D0848183

# HANDBOOK OF UNUSUAL NATURAL PHENOMENA

Compiled by

WILLIAM R. CORLISS

Illustrated by John C. Holden





Published and Distributed by

The Sourcebook Project Glen Arm, Md. 21057

Library of Congress Catalog Number 76-49382

ISBN 0-915554-01-1

NOTICE

This handbook is one of a series published by the Sourcebook Project, Glen Arm, MD 21057. The handbooks and the closely related sourcebooks may be purchased directly from the Sourcebook Project.

First printing: February 1977

Printed in the United States of America

# TABLE OF CONTENTS

# PREFACE

The primary objective of this handbook is to provide libraries and individuals with a comprehensive collection of reliable eye-witness accounts of unusual natural phenomena. To meet this goal, I have analyzed a substantial fraction of the geophysical literature published in English as well as the complete files of Nature (265 volumes) and Science (195 volumes). The result of this research is an incomparable assemblage of rare and curious phenomena. From these riches, I have selected the best and most representative for this book.

My criteria for selecting the "unusual" were that the phenomena be either: (1) beyond the reach of present scientific explanation, or (2) curious to me personally. Usually both criteria were satisfied simultaneously. The reader will soon see that secondary objectives of the book are to pose challenges to the scientific community and beguile the casual browser.

I make no claim of completeness because new phenomena appear constantly as the search of the literature continues. The complete master file of phenomenon reports is being published in stepwise fashion in a series of looseleaf sourcebooks. The Sourcebook Project welcomes inquiries concerning the sourcebooks, although some may find that these more exhaustive accumulations contain "more than they wish to know" about strange phenomena. For these individuals the carefully selected accounts appearing in the present handbook are ideal.

The looseleaf sourcebooks were, in fact, the first publications of the Sourcebook Project. Although thousands have been sold to libraries, feedback from librarians indicated that casebound books would be more acceptable. Such suggestions from librarians were the major factor in the decision to publish this casebound handbook.

My hope is that this handbook will become a standard reference work in the rather disorganized field of unusual natural phenomena. To this end, I have utilized reports taken almost exclusively from scientific journals. The screening provided by the editors and reviewers of these publications helps keep misidentifications and hoaxes at a minimum. The eyewitnesses in many cases are actually scientists who are well-experienced in the strange nuances of nature. Even with this conservative stance, the reader will find that many of the phenomena are very strange indeed and that reasonable explanations do not exist for the majority.

It is difficult to categorize and organize the unknown. I have used sensory stimuli as guides for organizing the chapters. Thus, we have chapters on luminous, acoustic, magnetic, and seismic phenomena and so on. Within each chapter are sections and subsections that divide the seemingly formless collection of observations into genera and species. For example, that perennial phenomenon, ball lightning, is categorized into several different types, as befits a most complex and curious manifestation of electricity. I might add that there is no classification labelled "UFOs" although the very thorough index will lead the reader to a number of UFO-like observations.

Some of the accounts reported herein are a century or more old, and odd spellings may be encountered on occasion. One also finds that English and American spellings differ in a few instances. No effort has been made to change the sources; they are as they were published originally.

Photographs and drawings are almost nonexistent in the older literature. For this reason, I engaged John C. Holden, a free-lance artist with considerable training in geophysics, to illustrate some of the phenomena from the published descriptions. He has also redrawn some illustrations from the original articles to conform with the handbook style. The 130 drawings constitute, I believe, one of the unique and most valuable features of this book. Not all drawings and photos referred to in the text are reproduced---only the most useful ones. Since several of the drawings portray humans involved in the phenomenon at hand, I should state that "no resemblance to anyone living or dead is intended."

Being that the bulk of this book consists of quotations, I hasten to acknowledge the many writers of papers, letters-to-the-editor, and sundry publications who have contributed their sightings and thoughts. Where lengthy quotations are taken from publications still protected by copyright, permissions have been obtained from the copyright holders.

William R. Corliss

Glen Arm, Maryland
December 5, 1976

# Chapter 1
# LUMINOUS PHENOMENA

## INTRODUCTION

The primary sensory channel for detecting new and unusual natural phenomena will always be the eye. Nothing catches the attention faster than a mysterious light. Phenomena may be self-luminous or made visible by the sun or moon. This chapter, however, deals only with the former; that is, those phenomena that make themselves visible through their intrinsic light-emitting capabilities. Nearly all are aerial or atmospheric, with the major exception being marine phosphorescent displays.

Self luminosity implies the release of internal energy or the conversion of energy entering the atmosphere into light. Lightning, ball lightning, electric discharge phenomena, and nocturnal lights seem to derive their energies from atmospheric processes, although these processes may be stimulated to some degree by extra-atmospheric forces. It has been suggested, for example, that ball lightning may be created by tiny antimatter meteorites and that earthquakes (and therefore earthquake lights) are influenced by solar activity. Extraterrestrial influences are less controversial when it comes to aurora-like and meteor-like phenomena. Here, the basic energy sources are almost certainly the sun and incoming meteors.

Genuine mysteries surround most of the phenomena of this chapter---otherwise they would not have been included. Most of the scientific problems center on the light-emitting processes, as in ball lightning and tornado lights. The actual light-emitting mechanisms are so ill-understood that some of the phenomena are denied objective existences by many scientists. Even more debatable are three other aspects of these phenomena:

- The coincident sounds and biological effects
- The prankish, uncanny "behavior" of some of the phenomena
- The role of extraterrestrial influences in stimulating the phenomena

All in all, strange luminous phenomena provide a fertile ground for scientific exploration. Unfortunately, the controversial aspects make it a risky field of research.

# REMARKABLE AURORAL PHENOMENA

"Normal" auroras inspired awe and legend in primitive man and even grip the modern beholder with their eerie draperies, arcs, beams, and flashing displays. Despite all our research with spacecraft, sounding rockets, and other scientific instruments, auroras have not yielded up all their mysteries. They are associated with solar activity and terrestrial magnetic storms, leading to the surmise that electrically charged particles emitted by the sun stimulate the auroral radiations high in the earth's atmosphere.

Beyond the reach of this explanation lie a host of peculiar luminous features and effects---seemingly auroral in nature---that must somehow be accommodated by geophysical theories. It may be, of course, that auroral displays have more than one explanation. The following classes of aurora-like phenomena will challenge any hypothesis-maker:

Auroral beams. Isolated, searchlight-like beams rising from the horizon well away from the normal auroral regions. Such phenomena have been seen prior to great earthquakes and also resemble the mountain-top glows described in the section on Electric Discharge Phenomena.

Auroral arches. Spanning the sky from horizon to horizon, often passing through zenith and south of it, these may be longer versions of auroral beams. They may encircle the entire planet.

Auroral "meteors". Isolated, well-defined patches of luminosity that move meteor-like across the sky.

Low auroras. "Conventional" auroras rarely descend below 50 km, yet ground-level observations of luminous displays with auroral overtones are not rare. Low auroras often seem to generate swishes, cracklings, and other sounds, which normal auroras located at 50-100 km would not seem likely to do. This controversial subject is explored farther in Chapter 4 . Low, noisy auroras may be associated with terrestrial electrical discharges.

Auroral odors. The very scarce reports of ozone connected with low-level auroral-like displays are also indicative of terrestrial electrical activity.

Artificial auroras. The artificial production of auroras with metallic arrays on mountain tops in northern latitudes also suggests electrostatic discharges as the sources of some auroras.

## BEAMS OF LIGHT ON THE HORIZON

### STRANGE AURORA
**Wagner, William H.; *Popular Astronomy,* 27:405, 1919.**

When returning to my home on the night of May 2, about 11:25 p. m. (Summer time), I noticed a bright beam of light spanning the sky almost directly overhead and reaching from the western horizon clear to the eastern in the form of an immense arch. It resembled the rays of a powerful searchlight, but had a dark rift running through it, dividing it into two parts. The light was very steady, with no evidence of flickering or rapid movement such as a searchlight or Aurora would have. The beam slowly changed its form, one side fading out, leaving a single beam. Later it split up again, only to resolve itself back into one again.

Then a strange thing occurred. The single beam slowly broke up into segments of different widths and brightness, as if broken up by some disturbing influence. Then it blended into a broad and fainter beam which slowly widened and finally faded out entirely.

In one hour it had disappeared completely. The night was clear and cloudless and when I first saw the phenomenon, it passed between the stars Gamma and Zeta in Leo, eastward through Arcturus to the horizon. The whole beam had a slow drift southward and when last seen it passed through Beta and Delta Sextantis eastward through Delta and Epsilon Ophiuchi, having moved this distance in about 40 minutes.

In appearance it resembled the tail of a comet, the Yerkes photograph of Halley's comet May 5, 1910, found on page 92 McKready's "Star Book" bearing a striking resemblance to what I saw.

It was perfectly transparent but grew very bright at times, having at one time almost blotted out the second magnitude star Gamma Leonis. Its width varied, but when first seen just filled the space between Gamma and Zeta Leonis, and just before it faded out it was fully twice as wide. (Popular Astronomy, 27:405, 1919)

## LIGHT RAYS AT NIGHT
**Groves, C. A.; *Marine Observer*, 7:65, 1930.**

The following is an extract from the Meteorological Report of S. S. Sheaf Mount, Captain C. A. Groves, Japan to Vancouver, B. C. Observer Mr. A. Macarthur, 2nd Officer.

"11th March, 1929, 10 p.m. A. T. S. in Latitude 49° 10' N. Longitude 134° 40' W.. Wind S. S. W. force 4. Barometer 29.80 in. Temperature Air 45°F., Sea 47°F. From behind Fracto Nimbus cloud stretching from horizon to a height of 11° and bearing S. 75° E. True, a shaft of light appeared, making an angle of 45° with the horizon and following a straight line 45' wide for a distance of 60°;

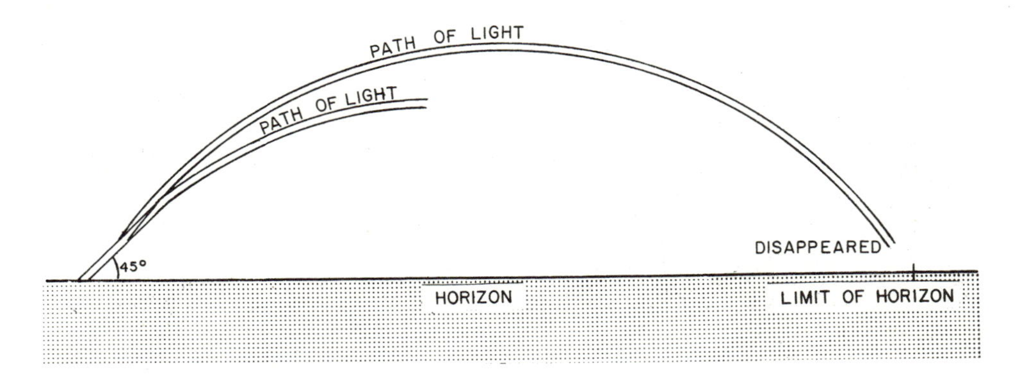


Split auroral beam over the Pacific

in the sky the light broke off into two curved sections passing through the zenith and finishing $3^{o}$ above the horizon bearing N. $75^{o}$ W. True. The straight portion of the phenomenon had the appearance of a gigantic searchlight. The whole streak of light was motionless over its arc of $166^{o}$ and commenced fading away at 10.20 p.m. and finally disappeared entirely ten minutes later. Fracto Nimbus and Cumulus clouds in S.W. portion of sky lit up by reflection of light."

Note.---It is possible that this was some kind of auroral ray. Another explanation is that it was the trail of a meteor, but if, as the description seems to imply, the ray was seen at first to progress across the sky, the bright head of the meteor would have been seen. (Marine Observer, 7:65, 1930)

## A CURIOUS PHENOMENON
**Noble, William; *Knowledge,* 4:173, 1883.**

Can any of my brother readers of Knowledge offer a feasible explanation of a very remarkable phenomenon which I witnessed at 10h. 35m. p.m. on Tuesday, August 28? I was just coming out of my observatory when, on the E.N.E. point of the horizon beneath the Pleiades, I saw a bright light. My first thought was that the moon was rising, but an instant's reflection sufficed to remind me that she would

Searchlight-like ray of light on horizon

not be up for the next two hours. As I watched the light becoming brighter and brighter, I saw that it threw a kind of radial illumination upward, the effect of which I have tried to reproduce in the accompanying rough little sketch. As will be seen, a few distant cumulo-stratus clouds, close to the horizon, crossed it. For a moment I imagined that I was viewing the apparition of a new and most glorious comet; but, as I watched, the "tail" disappeared and what would represent the nucleus flashed up brilliantly. Then I made up my mind that some distant house, barn, or haystack was on fire, and returned to the observatory for a 3 inch telescope, which I keep for looking over the landscape. Before I had time, however, to enter the door, every vestige of illumination disappeared as suddenly as it had come into view, and after waiting in vain for some time, I left the observatory and came into the house. I have diligently inquired if there was a fire anywhere in this part of Sussex on the night of which I am speaking, but there was none. (Knowledge, 4:173, 1883)

## A CURIOUS PHENOMENON
**Bradgate, W. K.; *Knowledge,* 4:207, 1883.**

The remarkable phenomenon which Mr. Noble described in No. 98 of Knowledge was also witnessed by me in Liverpool on Aug. 29, at 12 h. 40 m. a.m. I had just been looking at Saturn, when, for the first time, I saw a bright divergent cone of light about 7° above the horizon; the entire length of the cone was about 5°. The apex or nucleus displayed such a degree of concentration that I thought it was the planet Jupiter. I turned my telescope, a 2-inch, armed with a power of 30, on the point where the apex should be (it was now obscured by a cloud), with the expectation of being able to unravel the mystery, but was disappointed, as the cloud was too dense. I then ran my telescope along the major axis of the cone, and the field of view was so faintly illuminated that the brightest part could hardly be said to equal the lumiere cendree seen under similar conditions. It gradually faded from view, after having been visible for thirteen minutes. I continued watching the part of the heavens where it had disappeared, with the confident hope that it would return, but was at last obliged to give it up, as a great bank of clouds precluded all further observation.

It could hardly have been an auroral streamer, as the point where it appeared is 67° east of North. (Knowledge, 4:207, 1883)

## AURORAL APPEARANCE
**Bonnycastle, R. H.; *American Journal of Science,* 1:32:393-394, 1837.**

At a quarter past nine o'clock on Sunday night the eighth day of May, in the present year, my attention whilst regarding the heavens was forcibly attracted to the sudden appearance due east of a shining broad column of light.

At first, as my window overlooks the bay of Toronto and the low island which separates it from the lake, I took this singular pillar of light for the reflection from some steamboat on the clouds, but having sought the open air on the gallery which commands a full view of the bay and of Ontario, I was convinced that the meteor was an effluence of the sky, as I now saw it extend upwards from the eastern water horizon line to the zenith, in a well defined, equal, broad column of white strong light, resembling in some degree that of the aurora, but of a steady brightness and unchanging body, whilst there were few or no clouds. Ursa Major, then near the zenith, was situated with regard to this column, at a quarter past nine as below, the column passing nearly vertically between $\zeta$ and $\eta$.

There was no moon, as on that day it rose at 2h. 4m. consequently it was dark, and as the sky was not very cloudy the meteor was seen to the greatest advantage as the night wore on. It passed very slowly and bodily to the westward, continuing to occupy the space from the horizon to the zenith, until the upper part first faded slowly and then the whole gradually disappeared, after it had reached nearly to due northeast. I had unfortunately broken my thermometer and could therefore only

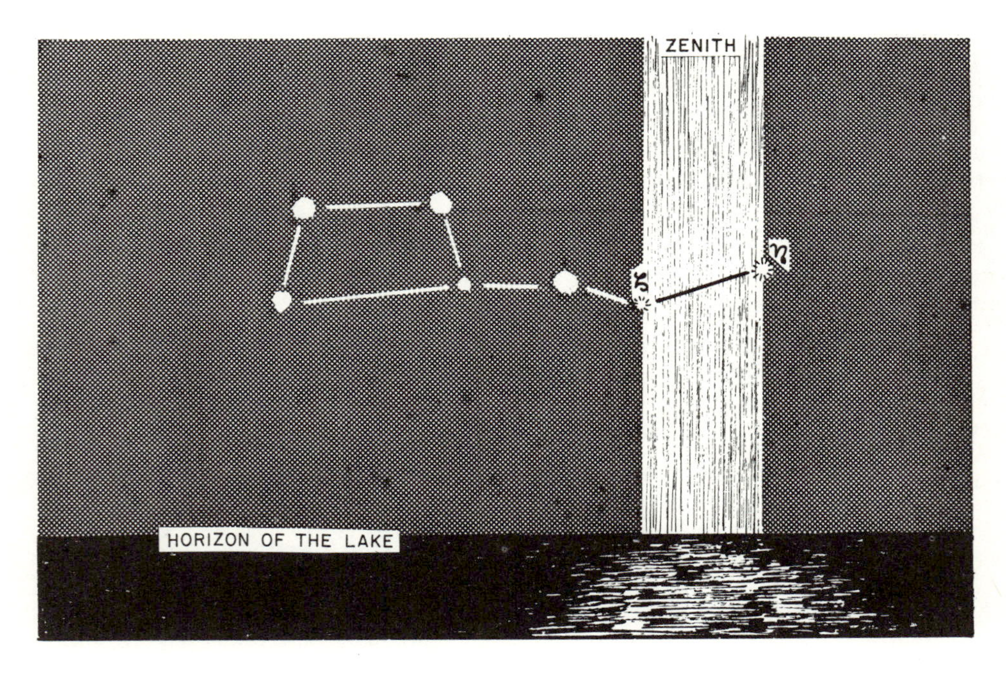


Pillar of light over Bay of Toronto

state that the weather was cold, and that there was no wind. At twenty five minutes past nine o'clock the pillar of light had vanished, but it immediately afterwards reappeared slightly in the horizon where it had been last seen, and in the mean time the constant auroral arch of the halos I have before mentioned became visible in the northern horizon, and increased very rapidly in brilliancy, and at ten minutes to ten gave so intense a glow to the sky that it was light enough to enable me to see the objects around distinctly as in pale moonlight. It was in short equal to the light of the moon at the end of the second quarter.

The auroral arch rose very high on this occasion and then flattened, and at ten the double arch, I have already described, was pecularly beautiful, the darkness under it being singularly grand. (American Journal of Science, 1:32:393-394, 1837)

## LIGHT RAYS ON HORIZON
**Edwards, W.; *Marine Observer*, 6:10-11, 1929.**

The following is an extract from the Meteorological Report of S. S. Mongolian Prince, Captain W. Edwards, Cape Town to New York. Observer, Mr. V. C. Palmer:---

"January 7th, 1928, 4.19 p.m., A. T. S., 1845 G. M. T., in Latitude 5° 19'N., Longitude 34° 53' W., observed rays of light emanating in the eastern section of the sky and extending to the southward bearing S 77° E true, and to the northward bearing N 35° E.

"The extreme altitude was 10° and they appeared in white and grey contrasts.

"1956 G. M. T., a broad ray appeared bearing N 50° E., true, and had the appearance of being light green in colour.

"At 2011 G. M. T., there were no rays visible. The computed time of moon rise U. L. was 6.26 A. T. S. and bearing N 67° E.

"January 8th, 1928, 5.45 p.m., A. T. S., 2022 G. M. T., in Latitude 8° 59' N., Longitude 37° 45' W. At sunset observed two rays of light in the eastern section of the sky emanating from a point on the horizon bearing N 65° E, true; one extended

Anomalous light rays on horizon

to an altitude of 45° and the other to an altitude of 60°. At the same time an arc was visible in the western section of the sky bearing S 50° W, true, and had an altitude of 50°, this was showing through Cirrus clouds. The colours visible were purple and green. This effect was visible till 2038 G. M. T.

"The sun's bearing at sunset was S 68° W, true. The computed time of moon-rise and bearing was, 7.20 A. T. S., N 70° E.

"The W/T Operator reported heavy static rushes." (Marine Observer, 6:10-11, 1929)

# SKY-SPANNING AURORAL ARCHES

## LUMINOUS ARCH
**Anonymous; *American Meteorological Journal,* 8:35, 1891.**

A curious phenomena, says the report, was witnessed near here last night by passengers on the north-bound passenger train on the Houston & Texas Central, which passes this point at 2:25 a.m. It was in the form of a luminous arch, possibly of an electrical character. The luminous mist was first observed by the engineer, when it was still several hundred yards ahead of the train, and thinking it a prairie fire, he slowed up, thus arousing the passengers, who, with the crew, crowded to the windows and on to the platforms to look at the vast, hueless rainbow spanning the heavens. As the arch was more closely approached its dim, white radiance was seen to be clearly defined against the sky as though painted there by the sweep of a brush dipped in white fire. The stars could be seen shining close against the rim of it, and all around and under the arch. It was in form the half of a perfect circle, one leg resting on the earth, while the other appeared to have been broken off near the base. It seemed to gradually increase in size.

The arch rose directly over the track, and as the train approached it seemed to gather a greater luster, as of the diamond or some clear, glittering star. The stars could be seen in close proximity to it. When the train passed directly under the bridge of light, the surrounding country spanned by it became plainly visible, appearing to be bathed in pale moonlight.

A curious feature of the luminosity was that while it gave all objects a weird, unreal aspect, the shadows which it caused them to throw were black and as clearly defined as silhouettes. In a few minutes after the train passed under the arch it seemed to fade away, melting gradually into the starlit sky. The night was fair and fogless. There was no moon, so the arch must have been self-luminous. (American Meteorological Journal, 8:35, 1891)

## A REMARKABLE PHENOMENON
**Campbell, Frederick; *Popular Astronomy,* 11:484-486, 1903.**

During the late evening of August 21st last, the writer, in company with several others, witnessed a celestial phenomenon surpassing in wonder anything that ever before had come to their attention. I was spending three weeks at Cranberry Lake,

St. Lawrence Co., N. Y., in the Adirondack wilderness. The latitude of this place is about 44 degrees and 12 minutes north, and the longitude 74 degrees and 45 minutes west. It is about 75 miles nearly east of the city of Watertown, N. Y.

When the Sun had set, the northern sky seemed remarkably bright; and, as the darkness deepened, it was apparent that the illumination was something more than that of evening twilight. We were being treated to a display of aurora borealis, or northern lights. The spectacle was more marked than is often seen even during winter nights. There were streamers that reached far beyond the Great Dipper; and occasionally, in the midst of the general radiance, certain spots would brighten to an intense glow and again fade somewhat. There was no color at any point, simply the clear pure light. But the glory was enhanced by occasional flashes of lightning near the northern horizon, but unaccompanied by thunder.

We gazed upon this with interest for some time, and then retired within doors and gave it very little more thought. At about half past nine, however, I was suddenly summoned out of doors to witness a spectacle such as I had never looked upon before nor had ever heard or read of. The heavens were spanned by two great bows of light, crossing each other near the zenith by a wide angle. The one was the familiar Milky Way or Galaxy, then in its glory, no Moon being present. The other was a remarkable archway of light, comparable with the Galaxy in width, but much brighter, and seemingly, stretching from horizon to horizon, though in the one direction its light, like that of the stars, faded near the horizon, and in the other it was obscured by a small cloudbank. I estimated the width of the arch to be three degrees or that of six full moons placed side by side. This bow stretched

Archway of auroral light crossing Milky Way

from a little north of west, to a little south of east. It passed just south of Arcturus in the west, and just north of Jupiter in the east, at 10:00 p. m. The arch remained in place for a long time, possibly half an hour or more. It changed very little in appearance and brilliancy. In places it had a fluted, wavy aspect, being completely broken only here and there, and then only by a very narrow interval. The waves appeared as if gently blowing in the direction of the arch, toward the northwest, like curtains waving in a breeze, or a little like steam being blown along. As the wind was actually blowing very freshly at the time, the illusion was not difficult. The entire arch was bright, and where brightest easily obscured all the stars in the immediate vicinity. Job's Coffin and Altair were in the arch.

The first thought was that this was a great streamer of the aurora borealis, seen earlier in the evening, and still continuing quite brightly; but it was apparent at a glance that it was something totally distinct; for one thing, it arched the entire heavens; again, it did not narrow to a point, but was equally broad across the entire sky; but most of all, it did not radiate from the north at all; it extended from a point nearer west than north, somewhere between northwest and west. Moreover, the next day I conversed with two reliable men who were together out-of-doors when the spectacle started up; and of them I learned that it did not originate at its more northerly or northwest extremity, but at its southeast end. They saw it rise from that direction and gradually extend itself along the sky in a generally northwest direction until the heavens were arched as when I first saw the sight. It seems altogether probable that the simultaneous appearance of the aurora borealis and of this archway of light indicates a common origin, even if the arch could not be regarded as a part or a streamer of the aurora. Evidently there was a wide-spread magnetic disturbance of the atmosphere which caused the aurora to appear first, and later threw this great bow across the sky. But why it should have manifested itself in this way, choosing its direction and regulating its width as it did, is a very great mystery. Something a little similar to this I recall seeing in the same part of the country several years ago, repeated two or three evenings; but at that time there was no arch, but a strange and persistent brightening of the heavens in a northwest direction, well up, and entirely detached from the horizon, with no accompanying aurora, as far as I now recall.

There was no possibility of this being a lunar rainbow, for there was no color, and there was no Moon. Borrelly's comet was near at the time, but there was no possibility of this being its tail, and doubt as to the comet's being in any way responsible. (Popular Astronomy, 11:484-486, 1903)

## AN UNUSUAL AURORAL DISPLAY
**Bobrovnikoff, N. T.; *Popular Astronomy,* 45:299-301, 1937.**

At about 8:30 E. S. T., April 27, 1937, a fine aurora was noticed by me on the northern horizon in Delaware, Ohio. There were low clouds in the north and northwest but otherwise the sky was clear. The aurora consisted of a low arch, which could not be seen distinctly on account of the clouds, and streamers showing their usual pulsation and motion from east to west. There was nothing especially remarkable about this aurora except perhaps its color which was distinctly reddish. Although auroras are seldom seen in our latitude, there would be no reason to describe this phenomenon in detail if it did not lead to another most remarkable display.

By 9:00 p.m. the aurora in the north apparently died out. The region from Polaris to the northern horizon was in clouds, but no particular illumination could be seen in the spaces between broken clouds. At this moment I noticed a strong narrow ray rising from the eastern horizon from a point some 5° south of the true east. The origin of the ray on the horizon could not be seen because of clouds. The ray, about 1° wide, spread in a few minutes across the sky, so that its center line was passing through δ Leonis a few degrees south of the zenith, and its western end through δ Geminorum. The western end was not sharp but was split into several branches apparently issuing from δ Geminorum and changing their position from time to time. The whole length of the ray was at least 150°.

The brightness of the ray was practically uniform, but somewhat greater near δ Leonis, probably because of atmospheric absorption in lower parts. Near the zenith the brightness was estimated by me to be about twice the brightness of the Milky Way in the Sagittarius cloud, but it may have been brighter as the sky was hazy and it was somewhat foggy. The color of the ray seemed to me perfectly white. The phenomenon lasted until 9:15 without noticeable change. At this moment I noticed that some 20° above the eastern horizon the ray terminated abruptly. I thought I saw the real eastern end of the ray but this was not so. I noticed soon that at the place mentioned there was simply a break in the intensity, the ray farther down toward the eastern horizon being much fainter. In two or three minutes the difference between these two parts of the ray nearly disappeared and the ray again regained its uniform brightness.

While I was watching these developments, the middle portion of the ray around δ Leonis suddenly burst into violent activity. Many sharp streamers seemed to issue from this star, following one another in their apparent motion from east through north to west. The streamers were about 10° long when east or west, but toward north they were shorter, some 5°, apparently the effect of foreshortening. There were no streamers of this kind on the south side of the main ray. The region around δ Leonis was rapidly changing its shape, now becoming narrow (1/2°), now wide (2°), and twisting in a most complicated fashion. Almost simultaneously with this there seemed to be a violent motion of masses from δ Leonis west, along the main ray, and on its northern boundary, parallel to it. The character of the motion can best be described by globes of light, some 1° in diameter moving about 1° per second for a distance of 10° from δ Leonis west toward δ Geminorum and gradually fading out. This phase of the auroral activity lasted not more than two minutes. At first I could not believe my eyes, but the motion was too evident to entertain any further doubts. Several people watching this phenomenon with me confirmed my impression. Pretty soon the activity in the ray ceased and it resumed its former appearance. At this time it seemed to me that all across the sky, south as well as north, there were faint streaks of light running perpendicular to the ray. At one time a fainter ray parallel to the main ray began to form to the south, but soon disappeared.

By 9:30 heavy clouds came from the west completely obscuring the sky for half an hour, except for a break in the east at about 9:40. At that time the eastern part of the ray was still in its former place. The phenomenon lasted therefore at least 40 minutes during which time it did not appreciably change its position among the stars. An hour later the clouds broke up, but the moon was already above the horizon, and there was no evidence of the ray, or of any aurora in the north.

Such auroral rays are seldom observed. Probably the most remarkable one was seen on August 21, 1903, from various parts of the country between Maine and Iowa (Pop. Ast., Vol. 11, 1903, several descriptions; Barnard, Ap. J., Vol 31, 210, 1910) giving evidence of the great height of such phenomena. According to Barnard (Williams Bay, Wisconsin) the ray of 1903 rose from the western horizon during an auroral display, was 80° long, and lasted 15 minutes. According to F. Campbell (Pop. Ast., Vol. 11, 484, 1903), however, the ray (northern New York state) rose from the eastern horizon, spreading toward the west. I could not find any description of unusual activity along this ray, but I do not doubt a careful search through records of aurorae would reveal similar observations. Formation of balls of light apparently moving along the streamers of ordinary aurorae has been observed many times before.

I have myself seen several such rays at various times, but they were quite faint and they never showed any activity. However, I thought it would be worth while to mention this interesting phenomenon as the sun-spot activity is increasing and, judging from the size of recent spots, promises to be higher than during the recent periods of sun-spot maxima. In fact a large group of spots was near the central meridian of the sun at the time of the aurora. I cannot but think of comets' tails in this connection. Various phenomena in cometary tails described loosely as coruscations, scintillation, flickering, etc., are generally, but unjustifiably, regarded as optical illusion and that in spite of a very considerable volume of evidence to the contrary. The auroral displays have many features in common with the tails of comets, which considering the mode of excitation of cometary material is not surprising. In the present connection I want to mention the observation of Comas Sola (A.N., Vol. 185, 75, 1910) of Halley's Comet in which he described globes of light constantly issuing from the nucleus and moving along the axis of the tail.

It is impossible to decide whether the motion of the luminous globes in aurora (as well as in comets) is actually transference of matter. In spite of a very definite impression to this effect, I think it was rather transference of luminescence. The velocity, however, assuming the height of the aurora to be 100 km would be only of the order of 2 km/sec. (Popular Astronomy, 45:299-301, 1937)

## PATTERN OF LIGHT RAYS
### Williams, J. O., and Craig, W. B.; *Marine Observer*, 32:64-65, 1962.

m.v. Waiwera. Captain J. O. Williams. Balboa to Auckland. Observers, Mr. B. J. Wardle, 2nd Officer, Mr. C. H. Swindells, 3rd Officer and Cadet M. Orr.

23rd June 1961. From 0815 to 0840 GMT a pattern of rays of light was observed bearing from 180° to 270°. They extended from near the horizon to an altitude of 55° when first seen but the angle of elevation increased until it was just past the zenith. In colour they were white, occasionally tinged with red; in structure they were wispy thin straight rays, resembling weak searchlight beams and sometimes merging into patches. The rays appeared to be related to a point about 10° below the horizon, bearing 260°.

There was continuous movement among the rays, some of which often faded out and reappeared. The moon, which was in its 9th day, was shining brightly through the ray pattern without noticeably weakening the strength of the latter. The brighter stars were also visible through the rays. A few fair-weather Cu. clouds were present but they did not in any way interfere with the observation.

Position of ship: 22° 38's, 123° 40'w.

Note. This is an observation of unusual interest. Although the description would suggest that a display of aurora was seen, Mr. J. Paton, of the Balfour Stewart Auroral Laboratory, University of Edinburgh, comments as follows:

"One can say quite certainly that this is not aurora. At this time the magnetic Kp index was I and we have no other reports of aurora. The sketch [not reproduced here] suggests crepuscular rays but the radiation point does not fit with the position of the sun." (Marine Observer, 32:64-65, 1962) m.v. Port Hobart. Captain W. B. Craig, Aden to Wyndham. Observers, Mr. G. J. Botterill, 3rd Officer, Mr. G. G. Blackler, 4th Officer and Mr. W. Knight. A.B.

1st June 1961. At 1800 GMT the upper of the two arcs illustrated appeared in the sky; 10 min. later, the lower narrower arc became visible. The upper one was very bright, while the lower was of slightly less intensity; both arcs attained maximum brilliance at 1820 and disappeared at 1832. They extended from 090°, through S, to 240°.

Another arc of different shape, extending from 205° to 270°, developed at 1845. It had a grey/white light which became brightest at 1852; it then gradually faded, finally vanished at 1900. There was a full moon, which during the display increased in altitude from 45° to 54°.

Only 2/8 of cloud, consisting of small Cu., Ac. and Ci., were present during the observation. Visibility was excellent.

Position of ship: 5° 06' N, 78° 30'E.

Note. Mr. J. Paton, of the Balfour Steward Auroral Laboratory, University of Edinburgh, comments:

"We can think of no satisfactory explanation of the remarkable arcs seen by observers in m.v. Port Hobart in the Indian Ocean south of Colombo around local midnight on the night of 1st June 1961. That they were auroral is unlikely, for geomagnetic disturbance at this time was no more than moderate. The change in shape at 1845 GMT suggests that they might have been caused by dust layers.

"Various sky phenomena, for example the twilight glow and crepuscular rays, may sometimes be so very like aurora that unless special instruments like the spectroscope or interference filter are available, the observer will find it impossible to decide if aurora should be reported. In these circumstances, most observers usually describe what is observed and add that aurora is suspected. We hope that observers will continue to follow this practice. When the report is later examined in the laboratory it is usually possible to ascertain from the level of magnetic disturbance at the time of aurora is likely to have been present; aurora is invariably associated with unusual magnetic disturbance." (Marine Observer, 32:65, 1962)

# AURORAL "METEORS"

## AURORAL "METEOR"
**Capron, J. Rand; *Nature*, 27:82, 1882.**

I happened to turn to the south, where the moon (with a very pronounced lumiere cendree on its dark part) was nearly on the meridian, when I saw a spindle-shaped beam of glowing white light, quite unlike an auroral ray, had formed in the east. As I looked this slowly mounted from its position, rose to the zenith, and passed it, gradually crossing apparently above the moon, and then sank into the west, slowly lessening in size and brilliancy as it did so, and fading away as it reached the horizon. The peculiar long spindle shape, slow gliding motion and glowing silver light, and the marked isolation of this cloud from the other portions of the aurora made it a most remarkable object, and I do not recollect in any former aurora to have seen anything similar. About 6 o'clock the aurora gradually died away, to revive again at 9 in the shape of a white semicircle of light in a point north by west, which did not last long. Owing to moonlight, but little could be done with the spectroscope with a wide slit on the most glowing parts of the red patches only the usual green line, with a faint continuous spectrum towards the violet could be made out. At times I thought I caught traces of other lines, but with no certainty at all. The spindle-shaped beam was also examined with the spectroscope, but only gave the green line. Even in the brightest parts of the red glow, the red line could not be made out. The peculiarity of the moving beam of light was its absolute southern position. Its apparent passage across the sky was only a few degrees above the moon, then at a comparatively low altitude. (Nature, 27:82, 1882)

## THE AURORA
**Anonymous; *Knowledge*, 2:419, 1882.**

At about five minutes past six a singular phenomenon was observed. A cloud of whitish light, shaped like torpedo, passed from the south-eastern to the north-western horizon (some accounts say from east to west, but the true direction was from a little east of south-east to a little north of north-west). It was of nearly

The spindle-shaped auroral meteoroid of 1882

uniform brightness. Its length was nearly ninety degrees. (One observer says that its size, compared with that of the moon (which was shining brightly at the time), was as a herring compared with a sixpenny-piece; but as he does not name the price of his illustrative herring, the illustration is not so satisfactory as could be wished: does he mean a penny herring, or one of those which may be obtained, we understand, at two a penny?) It passed across the heavens in about two minutes. It is described as a cometary body; but as it moved almost exactly at right-angles to the magnetic meridian, there can be very little doubt it was an electrical phenomenon. The appearance is akin to that sometimes seen in high latitudes, as the auroral streamers vary in position, aggregation, and length, seeming to throw folds of brightness, shaped like the folds of a curtain, athwart the auroral arch. (Knowledge, 2:419, 1882)

## AN UNUSUAL AURORA

**Howard, W. H.; *Science*, 22:39, 1893.**

On Saturday evening, July 15, there occurred an aurora which was unlike any the writer has ever seen, and a brief description of it may contribute something to the aggregate knowledge of those interesting phenomena.

The peculiar feature of this aurora was the movement of a series or succession of whitish flecks across the sky from east to west, resembling somewhat the waves of a body of water.

About 9.30, central time, my attention was first attracted to it. Flecks of white light were forming in the east at an altitude of about 45°, passing in regular succession westward, about 20° north of the zenith, and apparently accumulating in one larger band in the northwest, reaching at times from near the horizon to perhaps 80°. The white flecks or streaks were about 10° in length, strictly parallel north and south, and quite uniform in distance apart. They grew brighter and more distinct as they approached and passed the meridian. Their motion was very regular and quite rapid,---comparable to the swiftest apparent motion of light clouds. If they were as high as the electric theory would suggest, the velocity must have been enormous.

At times similar short bands, like strokes with a paint-brush, were stationary in the north, at about 45° altitude, for several minutes at a time.

A few minutes later a number, perhaps ten or twelve, white bands appeared north of the zenith, all converging towards a point some 10° south of the zenith, but vanishing before reaching the zenith. They remained only a few minutes. About 10 o'clock the moving flecks had disappeared, and one long, straight band extended from the northwest horizon, 50° or 60°, toward a point about 45° south of the zenith. Two or three other short flecks appeared parallel with the main band. About the same time the usual diffused glow appeared in the north horizon and continued till after 11 o'clock, but was not observable while the moving bands were seen. Many more gorgeous auroras have been seen in our latitude, but the rapidly-moving bands gave this one a new interest. (Science, 22:39, 1893)

# VERY LOW AURORAS

## THE AUDIBILITY AND LOWERMOST ALTITUDE OF THE AURORA POLARIS
**Chapman, S.; *Nature,* 127:342, 1931.**

Some of the accounts of low aurorae seen between the observer and terrestrial objects are very striking and circumstantial. One writer states that he and his party (members of a government radio station, in the winter of 1917-18) were enveloped in "a light mist or fog-like substance in the aurora"; a hand extended could be seen as if in a coloured fog and a kaleidoscope of colours was visible between the hand and the body. It was impossible to feel this visible fog or mist, and there was no dampness. By stooping close to the ground it was possible to see under this light, which did not go below four feet from the ground. The low-hung aurora lasted fifteen minutes, while great streaks and shafts of light came and went in the heavens. The occasion was unique in the writer's experience.

Aurora at earth's surface

Another writer, during a brilliant auroral display saw the light "play" down between himself and a steep glacial deposit bank about 125 feet away; he "stepped right into the aurora". Another saw an aurora between himself and a ten-foot bank no more than one half-mile away; there was "immense light", "on the very surface", and a shaft of light shot up to an immense height. Another in 1915 saw the familiar landscape "beautified by changing coloured lights which came and went rapidly, now bathing us, now withdrawing with a swish to a height of a mile or so"; "it came down in streamers, or again as a fog (only much faster than fogs come, only a matter of seconds)"; "we saw the tree trunks through the aurora", the colours being mostly, though not wholly, greens.

These letters make it difficult to deny that aurorae occur, very rarely, quite near the earth and are sometimes accompanied by noises. Almost all the reports come from a belt of country about three hundred miles wide, lying roughly along the auroral zone; the belt includes the Klondike region, where is situated the township of Dawson, with a population of several thousand; several reports come from this neighbourhood. The only other places along the auroral zone, likely to afford such favourable fields of inquiry, appear to be near Churchill, on Hudson Bay; the extreme south of Greenland; Iceland and the most northerly part of Norway.

Since very low aurorae seem to be very rare and to be confined to localities near the auroral zone, it is perhaps not surprising that Stormer, Vegard, and Krogness should have observed no such case, or that the Polar Year (1882-83) failed to provide evidence establishing their existence; with the better organisation of auroral observation which it is hoped to achieve during the proposed new Polar Year (1932-33), there is more chance that opportunities of critical examination of these appearances will occur.

These low aurorae must obviously be very different in character from those observed in the upper atmosphere, though connected with them. Inability to understand their physical nature is not a sufficient ground, in the present state of knowledge, for rejecting the possibility of such occurrences; such an attitude would, for example, forbid acceptance of the reality of globular lightning. The observations cited by Mr. Johnson constitutes at least a case for active further inquiry, and renders it highly desirable that auroral investigation near the auroral zone should include not only visual and photographic observations, but also atmospheric electric registration at a well-equipped observatory. (Nature, 127:341-342, 1931)

## AURORAL OBSERVATIONS IN THE ANTARCTIC
**Chree, C.; *Nature,* 101:115, 1918.**

In the daily logs [at Cape Royds] there are frequent indications of the observer's impression that the aurora was at no very great height, and that its form was influenced by Mt. Erebus when it lay in that direction. Thus, of an aurora on May 23 it is said:---"As it extended past the cone of Mt. Erebus, there appeared a local bend, curving outwards from the mountain....the lower border appeared to show below the summit of the mountain." Of a curtain on May 31 it is said:---"It appeared to be very low over Mt. Erebus, and to touch the....crater. At one stage it ringed the crater." On June 21, we are told, "a strong luminous nebula appeared on the N. flanks of Mt. Erebus....The luminous nebula stood out brightly between us and the slopes of Mt. Erebus." (Nature, 101:114-115, 1918)

## LOW AURORAS
**Simpson, G. C.; *Royal Meteorological Society, Quarterly Journal,* 59:189, 1933.**

One afternoon I was walking down John river on the smooth glare ice. I came to a place where the water was bubbling up through the ice, due to great pressure. I hurried away from it as soon as possible, rounded the bend and went to my cabin.

That night about 8 o'clock I went out of doors, and at this very spot that I had crossed over in the afternoon there was an aurora, between me and a ten-foot bank, not to exceed one-half mile from where I was standing in a south-east direction. What interested me most of all was the immense light. The light was on the very surface. And a shaft or column of light shot up to an immense height, the streamers seemed to be on the south-east side of the shaft of light, and as they went up seemed to flatten out and at great height was carried away on the air. The light on the surface extended out for considerable distance---one-quarter of a mile. There was no noise, no sound, no swishing. The night was cold, still, dark, temperature about 40 degrees below zero. (Royal Meteorological Society, Quarterly Journal, 59:189, 1933)

## A LOW AURORA AND ITS EFFECT ON A RADIO RECEIVER
**Anonymous; *Nature,* 127:108, 1931.**

Though no photographic determination of the height of aurorae has ever yet yielded reliable values below 80 km., cases of aurorae coming much lower, even to the ground, are occasionally reported. Mr. A. C. Cummings, Room 1111, 180 Varrick Street, New York, sends us such a report---which also has other unusual features ---of an aurora witnessed by his brother at Norwood, Ontario, Canada, early in the winter of 1929-30. He observed that his radio set was 'dead', though the valves were alight and the aerial and ground wires were connected; no reason for the absolute quietness of the set was apparent. Looking out, he saw that a bright aurora was in progress; on going outside to get a better view he found, "to his astonishment, that a curtain of streamers extended from the sky to the ground and completely surrounded the house but at a distance of several feet". The curtain was many-coloured and unsteady, its scintillations being accompanied by "visible sparking" and by "snapping sounds". (Nature, 127:108, 1931)

## SOME NEGLECTED ASPECTS OF THE AURORA
**Botley, Cicely, M.; *Weather,* 18:217-219, 1963.**

From the published bibliographies it would seem that never in the 350 years during which it has been scientifically studied has there been so much intensive investigation of the Aurora. Yet one of the most interesting aspects, once well known, is now, with a few exceptions, practically ignored.

Ever since the time of Gassendi's description of the great display of 12 September 1621, with 'most brilliant white fumes' (candissimi fumi), and perhaps before (e.g. Stow's 'divers impressions of fire and smoke' and certain broadsheets) it has been noticed that after a major aurora the sky clouds over, loses its clearness, and is covered with a light veil of cirrostratus giving rise to lunar halos. Observations of this kind occur in the literature from J. J. de Mairan (1754) through the records of the great Polar expeditions of the last century, and the observations of devoted amateurs like H. Weber of Peckloh, to those of the Second Polar Year at Scoresby Sund. A typical quotation would be that from von Wrangel (1844) from Siberia 1820-24.

"When the streamers extend to the zenith or nearly so, they sometimes resolve themselves into small luminous and cloud-like patches of a milk white colour, which not infrequently continue to be visible on the following day in the shape of white, wave-like clouds."

Also, he noted that when the moon was reached by the rays, a halo of $20^{o}$-$30^{o}$ radius developed, which lasted some time.

There are other interesting observations: the development of clouds in the clear part of the sky, south of the aurora, clouds which formed an arch, which lasted for some time, and then dissipated. The thickness of certain arches is remarkable: notably the red band observed to the south from Geneva and Montpellier on 15 February 1730 (which must have formed well inside the minauroral limit at magnetic latitude $45^{o}$N), through which the stars showed dull, and which even occasionally obscured Jupiter (Mairan 1754). Appearances of aurora below or issuing from cloud, often dismissed as erroneous, have been recorded by observers of the front rank such as Lemstrom and Weber. The latter noted aurora which began in the region of the Polarbanden (never above 5 miles) and then descended.

Also, in association with aurora, halos of the unusual radii $10^{o}$ and $15^{o}$ were noticed by Richardson (1855) on the 1820 Franklin expedition. Other halo observations include a greenish halo noted at Uppsala on 4 February 1874, a 'peculiar' corona recorded on 3 March in the same year at Peckloh in 'currus', of radius $20^{o}$. A particularly interesting observation was made at Bossekop on 2 November 1838 when an auroral ray reached the ordinary halo of $22^{o}$ radius; 'this is wider and its light appears more condensed'.

During the Polar Year expeditions to Scoresby Sund, 1932-33, Dauvillier recorded 22 lunar halos, along with auroral 'fog'. It was suspected that these 'halos' were not of the ordinary refraction type formed in ordinary cirrus or cirrostratus. Toppler sought to explain them as diffraction coronae formed in meteoric dust, then held to have some connection with aurorae.

Recently Dauvillier (1962) has put forward a view which could explain not only the halos but the appearance of aurora below clouds, which now is regarded as physically impossible. According to Dauvillier the auroral clouds, fog, etc., are formed in the ionosphere in a microcrystalline aerosol made of ammonium nitrite, according to Thenard's reaction---

$$N_2 + 2H_2O = NH_4NO_2$$

The necessary constituents, nitrogen, hydrogen, oxygen, and the molecules OH and NO, are known to exist above 70-90 km. The energy is supplied by the aurora in the polar regions, and the airglow in lower latitudes. Dauvillier has attributed to similar causes the 'ashen light' often seen on the dark side of Venus.

Auroral halos are regarded by Dauvillier and the author as examples of the very large corona known as Bishop's Ring, named after S. Bishop of Honolulu who reported it in the dust from Krakatoa in 1883. It appears as a circle, bluish inside and reddish-brown outside---the red of a halo is inside---though at night the feebleness of the light may prevent the colour being seen, thus increasing the chances of confusion with the $22^{o}$ halo. Unlike the halo, it can be of variable dimensions, which would explain the anomalous rings seen by Richardson. As Dauvillier points out, the size of particle about 2 microns given by the radius $20^{o}$-$30^{o}$ corresponds to that deduced by Lyot by his studies of polarization for the clouds of Venus. Bishop's Ring is certainly formed at high altitudes as ordinary cirrus clouds have been seen to pass before it. Also there seems to be some relation between its appearance and solar activity.

Another possible relation is with noctilucent clouds. As Paton recently remarked, these seem to resemble cirrus and yet it is difficult to imagine how enough water exists to form them at 80 km.

Again, after aurora, with a delay of about 48 hours, normal cirrus cloud appears, which of course may yield normal $22^{o}$ halos. The author recorded a halo on 5 September 1958 after auroral activity beginning on the 3rd. Readers of

Weather may remember the paper by Needham and Ho on the great apparent knowledge of halos in medieval China; is there a connection between this and the recorded great number of aurorae at that period? It is certain that the connection was noted in his journal by Tycho Brahe, at Uraniborg (hveen), Denmark, 1583-1596, during a period of great auroral activity, and that 300 years later, when the aurora was also prevalent, despite difficulties caused by weather and bright summer nights, Tromholt was able to show from records in his father's diary, kept in the same part of Europe in 1859-1873, excellent correlation. The years of maxima correspond exactly for both phenomena; as to monthly frequency, the spring maximum agrees but there is a difference of a month in the autumn, perhaps accidental.

Dauvillier holds that the shower of secondary electrons exciting auroral radiations could produce negative ions at 80 km which could be carried downwards by descending currents to form condensation-nuclei at cirrus levels. A pointer to downward transfer of actual auroral matter is that in Iceland, despite the lack of thunderstorms, which are its usual source, the deposit of ammonium nitrite reaches 1.3 kg/hectare (about 1.2 lb/acre); this deposit oxidizes quickly. Iceland is in the region of continuous aurora, unaffected by the solar cycle.

Dauvillier also considered that the ionospheric disturbance transmitted downwards could cause the lower atmospheric movements shown by increased twinkling of stars during auroral and magnetic activity. He also holds that though the mass of the ionosphere is negligible as compared with that of the lower regions, yet if these are in an unstable state trigger action is possible. He points out that the Icelandic depressions which concern the inhabitants of western Europe are born in the auroral zone!

Some lines for investigation may be suggested:

1. Measurements of the radii of lunar halos associated with aurorae.
2. Careful note of auroral glows apparently seen against clouds.
3. Estimation of the delay between the appearance of aurorae and of true cirrus clouds showing refraction halos of 22°.
4. If practicable, measurements of the height of daytime polar clouds, their connection with magnetic activity, their alignment with magnetic bearings, and any change at night into aurora. Also whether they become an aurora, or replace it (this latter point is valid outside the polar regions, for apparent cases have been observed as far south as Rome).
5. Observations of any connection between aurora and subsequent weather, with relation to local views, such as the Scottish saying that the first great aurora of autumn is followed in two days by a storm.
6. Investigation of the luminous glows which have been recorded mainly from high latitudes apparently at very low levels, even to the sea, of the type of outside St. Elmo's Fire seen on mountain peaks, * and which cannot be explained entirely by mistakes, illuminated mists, etc. For instance such distinguished explorers as Sabine and John Ross actually sailed through such a mist.

Observers need not be discouraged by the fact that at present solar activity is on the wane, and that 1964 is to be observed internationally as the Year of the Quiet Sun. There is no closed auroral season inside geomagnetic latitude 58° (Scotland, Canada and south polar stations), also Sol may spring a surprise or two! (Weather, 18:217-219, 1963)

* Known as the Andes Lights, in that chain it reaches extraordinary development, but has been seen in the Alps, Teneriffe, the famous Mount Ida in Turkey, and Lapland. The popular explanation in South America of volcanic activity cannot, from the facts, be maintained; the glow is probably intense point discharge, with increase of height.

# AURORAS AND OZONE ODOR

## LOW AURORAS AND TERRESTRIAL DISCHARGES
**Beals, C. S.; *Nature,* 132:245, 1933.**

Investigations by Tromholt, Chant, Johnson and myself have produced evidence which points to the existence of sounds of a hissing, swishing or crackling nature associated with brilliant displays of the aurora. Evidence has also been presented which suggests the existence of luminous effects at low levels accompanying occasional displays. In explanation of these phenomena, the hypothesis has been advanced that there is sometimes present, during auroral displays, an unusual electrical condition of the atmosphere which results in an electrostatic discharge close to the earth's surface. Such a discharge, if real, is probably of a 'secondary' nature induced in some way by processes which take place at great heights and are responsible for the more conspicuous and familiar aspects of the aurora as investigated by Stormer and his collaborators.

Recently a number of reports have been received from observers in northern Canada who have detected the odour of ozone during auroral displays when the displays were accompanied by sounds. These observations would appear to offer in some degree confirmation of the idea that the sounds and the low level displays are due to electrostatic discharges close to the earth's surface. (Nature, 132:245, 1933)

# ARTIFICIAL AURORAS

## AN ARTIFICIAL AURORA
**Lemstrom, Prof.; *Symons's Monthly Meteorological Magazine,* 18:51-55, 1883.**

Amid the almost countless forms presented by aurorae, there is one of special interest as indicative of its origin. This form, which was observed during the Swedish Polar expedition of 1868, consists in thin flames or a phosphorescent glow surrounding elevated objects, and especially mountain tops. This characteristic is so marked that it needs no training to observe it, for it was observed by our celebrated philologist, M. A. Castren, during his voyages in Siberia, and by him very tersely and accurately described.

"My first experiments in 1871 were suggested by what I had myself observed in Spitzbergen, and were directed towards ascertaining whether a similar appearance could not be produced, or at least, developed by suitable methods.

"Starting with the hypothesis that aurorae in general, and especially the above-named variety thereof, are due to electrical currents in the atmosphere, I erected on the summit of Luosmavaara, about 560 feet above Lake Enare [in Lapland] an apparatus composed of a corona of numerous fine points of copper wire united to a single copper wire, this coronet, about 8 in. square, was placed on the top of a tall pole. From it an insulated copper wire (0.015 inch in diameter) led to a galvanometer, which was placed in a room at the rectory of Enare, about 2-1/2 miles distant; another copper wire was led from the galvanometer to a plate of platinum buried in the earth. When the connection between the coronet and the earth was broken the needle of the galvanometer was appreciably, though but slightly, deflected. The very night (November 27th, 1881) after the apparatus had been erected an aurora began, the first manifestation being a single ray perpendicularly over Luosmavaara. This ray gave in the spectroscope the true greenish yellow auroral line, but it was not certain whether the ray was truly vertical over the mountain or behind it. Subsequent observations, which I will presently mention, render it nearly certain that the ray was produced by the apparatus.

"During the same expedition, but a few days earlier (November 21st, 1871) while studying the spectra of the flames on surrounding summits which, on that day, were somewhat brighter than usual, we found that the spectroscope gave the greenish yellow line from nearly all the surrounding objects, the ice of a pond, the roof of a house, and even (but very feebly) from the immediate vicinity of the observer. These observations led me to believe that I was then in the midst of a feeble aurora, and from that time onwards it has always been my wish to subject aurorae to further practical tests. This desire was also strengthened by an accident, viz., the appearance of light in a Geissler tube, which was not in direct connection with any electrical machine. For this accident led to researches which proved that the current which leaves one pole of an electrical machine, while the other pole is connected to earth, may pass obscurely through a stratum of air of ordinary density, but produce light in a stratum of diminished density."

. . . . . . . . . .

Prof. Lemstrom then explains that various causes conspired to delay his operations, and that, finally, he had to start with somewhat makeshift apparatus. His first experiments made upon a clocktower at Sodankyla were unsuccessful. He therefore removed his apparatus to the top of Oratunturi, a mountain 12-1/2 miles from Sodankyla and in lat. 67° 21' N, and lon. 27° 17'E. The summit is about 1,798 feet above the sea, and 971 feet above the village of Sodankyla. His so-called discharging apparatus, placed on this summit, consisted of a stout copper wire, 0.08 inch in diameter, supported on insulators fixed to posts 8 feet high. The wire was arranged in a rectangular spiral, somewhat as shown in the annexed diagram, the posts being at the angles,. and no two lines of the wire at a less distance than 3 feet; the wire was studded with brass needle points at each 20 inches. Judging from the description, the coronet must have contained about half-a-mile of wire, and been studded with more than 2,000 points. An insulated wire was led from the interior extremity of the spiral to a rough shed about 600 feet lower down the mountain, in which it was connected to a galvanometer, whence another wire was carried to "earth" in a small stream.

This apparatus was completed on December 5th, 1882, and thereafter there was nearly every night a yellowish-white light over the summit, though none was visible on any adjacent mountain. The light was a flashing one, that is to say, very variable in intensity, and always in motion. Examined on three occasions by the spectroscope at a distance of 2-1/2 miles, it gave a feeble continuous spectrum from D to F, upon which could be observed the auroral line $\lambda = 5,569$ faint, and varying in intensity. The galvanometer always showed a positive current from the coronet to the earth. The temperature was, however, very severe, generally about -20°F (52° below freezing), and the apparatus was constantly being deranged and damaged by the masses of hoar frost which formed upon it, destroying the insulation, and frequently breaking the insulated wire between the coronet and the cabin.

Subsequently Prof. Lemstrom erected a somewhat similar apparatus on Pietarintunturi, a mountain near "Kultala (lat. 78° 32' N. lon. 27° 17' E.)" [There is some mistake here, probably this a place shown on the U. K. map as Kustala in lat. 67° 50' N. and lon. 25° 35' E. and about 40 miles N. W. of Sodankyla]. Here experiments were possible on only two days, but on the first of them, December 29th, 1882, an auroral streamer about 400 feet long was formed, which observations from several azimuths proved to be exactly over the apparatus. (Symons's Monthly Meteorological Magazine, 12:53-54, 1883)

## LEMSTROM'S ARTIFICIAL AURORA
**Anonymous; *Science*, 4:465-466, 1884.**

The effect produced by Mr. Selim Lemstrom in his experiments on the artificial production of the aurora is well shown in the illustration on the next page, from a drawing by Mr. Lemstrom. On the top of the mountain a wire was stretched on poles, and furnished at every foot or two with brass points. This wire was several miles in length, and was carried over the top of the hill in the form of a spiral. The poles were supplied with sulphuric-acid insulators in order to prevent, as far as possible, the creeping of electricity over their surfaces. From this insulated system of points a wire was run to a ground-connection at a lower level, perhaps from five hundred to a thousand feet below. If the ground-connection had been made on the same level, no current would have been observed; but when there was a difference of level, even if not more than thirty feet, a current was always observed from the higher point to the lower. The luminous effects portrayed in the cut were only visible to the naked eye when there was a marked display of northern lights; although by the aid of the spectroscope, which would show the peculiar spectrum of auroral light, the existence of the streamer could often be proved. (Science, 4:465-466, 1884)

"Artificial aurora" over Lemstrom's system of wires in Lapland

# GEOGRAPHICALLY DISPLACED AURORAS

## THE AURORA OF MARCH 30, 1894
**Veeder, M. A.; *American Meteorological Journal*, 11:77-78, 1894.**

The aurora of Friday evening, March 30, was first seen at Lyons, N. Y., at 7.40 p.m. It then consisted of flickering, irregular patches which filled the entire heavens from the zenith down to the southern horizon, there being nothing whatever to be seen in the usual location of auroras toward the north. This distribution of the display continued with some variations in brightness until 8.38 p.m., standard time, when streamers began to form northward for the first time during the evening, and the heavens became luminous in that direction, presenting an appearance similar to that previously seen toward the south. Fortunately, quite a large number of observers are making concerted observations of the aurora from South Carolina northward in the vicinity of the 77th meridian into Canada. From the reports thus far received, it appears that at the hour above indicated, the aurora was seen toward the south exclusively at stations directly north of Lyons, while southward in Pennsylvania the entire sky was more or less covered with flickering patches and masses of light, and still further south the aurora was seen toward the north exclusively, at Charleston, S. C., rising about 25° above the northern horizon. Thus it appears that the central point of luminous mass at the longitude named was probably within the bounds of the State of Pennsylvania. A similar distribution in latitude has been noticed in some other instances, heretofore, but has not been so strongly defined. (American Meteorological Journal, 11:77-78, 1894)

# ANOMALOUS AURORAS

## PECULIAR AURORAL PHENOMENON
**Anonymous; *Nature*, 26:160-161, 1882.**

A peculiar and interesting auroral phenomenon witnessed from the steamship Atlantis off the Newfoundland coast on September 12 last year, has been described by Mr. Engler to the St. Louis Academy. While an aurora of normal type was clearly seen in the northern sky, there appeared in the southeast, about 30 to 35 deg. above the horizon two horizontal streaks of light, about 5 deg. apart, and 15 or 20 deg. in length. Their pale hazy light resembled moonlight. From the upper streak were suspended, by small cords of light, a number of balls, brighter than either of the streaks, which were continually jumping up and down in vertical lines, much like pith-balls when charged with electricity. Above the upper streak was a bright gauze space with convergent sides, seemingly composed of streamers of light, the brightness diminishing from the streak outwards. From the lower streak extended a similar mass, differing only in a greater inclination of the streamers. The balls and cords gradually disappeared first, then the streamers, then the streaks; and the whole phenomenon lasted about half an hour. No explanation is offered. It is noteworthy that on the same evening and at the same hour, a most remarkable band of white light was seen at Albany, N. Y., Utica, N. Y., Hanover, N. H., Boston, Mass., and elsewhere in the North Atlantic States, spanning the heavens from east to west near the zenith. (Nature, 26:160-161, 1882)

# PHENOMENA COINCIDENT WITH AURORAS

Thunderstorms are manifestly electrical in nature, and we know that they sometimes penetrate far into the ionosphere. Earthquakes likewise generate "disturbances" that reach into the upper atmosphere and ionosphere and which therefore may be electrical in nature. In concept, then, thunderstorms and earthquakes could create low conductivity paths for earth-to-space electrical discharges. Meteors, as is well-known, leave long ionized trails at high altitudes and could therefore interact with "other" aurora-generating mechanisms. The section on Electric Discharge Phenomena has many auroral nuances as one might suspect from the observations recorded below. Mountain-top glows and some earthquake lights, for example, seem related to auroral phenomena.

## AURORAS AND THUNDERSTORMS

### THUNDERSTORMS AND AURORAL PHENOMENA
**Davidson, J. Ewen; *Nature*, 47:582, 1893.**

I am residing in tropical Queensland, lat. 21$^{o}$S., and consequently am not likely to see any auroral phenomena, particularly in the middle of our hot and rainy season; but last night between 8 and 9 p.m. there occurred the following remarkable appearances, which were seen by me and several others.

There was a sharp thunderstorm with incessant lightning visible on the southern horizon, occupying a width of 10$^{o}$ and an altitude of from 5$^{o}$ to 10$^{o}$ above the horizon, probably from 80 to 100 miles off.

But for the distant thunderclouds the sky was clear and starlight, with a few light cirrhus clouds drifting before the north wind.

I was sitting on the lawn watching the distant flashes, when suddenly a patch or cloud of rosy light---5$^{o}$ to 6$^{o}$ in diameter---rose up from above the thunderstorm and mounted upwards; disappearing at an elevation of from 40$^{o}$ - 45$^{o}$. There were about twenty to twenty-five of these patches in the course of half an hour, sometimes three or four in quick succession; they took from one to two seconds to mount, and were not associated with any particular flash; the rosy colour contrasted strangely with the silvery light of Nubecula Major just above. There were also occasional streamers, sometimes bifurcated, of 2$^{o}$ in breadth, which shot up in the same way as the auroral streamers, which I have seen both in the arctic and antarctic zones.

Auroral phenomena are known to be electrical manifestations, but here were the same phenomena exhibited in connection with a thunderstorm in the tropics. Thinking this phase of electrical action worthy of note, I send you this account and enclose my card.

P.S.---The thunderstorm, patches of light, and streamers were distinctly connected; it was not a case of an ordinary aurora, with a thunderstorm interposed. (Nature, 47:582, 1893)

## AURORA AND REMARKABLE LIGHTNING
**Riley, F. N.: *Marine Observer*, 23:81-82, 1953.**

M. V. Melbourne Star. Captain F. N. Riley, D. S. O. Adelaide to Southampton. Observers, the Master, Mr. R. B. Escreet, 4th Officer, and Mr. C. T. Whitaker, 3rd Officer.

27th June, 1952, 1800 G. M. T. The evening was dark but fine, and exceptionally clear, with 1/8 Cu at 4, 000 ft. The first signs of a coming display of aurora had the appearance of distant lightning. By 1900 the aurora was more pronounced and appeared as ribbons of flickering light in an arc extending from about 180° to 290°(T), with the lower edge sharply defined in a brilliant golden orange colour, while the upper edge was more whitish. The lower edge slowly increased in altitude until it reached about 18°. The width of the arc, measured at the horizon, was 29°. It was observed to the NW that diffused patches of light were scattered across the horizon; these illuminated the lower cloud (Cu) most vividly with a pale bluish-white colour.

At 2015 a brilliant red rod shot upwards from the horizon to about 15° and then spread out about 15° on either side in a vivid pale pink. By 2030 the aurora had attained such brilliancy that the evening resembled sunrise, but soon the sky became overcast, with a rising wind and falling barometer. At 2130 the colourful scene was fading out.

During the whole display no exceptional errors in the magnetic compasses were observed, although around 2030 the cards were swinging 5° or 6° either way.

At 2130, following on the display of Aurora Australis, the sky became overcast at 1, 500 ft and lightning became intense, the flashes being about 10° above the horizon and circling around the sky. There were also streaks which commenced at about 170°(T) and travelled through an arc of 130° over the zenith and terminated with a crack and a slight rumble of thunder. The colour of the streaks was a whitish-blue, while intense rain storms followed at intervals of 7 to 10 min. Lightning of a general character developed, while single cracks of lightning were observed at 160° (T) at an altitude of approximately 8°. These were light blue in colour and left a crimson streak in the sky for about 6 sec.

At 2330 a light-blue streak was observed in the form of an arc between bearings 280° and 310°(T), with a thickness of 4° to 5° and altitude of 10°. The edges were of a bright crimson with a centre of pale blue. This streak increased in altitude to about 20°, in a vertical direction towards the zenith and remained 3 to 5 min before it disintegrated. At 0030 (28th) the sky was 3/8 clouded and no more lightning was observed.

Position of ship at 1800 on 27th: 33° 35'S, 46° 50'E.

Note. This is an interesting observation as the thunderstorm which followed the auroral display seems to have shown abnormal forms of lightning. The crimson streaks left in the sky by some of the flashes were probably an effect of persistence of vision, coupled perhaps with colour fatigue of the eye due to the brilliance of the original flashes. An ordinary lightning flash, where the track of the flash is seen in the sky, is a result of persistence of vision, but this does not normally last for more than half a second or so.

The phenomenon seen at 2330 is more difficult to explain. It is not clearly stated that the sky remained completely overcast at this time, in which case the phenomenon might have been a bi-coloured auroral ray. This is the most likely explanation as it does not seem possible for a lightning flash to persist for three to five minutes, but the bearing on which it was observed seems to argue against it being of auroral origin. (Marine Observer, 23:81-82, 1953)

### AURORAS VERSUS THUNDER-STORMS
**Veeder, M. A.; *Science,* 20:221, 1892.**

During September just past sun-spots were very numerous and large. Nevertheless, auroras during the month were without exception comparatively inconspicuous. In this case certainly large sun-spots have not been attended by bright auroras, as some have held to be the rule. The explanation of this anomaly, which appears to be justified by systematic records in my possession, is that thunder storms took the place of auroras. It has been found that not unfrequently thunder-storms become widely prevalent upon dates upon which auroras should fall in accordance with their periodicity corresponding to the time of a synodic revolution of the sun. When this happens, it robs them of their brightness, wholly or in part. The relation between these two classes of phenomena appears to be reciprocal or substitutive, the one taking the place of the other under conditions which are only just beginning to be understood, and which are in process of investigation. (Science, 20:221, 1892)

## AURORAS AND EARTHQUAKES

### AURORA
**Veeder, M. A.; *Nature,* 35:54, 1886.**

Last evening (November 2), between the hours of seven and eight o'clock, a bright aurora was visible in this vicinity. At intervals later in the evening, patches of cirrus clouds in the northern sky became luminous. The disturbance of the suspended magnet was at its height early in the evening, when the aurora was brightest. It is interesting to note the fact that this aurora was twenty-six days removed from that of October 7 and 8, corresponding to the time of the revolution of the sun on his axis. It is noteworthy, also, that very near to the time of the appearance of each aurora there was a slight renewal of earthquake activity in South Carolina and other localities. (Nature, 35:54, 1886)

### ON CERTAIN METEORIC PHENOMENA
**Clarke, W. B.; *Magazine of Natural History,* 7:300-301, 1834.**

The year 1750 is celebrated for shocks of earthquakes throughout the earth. The Rev. Mr. Sedden of Warrington has recorded that, at the moment of the shock there (April 2.), he saw an infinite number of rays of light, proceeding from all parts of the sky, to one point near the zenith. The rays were at first yellow, afterwards blood-red, and were visible for twenty minutes. Dr. Doddridge of Northampton relates that, at 4 a.m. of Sept. 30 (the day of the shock at that place), a ball of fire was reported to have been seen; that on the following night the sky was red as blood, and on the night after appeared the finest aurora he ever saw. Several days before the shock at London (March 2), there were reddish bows in the air, which took the same direction as the shock. The aurora also accompanied the earthquake of Aug. 23 at Spalding. Sir W. Hamilton mentions that for many hours after the eruption of Vesuvius, in 1799, the air "was filled with meteors, such as are vulgarly called falling stars: they shot generally in a horizontal direction, leaving a luminous trace behind them, but which quickly disappeared. The night was remarkably fine, starlight, and without a cloud." Ashes fell that night at Manfredonia, 100 miles from Vesuvius, two hours after the eruption. (P. T., 1780.) The volcanic lightning and electrical phenomena, on that occasion, are mentioned as truly astonishing. Mr. Poulett Scrope has a similar remark respecting the eruption of 1822. (Considerations on Volcanoes, p. 81.) During the eruption of 1794, it is stated that "out of these gigantic and volcanic clouds, besides the lightning, both during this eruption and that of 1779, I have, says Sir W. Hamilton, with many others, seen balls of fire issue, and some of a considerable magnitude, which, bursting in the air, produced nearly the same effect as that from the air-balloons in fireworks, the electric fire that came out having the appearance of the serpents with which those firework balloons are filled." (P. T., 1795.) Sir D. Brewster witnessed an aurora in combination with a thunderstorm at Belleville, Inverness-shire, on Aug. 29, 1821, 9-1/2 p.m. (Encyc. Brit., viii. 623., 7th ed.) (Magazine of Natural History, 7:301, 1834)

# AURORAS AND METEORS

## SPECTRA OF METEOR WAKES
**Botley, Cicely M.; *Nature,* 200:1194, 1963.**

With regard to the interesting communication by Rao and Lokanadham I would direct attention to a possible consequence.

Hoffmeister has collected various instances of auroral-like glows connected with meteoric showers, Perseids, Leonids, 1872 Bielids. Such was also noted in Germany and England during the famous Leonid shower of November 13-14. It was indeed remarked it might be possible that "brilliant meteor streams electrify the higher regions of the atmosphere and produce phenomena analogous to the aurora borealis."

The detection of H and the red line of NI is interesting in view of the observation by von Wrangel in Siberian waters 1820-24 that dark areas of aurora traversed by meteors lighted up and continued to glow. (Nature, 200:1194, 1963)

## AURORA AND METEOR DISPLAY OF SEPTEMBER 14
**Packer, D. E.; *English Mechanic,* 88:211, 1908.**

1908. Sept. 14, 8.40 p.m.---Sky clear generally, and faintly luminous, owing to moonrise in N. E. A bright, horizontal cloud observed stretching across the fore part of Ursa Major, stretching across the fore part of Ursa Major, and superposed upon a dark cloud or darkness, so intense as to appear almost black in the general twilight. The bright cloud was in a constant state of fluctuation, flashing up and fading with almost the same rapidity as the well-known scintillation effect observed in bright stars. The colour was a pale white, and, as far as could be judged, the cloud was stationary over the magnetic meridian during the whole period of observation. At 8.45 a meteor from Perseus approached it, and on entering the cloud at its eastern extremity burst out with extraordinary splendour, far surpassing Venus in lustre, and exhibiting a string of many smaller nuclei on each side of the main nucleus. At 8.50 the cloud at this point had thrown out a streamer, which was traceable between Auriga and Perseus into Aries. At 9 p.m. the streamer had faded, and the cloud showed signs of disruption into two separate masses, united by a narrow neck; also there now appeared evidences of other clouds forming in the neighbourhood, less luminous than the main cloud, and more difficult of observation, owing to the increasing moonlight. Subsequently the main cloud became more compact as at first, the fluctuations became less pronounced, and by 9.30 no traces of the phenomenon were visible. (English Mechanic, 88:211, 1908)

# AURORAS AND BRIGHT CLOUDS

## THE AURORA OF SEPTEMBER 9, 1898
**Purnell, Chas. W.; *Nature,* 59:320, 1899.**

I observe, from Nature, that an auroral display was visible in the South of England on the evening of September 9. It may interest some of your readers to know that an aurora was seen here on the evening of September 10. The display began at about a quarter to eight o'clock, and lasted for an hour or so. The whole southern heavens at first became suffused with a bright orange light low down upon the horizon, from which a few streamers issued from time to time, rising (judging by the eye) to a height of, say 45 degrees above the horizon. When both glow and streamers had faded away, I noticed three luminous clouds, one at the zenith. The largest of these clouds increased in size, and shot forth a few streamers of light, both upwards and downwards, and all then disappeared. I have witnessed several auroral displays at Ashburton, but none like that of September 10, the distinguishing features of which were the orange glow and the luminous clouds.

On the following day, my telephone, which had never failed me before, worked irregularly, and some of the other telephones in the town were similarly affected. (Nature, 59:320, 1899)

## AURORAL PHENOMENA
**Ingleby, C. M.; *Nature*, 23:363, 1881.**

It is perhaps worth a note that my daughter saw at Folkestone a very unusual phenomenon on the evening of January 25, a little before 6.30. Some distance to the left of Orion (for the night was clear and starry) she observed a small cloud of a bright golden hue, from which streamers of great brilliancy darted in various directions, the cloud alternately paling and brightening. She describes the streamers as like small meteors, leaving trails of light behind them. (Nature, 23:363, 1881)

Bright cloud with radiating streamers

# SKY GLOWS, SKY BRIGHTENINGS, AND WEATHER LIGHTS

Typical auroras are confined to one portion of the sky and identify themselves through their distinctive structures. Much more rarely, the entire sky brightens well above its usual nighttime level. The glow may last hours or just a fraction of a second. The long-lived glows may arise from dust suspended throughout the high atmosphere after a major volcanic eruption or a collision with an astronomical body. For example, the 1908 sky glows reported below have been attributed to the famous Tunguska Event, also known as the Siberian Meteor. (It may have been a small comet!) Transient sky brightenings, however, are more mysterious. Could they be all-sky auroras stimulated by the sudden influx of extraterrestrial radiation that for some reason is not deflected to the polar regions by the earth's magnetic field? No answers are at hand.

Localized sky glows and brightenings on the horizon are reputed to presage weather changes and are sometimes called "weather lights." Weather lights have the curious habit of moving around the horizon, indicating that they are probably not bright cloud tops. Since these glows are quite steady, lightning also seems out of the question. Localized auroral activity and massive terrestrial electric discharges, both of which may be associated with weather changes are other possibilities. For example, the section on Electric Discharge Phenomena describes columns of light seen during some tornadoes.

## NIGHTTIME SKY GLOWS

### AURORA IN 1886
**Fleming, David; *Popular Astronomy,* 29:448-449, 1921.**

The description of the auroral effect observed to the southeast from Riverside Drive, New York City, on the night of May 14th ult., published in your June-July number, induces me to record the account of a peculiar phenomenon observed in Florida about the year 1886, by the late Col. A. F. Wrotnowski, who was with James B. Eads and E. L. Corthell on the Mississippi jetties work.

The strange character of the occurrence so impressed me that I am able to recall almost verbatim his words in the description given to me by him in 1910, when I met him in Mexico.

He said, "I was making some preliminary surveys for canals, and was camped in a house boat at the middle of a large lake in Southeastern Florida. We were far from any settlement.

"I awoke a little before 3:00 a.m. and noticed that it was very light, and as there was no moon, got up and went out on the deck, and to my surprise saw everything illuminated with a pale greenish light so intense that we could read by it.

"I woke up the rest of the party, some 8 or 10 men and we all watched the phenomenon with great astonishment.

"The atmosphere was hazy, and we could not make out the stars, and the light seemed to be general and from no particular direction. I remember we got out a newspaper and the print could be easily read.

"The illumination lasted fully an hour and gradually faded away to normal.

"We discussed the occurrence for days, but never reached any plausible solution of it."

Col. Wrotnowski was a man of the highest character and of unquestionable veracity, and the similarity of the green color described, and that noted by Mr. Wilson in New York in the direction of Arcturus, leads me to believe that while the New York effect was local to that region, the above occurrence, though so far south, must have been from an aurora. (Popular Astronomy, 29:448-449, 1921)

## BRILLIANT SKY GLOWS, JUNE 30 AND JULY 1, 1908
**Anonymous; *Royal Meteorological Society, Quarterly Journal,* 34:202, 1908.**

Brilliant sky glows were observed in many different parts of the United Kingdom on the night of June 30 and on several succeeding nights, the phenomenon being apparantly at its maximum intensity on the night of July 1. The whole of the northern part of the sky, from the horizon to an altitude of about 45$^{o}$, and extending to the west, was suffused with a reddish hue, the colour varying from a pink to an Indian red, whilst to the eastward of north the colour was distinctly a pale green. No flickering or scintillation was observed on the reddened sky, nor was there any tendency to the formation of the streamers or luminous arch characteristic of aurora. Cirro-stratus clouds near the horizon were tinged with the same colour as the surrounding sky. A special feature in connection with the phenomenon was the prolongation of twilight, extending almost to the following daybreak, and from the experience cited by many observers in various parts of Great Britain the light at midnight was sufficient to allow of fairly small print being read without any aid from artificial light. (Royal Meteorological Society, Quarterly Journal, 34:202, 1908)

## AN EXCEPTIONAL NIGHT SKY
**Anonymous; *Nature,* 127:871, 1931.**

Lord Rayleigh, in the May number of the Proceedings of the Royal Society, describes an unusual night sky which was watched by him at Terling (52$^{o}$N.) on Nov. 8, 1929. The light was much the same in constitution as on ordinary nights, extending uniformly over the sky, but was very much brighter than usual, and of constant intensity over a period of hours. Its spectrum seemed continuous, and the green auroral line ($\lambda$ 5577) was not definitely seen. It was evidently of a totally different nature from the polar auroral light, and, confirming this distinction, the state of the earth's magnetism was steady. The negative bands of nitrogen could not be searched for on this occasion, as they were outside the range of the spectroscope used, but Lord Rayleigh says that in his experience the most striking distinction between the ordinary night sky and the polar aurora is the absence of the bands from the former; a contrary conclusion has, however, been arrived at by Sommer, who finds that these bands are present in the normal night light at Gottingen. Lord Rayleigh has remeasured the two unidentified lines (or band heads) in the night sky spectrum at $\lambda$ 4419 and $\lambda$4168. (Nature, 127:871, 1931)

## THE ATMOSPHERE NEVER DARK ON A WINDY NIGHT
**Anonymous; *Scientific American,* 3:226, 1848.**

Several years since, says a writer in the Magazine of Natural History, when travelling by night in the mail coach, in the depth of winter, and during the absence of the moon, I was surprised to observe, that though dense clouds covered every part of the horizon, and not a single star could be seen, yet the night was far from being dark, and large objects near the road side were easily discerned. On expressing my surprise to the driver, he replied, "The wind is very high, and during a great many years that I have been upon this road I never knew it to be dark on a windy night." The observation was at that time new to me; but subsequent experience has convinced me that it was true. (Scientific American, 3:226, 1848)

## "FALSE DAWN"
**Cumming, J. H.; *Meteorological Magazine,* 71:189-190, 1936.**

I write to describe a phenomenon which much interested me. On the night January 6th-7th, 1933, I had occasion to proceed from London to Torquay. I left London at 1 a.m. precisely and travelled by car by the main road, i.e. A. 30. The night was very dark and at Bagshot I ran into a storm area, heavy rain and gusts of wind. I reached Shaftesbury, 101 miles, between 4.30 and 5 a.m. I cannot give exact times but I have driven over this road many times and have often had to plan my departure from London so as to reach certain places in Somerset and Devon at certain times, so that in actual fact the times will be only a few minutes out, one way or the other. Descending the hill from Shaftesbury, where in daylight you get such a magnificent view westward, I could see nothing of the surrounding country owing to the darkness. I proceeded for a few miles and then stopped for a short halt, and whilst I cannot name the exact spot it was a small village and geographically would correspond with East Stour, i.e. between Shaftesbury and Milbourne port.

I got out of the car and found the rain had ceased and strolled up and down for a few minutes, when I suddenly realised that it was getting light, and at first I thought it was the dawn, but knowing it was only about 5 a.m. I realised I was mistaken, but very unfortunately I cannot give exact times. But the time was almost certainly about 5.10 a.m. It became so light that I was able to see the surrounding countryside. I could see the Dorsetshire hills sufficiently plainly to see that the clouds lay on their tops and that they were clear below, and I was able to turn off my headlights. The light was a general diffused light with no apparent focal point but I should say it was lighter towards the south and east, although I could quite well see towards the north and west---and here again I failed to note times; but this condition lasted probably for about 20 minutes, after which period, and about as suddenly as it appeared, it became quite dark again. I had to switch on my headlights and keep them on until Fairmile, which is 10 miles short of Exeter, at about 7.30 a.m., when the true dawn commenced. The other I can only call a "False Dawn". I have many times read of it, chiefly in seafaring stories, and have asked many people but found none who have seen it. There was nothing remarkable about it beyond the fact that it became light about two hours before dawn was due, as at that time sunrise is about 8.4 a.m., and then became quite dark again. (Meteorological Magazine, 71:189-190, 1936)

# TRANSIENT SKY BRIGHTENINGS

## FLASH OF LIGHT COVERS NIGHT SKY
**Hopkins, L. J.; *Marine Observer,* 27:209, 1957.**

S.S. Gothic. Captain L. J. Hopkins. Wellington to Balboa. Observers, Mr. T. I. Oliver, 3rd Officer, and Quartermaster Tavener.

5th December, 1956. At 0643 G.M.T. a harsh, brilliant flash, bluish-white in appearance, covering the whole of the night sky, appeared, lasting for approximately 1-1/2 sec. The sky was completely covered by a thick layer of Sc.

The possibility of a meteor was ruled out as the cloud would not have permitted it to be visible, and the character and intensity of the flash did not conform to lightning or other electric phenomena. The forecastle head was plainly visible from the bridge and the horizon clearly defined. The source of the light appeared to issue from the NE. sky at an altitude of approximately 60°. The lookout confirmed this. The radio officer did not hear any static at the time. Bar. 1022 mb. Air temp. 63°F, wet bulb 58°, sea 65°. Wind E., force 5. (Marine Observer, 27:209, 1957)

## PURPLE LIGHT AT NIGHT
**Furneaux, S. J.; *Marine Observer,* 9:93, 1932.**

The following is an extract from the Meteorological Log of S.S. Nova Scotia, Captain S. J. Furneaux, Liverpool to St. John's, N. F. Observer, Mr. J. E. Wilson.

"May 24th, 1931. At 0135 G.M.T. Sea and sky were suddenly quite brilliantly lit for about three seconds with a flickering purplish light, which did not appear to emanate from any particular point. Weather at the time was light W.N.W. wind---smooth sea; cloud A-St. amount 10. Shortly before this the wind has been S.W., light and the sky perfectly clear. (Marine Observer, 9:93, 1932)

## BRIGHT BLUE LIGHT FROM CLOUDS
**Williams, J.; *Marine Observer,* 30:194, 1960.**

M. V. Trevean. Captain J. Williams. Curacao to London. Observer, Mr. D. C. Penberthy, Chief officer.

1st December, 1959. About 2000 G. M. T. when low cloud was passing across, a bright blue diffused light suddenly grew, as from the clouds overhead. It was thought at first to be an electrical fault in the wheelhouse and I looked inboard for it: a few sec later I stepped out on to the open deck, when the light faded and went out. No actual flash was seen---it was as though a diffused blue light had been gradually unshaded behind the clouds and then gradually covered again. Both the lookout man and the man at the wheel were in agreement about the appearance and intensity of the light. The wind at the time was NW., force 6-7.

Position of ship: 39° 45'N., 17° 40'W. (Marine Observer, 30:194, 1960)

## STRANGE PHENOMENON
**Rogers J.; *Knowledge,* 5:171, 1884.**

Whilst walking in the country last Monday evening, at about 10.20, I saw the following strange phenomenon. The night was dark and calm, and a drizzling misty rain was falling, when suddenly the whole heavens became illuminated for about two seconds, and then all was darkness again. The light was not vivid, but soft, and bright enough to show clearly the surrounding objects. (Knowledge, 5:171, 1884)

## NOTICE OF A REMARKABLE METEOR WHICH APPEARED ON THE 2D OF DECEMBER, 1814
**Wallis, John; *Annals of Philosophy,* 5:235-236, 1815.**

On Friday night, the 2d of Dec. at about 20 min. before 11, I was walking in an open part of the village of Peckham, about four miles S. S. E. of London. The night was cloudy and dark, the lower part of the atmosphere clear and calm, a very slight wind blowing from the E. Suddenly I was surrounded by a great light. I remember that at the instant I shrunk downward and stooped forward; as I was apprehensive of some danger behind me, I instantly ran a few paces. I turned about in a few seconds to the N. E.; for I was certain the light came from that part of the heavens (as it brilliantly illuminated some houses to the S. W. of me); and I think at a considerable height from the horizon. But I saw nothing to cause this light. It did not give me the idea of the force and intensity of lightning; its brilliancy was not so instantaneous and fierce; but it was a softer and paler kind of light, and lasted perhaps three seconds. I could discover no noise, though immediately I expected an explosion.

The strength of the light was nearly equal to that of common day-light; all near objects were distinctly visible. The light very much resembled that of the luminous balls thrown from a sky-rocket when it finally explodes. It was not as vivid and blue as that of a strong flash of lightning at night. None of the persons I met that night thought it to be lightning, though none of them saw anything but the light. Unless it had been very high, if it was in the direction I have supposed, it would be concealed from them by a high wall and some houses.

A relation of mine, who resides at the northern extremity of Tottenham, saw the light of it as he lay in bed, through a window facing the west. He describes it to have been as light as day. (Annals of Philosophy, 5:235-236, 1815)

# LUMINOUS PATCHES ON THE HORIZON (WEATHER LIGHTS)

## WEATHER LIGHTS
**Conrad,Cuthbert P.; *American Meteorological Journal,* 1:81, 1884.**

On the evening of February 26, 1884, a phenomenon known locally as "weather lights," was the precursor of a violent snow storm. I have watched these "weather lights" in this locality for four years, and while I have lived in five states east of the Alleghanies, have never witnessed similar phenomena elsewhere. The appearance is that of a rosy red to white light appearing above the horizon, $5^{o}$, $10^{o}$, and even $30^{o}$, and all the way from northeast around to the southwest---sometimes only in the northwest (the most frequent quarter); sometimes first in the northeast, fading out and appearing in the northwest, west, or southwest. These lights invariably precede a change in weather---either rain or snow (i.e., a change invariably follows), but I have not been able to fix upon any definite interval of time. (American Meteorological Journal, 1:81, 1884)

## PATCHES OF LIGHT MOVE AROUND HORIZON
**Porter, L.; *Marine Observer,* 25:213, 1955.**

S.S. Perim. Captain L. Porter. Adelaide to Aden. Observer, Mr. G. S. Forbes, Cadet.

8th October, 1954, 1800 S.M.T. (1445 G.M.T.). Just after the sun had disappeared below the horizon bright patches of light were seen to travel round the horizon in a clockwise direction as shown in the sketch. Each wedge was seen to sink into the horizon and start again.

Position of ship: $12^{o}$ 06'N, $60^{o}$ 38'E.

Note. This is a very remarkable observation, depending on some unusual form of abnormal refraction. (Marine Observer, 25:213, 1955)

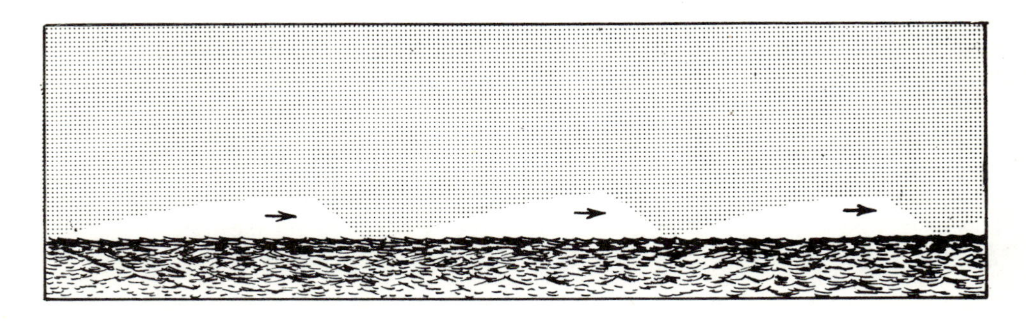

Patches of light move around horizon in Arabian Sea

## AN UNUSUAL AURORA
**Newcomb, Simon; *Science*, 8:410-411, 1898.**

Last evening (September 11th) I witnessed at this place what I suppose to be an aurora, and which, if such, showed features so unusual as to seem worthy of record. The air was remarkably clear for the climate of this region and no perceptible wind was blowing. At 7h 50m E. S. T. I walked out to a good point of view, free from artificial lights, to look for the zodiacal light and the 'Gegenschein.' I soon noticed in the south what I supposed at first to be a white cloud, which, however, soon disappeared. Later the supposed cloud repeatedly reappeared and disappeared in so unusual a way that I watched it more closely. It was oval in form, the longer axis parallel to the horizon, bright in the central part and fading out gradually at the border. It filled the comparatively vacant space to the east of a Capricorni, and was perhaps five or six degrees in length. After some time I satisfied myself that it could not be a cloud from the facts that it did not obscure the stars, one or two of which were on its boundary; that it was, at brightest, twice as bright as the Milky Way; that it brightened up and disappeared again too rapidly, and was apparently almost fixed in position. In the latter feature and in its regularity of outline it also differed from any aurora I have ever seen. Toward the close of the exhibition it moved a little to the west, so that its last appearance was nearly central over a Capricorni. It last showed itself about 8h 30m. It must, therefore, have lasted in all at least 40 minutes, during which time it brightened up and nearly or quite disappeared again perhaps ten or twenty times. A noteworthy feature was that there was nothing like an auroral streamer and no aurora elsewhere, unless an extremely faint, fixed illumination of the sky along the north horizon was such.

Quite likely it was an auroral beam seen end on. If so, it affords one of the best opportunities that have ever occurred to determine the height and length of such a beam. I, therefore, describe the phenomenon in the hope that it may have been seen and its position noted in other parts of the country. (Science, 8:410-411, 1898)

## UNIDENTIFIED LIGHTS SEEN AT SEA
**Crane, T. C.; *Marine Observer*, 28:81, 1958.**

S. S. Corrales. Captain T. C. Crane. Santa Isabel to Liverpool. Observers, the Master, Mr. J. H. Crossley, 3rd Officer, Mr. A. Campbell, Senior Radio Officer, and the Lookout.

7th August, 1957. At 2150 G. M. T. what appeared to be the loom of three lights appeared on the eastern horizon. The central loom was the most distinct and bore $098^{o}$, the others were $2^{o}$ on each side of it. It was thought at first that the phenomenon was due to the reflection of moonlight from the clouds, but the source of light appeared to be below the horizon and was not steady. The moon bore $168^{o}$.

No echoes showed on the radar screen on this bearing, though the echo of a ship at a distance of 15 miles showed clearly, and also those of rain squalls over 30 miles away, so it would appear unlikely that the source of the lights was associated with a ship, or ships. As the nearest land, Cape St. Vincent, was 200 miles distant, it was difficult to relate the lights to anything on the land. The central light persisted for 40 min, until 2230; its bearing became more N'ly and was $080^{o}$ when it disappeared. The lesser lights lasted 30 min, until 2220. Air temp. $71^{o}$F, wet $68^{o}$, 7/8 Sc. (Marine Observer, 28:81, 1958)

## UNUSUAL SKY APPEARANCE
**Abbott, C. G.; *Science,* 81:294, 1935.**

A correspondent from Vienna, Va., writes that on either January 22 or 23, about 8 o'clock in the evening, she saw a light flashing in the southwest something like lightning. It would flare up several times then die down. As she watched it, it became very vivid till it seemed to come from a great blazing light, almost a ball of fire. All this time it was moving around the horizon from the southwest until it had almost reached the starting point. She thought it perhaps more vivid when in the north, and that it seemed to be dying away in the southeast. It appeared to be very low, just showing above the foothills.

I myself was driving along Wisconsin Avenue in Washington on the evening in question, with my wife and we were startled by what was probably the same appearance. It resembled what is called "heat lightning," only that it seemed to be very near indeed and not associated with any noise. The night, as I recall it, was very cold and dry, and I believe on the turn between two contrasting types of weather. (Science, 81:294, 1935)

## THE AURORA AND THE WEATHER
**Botley, C. M.; *Weather,* 20:117-118, 1965.**

Amongst peoples living in the auroral zone there has always been some association between the aurora and the weather, much of it confused and contradictory, but none the less with what seems to have been sound observation.

A Finnish name for the aurora is Vindlys = Windlight.

Scottish farmers in the last century held that the first great storm of the year was heralded by an aurora. Observations between 1843 and 1870 at Uckfield, Sussex (lat. 51°N), made during an aurorally 'rich period', showed that after the brilliant displays occasionally seen, very stormy weather followed almost invariably after 10 to 14 days.

The Southern Cross Expedition, which wintered at Cape Adare, Ross Dependency (72°S), in 1899-1900 found that brilliant and active aurora, observed to the north, was followed, too often for coincidence, by a violent storm from the southeast.

At Ivigtut, west Greenland (72°N), about 1881, it was observed that there seemed to be a curious connection between aurora, the fiord (Arsuk) and the weather. The aurora seemed to begin at the edge of the sea, spreading eastward along the fiord, reaching full development 'as a rule' in fine weather, and about 2 hours after sunset, when it appeared as a long band with luminous streaks sometimes showing the colours of the rainbow, and bright enough to enable one to read small print. But when the fohn blew from the hills at the head of the fiord such bands did appear often 'but then they were restless and torn'.

It is not uncommon, 48 hours after a great aurora, for cirrus clouds to develop.

Sir W. Herschel discovered in the 18th century that aurora made 'the stars to undulate' and increased twinkling, a result confirmed by the observations of Montigny between 1880 and 1886. 'In a word the turbulence of the upper atmosphere is increased by solar activity' (Dauvillier 1962).

The association of the aurora and the weather was a regular feature of 19th century books, such as the standard works of Fritz and Angot and in our time the matter has been revived by the French savant quoted above, who has also remarked

(1954) that the auroral zone in the Atlantic coincided with the area of formation of Icelandic depressions 'which concern the inhabitants of western Europe'. He postulates a kind of relay system from the ionosphere.

More recently observations in the western hemisphere have shown a connection 'between trough development in the Gulf of Alaska and aurorae. Since the Gulf of Alaska is a breeding ground for troughs which later move eastwards a connection with widespread meteorology is possible'.

The Gulf is actually somewhat to the south of the auroral zone, but this does not seem to be fixed, and moves southward at sunspot maximum.

The paper quoted above also mentions some relation between the circulation at the 500 mb and the 300 mb level and magnetic disturbance.

Furthermore in the same symposium (Link 1964) it is remarked 'the influence of solar activity on the climate is betrayed by the periodic displacements of the principal zones of depressions whose latitude is in indirect relation with solar activity', and that 'there are reasons to suppose that these displacements are controlled by the extension of the auroral zone, variable with the solar activity, therefore by the incidence of corpuscular radiations in the atmosphere. (Weather, 20: 117-118, 1965)

## AURORAL DISPLAY
**Campbell, John T.; *Science*, 20:66-67, 1892.**

On Saturday night, July 16, 1892, I was returning to my home in Rockville, Indiana, from Clinton, Indiana, sixteen miles southwest. Mr. Harry McIntosh, a young man of this place who had been helping me make a survey near Clinton, was riding with me in my buggy. We amused ourselves looking at a most beautiful sunset as we rode over the Lafayette and Terre Haute road, along the foot of the high hills east of the Wabash River.

When we turned eastward, over the hills toward Rockville, it began to grow dark, and most of the clouds that showed up so beautiful at sunset began to vanish, till only a few streaks of stratus clouds remained. As we were descending the west hill at Iron Creek, five miles south-west of Rockville, we saw in front of us what we supposed was the new electric light at Rockville, thrown upward and reflected from a cloud or mist. As we were ascending the hill on the east side of the creek and near its summit, we saw in our front the reflection of a great light from behind us. It was so noticeable as to cause us both to turn about on our buggy seat and look backward. There, at a bearing S. 60$^{o}$ W. (that is the bearing of the road, with which the light was in alinement), we saw a great white light radiating from a point at the horizon where it was brightest, right, left, and upward to a height of 10$^{o}$ to 15$^{o}$, weakening in brilliancy as it radiated and terminated in a dark band or segment of rainbow shape, some 10$^{o}$ wide. The light seemed to radiate from a point a half-radius above the centre of the circle which the black segment would indicate. Above the dark segment another segment or band of light, not so bright as the one at the horizon, formed a rainbow, or arch, some 10$^{o}$ to 15$^{o}$ wide. Above that second band of light was a light haze, or mist, through which the stars could be easily distinguished. Some 10$^{o}$ up in that mist, and directly over the centre of the light at the horizon, was a light about as large as a man would appear to be if suspended from a balloon a thousand feet distant. It was about four times as long vertically as wide horizontally. Young McIntosh saw it first and called my attention to it, as I was watching the bright light at the horizon. When I first caught sight of it, it had the appearance of the head of a comet, only it was long vertically. When young McIntosh first saw it, it seemed to be a blaze such as a large meteor appears to carry at its front. We halted and watched it about ten minutes, during which time it (the small light) slowly faded till only its locality could barely be noticed, then suddenly loomed bright almost to a white blaze, then slowly faded as before. It would loom up in five seconds, and consume five minutes in fading away. It kept the same position all the time, for we watched its position with relation to the stars to see if it moved. At this second appearance I decided to commit the general appearance to memory so I could sketch it afterward. This little light loomed up and faded four times when the big light under it faded also and made it dark there.

I am not sure we saw this light the first time it appeared, but think we did. The small light above looked as the moon does when shining through a thin cloud, except as to the oblong shape vertically.

When the first or south-western light faded nearly out, a light at the horizon in the south loomed up, but not so bright as the first, nor had it any of the upper characteristics of the first, nor did it last over five minutes. When this second light faded a third loomed up in the north, quite as bright at the horizon as the first, but it was obscured or cut off from our view by a stratus cloud. This cloud was about $10^{o}$ above the horizon, at its underside (which, by the way, was its most northern limit). This limit, I judge from my frequent observation of clouds, was fully twenty-five miles north of us. We could see the light through one hole in the cloud near its bottom (or distant) side, and also through several thin places, but could not determine its upper shape. This third light (counting the southwestern light as the first) lasted about five minutes, when a fourth light loomed up in the north-west, and, very bright at the horizon, reached upward about $15^{o}$, lasted a few minutes, and faded out as did the others. Then one appeared in the north-east, in the direction of Rockville; but we were so near the town we were sure it was the new electric light (we had been gone a week), but on entering the town found the old gasoline lamps still doing service.

On the first appearance of these lights at the horizon, I thought I saw a flash of light, not as a blaze, but as if a mirror had been turned so as to flash the light into my face, then away so quick I could not be certain what I saw. Young McIntosh thought he saw the same flashes of light when the great lights first made their appearance.

I saw this same electrical storm (if that is what it is) in the summer of 1884, from the town of Clinton, Indiana, and in July, I think. It had all the features I have given of this, except the one in the south-west with its three lights and dark segment herein described. The Clinton display was watched by apparently the whole population of the place, and was described by the Clinton Argus at the time. I reported it to the U. S. Signal Office at the time, as I was then making voluntary observations for that office.

The small light I have described as seen in the south-west, in the first light last Saturday night, is a new feature, so far as I know or can learn from my authorities. These lights occurred from about half past nine to half past ten o'clock at night. (Science, 20:66-67, 1892)

# ELECTRIC DISCHARGE PHENOMENA

The varieties of electric discharge phenomena encompassed by this category are "slow;" that is, the flow of electricity is gentle and rather quiet in contrast to lightning with its accompanying thunder. While not as vivid as lightning, slow electric discharge phenomena are still luminous and sometimes strangely beautiful, much like the auroras. The best known phenomena in this category are St. Elmo's Fire and coronas surrounding sharp points. We skip these and concentrate on more challenging and mysterious fare.

Mountain-top glows form the first example. Since some mountains are sharp, projecting surfaces, the existence of slow electrical discharges from their peaks is not particularly surprising. Nevertheless, the aurora-like beams and sheets of flame arising from this process can be spectacular and visible for hundreds of miles. Where does this mountain electricity go? To outer space? And why are mountain-top glows enhanced during earthquakes?

These questions bring us to the subject of earthquake lights that are seen on occasion near epicenters, and which resemble the aurora-like mountain-top glows. Recent observations indicate that earthquake shock waves can perturb the atmosphere and even the ionosphere, possibly establishing low-conductivity paths for earth-space discharges. Earthquake lights also include balls of fire (ball lightning?), sheets of flame issuing from the ground, and other luminous displays. The origins of these phenomena are unknown at present.

Volcanoes and tornadoes are also accompanied by unusual "lights". Lightning and ball lightning are expected from tornado activity but whence the peculiar shafts of light and flame-like displays? The immense shafts of light seen in the cores of some tornadoes resemble huge fluorescent lights and may be explained as glow discharges. The possibility exists that tornado and volcano activity penetrates into the ionosphere, paving the way for electrical discharges from the surface of the earth.

On a smaller, less violent scale, patches or waves of electrical activity sometimes travel along the ground engulfing people and, in the process, engendering peculiar effects and even stranger stories. In some ways, these surface displays seem related to luminous marine displays, such as the phosphorescent wheel.

Much more mysterious are the rare instances of luminous, balloon-like spheres that resemble neither ball lightning or nocturnal lights. They are in a class by themselves and may not be electrical in origin. Even if these floating, luminous bubbles are electrical, modern theories do not begin to explain them.

## MOUNTAIN-TOP GLOWS

### MOUNTAIN-PEAK POTENTIAL-GRADIENT MEASUREMENTS AND THE ANDES GLOW

**Markson, Ralph, and Nelson, Richard; *Weather,* 25:350, 1971.**

'Andes glow' or 'Andes lights' are terms used to describe illumination seen at night in the vicinity of certain mountain peaks. While the majority of reports have come from the Andes mountains of Bolivia, Chile and Peru, this phenomenon has also been reported in the European Alps, Mexico and Lapland and presumably could occur in many mountainous regions under favourable conditions. While sometimes thought to be lightning, for lack of a more obvious explanation, the interesting property of these light displays is that they can occur under cloudless skies. Sometimes they are but one single flash, while at other times they may persist intermittently for hours. On occasion, a periodicity has been noted in the time between flashes. At their most spectacular they have been described as " . . . not only clothing the peaks, but producing great beams, which can be seen miles out at sea." They seem to favour particular mountain peaks where often they can be seen during the dry season. Very low humidity is frequently mentioned as the dominating meteorological factor when Andes lights are seen. Rather than lightning, the Andes lights may be a large scale Saint Elmo's Fire with many points across the mountain top going into corona discharge. (Weather, 25:350, 1971)

## MOUNTAIN-TOP DISCHARGES AT MADEIRA
**Robson, G.; *Marine Observer*, 25:95-96, 1955.**

S.S. Tribulus. Captain G. Robson. Las Piedras to Algiers. Observer, Mr. R. W. Lumsden, Chief Officer.

7th June, 1954, 1800 G.M.T. On approaching Madeira from SW the island was completely covered with low cloud, St and Sc. On arriving within 16 miles of the island the cloud rapidly lifted and numerous brilliant white flashes were observed at frequent intervals on various mountain peaks. At the time of these occurrences

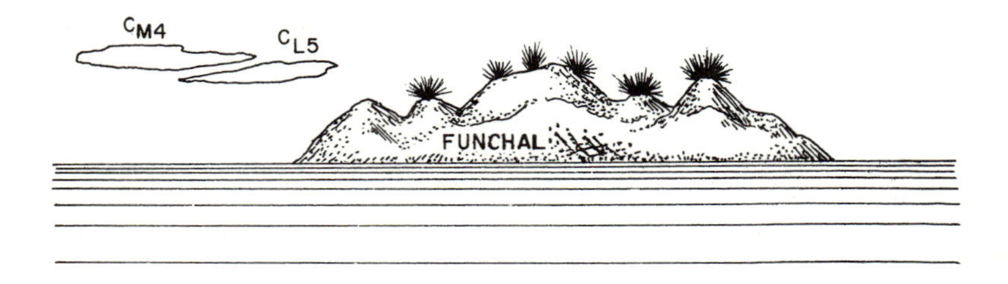


Electric discharges from mountain tops seen at sea off Madeira

the cloud was clear of the island, although there was some Sc to NW. After the flashes had continued for some 20 min a low rumbling was heard like distant thunder. Note. The brush-like discharge of electricity from objects on the earth's surface is normally a small-scale phenomenon visible only at relatively small distances, such as when it is seen on ships' masts or aerials, etc., from the deck. There are several observations on record, however, of it having been seen on clouds, both at night and by day, in which case the phenomenon must have been on a larger scale to be visible at cloud distance. We can think of no other explanation of the above remarkable observation. The discharges must have been very large and bright to be visible at 16 miles, and if it really was St. Elmo's Fire the observation is probably unique. (Marine Observer, 25:95-96, 1955)

## CURIOUS LIGHTNING IN THE ANDES
**Anonymous; *Scientific American,* 106:464, 1912.**

Dr. Walter Knoche, the German director of the Chilean meteorological service, has begun an investigation of the remarkable displays of so-called "heat lightning" which are often observed along the crest of the Andes, and are sometimes visible far out at sea.

Thunderstorms are rare in Chile, and this fact may possibly be explained on the assumption that the Andes act as a gigantic lightning-rod, between which and the clouds silent discharges take place on a vast scale. The visible discharges occur during the warm season, from late spring to autumn, and appear to come especially from certain fixed points. According to Dr. Knoche they are confined almost exclusively to the Andes proper, or Cordillera Real, as distinguished from the coast cordillera. Viewed from a favorable point near their origin there is seen to be, at times, a constant glow around the summits of the mountains, with occasional outbursts, which often simulate the beams of a great searchlight, and may be directed westward so as to extend out over the ocean. The color of the light is pale yellow, or rarely reddish.

One striking feature of these discharges is that they are especially magnificent during earthquakes. At the time of the great earthquake of August, 1906, throughout central Chile the whole sky seemed to be on fire; never before or since has the display been so brilliant. The natives regard these lights as the reflection in the sky of the glowing lava in the craters of volcanoes; but there seems to be no doubt that they are electrical discharges.

It is planned to make spectroscopic observations of this singular phenomenon, and also, if possible, measurements of the electrical state of the atmosphere in the high Andes where it appears to have its origin. Possibly the result may be to connect up "Andes lightning" with a peculiar form of aurora which has been observed by Lemstrom over mountain summits. (Scientific American, 106:464, 1912)

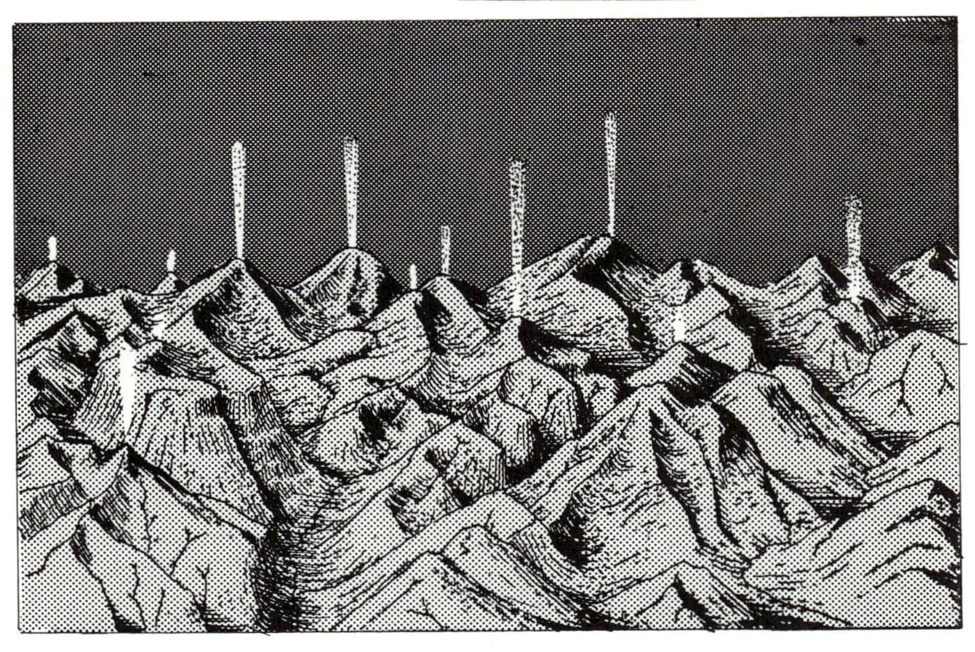

## SUMMER LIGHTNING
**Geikie, Arch.; *Nature,* 68:367-368, 1903.**

Although a good deal has been written on the subject of "summer lightning," it may not be superfluous to describe a display of the phenomenon which occurred here last evening on a scale far surpassing anything which it had been my good fortune to witness before. There had been several thunderstorms in the district during the previous five or six days, and a few peals were heard and heavy rain fell in the early afternoon of the day before (August 13). But the sky cleared rapidly thereafter, and the evening and night of that day were cloudless, every peak and crest standing out sharply defined in the clear air. Yesterday was still fine, but warmer and less bracing than visitors here expect. Late in the afternoon wisps of white mist began to gather round the summit of the Jungfrau, and streaks of thin cloud took shape in the higher air above the great mountain ridge that extends from the Silberhorn to the Breithorn. About 8 p.m. I noticed a faint quivering light overhead, supplemented by occasional flashes of greater brilliance and different colour. These manifestations rapidly increased in distinctness, and continued to play only along the opposite mountain-ridge, not extending into the regions beyond, so far as these could be seen from here, though I have since learnt that an independent series of flashes was seen around the Schilthorn on this side of the valley. Not a single peal of thunder was at any time audible. A long bank of cloud formed at a higher level than the summits of the mountain-ridge, and at some distance on the further side of it, so that the stars, elsewhere brilliant, were hidden along the strip of sky above the crest.

As one watched the display it was easy to distinguish more definitely the two kinds of discharge. One of them took the form of a faintly luminous reddish or pink light, which shot with a tremulous streamer-like motion in horizontal beams that proceeded apparently from left to right, as if their starting point lay somewhere about the back of the Jungfrau. These streamers so closely resembled the aurora borealis that, had they appeared alone, one would have been inclined to wonder whether the "northern lights" had not here made an incursion into more southern latitudes. So feeble were they when they sped across the clear sky that the stars were clearly visible through them. Sometimes they quivered on the far side of the cloud, lighting up its edges and shooting beyond it across the still unclouded blue. At other times they appeared on this side of the cloud, and showed the dark outline of the mountains in clear relief against the luminous background. They so rapidly succeeded each other that they might be said to be continuous, a faint pinkish luminosity seeming to remain always visible, though pulsating in rapid vibrations of horizontal streamers.

The brighter discharges were not only far more brilliant, but much more momentary. They had a pale bluish-white colour, and came and went with the rapidity of ordinary lightning. But they were clearly connected with the mountains, and not reflections from a series of distant flashes. Sometimes they arose on the other side of the great ridge, allowing its jagged crest to be seen against the illuminated surface of the cloud beyond, but leaving all the precipices and slopes on this side in shade. In other cases they clearly showed themselves on this side of the mountains, lighting up especially the snow-basins and glaciers with the dark crags around them. Nothing of the nature of forked lightning was observed among them. In one instance the flash or horizontal band of vivid light, a mile or two in length, seemed to shoot upward from the slope at the base of the precipices of the Silberhorn, as if it sprang out of the ground, having a sharply defined and brilliant base, rapidly diminishing in intensity upward, and vanishing before reaching halfway up to the crest.

But the most singular feature of the more brilliant white discharges was to be seen when one of the great couloirs of snow or a portion of a glacier remained for a minute or two continuously luminous with a faint bluish-white light. After an interval the same or another portion, perhaps several miles distant, would gleam out in the same way. My first impression was that this radiance could only be a reflection from some illuminated part of the cloud. But I could not satisfy myself of the existence of any continuously bright portions of the cloud. Moreover, the luminosity of the snow and ice remained local and sporadic, as if the beam of a search-light had been directed to one special part of the mountain declivity, and then after a while to another. While watching one of these patches of illumination, I noticed a bright point of light at the top of one of the basins of neve on the slopes of the Mittaghorn. It quickly vanished, but soon reappeared, and then as rapidly was lost again. I thought that it was probably a star briefly exposed through rifts in the cloud, though its position seemed rather below that of the mountain-crest. Half an hour later, however, a similar bright light appeared about the same place, more diffused than the first, and having a somewhat elongated shape. Whether it was really a star seen through the distorting medium of a wreath of mist, or a form of St. Elmo's fire clinging to some peak on the precipice, could not be ascertained from its momentary visibility.

I learnt this morning that other observers who could watch at the same time the mountain ridges on each side of the Lauterbrunnen valley noticed that sheet-lightning was also playing about the Schilthorn, but quite independently of that on the Jungfrau range, the one mountain being dark, while the other was illuminated. The distance of the two electric centres from each other is between five and six miles. The whole display last evening afforded an admirably complete demonstration of the erroneousness of the notion formerly prevalent that summer lightning is only the reflection of distant ordinary lightning, and of the truth of the more recent views as to the nature of the phenomenon.

I may add that, as the lightning increased, the air, which had previously been nearly calm, freshened into a strong breeze, which blew from the south-west down the valley, but died down after the illumination faded away. The cloud above the mountain began to assume irregular dark cumulus shapes, and the sky became generally overcast. Early this morning rain was falling heavily. The mountains have been all day shrouded in dripping cloud, and the deluge still continues. (Nature, 68:367-368, 1903)

## ST. ELMO'S FIRE IN EGYPT

**Botley, Cicely M.; *Meteorological Magazine*, 73:96, 1938.**

In view of Mr. Sutton's note in the March issue of the Meteorological Magazine the following extract from "A Search in Secret Egypt" by Paul Brunton (Rider & Co. 1936) may be of interest. A footnote to page 77 reads:

"Dr. Abbate Pacha, Vice-President of the Institut Egyptien spent a night in the desert near the Pyramids, together with Mr. William Groff a member of the Institut. In the official report of their experiences the latter said: 'Towards eight o'clock in the evening, I noticed a light which appeared to turn slowly around the Third Pyramid almost up to the apex; it was like a small flame. The light made three circuits round the Pyramid and then disappeared. For a good part of the night I attentively watched the Pyramid; towards eleven o'clock I again noticed the same light, but this time it was of a bluish colour; it mounted slowly almost in a straight line and arrived at a certain height above the Pyramid's summit and then disappeared.'"

The account goes on to say that Mr. Groff found that this phenomenon was well-known to the Arabs who put it down to spirits. Unfortunately the weather conditions do not appear in the excerpt as cited. (Meteorological Magazine, 73:96, 1938)

# INTER-MOUNTAIN DISCHARGES

## ELECTRIC PHENOMENA IN THE EUPHRATES VALLEY
**Huntington, Ellsworth; *Monthly Weather Review,* 28:286-287, 1900.**

During a recent ten days' geological trip through an almost unvisited part of the Taurus Mountains to the south of Harpoot I heard of a phenomenon which I should be glad to have you explain, either by letter or through the columns of the Review. Before leaving Harpoot I was told by a man from Aivose that Keklujek Mountain, near his village, fought with Ziaret Mountain, on the other side of the Euphrates River. The weapons were balls of light, which the mountains threw at each other. As the region was one of volcanic activity in comparatively recent times, and as hot springs at extinct craters are still to be seen, I thought at first that this must be a traditional account of a volcanic eruption. Subsequent investigation, however, showed that the story had its origin in a meteorological phenomenon. At first I was skeptical as to the truth of what follows. After hearing substantially the same story from ten or twelve men whom I saw in five different places separated by an extreme distance of 40 or more miles, I became thoroughly convinced of its truth. It may be a common occurrence, but I have never heard of it and can find no account of it in the few books at my command.

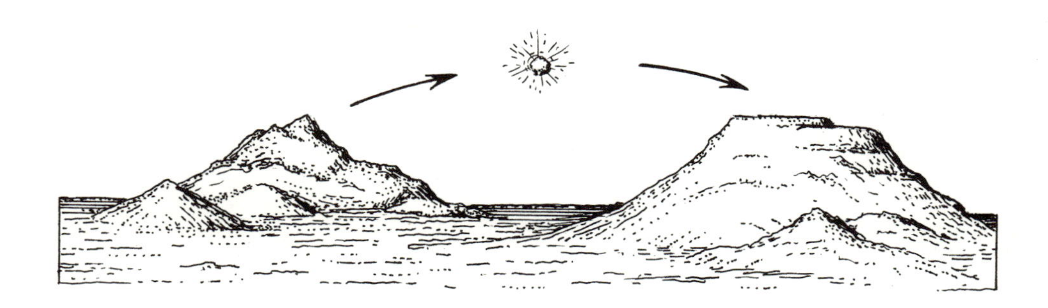

Globe discharge between mountains in Euphrates Valley

The facts, upon which all agree, are as follows: A ball of fire is sometimes seen to start from one mountain and to go like a flash to another. At the same time there is a sound like thunder. This occurs by day or by night, although by day no light is seen. It always occurs when the sky is clear and never when it is cloudy. It sometimes happens two or three times in a year, and then again is not seen for several years. For the last two years it has not been seen. It is most common (or possibly never happens except) in the fall at the end of the long, dry season of three months. The mountains show no special features different from other mountains. I visited one of them, Karaoghlon (Black Son) Mountain, and found it to be composed of metamorphic schistose shale of cretaceous age. Its height is 7,350 feet and the top is comparatively flat. One observer said that a glow remained after the flash, but all the rest contradicted this. Another said that the ball of fire was first small, but grew larger as it passed over, and then grew smaller again. He evidently was between the two mountains. (Monthly Weather Review, 28:286-287, 1900)

# UNUSUAL BRUSH DISCHARGE AND ST. ELMO'S FIRE

## ELECTRICAL PHENOMENON
**Watt, J. B. A.; *Nature*, 32:316-317, 1885.**

About ten o'clock in the evening of July 23 a party of four of us were standing at the head of the avenue leading to this house [Midlothian, England], when we saw a feebly-luminous flash appear on the ground at a distance of some thirty yards down the avenue. It rushed towards us with a wave-like motion, at a rate which I estimate at thirty miles an hour, and seemed to envelop us for an instant. My left

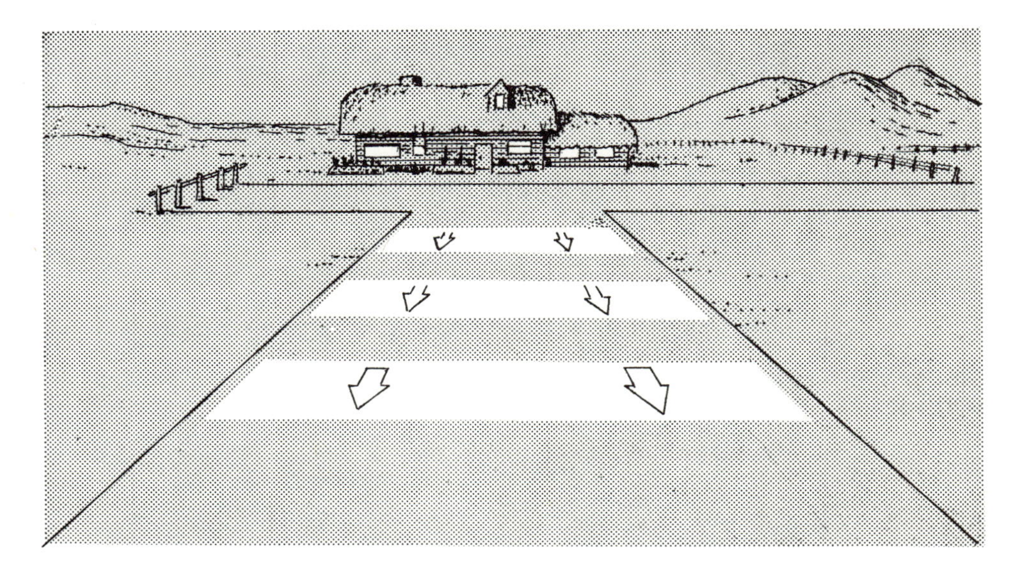

Luminous waves envelope pedestrians

hand, which was hanging by my side, experienced precisely the same sensation as I have felt in receiving a shock from a weak galvanic battery. About three minutes afterwards we heard a peal of thunder, but, though we waited for some time, we neither saw nor heard anything further.

The gardener, who was one of the four, thus describes what he saw:---I thought it was a cloud of dust blowing up the avenue, and before I could think how that could be when there was not a breath of wind, I saw you three gentlemen covered for a second in a bright light, and that was all. Another of the party says that he observed what seemed to be a luminous cloud running up the avenue with a wavy motion. When it reached the party it rose off the ground and passed over the bodies of two of them, casting a sort of flash on their shoulders. The distance traversed was about twenty yards, and the time occupied between two and three seconds. (My own estimate of distance and velocity makes the time occupied almost exactly two seconds.) The day had been extremely hot and sultry, as also had the preceding day been, the the rmometer readings being sometimes 80°F. in the shade.

On asking the gardener for further particulars, he tells me that the distance traversed by the luminous cloud was about forty yards, and that, when it had gone about half the distance, he saw a flash of lightning in the direction of it, but sideways; also that the top of the cloud seemed to be three or four feet from the ground, and it gradually rose higher as it came along. When the cloud reached the party he saw one of them distinctly by its light, the night being otherwise quite dark at the time; and, lastly, that the cloud went a few yards beyond the party into the open space in front of the house, and then disappeared. (Nature, 32:316-317, 1885)

## ELECTRICAL PHENOMENON IN MID-LOTHIAN
**Lucas, Robert; *Nature,* 32:343, 1885.**

I have observed in a daily contemporary a communication quoted from your journal with reference to this occurrence on the 23rd ult.

For the information of those of your readers who are interested in such matters perhaps you will kindly allow me to observe that I also witnessed a similar, or the same, phenomenon that evening.

When driving home from a professional visit in the country, and a mile south of this town, about ten o'clock I was suddenly startled by a peculiar sensation or slight shock, and immediately perceived, ten yards in front, on the road, a bright opalescent luminosity which travelled deliberately away in a northerly direction. This cloud or wave of light covered the whole breadth of the road, and was distinctly visible for some seconds. It seemed to rest entirely on the ground, and in character reminded one somewhat of the illumination resulting from the electric light. I should imagine it was travelling at the rate of twenty miles an hour, as it was going much in the same direction I was, but of course much faster. The part of the road where it showed itself is lined by high trees on both sides in full foliage. I heard no thunder and saw no lightning or meteor to account for the strange and weird-looking light.

The interesting question then arises, What was the nature of this phenomenon?

It will be remembered that the thermometer was for several days at that time above 80°F. in the shade. Might it not be possible, therefore, for a certain volume of air to become electrified, and then, perfectly insulated by the dry surrounding atmosphere, show its existence in this manner as a luminous cloud rushing along the ground?

I may mention in conclusion that my groom, who was driving me at the time, also witnessed the occurrence. (Nature, 32:343, 1885)

## AN ELECTRICAL "BATH" IN YELLOWSTONE
**Sanborn, William B.; *Natural History,* 59:258-259, 1950.**

In early September of 1949 I had occasion to drive from Madison Junction to Mammoth Hot Springs in the Yellowstone Park, and dusk was gathering when I drove onto the stretch of straight highway that parallels Swan Lake. A violent electrical storm was centered over Electric Peak, extending several miles eastward into the Beartooth Range. Heavy bolts of lightning frequently stabbed downward, and the clouds themselves seemed to be constantly flickering with an orange light. The display was so spectacular that I pulled off the road near the north end of Swan Lake and stepped out of the car to watch it.

It was then that I noticed a bluish light coming from over the low ridge to the west of Swan Lake. My first thought was of a fire, perhaps caused by lightning. I watched the ridge for a moment and was amazed to see what can best be described as a hazy patch of blue light coming over the ridge and moving down the hill slope toward the flats around the lake. It was then that I observed a very low lead-gray cloud moving swiftly above the patch of light. The patch moved through the marshy north end of Swan Lake and caused several waterfowl to rise in hurried flight.

The patch of light moved off the lake and onto the flats at a steady rate and proceeded directly toward my viewpoint. When the patch was but a few yards away, I noted a sudden calm in the air and a marked change in temperature, as well as what I believe was the odor of ozone. It was then that I realized that the display before me was some manner of static electricity, comparable perhaps to St. Elmo's fire and directly controlled by the low cloud moving above.

The patch, which actually was a static field, enveloped my immediate area. To describe the weird feeling caused by viewing the progress of this phenomenon is difficult. It kept low to the ground, actually "flowing around" everything that it came in contact with, coating it with a strange pulsating light. Each twig on the sagebrush was surrounded by a halo of light about two inches in diameter. It covered the automobile and my person but did not cover my skin. There was a marked tingling sensation in my scalp, and brushing my hair with the hand caused a snapping of tiny sparks.

The most unusual aspect of the disturbance occurred when I threw a rock along the ground in much the way that one skips a stone over a pond or lake. Every time the rock hit the ground or a piece of brush it gave the appearance of "splashing." The light would momentarily disappear from the contact spot but immediately build up again. I noted that the static could be brushed off the car surface, and it also would establish the light again in a matter of seconds. I obtained no shock from touching any object on the ground or the outside of the car.

My estimate of the speed of the disturbance was about three or four feet a second, or faster. The field seemed to be about 50 yards wide and perhaps some 250 yards in length, roughly comparable to the low cloud above. (Natural History 59:258-259, 1950)

## ELECTRIC FLUID ENVELOPES GIRL
**Anonymous; *Nature,* 22:204, 1880.**

The Times Geneva correspondent writes under date June 20 that a remarkable electrical phenomenon occurred at Clarens on the afternoon of Thursday last. Heavy masses of rain-cloud hid from view the mountains which separate Fribourg from Montreux, but their summits were from time to time lit up by vivid flashes of lightning, and a heavy thunderstorm seemed to be raging in the valleys of the Avants and the Alliaz. No rain was falling near the lake, and the storm still appeared far off, when a tremendous peal of thunder shook the houses of Clarens and Tavel to their foundations. At the same instant a magnificent cherry-tree near the cemetery, measuring a metre in circumference, was struck by lightning. Some people who were working in a vineyard hard by saw the electric "fluid" play about a little girl who had been gathering cherries and was already 30 paces from the tree. She was literally folded in a sheet of fire. The vine-dressers fled in terror from the spot. In the cemetery six persons, separated into three groups, none of them within 250 paces of the cherry-tree, were enveloped in a luminous cloud. They felt as if they were being struck in the face with hailstones or fine gravel, and when they touched each other sparks of electricity passed from their finger-ends. At the same time a column of fire was seen to descend in the direction of Chatelard, and it is averred that the electric fluid could be distinctly heard as it ran from point to point of the iron railing of a vault in the cemetery. The strangest part of the story is that neither the little girl, the people in the cemetery, nor the vine-dressers appear to have been hurt; the only inconvenience complained of being an unpleasant sensation which was felt with more or less acuteness for a few hours after. The explanation of this phenomenon is probably to be found in Prof. Colladon's theory of the way in which lightning descends, as described in Nature, vol. xxii, p. 65. The Professor contends that it falls in a shower, not in a perpendicular flash, and that it runs along branches of trees until it is all gathered in the trunk, which it bursts or tears open in its effort to reach the ground. In the instance in question the trunk of the cherry-tree is as completely shivered as if it had been exploded by a charge of dynamite. (Nature, 22:204, 1880)

## LUMINOUS FOREST
**Anonymous; *Nature,* 20:423, 1879.**

The Times Geneva correspondent writes, under date August 22:---"On the evening of August 5, six persons who were standing in the gallery of a chalet in the Jura, above St. Cergues, witnessed an atmospheric phenomenon equally rare and curious. The aspect of the sky was dark and stormy. The air was thick with clouds, out of which darted at intervals bright flashes of lightning. At length one of these clouds, seeming to break loose from the mountains between Nyon and the Dole, advanced in the direction of a storm which had, meanwhile, broken out over Morges. The sun was hidden and the country covered with thick darkness. At this moment the pine forest round St. Cergues was suddenly illuminated and shone with a light bearing a striking resemblance to the phosphorescence of the sea as seen in the tropics. The light disappeared with every clap of thunder, but only to re-appear with increased intensity until the subsidence of the tempest. M. Raoul Pictet, the eminent chemist, who was one of the witnesses of the phenomenon, thus explains it in the last number of the Archives des Sciences Physiques et Naturelles:---'Before the appearance of this fire of St. Elmo, which covered the whole of the forest, it had rained several minutes during the first part of the storm. The rain had converted the trees into conductors of electricity. Then, when the cloud, strongly charged with the electric fluid, passed over this multitude of points, the discharges were sufficiently vivid to give rise to the luminous appearance. The effect was produced by the action of the electricity of the atmosphere on the electricity of the earth, an effect which, on the occasion in question, was considerably increased by the height of the locality, the proximity of a storm-cloud, and the action of the rain, which turned all the trees of the forest into conductors.'" (Nature, 20-423, 1879)

## PHENOMENON WITNESSED IN NEW ZEALAND ON FEBRUARY 29TH, 1936
**Laurenson, M. D.; *Meteorological Magazine*, 71:134-136, 1936.**

Travelling alone by car from Hamilton to Tauranga, I reached the top of the Kaimai Road, and was greatly struck with the vivid lightning display over the Bay of Plenty. As this display gradually increased in intensity, I stopped my car where I had a clear view to the east, at a spot 13 miles from Tauranga on the Kaimai Hills, about 500 ft. above sea level.

Having stopped the engine and extinguished the lights, I noticed that the major flashes came from behind a cloud lying practically due north. After a major flash from the north there would be a pause, and then from a point or two to the east of where the major flash had originated, forked lightning would run across the sky, through east to south. At about 10.10 p.m. I noted, due east, a faintly glowing light. At first I thought this came from a house window, or the headlights of a car. I subconsciously decided the former surmise was correct, as although it was definitely a light itself (and not the reflection of one) it was a bit faint for a car. All the same, I mentally registered the impression that it was a bit high up for a house or a car. (I am well acquainted with the district, so that I am fairly sure as regards localities and directions.)

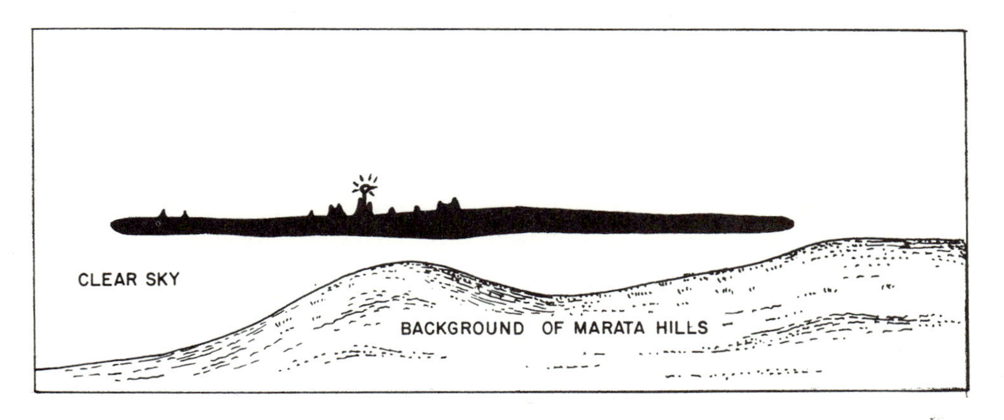


Intense white light atop cloud deck in New Zealand

Some minutes later, a bright flash from the north lit up the sky, and I was more than amazed to realise that the glowing light actually proceeded from the upper surface of a black bank of cloud. The time was then 10.15 p.m. and the glowing light had been practically steady for five minutes. By steady, I mean it had stayed in the same position, no movement right or left, up or down. It had waxed and waned, only very slightly, which had made me think, hitherto, that it was a house light, and that rain had drifted past, causing it to alter in apparent intensity.

At 10.15, as stated, I realised the glowing light was proceeding from a cloud. Before I had time to conjecture what that could mean, I witnessed one of the most weird and uncanny sights I have ever seen. It suddenly seemed to pulsate, it took definite shape as a molten ball of soft light, but although in itself not dazzling to the eyes, threw off an indescribably bright, greenish white light, or rather radiance. This radiance lit up the whole of the upper surface of the cloud bank and showed the ball of light balanced on a finger of cloud. On either side of this finger were ugly looking black peaks, and all these were silhouetted against the radiance. From being a small ball of light, it instantly became larger, until it was the size of a half-crown, seen from a distance of 18 ft. The radiance became brighter and lit up all the landscape and countryside. The ball itself, although molten in appearance, did not hurt or dazzle the eyes. It held this size for about 15 seconds (I was far too interested in watching it, to take the time). Then it pulsated again (seeming to slightly contract and expand once or twice) and almost immediately became much enlarged. It appeared to be the size of a large orange seen from 18 ft. away. This time the radiance was terrific. Thousands of searchlights would not have equalled the intensity. But still the focal point or ball was not blinding.

My own feelings, although not scientific evidence, were interesting, to say the least. I had no time to be scared for it all was so absorbingly interesting. I felt literally that anything might happen, and nothing could have surprised me. Situated as I was, on an absolutely deserted road, I might have been watching the creation of a new world, or the extermination of an old world.

This phase could have lasted no longer than 15 seconds, but of course, time is only relative. Then the ball of light commenced to contract very quickly, and went back to its original glow. It disappeared momentarily, but for certainly ten more minutes kept on appearing and disappearing. For the 15 minutes that I saw it, it did not move in the sky. It did not again come into prominence, although I waited for at least half an hour. All this time the lightning display continued.

I also noticed a bright and constant radiance, somewhat similar to the one described above. This was approximately in the direction of Rotorua, but it would have needed a city the size of London to give the same effect as a reflection. I could not see the focal point of this radiance, owing to the tops of the hills which were outlined. This glow remained practically steady for 15 minutes.

One other detail I noted with regard to the phenomenon detailed at length above. After the glowing light had finally disappeared, one particular display of lightning emanating from the northern cloud flashed across the sky. One jagged fork was travelling horizontally above the spot where the ball had been. Suddenly this jagged fork resolved itself into an absolutely straight line, and drove itself directly into the spot where the ball had been. Its angle of flight was from 11 o'clock to 5 (i.e., slightly off the perpendicular).

I am writing this on the day following, while it is all fresh in my memory. I have heard no mention of any similar experience over the air, although I have been told that others have witnessed the same thing. Nevertheless, I have purposely refrained from discussing it with any of those people as I do not want to risk having my impressions varied by the experiences of others until I have put my record on paper. (Meteorological Magazine, 71:134-136, 1936)

# ELECTRICAL EFFECTS IN DUST AND SNOW STORMS

## VISUAL OBSERVATION OF ELECTRIC SPARKS ON GYPSUM DUNES

**Kamra, A. K.; *Nature,* 240:143, 1972.**

Electrification in dust storms has been studied at White Sands National Monument in southeastern New Mexico. In this region of New Mexico gypsum dunes, some of

Electric sparks from gypsum dunes

which are up to 10 m high, covered an area of about 690 km and constantly change shape with the prevailing southwest winds.

The observations reported here consisted of measurements of atmospheric electric potential gradient with a radioactive probe at ground level, space charge by the filtration technique suggested by Obolensky and Moore et al., and wind speed with a three-cup anemometer, at 1.25 m above the ground. Atmospheric temperature and relative humidity were measured with a psychrometer at ground level. The apparatus was installed on top of the dunes or on the plane rugged areas between the dunes.

I made a fascinating observation of electrical activity on May 11, 1971, when a thunderstorm passed over the site. Strong winds were blowing large quantities of sand into the air, and atmospheric temperatures were lower and relative humidities were higher than usual for the month of May in that region.

The apparatus was set up on a rugged surface about 10 m downwind of an 8 to 10 m high dune. The recording started at 1510 MDT (Mountain Daylight Time) when the strong winds were already blowing sand into the air. Electric sparks were observed extending from the top of the sand dunes up into the air terminating at a height of a few metres. At least five sparks, all on different dune tops, were observed within the period 1510 to 1600 MDT. These sparks extended straight up, and no branches were observed. The atmospheric potential gradient record made during this period shows some very rapid field changes, which are both positive and negative. One of the sparks that occurred at 1535 MDT corresponded to a positive potential gradient increase of 1,500 V $m^{-1}$. Corresponding rapid changes in space charge density observed during this period are presumably due to some displacement current through the space charge tube. (Nature, 240:143, 1972)

## LOCAL ELECTRICITY IN A WINDSTORM IN WYOMING
**Anonymous; *Monthly Weather Review,* 22:509, 1894.**

Mr. John Hunton, voluntary observer at Fort Laramie, Wyo. (N. 42° 12', W. 104° 31', altitude 4,519 feet), reports as follows:

On December 20, at 1.30 a.m., observed vertical red streaks above and below moon. Each about one and one-quarter diameters of moon in width and each about one diameter of moon from it. One streak extended below the horizon, the other extended upward about 20 diameters of moon in length. At 10.15 a.m., same date, electrical wind storm of great velocity commenced and continued until 7 p.m. Maximum velocity of wind obtained at about 1.30 p.m., when three houses were unroofed and substantial steel tower windmill was blown down, causing a loss of about $700.

The wind was phenomenal in that it would remove and break into pieces the most solid wood work of buildings and leave adjoining frail parts of the wood work undisturbed. Reliable information, obtained from over a section of country extending 12 miles south and 20 miles east, states that the electrical current was freely felt in many localities embraced in that area.

Mr. Silas Doty, a very reliable man living 8 miles south, reports that he discovered a fence post on fire 200 yards north of his house, and upon going to it to extinguish the fire found it burnt more than half through where one of the fence wires was fastened, and partly burned where another wire was in contact with it. Other fence posts were slightly marked by the electric sparks.

Mr. John F. Barnes, another reliable man, living 11 miles south of here, states that in going from his stable to his house he caught hold of a fence wire to assist him in walking against the strong wind and received a severe electric shock from which he was some time recovering. He had a strong healthy cow in a lot inclosed with wire fence. The wind drifted the cow into a corner against the fence and held her there, where Mr. Barnes found her dead late in the evening. He thinks that long contact with the heavily charged wires killed her, as there was no mark of violence or internal derangement to cause death.

Mr. E. B. Hudson, 20 miles east of here, reports that two of his employees when going from work to house, soon after the commencement of the wind-storm, were severely shocked when crossing a wire fence. One of the men caught hold of a wire and received a shock which numbed and weakened his hand to such an extent that he was unable to take it from the wire but had to release it with his other hand. The injured hand and arm remained nearly helpless for several moments. Mr. Hudson, as well as the other two men referred to, is a reliable, truthful man. (Monthly Weather Review, 22:509, 1894)

## ELECTRIC TREES IN DAKOTA
**Anonymous; *Symons's Monthly Meteorological Magazine,* 32:27, 1897.**

On January 4th, during the worst of the great wind and snowstorm at Huron, the air was heavily laden with electricity. The cottonwood trees in front of the Chicago and North-western offices presented a very strange and novel appearance. The trees were buried in snow almost to their tops, but at the end of each twig on every branch in sight was an electrical spark about as large as a common field Pea. On taking hold of a twig the spark extinguished, but on withdrawing the hand the spark reappeared. Dispatcher Wilson, who wore a glove with a hole in the thumb, took hold of a twig, and the spark transferred itself to his thumb and back to the twig when he let go. There was no shock experienced, says an American contemporary, in handling the twigs, and the light did not waver or tremble, but was quite steady. The trees looked as if a colony of fireflies had settled upon them for the purpose of an illumination. (Symons's Monthly Meteorological Magazine, 32:27, 1897)

## ON LUMINOUS APPEARANCES IN THE ATMOSPHERE
**Allen, J. A.; *American Journal of Science,* 1:4:341-342, 1822.**

On the evening of the 18th of January 1817, during a rapid fall of moist snow, attended with repeated claps of thunder; lights or luminous appearances were seen in the atmosphere in many places on the Green Mountains.---The lights were observed by different persons in the towns of Andover, Jamaica, Wardsborough, Dover, Somerset, Stratton and New Fane.

In all these places the lights were described as having the same appearance. They were observed on the tops of bushes, fences, houses &c. Some persons represented them as appearing like the blaze of candles, but all agreed that they were luminous spaces which appeared to rest on pointed or elevated substances--- In some instances, persons who were travelling, suddenly observed a light surrounding their heads; in others they were completely enveloped in a light but little less than the ordinary light of the sun---Several persons found when they elevated their hands, that the light appeared to stream from their fingers. This fact was particularly noticed by J. Deming, Esq. of Andover.

Such phenomena as these it is believed, have seldom been observed in this vicinity, probably this is the first instance since the settlement of the country. But in other parts of the world they have long been witnessed---though not very frequently. (American Journal of Science, 1:4:341-342, 1822)

## AN UNUSUAL TYPE OF LIGHTNING
**Matthews, J. Brian; *Weather,* 19:291-292, 1964.**

On 3 March 1964 Tucson, Arizona, suffered a snow-storm, an unusual event for this locale, which exhibited an even more unusual form of electric discharge. The observed phenomenon appears to be sufficiently rare to merit this brief note.

A moderately strong cold front passed through Tucson about 8 pm and was followed by precipitation which later turned to snow. Between 8 and 8.30 pm, before the precipitation commenced, several flashes of lightning were observed over the surrounding mountains. Precipitation in the form of rain began about 9 pm and turned to snow about 9.30 pm.

From 10 to about 11.30 pm, the snowfall was heavy and composed of large, wet flakes (up to 2 in. dia.). The wind was north-east at about 15 mph but backed to north at 25 mph about 11 pm. The air temperature remained at 33°F. The author and Dr. D. O. Staley, a meteorologist of the Institute of Atmospheric Physics, watched the snowstorm from the 80-foot high observation room of the Institute. Short flashes of 'lightning' were seen to be occurring at intervals estimated to be about 15-20 seconds and at random places around the town. The flashes continued to occur throughout the storm, until the snow ceased at about 11.30 pm. Visibility was about 1 mile at first, but later decreased during the heaviest snowfall to about 200 yards.

The flashes of 'lightning' contained several unusual features: they appeared to be single, short flashes of light without the 'flicker' normally associated with lightning. They were less intense than normal lightning and were not observed to case sharp shadows. Moreover, thunder was not heard at any time, nor could correlation between the flashes and 'static' over the radio be found.

The 80-foot high observation tower affords a good view over the surrounding town, which is mainly occupied by single-story dwellings. From this elevation, the lightning was seen to originate from points at or very close to the ground. No brilliant, illuminated track such as is usually associated with lightning could be detected, although several discharges as close as 200 yards were observed. The illumination seemed to emanate from a single point which lit up the falling snow and surrounding cloud. The phenomenon was observed by several trained meteorological observers, in addition to Dr. Staley, but it was seen in much greater detail by the latter and the observers in the observation tower.

Several non-meteorological causes for the phenomenon were considered but had to be rejected. For example, it was suggested that car headlights sweeping the sky as the vehicle passed through a dip in the road could be mistaken for this 'lightning' However, by direct comparison with passing cars, this could be seen not to be the case. Arcing from H. T. power cables was also considered as a possible cause. However, the local power company reported no arcing from its overhead H. T. lines, though their engineer did report flashes which he attributed to lightning. Moreover, a flash was seen by the author when at a distance of at least two miles from the nearest house or power line.

A brief search of the literature does not reveal reports of similar phenomena, although normal lightning from snow is occasionally cited, e. g. Tucker, Hancock. The present phenomenon seems to be unusual in the form of the discharge and the type of precipitation. Hodkinson noted a discharge from a cloud to a tethered balloon at a height of 2, 750 feet above ground during a sleet storm. Though this flash was undoubtedly directly caused by the balloon, it does indicate the presence of a small quantity of space charge in a sleet storm.

The only possible explanation of the present phenomenon seems to be that isolated, small pockets of electric charge were carried to earth on the wet snow. The existence of regular thunderstorm activity seems to be indicated by the lightning seen over the surrounding mountains before the storm reached the town. It is perhaps not inconceivable that the final dissipation of space charge from a mature thunderstorm is the cause of the phenomenon here reported. Clearly, this leaves many unanswered questions and much more evidence is needed before any tentative conclusions may be drawn. (Weather, 19:291-292, 1964)

# PUZZLING SPHERES OF LIGHT

## NOTE ON A MANIFESTATION OF ELECTRICITY AT RINGSTEAD BAY

**Eaton, H. S.; *Royal Meteorological Society, Quarterly Journal,* 13:305, 1887.**

A remarkable display of electricity was witnessed at Ringstead Bay, on the Dorset Coast, on August 17th, 1876, by Mrs. and Miss Warry, of Holwell, near Sherborne.

North and east of the Bay the cliffs and steep declivities of broken ground, so conspicuous in the view from the Esplanade at Weymouth, descend from the Downs as a kind of undercliff. On the western side of the Bay the cliff diminishes, ending near Ringstead Ledges. About half way down the hill the coast-guard path from Whitenose to Osmington Mill runs close by a small sandstone quarry, on the brow of the cliff where the Upper Green Sand is faulted against the Kimmeridge Clay. The clay extends from the quarry to the foot of the hill, and falls sideways rather abruptly to the beach. The grassy, treeless brink of this clay cliff, for a distance of some 200 or 300 yards from the quarry downwards, and over a width of 3 or 4 yards from the edge of the cliff, was the place where the phenomenon was observed, the height above sea-level being probably from 150 to 250 feet. It is not known how much further the display extended along the edge of the cliff up or down the hill.

The weather for some weeks had been hot and dry, and the fields were parched by the drought; but at last a change was evidently approaching. The day was dull, sunless, and very sultry; from an early hour sheet lightning had been flashing frequently in various directions, unaccompanied by thunder. Between 4 and 5 o'clock in the afternoon the above-mentioned ladies, longing for a breeze, strolled to the edge of the cliff some distance below the quarry from a neighbouring cottage. The heat was oppressive; hardly a breath was astir, but faint puffs of sweltering air surged fitfully upwards from the ground. Mrs. Warry's first impression, on gaining the brink of the declivity facing the sea, was that the heat had affected her sight.

Over the crest of the ground, surrounding them on all sides, and extending from a few inches above the surface to 2 or 3 feet overhead, numerous globes of light, the size of billiard balls, were moving independently and vertically up and down, sometimes within a few inches of the observers, but always eluding the grasp; now gliding slowly upwards 2 or 3 feet, and as slowly falling again, resembling in their movements soap bubbles floating in the air. The balls were all aglow, but now

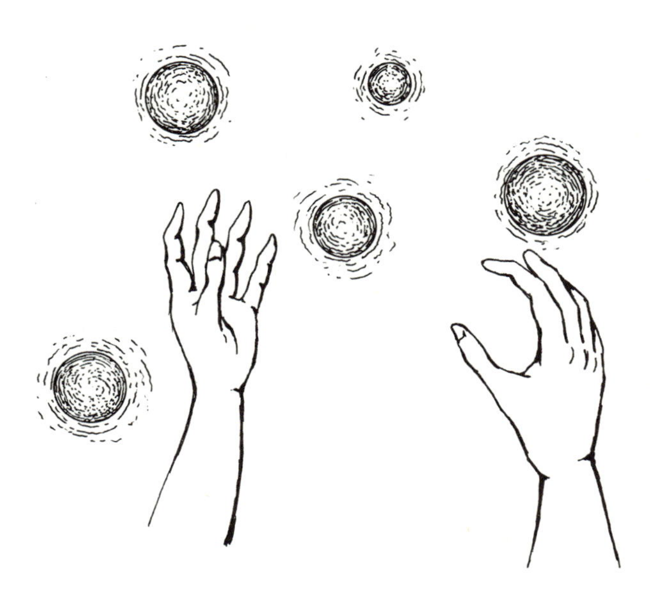

Spheres of light elude grasp at Ringstead Bay

dazzling, with a soft, superb iridescence, rich and warm of hue, and each of variable tints, their charming colours heightening the extreme beauty of the scene. The subdued magnificence of this fascinating spectacle is described as baffling description. Their numbers were continually fluctuating; at one time thousands of them apparently enveloped the observers, and a few minutes afterwards the numbers would dwindle to perhaps as few as twenty, but soon they would be swarming again as numerous as ever. Not the slightest noise accompanied this display.

The ladies sauntered up to the quarry and down again several times along the edge of the cliff, viewing the phenomenon for upwards of an hour with vague and increasing apprehension, and returned to the cottage, leaving the display in active operation. How long it continued, and when it began, is unknown.

About 10 p. m. a severe thunderstorm, attended with torrents of rain, came up from the sea; and on the next day, about 5 p. m., a waterspout was seen off Whitenose. (Royal Meteorological Society, Quarterly Journal, 13:305, 1887)

## NOTE ON AN APPEARANCE OF LUMINOUS BUBBLES IN THE ATMOSPHERE

**Bonney, A.; *Royal Meteorological Society, Quarterly Journal,* 13:306, 1887.**

On a day in the month of January 1871, a lady residing at Park Place, Remenham, observed the following phenomenon shortly before noon.

The weather was intensely cold, snow was lying after a fall some days previously. The sky was dull grey, with "rusty" clouds hanging rather low, the sun just showing itself, and the air was perfectly still.

The wall paper and furniture of the room in which the lady was sitting were suddenly flushed with rose colour, which gradually deepened into crimson, passing through bright gold into orange, lilac and deep violet.

It was then seen that from the centre of the level space of snow within view, a group of air bubbles, of the shape and apparent size of the coloured India rubber balls sold in the streets, rose to a considerable height and then began to move up and down within a limited area, and at an equal distance from each other, some ascending others descending.

The appearance lasted about two minutes, at the expiration of which the balls were carried away by a current of wind to the eastward and disappeared.

Another group of balls arose subsequently from the same spot, and the phenomena were precisely reproduced.

It was remarked that the balls assumed in succession the tints which had been observed on the walls of the room.

The appearance was also witnessed by a maid servant, who, on entering the room, at once exclaimed, "Oh, look at those little balls going up and down."

The above particulars were noted down immediately after a conversation with the lady who saw the bubbles, but she had frequently referred to the matter previously. Though rather advanced in years, she is in full possession of her faculties and is rather unusually keen and observant. (Royal Meteorological Society, Quarterly Journal, 13:306, 1887)

## MYRIAD OF LUMINOUS BODIES CROSS SKY
**Anonymous; *Nature,* 22:64, 1880.**

A remarkable phenomenon was observed at Kattenau, near Trakehnen (Germany), and in the surrounding district, on March 22. About half an hour before sunrise an enormous number of luminous bodies rose from the horizon and passed in a horizontal direction from east to west. Some of them seemed of the size of a walnut, others resembled the sparks flying from a chimney. They moved through space like a string of beads, and shone with a remarkably brilliant light. The belt containing them appeared about 3 metres in length and 2/3 metre in breadth. (Nature, 22:64, 1880)

## AERIAL BUBBLES
**Swinnerton, Henry U.; *Science,* 21:136, 1893.**

The account of "snow-rollers" in your recent issue recalls an atmospheric phenomenon which was beheld here by two witnesses of unimpeachable character several years ago, of which no account has ever been published. Towards sunset, late in April, 1886, on a warm, thawing day, the snow rapidly disappearing, two men, Capt. John E. Hetherington and Mr. Marcus Sternberg, as they rode down the long hill towards this village from the east, saw what appeared to be innumerable spherical bodies floating in the air like soap-bubbles. Both men saw and wondered at the appearance for some moments before either spoke. Capt. H. then said, "I wonder whether I am dreaming?" The other rubbed his eyes and echoed the sentiment. "Well," said the captain, "I wonder if you see what I see; what do you see?" They questioned each other, and both agreed as to their impressions. An orchard lay along the lower and northwesterly side of the road, and all in among the apple-trees were thick, gently descending multitudes of these bubbles, pretty uniform in size, say, 8 or 9 inches in diameter, apparently; none less than six; no small ones being observed.

The two observers state that they carefully fixed their attention on particular bubbles, in order to compare notes, and saw them seem to rest on the bough of a tree, or the top board of the fence, and then gently roll off and disappear or go out of sight. The sun was sinking and dropped below the opposite hills as they reached the foot of the long descent and entered the village, and the appearance came to an end. But up to this time the air seemed to be filled with these transparent floating spheres. The position of the observers with regard to the light seems to have made some difference as to seeing well this or that large aggregation or swarm that one or the other pointed out. The bubbles were highly colored, iridescent, gave the same sort of reflections as soap-bubbles, and apparently vanished individually in much the same way. All these points I have ascertained by repeated conversations.

Captain Hetherington (Lieutenant Colonel by merit) is widely known for his extensive apiaries, the largest in the country, and is an exceptionally good observer. Mr. Sternberg also is a gentleman of intelligence and careful observant character.

The only theory I have been able to form to account for such a phenomenon is, that if a certain kind of dust floated off in the air, each particle composed of some sort of saponaceous envelope, enclosing a highly expansible centre or core, like ammonia,---particles of this character expanded by the warm air, and at the same time moistened, might, under very nice conditions, produce such an effect. (Science, 21:136, 1893)

# EARTHQUAKE LIGHTS

## EARTHQUAKE LIGHTS: A REVIEW OF OBSERVATIONS AND PRESENT THEORIES

**Derr, John S.; *Seismological Society of America, Bulletin,* 63:2177-2187, 1973.**

The problem of earthquake lights (EQL), or luminous phenomena, as noted by Byerly (1942), has always been the darkest area of seismology. Very few scientists have ever worked on the problem, and few today are willing to tackle it because most of the reports are personal observations of untrained observers, and, until recently, there were no "hard data" which could be subjected to scientific analysis. Observations, however, have been made for many years, as suggested by an old Japanese haiku, quoted by Finkelstein and Powell (1971).

The earth speaks softly
To the mountain
Which trembles
And lights the sky

Recently, popular interest in EQL has been raised by press reports and the search for methods for earthquake prediction. The data which are now available consist of pictures of luminous phenomena taken at Matsushiro, Japan, from 1965 to 1967. This paper includes descriptions of the phenomenon, and reviews several theories advanced to date as possible explanations.

The first known investigations which led to significant interpretations and conclusions were done in the early 1930's by two Japanese seismologists, Terada (1931, 1934) and Musya (1931, 1932, 1934), and were described by Davison (1936, 1937). Musya collected some 1,500 reports of EQL from the Idu Peninsula earthquake of November 26, 1930, at 4:30 a.m. "The observations were so abundant and so carefully made that we can no longer feel much doubt as to the reality of the phenomena and of their connection with the shock. In most of them, the sky was lit up as if by sheet lightning, and nearly all the observers agree in estimating the duration of a single flash as decidedly longer than that of lightning. At one place on the east side of Tokyo Bay, the light resembled auroral streamers diverging from a point on the horizon. Beams and columns of light were seen at different places, several observers comparing the beams to those of a searchlight. Others describe the lights as like that of fireballs. Some state that detached clouds were illuminated or that a ruddy glow was seen in the sky. At Hakona-Mati, close to the epicenter and to the northeast, a flash of light was seen, now in one spot, now in another, and, when the earthquake was at its height, a straight row of round masses of light appeared in the southwest. According to most of the observers, the colour of the light was a pale blue or white or like that of lightning, but a large number state that it was of a reddish or orange colour. The light is said to have been so bright in Tokyo that objects in a room could be seen. At another place, about 30 miles from the epicentre, it was brighter than that of moonlight....

"The lights were seen to a distance of 50 miles to the east of the epicentre, nearly 70 miles to the northeast, and more than 40 miles to the west. They were seen both before and for some time after the earthquake, but were most conspicuous during the middle of the shock. The directions in which the lights were seen point, as a rule, to the neighborhood of the epicentre, that is, to the northern part of the Idu Peninsula. The light was, however, seen in other directions, sometimes in the direction of the sea....

"During the year following the Idu earthquake, Mr. Musya studied the luminous phenomena attending four other Japanese earthquakes. The reports that he received were most numerous for the South Hyuga earthquake of November 2, 1931. With this earthquake, the lights were usually described as beams radiating from a point on the horizon, as like lightning or a searchlight turned to the sky, and as of a blue or bluish colour. They were seen before the earthquake by 26 observers, during it by 99, and after it by 22." (Davison, 1937).

Radiating beams; one type of earthquake light

Terada and Musya reached no conclusions as to the possible causes of earthquake lights. Terada (1931) made some calculations on potential differences in the Earth, which McDonald (1968) considers to be in error. Nevertheless, he made some perceptive comments about the quality of testimonies of witnesses under stress, which are quite relevant to the problem of collecting subjective data during an earthquake.

"With regards to all these testimonies of witnesses, it must be always kept in mind that people are naturally alive to all kinds of phenomena observed at the time of a severe earthquake and apt to regard them as something connected with the catastrophal occurrence, while they forget to consider that the same phenomena are frequently observed on many other occasions not at all connected with earthquake. On the other hand, we learn from the results of investigations by phychologists in what a ludicrous manner the testimonies of people, otherwise quite normal in mentality, may appear distorted when compared to the bare truth."

Another assessment of the problem of earthquake lights is given by Byerly (1942). In addition to his general description, he documents observations of earthquake lights observed at sea. If these lights have the same cause as those observed on land, severe restrictions are placed on the mechanism of their generation.

"Occasionally during an earthquake shock or immediately before or after, observers report luminous phenomena in the heavens. All types of lights are reported seen, although it is rarely that two observers see exactly the same. There are steady glows, red and blue, and white: there are flashes, balls of fire, and streamers.

"At the time of the earthquake off the coast of northern California in January, 1922, one observer reported a glow at sea which he at first took to be a ship on fire. At the time of earthquake of October, 1926, centering in Monterey Bay, an observer reported a flash at sea which resembled 'a transformer exploding.' During the Humboldt County (California) earthquake of 1932 an observer reported, 'Several of my friends and I saw to the east what appeared to be bolts of lightning travel from the ground toward the sky. The night was clear.' It has been customary to attempt to ascribe earthquake lights to secondary phenomena, since we know of no source of such lights in the original earthquake action. True, movement on a fault would generate considerable heat, and, after the Sonora earthquake of 1897, trees overhanging the fault were scorched. Such would scarcely produce flashes in the sky, particularly over the ocean. In modern times, the prevalence of electric power lines enables one to explain away many such observations as due to probably breaks in such lines; but many may not be so dismissed. Landslides in mountains may generate great heat by friction. In the Owens Valley earthquake of 1872, fires were started in the mountains by such sources. In some cases thunderstorms may happen to accompany earthquakes, and then lightning may be called upon to explain flashes. Lights over the sea have been attributed to luminous marine organisms excited by the earthquake vibrations."

More recently, research into observations of luminous phenomena in Japan has been done by Yasui (1968, 1971, 1972), who collected and studied pictures, taken by various other observers, of earthquake lights observed during the Matsushiro earthquake swarm of 1965 to 1967. He has also studied reports of other sightings in Japan. Of the approximately 35 sightings, any pictures which might have recorded unrelated phenomena---distant lightning, meteors, twilight, zodiacal light, arcing power lines--were eliminated. At least 18 separate sightings remained unexplained. He concludes that luminescence over a mountain area lasting several tens of seconds on a clear and calm winter night is not a known phenomenon. He thinks it to be an atmospheric electrical phenomenon, but the earthquake trigger mechanism is unknown.

There are five general characteristics of the phenomenon as observed by Yasui (1968):

1. The central luminous body is a hemisphere, diameter about 20 to 200 m, contacting the surface. The body is white, but reflections from clouds may be colored.
2. The luminescence generally follows the earthquake with a duration of 10 sec to 2 min.
3. The luminescence is restricted to several areas, none of them the epicenter. Rather, they occur on mountain summits in a quartz-diorite faulted rock.
4. Sferics generally follow the luminescence and are strongest in the 10 to 20 kHz range. The luminescence occurs frequently at the time of a cold frontal passage.
5. There was no indication on the magnetometers at the local observatory.

These observations are not consistent with explanations based on auroras or other ionospheric phenomena, noctilucent clouds, or rapid spark discharges. However, the observations near mountain tops and along faults are consistent with large atmospheric potential gradients and with an unusual increase of radon gas in the vicinity.

Yasui believes that ionization in the lowest atmosphere becomes unusually large at the time of an earthquake and causes the luminous phenomena at the place where the electrical potential gradient is highest. The electric field is not expected to be large, as it is under a thunderstorm, nor is the atmospheric conductivity expected to be high. Therefore, some action of the earthquake must contribute to triggering luminescence, e.g., violent atmospheric oscillation, but the process is still unknown.

An unusual report of EQL near Hollister, California, was given by Nason, personal communication, (1973). In this case, the lights were seen as discrete sources against a hill, also noted by Davison (1937), rather than as the more commonly observed general sky luminosity. Mr. Reese Dooley, a poultry rancher living south of Hollister, observed the EQL in 1961. There were two earthquakes about 2-1/2 min apart. It was dark when he felt the first one, which was strong enough to make him want to go home to check on his family. Just as he reached his car, the second one started. As he looked west toward a hill, he saw a number of small, sequential flashes from different, random places on the hillside. Nason inspected the hillside and found no electric wiring or any other conventional explanation for the lights. Clearly, Mr. Dooley was very close to the source of the lights. This observation suggests that the extensive EQL observed in Japan which lit up most of the sky could be caused by a great number of small, random point discharges over part of the epicentral area.

Lomnitz, personal communication (1972) agrees with the author's hypothesis that a whole range of precursory phenomena including lights, sounds, and animal reactions, are probably caused by electromagnetic effects. For example, Lomnitz reports his personal observation in Mexico City of extremely unusual behavior in a dog, at least 1 min before the August 2, 1968 earthquake near the coast of Oaxaca. He also notes that luminous effects were widely observed in Mexico City at the time of the 1957 earthquake near Acapulco. Thus, any theory of EQL would have to account for the occurrence of electromagnetic effects at distances of $3^{o}$ to $4^{o}$ from the epicenter of a shock of magnitude 6.5 or greater.

Most recently Yasui (1972) has commented on observations of EQL during the October 1, 1969 earthquake at Santa Rosa, California (Engdahl, 1969). The lights were seen extensively over the Santa Rosa area and described in terms of lightning or electric sparks, Saint Elmo's Fire, fireballs or meteors. Some people also heard sounds like explosions. Just how many reports are genuine EQL and how many are caused by earthquake effects on man-made objects cannot be determined. From the published description, however, the Santa Rosa observations did not include what was described by Davison (1937) as appearing to be auroral streamers diverging from a point on the horizon, a description which does fit observations, for example, in Chiba prefecture, Japan, January 5, 1968, as sketched by Yasui (1971). (Seismological Society of America, Bulletin, 63:2177-2187, 1973)

## FLASHES OF FIRE DURING LANDSLIDE
**Anonymous; *Scientific American*, 85:35, 1901.**

A curious phenomenon was observed at the village of Le Ghazil, in the French Alps, recently. One day toward evening the inhabitants were disturbed by a loud rumbling in the vicinity of Mont Farand, which increased in intensity. Looking toward the scene of the disturbance, the villagers were further startled by seeing bright flashes of fire. At first the unusual spectacle was attributed to volcanic agencies, and a party of civil engineers set out to examine the cause of the phenomenon. They discovered that the intense dry heat had caused the chalk rocks on the summit of the mountain to crack and to break away in all directions. These rocks had descended the mountain like an avalanche, and being thickly veined with silex, in descending they had struck one another with terrific force, scattering brilliant showers of sparks in all directions, with such rapidity that they resembled one single sheet of flame. (Scientific American, 85:35, 1901)

# VOLCANO LIGHTS

## THE FLASHING ARCS: A VOLCANIC PHENOMENON
**Perret, Frank A.; *American Journal of Science,* 4:329, 1912.**

On the afternoon of April 7, 1906, the present writer, in company with Professor Matteucci, was skirting the southern flank of Vesuvius on a trip to the main source of the lava at the Bosco Cognoli. The volcano at this time was entering one of those paroxysmal phases by which the eruption---already three days old---worked progressively up to its great culmination, which occurred, it will be remembered, between this and the following day. The ejected detritus was of a mixed nature, viz., the fresh lava, clear red in full daylight, being mingled with old material from the upper portions of the cone, then in process of rapid demolition. The frequency of the explosions varied from approximately one every three or four seconds to at least three per second. Although powerful, they were very sharp and sudden in their nature, and at the instant of each---but before it could be sensed by the eye or ear---a thin, luminous arc flashed upward and outward from the crater and disappeared in space. Then came the sound of the explosion and the projection of gas and detritus above the lip of the crater. The motion of translation of the arcs, while very rapid in comparison with that of the detritus, was not above the limits of easy observation and there could be no doubt as to the reality of the phenomenon, which was repeated some hundreds of times. (American Journal of Science, 4:329-335, 1912)

Flashing volcanic arcs

**NOTES ON THE RECENT ERUPTIONS OF MOUNT PELEE**
**Nicholls, H. A. Alford; *Nature,* 66:638, 1902.**

At 11 p. m., lightning shot out from the mountain [Mt. Pelee] in all directions, zig-zagging and flickering flashes alternating with, or being accompanied by, reddish globes, which ascended and exploded, and shot out stars and long rays. Away towards the south-west was another large focus of electric energy, which appeared to me to have a distinct relation to the volcanic electric discharges from Mont Pelee. This spot was, I reckoned, at least forty miles from the volcano, from which it bore almost west two points north. This latter electrical display was similar, but less extensive than that from Mont Pelee, and it was accompanied by curious glowing globes, which burst and shot out tongues of lightning. The most curious part of the magnificent sight, however, was that occasionally long rays of light, very like to the rays of a searchlight, shot out from the direction of Mont Pelee downwards to the secondary and distant electrical display, and on this broad ray reaching the western focus, the lightning there became more vivid, intense and extensive. (Nature, 66:638-639, 1902)

## TORNADO LIGHT COLUMNS

**LUMINOUS PHENOMENA IN NOCTURNAL TORNADOES**
**Vonnegut, B., and Weyer, James; *Science,* 153:1213-1220, 1966. (Copyright 1966 by the American Association for the Advancement of Science)**

The twister came through here up high. [Toledo, April 11, 1965] From southwest to northeast.... There was hard wind. I was outside. Suddenly there was no wind. My eardrums felt like they would burst, very intense. I heard a far away roar and at the same time it was all around me.... Then a great wall of white came. There was hard wind and all white. I could not see through the white. There was very little damage here.

..........

Just after the tornado struck, I was inside of the house looking out, I saw something very bright about the size of a basketball about six feet away from me and about five feet off the ground. It was white, blue and yellow in color and coming slowly toward me.... at less than the speed a person would walk.... when it seemed to hit the door, it made the door sound like it was singing.

..........

We were shaken up and our trailer along with others was dented badly from hail the size of baseballs. The beautiful electric blue light that was around the tornado was something to see, and balls of orange and lightning came from the cone point of the tornado. The cone or tail of the tornado reminded me of an elephant trunk. It would dip down as if to get food then rise up again as if the trunk of an elephant would put the food in his mouth. While the trunk was up the tornado was not dangerous, just when the point came down is when the damage started. My son and I watched the orange balls of fire roll down the Race Way Park then it lifted and the roof came off one of the horse barns.

..........

We thought we saw searchlights all around us, but there were no light beams shooting up to the clouds from the ground. The lights darting around in the clouds were sort of luminous and appeared to be more round in shape than anything else, also they were quite large. The lights were not as bright as a stroke of lightning, but they were above a dense layer of clouds, and bluish white in color....they were shooting around. We could see the lights in the west, northwest, north, and northeast from 8:00 p.m. on.... We did not have much stroke lightning and I do not recall hearing thunder. Our electricity was knocked out at 9:32 p.m., and that is when we saw one black funnel. The reason we could see it is because there was a slight glow coming from the sides of it. It came from due west.

..........

I was standing in my backyard about 9:00 p.m. looking at the sky toward the west to north about 15°. The sky was really black.... All at once a big hole opened up in the sky with a mass of cherry red. The opening looked about 1/4 mile long and about the same height. It had a yellow tinge in the center and the edges were darker cherry red with black spots in the edges. This opening opened up complete in about 6 seconds, stayed open about 10 to 12 seconds then closed in 6 seconds. The sky was completely black again. It was about 30° to 35° up. The black spots that I referred to were small portions of dark clouds like balls, just a few on the right of the sketch and the lower left. They were rolling some, always working toward the outer sides. The cherry red had some vibration to it, very little. The upper center....had a very fast quiver. Also a motion like hot steel melting in a pot. This motion was very small but fast. (Science, 153:1213-1220, 1966)

## EYEWITNESS TORNADO OBSERVATIONS OBTAINED WITH TELEPHONE AND TAPE RECORDER
**Ryan, R., and Vonnegut, B.; *Weatherwise*, 23:129, 1970.**

At about 2100 [May 15, 1968] the tornado struck Tuckerman, Arkansas, which lies 17 miles ENE of Oil Trough and 30 miles WSW of Jonesboro. Mr. Gene Elkins and Mr. Sam Smith observed the storm where it first touched down near Mr. Elkins' store on the southwestern side of Tuckerman. Mr. Elkins described what he saw:

It was raining real hard, and then the rain turned into hail, and it hailed for about 2 or 3 seconds, and then objects started flying out of there (the tornado), pieces of tin and pieces of metal. The funnel was going around and there was a lot of material in the air....I was looking to the east and all this stuff was going from my right to my left, and it was going in an upward direction....I'd say this was about 75 feet from me....the funnel it came from behind, and it was kind of a green glowish light and evidently the funnel was just right around us, but it appeared to be on a hill and just one edge of the funnel was touching ground....the other side of the funnel was up in the air. The north edge was off the ground, but the south edge was on the ground. It was a pale green neon light....in the funnel itself. I could see it (objects) by a light green light. There were 3 people in the store here with me and we could all see this stuff in the air, and it was just a light colored green light, and I don't know if that was kind of a solid bolt of lightning or where that light was coming from. It lasted 2 or 3 seconds...I could see stuff turning over up there. I believe it was just kind of constant, cause I could just keep seeing....It wasn't up real high because I remember the stuff going out of sight up higher in the cloud....I noticed my ears hurt....it pulled the glasses out of the cars, but it didn't hurt the building. The noise was really terrific; it sounded like a jet. (Weatherwise, 23:129, 1970)

## ELECTRICAL ACTIVITY ASSOCIATED WITH THE BLACKWELL-UDALL TORNADO

**Vonnegut, B., and Moore, C. B.; *Journal of Meteorology,* 14:284-285, 1957.**

Eyewitnesses to the tornado in Blackwell also have reported evidence of intense electrical activity. Montgomery, who viewed the tornado from a distance of about 3000 ft, reports: "As the storm was directly east of me, the fire up near the top of the funnel looked like a child's Fourth of July pin wheel." "There were rapidly rotating clouds passing in front of the top of the funnel. These clouds were illuminated only by the luminous band of light. The light would grow dim when these clouds were in front, and then it would grow bright again as I could see between the clouds. As near as I can explain, I would say that the light was the same color as an electric arc welder but very much brighter. The light was so intense that I had to look away when there were no clouds in front of it. The light and the clouds seemed to be turning to the right like a beacon in a lighthouse."

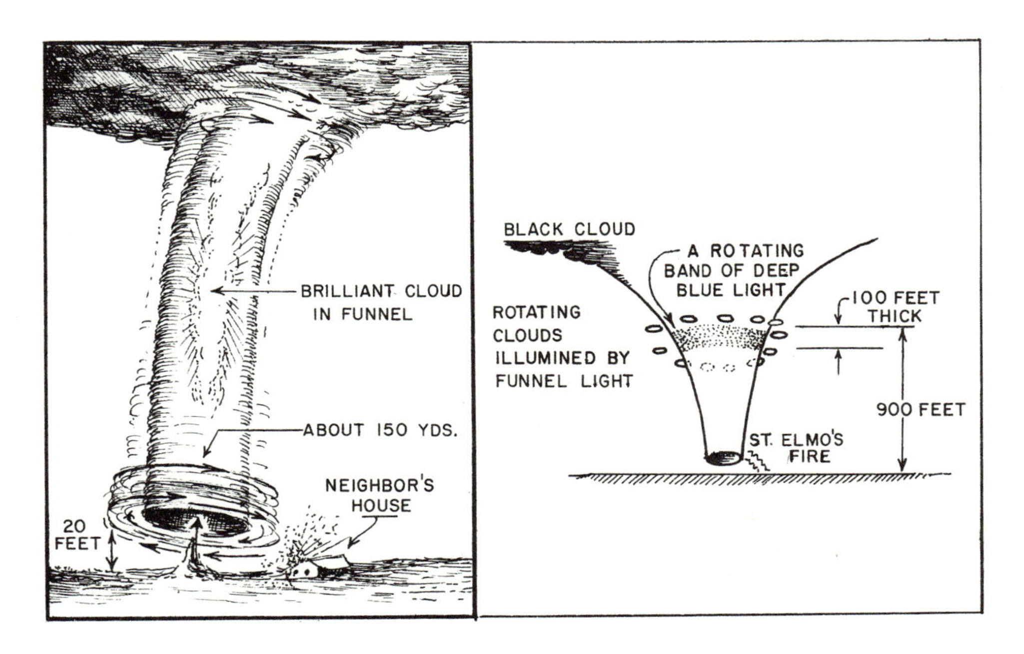


Luminous phenomena of the Blackwell tornado

Montgomery also tells of other eyewitnesses who had different views of the phenomenon, such as Mrs. Carl Sjoberg whose house in the direct path was completely demolished: "She saw lightning coming up from the ground two or three feet high and about half as wide as adding machine tape. It was a deep blue and forked on the end like a 'Y' or like a snake's tongue."

According to Montgomery, Lee Hunter, who was 4 mi north of Blackwell, described the tornado as follows: "The funnel from the cloud to the ground was lit up. It was a steady, deep blue light---very bright. It had an orange color fire in the center from the cloud to the ground. As it came along my field, it took a swath about 100 yards wide. As it swung from left to right, it looked like a giant neon tube in the air, or a flagman at a railroad crossing. As it swung along the ground level, the orange fire or electricity would gush out from the bottom of the funnel, and the updraft would take it up in the air causing a terrific light---and it was gone! As it swung to the other side, the orange fire would flare up and do the same." (Journal of Meteorology, 14:284-285, 1957)

# BALL LIGHTNING

Ball lightning cannot be ignored. It exists. Thousands of people have seen it and hundreds of scientists have written about it. Nevertheless, it remains as inscrutible as ever.

One reason ball lightning resists explanation is that it is so variable. It may be no larger than a pea or it may rival a house in size. It may be violet, red, yellow, or change colors during the few seconds it exists. The shape of ball lightning is usually spherical but rods, dumbbells, spiked balls, and other shapes have been reported. The phenomenon may glide silently and even inquisitively around a room and then quietly dematerialize. Most ball lightning, though, seems to explode violently; or it might be better to say that its disappearance is accompanied by an explosion. One has to hedge because the actual explosions sometimes seem to occur elsewhere---outside the house in which the ball has appeared, for example. This fact plus the frequent materialization of ball lightning in closed rooms and metal aircraft suggests that electrical induction may be involved. In other words, electrical forces from a surrounding storm may create a glowing ball of plasma in the air after the fashion of St. Elmo's Fire. When the electrical forces disappear (with a detonation), so does the ball lightning. Observers, however, have no doubt that something palpable has visited them because the room is usually filled with a sulphurous stench and considerable material damage.

Strangely, ball lightning rarely hurts people even though it may follow and hover about them. Ordinary lightning, in contrast, kills many. The listing of ball lightning's peculiar attributes could fill many more pages; several hundred examples have been gleaned from the scientific literature. It is best to let the observers themselves tell what they saw in some selected cases. Many were shocked that such a weird apparition exists and scared, too, because of the uncanniness of the phenomenon.

No reasonable scientific explanation exists for ball lightning. Electromagnetic plasmas, chemical reactions in the air, antimatter meteorites, nuclear reactions, and many other ideas have been rendered and found wanting. Perhaps several different phenomena are actually present. It is even possible that intense electrical forces may somehow distort our space-time continuum and provide a fleeting window onto some unknown cosmos.

## "ORDINARY" BALL LIGHTNING

### BALL LIGHTNING
**Stenhof, Mark; *Nature,* 260:596-597, 1976.**

A dramatic and characteristic ball lightning event, important because it may reveal details of the energetics of the phenomenon, was reported by a housewife to have occurred during a vigorous thunderstorm in the Midlands area of England on August 8, 1975. During this storm, a number of houses in the area were struck by lightning.

The witness was in the kitchen of her house in Smethwick, Warley, at about 1945, when a sphere of light appeared over the cooker. The ball was ~ 10 cm across and surrounded by a flame-coloured halo; its colour was bright blue to purple. The ball moved straight towards the witness at an estimated height of 95 cm from the ground. Burning heat was felt, and there was a singeing smell. A sound something like a rattle was heard. The ball was in sight for only a second or so because its lifetime was cut short:

"The ball seemed to hit me below the belt, as it were, and I automatically brushed it from me and it just disappeared. Where I brushed it away there appeared a redness and swelling on my left hand. It seemed as if my gold wedding ring was burning into my finger."

Where the ball struck her, the clothing of the witness was damaged. A hole was produced in her dress and her tights. Her legs were not actually burned, however, but became red and numb. The dress has been secured for examination, and tests are currently being carried out to try to establish the nature of the damage. There is a central area where the synthetic fibre of the dress has disappeared altogether, and the shape of the hole is in agreement with the witness' statement that she brushed the ball downward a couple of inches. The hole is $\sim$ 11 x 4 cm. Around its perimeter, the material has shrivelled, but is not charred. In addition, there is a secondary area where the fibre has shrivelled, and the pattern printed on the fabric has faded noticeably. The position of the primary damage is $\sim$ 20 cm from the hem of the dress, so it is estimated that the ball struck at $\sim$ 95 cm from the ground.

Although thunder and lightning were going on at the time, the witness is not sure whether or not the manifestation was associated with a thunderclap. The ball exploded with a bang, but she is not certain whether it exploded at the exact time at which she touched it; she felt this to be the case, however. Before my contacting her, the witness had had no acquaintance with the phenomenon of ball lightning. (Nature, 260:596-597, 1976)

## CAT-LIKE BALL LIGHTNING
**Anonymous; *Comptes Rendus,* 35:2-3, 1852.**

The following report to the Academy concerns a case of ball lightning which the Academy had entrusted me to investigate several years ago. Not on the downward stroke but rather during its return stroke, so to speak, the lightning in question struck a house in the Rue St. Jacques next to the Val-de-Grace church; in fact, this took place in a proximity which would have led one to think that the house should have been protected against this sort of mishap by the high lightning rod mounted on the dome of the Val-de-Grace church. The following is a condensed version of the account given by the tailor in whose room the ball lightning descended and then lifted upward again.

Somewhat after a very powerful lightning stroke, the tailor, who was sitting next to his table after having finished his meal, suddenly saw how the paperlined frame covering the fireplace opening fell down, as if it had been knocked over by a strong gust of wind, and a fireball the size of a child's head gently entered via the opening and slowly wandered about the room at low level over the floor-tiles. According to the tailor, the fiery ball appeared like an average-sized young cat which had rolled itself into a ball and moved without using its paws for support. The fireball was shiny and luminous rather than hot and inflamed, and the tailor experienced no heat sensation. The ball approached his feet like a young cat, which wanted to play and, as is customary for such animals, to rub itself along the leg of its master. The tailor nevertheless withdrew his feet and by means of several careful, or in his own words, very gently performed movements, he avoided making contact with the phenomenon. The latter seemed to pause several seconds beside the feet of the seated tailor, who, bending somewhat forward, watched it attentively. After the ball had carried out a few movements in different directions, but without veering from the center of the room, it lifted straight upward to the level of the tailor's head, who sat upright again and sank back into his chair in order to avoid being hit in the face and at the same time to be able to follow the course of the phenomenon. When the fireball had reached a height of about 1 m above the floor, it became somewhat elongated and made its way along a slanted path toward a hole about 1 m above the upper cornice of the fireplace. This hole served to accommodate the pipe of a stove, which the tailor used in winter. However, the lightning could not, in the tailor's words, see the hole, since the latter was covered with paper. Nevertheless, the fireball went straight toward this hole, peeled off the paper without tearing it and ascended the chimney. Thus, after it had taken the time (the mode of expression used by several eyewitnesses) to climb up the chimney at the same speed, i.e., fairly slowly, as when it had come and had reached the mouth of the chimney, which was at least 20 m above the ground, it produced a dreadful explosion, which destroyed part of the chimney and hurled the broken pieces into the yard. The roofings of several smaller buildings caved in, but fortunately without any disastrous consequences. The tailor's dwelling was situated on the third floor, i.e., not at midheight of the house. The lightning did not reach the ground floor. The motion of the luminous ball was at all times slow and steady. Its lustre was by no means brilliant. Furthermore, it did not radiate any noticeable heat. The ball did not seem to have any tendency to follow conducting bodies or to veer away from air currents. (Comptes Rendus, 35:2-3, 1852)

## BALL LIGHTNING
## Clark, R.; *Weather,* 20:134, 1965.

Some years ago, while I was walking along a road near Denby Dale, West Yorkshire, I was overtaken by a thunderstorm. When the storm was at its height and the lightning and thunder very close, I sheltered by some lodge gates which opened off the tree-lined road. After one lightning discharge there appeared a ball of light, 'electric' blue in the centre with an ill-defined yellow white fringe, the whole being about 18 in. in diameter. It first appeared from beneath the tree branches on my side of the road, perhaps 20 yards away, and approached slowly with a staggering undulating motion, but keeping approximately to the same height under the trees. As it approached me, it crossed the road gradually and then, when almost opposite me, turned back across the road and slightly backwards, passed through the lodge gates where I lost sight of it.

I estimate it moved at a moderate walking pace for 30 yards before I lost view of it and in this time, no changes in size or brightness were noted.

In the same storm a second 'ball' appeared at about the same place, but disappeared after moving four or five yards. In colour, size and speed of motion, it was much as the first one described. (Weather, 20:134, 1965)

## EXTRAORDINARY ATMOSPHERIC PHENOMENON
**Hannay, J. B.; *Nature,* 25:125, 1881.**

Those on board the Campbelton Steamer Kinloch (Capt. Kerr), which left Greenock on its usual run about half-past eleven o'clock on Tuesday morning after the storm that raged during the night, had a somewhat extraordinary experience while passing down the Firth. The vessel was enveloped in a dense shower of hail, and for some time it was awfully dark, and occasionally the vessel was lit up by vivid flashes of lightning. One of the flashes was very bright, and its shape was something like that of the arteries of the human body, with a central column all shattered and broken. About noon, while opposite the Cloch Lighthouse, and not far from the shore, the captain observed immediately over the ship what appeared to be a series of clear balls of lightning, each about a foot in length, and resembling a chain, except that they were disconnected. This phenomenon was quickly succeeded by an explosion in the funnel of the steamer, and several balls of fire upon the bridge running about, and then bounding off into the water. The first impression of the spectators was that something had exploded on board, but on inquiry it was found that this was not the case. The mate stated, however, that a ball of lightning had almost struck him where he stood. A fireman rushed upon deck to see what had happened, as the engine-room was filled with smoke, and a choking sensation was experienced below. The explanation appears to be that a portion of the lightning had passed down the funnel until its force was spent by the fire, and the sudden recovery of the draught of the funnel afterwards accounted for the loud report that was heard. The captain, in his long experience at sea, never encountered such a phenomenon before, and it may be taken as an indication of the extraordinary atmospheric forces which had been at work during the storm, and which seemed to centre in this locality. (Nature, 25:125, 1881)

## GLOBULAR LIGHTNING
**Anonymous; *Science,* 10:324, 1887.**

The following report from the Hydrographic Office relates to one of the rarest and most inexplicable forms of lightning. Can any of the readers of Science give any information on the subject? A globe of fire floats leisurely along in the air in an erratic sort of a course, sometimes exploding with great force, at other times disappearing without exploding. On land it has been observed to go into the ground and then reappear at a short distance, and where it entered the soil it left a rugged hole some twenty feet in diameter. Although there is no doubt as to the facts regarding the phenomenon, no satisfactory explanation of the cause has ever been given. It is, of course, entirely different in character from St. Elmo's fires, so often seen on board vessels during thunder-storms: these remain stationary at the yard-arms and mast-heads, and are analogous to the 'brush discharge' of an electric machine. Captain Moore, British steamship 'Siberian,' reports, "Nov. 12, midnight, Cape Race bearing west by north, distant ten miles, wind strong south by east, a large ball of fire appeared to rise out of the sea to a height of about fifty feet, and come right against the wind close up to the ship. It then altered its course, and ran along with the ship to a distance of about one and one-half miles. In about two minutes it again altered its course, and went away to the south-east against the wind. It lasted, in all, not over five minutes. Have noticed the same phenomenon before off Cape Race, and it seemed to indicate that an easterly or south-easterly gale was coming on." (Science, 10:324, 1887)

## A FIRE BALL?
**Matts, E.; *Weather*, 19:228, 1964.**

It happened on 10 November 1940 about 12.15 pm: I was working at the far end of my garden, the weather was normal, no rain, no signs of thunder. Suddenly, I seemed to be in the centre of intense blackness and looking down observed at my feet a ball about 2 ft across. It was of a pale blue-green colour and seemed made of a mass of writhing strings of light, about 1/4 in. in diameter.

It remained there for about 3 seconds and then rose, away from me, just missing a poplar tree about 8 ft away. It cleared the houses by about 20 ft and landed at the rear of the Weavers Arms on the Bell Green Road, a distance of about 1/4 mile. There was a loud explosion and much damage was done to the public house.

As a matter of interest, I felt no alarm whatever, but this may be explained by the fact that we had experienced considerable bombing in Coventry at that period. (Weather, 19:228, 1964)

## RARE ELECTRICAL PHENOMENON AT SEA
**Anonymous; *Monthly Weather Review*, 15:84, 1887.**

Capt. C. D. Swart, of the Dutch bark "J. P. A.," makes the following report of a remarkable phenomenon observed by him at 5 p.m. March 19, 1887, in N. 37° 39', W. 57° 00':

During a severe storm saw a meteor in the shape of two balls, one of them very black and the other illuminated. The illuminated ball was oblong, and appeared as if ready to drop on deck amidships. In a moment it became as dark as night above, but below, on board and surrounding the vessel, everything appeared like a sea of fire. The ball fell into the water very close alongside the vessel with a roar, and caused the sea to make tremendous breakers which swept over the vessel. A suffocating atmosphere prevailed, and the perspiration ran down every person's face on board and caused everyone to gasp for fresh air. Immediately after this solid lumps of ice fell on deck, and everything on deck and in the rigging became iced, notwithstanding that the thermometer registered 19° Centigrade. The barometer during this time oscillated so as to make it impossible to obtain a correct reading. Upon an examination of the vessel and rigging no damage was noticed, but on that side of the vessel where the meteor fell into the water the ship's side appeared black and the copper plating was found to be blistered. After this phenomenon the wind increased to hurricane force. (Monthly Weather Review, 15:84, 1887)

# SPIKED BALL LIGHTNING

## GLOBULAR LIGHTNING
**Gilmore, G.; *Nature*, 103:284, 1919.**

On the night of May 14 a thunderstorm took place over Dublin. A shower of rain fell after 9 p.m., but between about 9.25 and 9.40 there was practically no rain, only a few drops falling. At about 9.50 I went outside, and when I had gone about

Spiked ball lightning in street

two steps from the door I suddenly saw a luminous ball apparently lying in the middle of the street. It remained stationary for a very brief interval---perhaps a second---and then vanished, a loud peal of thunder occurring at the same time. The ball appeared to be about 18 in. in diameter, and was of a blue colour with two protuberances of a yellow colour projecting from the upper quadrants. It left no trace on the roadway. The street is about eight yards wide from footpath to footpath, with houses on both sides, the total distance across the street between the houses being about twenty yards. There are no tramlines on the street. When I observed the ball its distance from me was about ten yards. The thunder was heard just at the disappearance of the ball, but the sound seemed to come from overhead rather than from the place where the ball was. This was the first peal of thunder that I heard, and there was no more thunder or lightning until after 10.15. From 19.40 onwards the thunderstorm was rather violent and the rain heavy. The rain ceased about 12 midnight, but sheet lightning continued to play over the sky. I was looking towards the north at about 12.15, where the sky was fairly clear, with small white clouds scattered over it, when I saw a yellow-coloured ball which appeared to travel a short distance and then disappear. This ball was high up in the sky, and appeared smaller than the first ball described above. (Nature, 103:284, 1919)

# RAYED BALL LIGHTNING

## VIOLET BALL SURROUNDED WITH RAYS
**Anonymous; *Nature,* 42:458, 1890.**

The Caucasus papers relate an interesting case of globular lightning which was witnessed by a party of geodesists on the summit of the Bohul Mountain, 12,000 feet above the sea. About 3 p.m., dense clouds of a dark violet colour began to rise from the gorges beneath. At 8 p.m., there was rain, which was soon followed by hail and lightning. An extremely bright violet ball, surrounded with rays which were, the party says, about two yards long, struck the top of the peak. A second and a third followed, and the whole summit of the peak was soon covered with an electric light which lasted no less than four hours. The party, with one exception, crawled down the slope of the peak to a better sheltered place, situated a few yards beneath. The one who remained was M. Tatosoff. He was considered dead, but proved to have been only injured by the first stroke of lightning, which had pierced his sheepskin coat and shirt, and burned the skin on his chest, sides, and back. At midnight the second camp was struck by globular lightning of the same character, and two persons slightly felt its effects. (Nature, 42:458, 1890)

# ROD-SHAPED BALL LIGHTNING

## GLOBULAR LIGHTNING
**Hare, A. T.; *Nature,* 40:415, 1889.**

On Monday, the 5th instant, at midday, this district was visited by a violent storm of rain, which lasted half an hour, and was accompanied by thunder and lightning. When the storm had passed over and the sky was getting bright, a rod-like object was seen to descend from the sky. It is described as being of a pale yellow colour, like hot iron, and apparently about 15 inches long by 5 inches across. These dimensions are given by an observer who estimated its distance (about correctly, as it subsequently appeared) at 100 yards, and are not therefore affected by the uncertainty attaching to estimates of the sizes of objects whose distance is quite unknown. This object descended "moderately slowly," "not too fast to be followed by the eye" and quite vertically.

On reaching a point about 40 feet from the ground, and in close proximity with the chimney-stack belonging to a house in Twickenham Park, the object seemed to "flash out horizontally as if it burst," showing an intensely white light in the centre and a rosy red towards the outer parts. At the same instant a violent explosion was heard, and soon afterwards a strong smell was perceived, which is described by the observers as "resembling that of burning sulphur," for which the smell of ozone and nitric oxide might easily be mistaken.

..........

"The explosion filled the kitchen with smoke and soot. The dining-room also was filled with smoke and soot, though no fire was burning in it.

"The master of the house was just coming into the dining-room from the conservatory when he heard the detonation and simultaneously saw a bright flash of light. He staggered back a moment, and then ran through the smoke and soot to the hall, and called out to know if anyone was hurt. Finding all safe he returned into the dining-room. A Japanese umbrella set open as an ornament in the empty grate, but not fixed in any way, was undisturbed, though the hub of it was hot to the touch. Piles of soot spread out in a semicircle to the centres of the side walls of the room, and an arm-chair, which had been standing close to the fire-place, was 6 feet from its previous position, and had evidently been turned round and thrust against the wall. In the bed-room, on the first floor, soot was on the floor and in the fire-place. The slab of marble forming the architrave under the mantel-shelf, and extending the whole width of the fire-place, had been thrust out from its setting, and was, with a number of bricks, lying 6 feet away on the floor. (Nature 40:415, 1889)

## A POSSIBLE CASE OF BALL LIGHTNING
**Alexander, William H.; *Monthly Weather Review*, 36:310, 1907.**

I was standing on the corner of Church and College streets, just in front of the Howard Bank and facing east, engaged in conversation with Ex-Governor Woodbury and Mr. A. A. Buell, when, without the slightest indication or warning, we were startled by what sounded like a most unusual and terrific explosion, evidently very near by. Raising my eyes and looking eastward along College Street, I observed a torpedo-shaped body some 300 feet away, stationary in appearance and suspended in the air about 50 feet above the tops of the buildings. In size, it was about 6 feet long by 8 inches in diameter, the shell or cover having a dark appearance, with here and there tongues of fire issuing from spots on the surface resembling red-hot unburnished copper. Altho stationary when first noticed this object soon began to move, rather slowly, and disappeared over Dolan Brothers' store, southward. As it moved, the covering seemed rupturing in places and thru these intensely red flames issued. My first impression was that it was some explosive shot from the upper portion of the Hall furniture store. When first seen it was surrounded by a halo of dim light, some 20 feet in diameter. There was no odor that I am aware of perceptible after the disappearance of the phenomenon, nor was there any damage done so far as known to me. Altho the sky was entirely clear overhead, there was an angry-looking cumulo-nimbus cloud approaching from the northwest; otherwise there was absolutely nothing to lead us to expect anything so remarkable. And, strange to say, altho the downpour of rain following this phenomenon, perhaps twenty minutes later, lasted at least half an hour, there was no indication of any other flash of lightning or sound of thunder.

Four weeks have passed since the occurrence of this event, but the picture of that scene and the terrific concussion caused by it are vividly before me, while the crashing sound still rings in my ears. I hope I may never hear or see a similar phenomenon, at least at such close range.

Mr. Alvaro Adsit says:

I was standing in my store door facing the north; my attention was attracted by this "ball of fire" apparently descending toward a point on the opposite side of the street in front of the Hall furniture store; when within 18 or 20 feet of the ground the ball exploded with a deafening sound; the ball, before the explosion, was apparently 8 or 10 inches in diameter; the halo of light resulting from the explosion was 8 or 10 feet in diameter; the light had a yellowish tinge, some-

Torpedo-shaped ball lightning at Burlington, Vt.

what like a candle light; no noise or sound was heard before or after the explosion; no damage was done so far as known to me. (Monthly Weather Review, 36:310, 1907.

# DUMBBELL-SHAPED BALL LIGHTNING

## DUMBBELL-SHAPED BALL LIGHTNING
**Anonymous; *Meteorologische Zeitschrift*, 29:384, 1912.**

A remarkable example of ball lightning was observed on May 12, 1912, at 6.45 p.m. from the 200-m-high section of Plaven near Dresden on the adjacent side of the narrow Weisseritz Valley. Although the observation site itself lay beyond the range of the thunderstorm (the nearest of the three thunderclouds was at a horizontal distance of 3 km), suddenly, without there preceding a lightning of some other form, there appeared at a height, which must have been less than 100 m above the ground, a luminous spherical mass at a distance of approximately 1 km from the observation site. The instant that one of us first noticed the luminous phenomenon there was a second, smaller ball about 1.5 m below the first which was connected to the upper ball by a fine strip similar to a string of beads. The color of the balls was reddish yellow (about 600 mμ). Both moved slowly one directly above the other in the direction northeast, whereby the upper ball maintained a constant height above the ground, while the lower one slowly descended as the connecting luminous strip faded. At the observation site the wind came from the north. As soon as the lower ball was to within a short distance from the ground it continued to move in the same way as the upper one.

Both balls had a velocity which could hardly have exceeded 1 m/sec. The lower luminous ball soon vanished behind a bush; from its height, distance and angle, under which the upper edge of the bush could be seen, we concluded that the diameter of the lower ball could not have been greater than 1 m. The upper ball suddenly faded shortly after it, and likewise started to follow a downward sloping trajectory. No noise was perceived, and the total duration of the event from the moment the phenomenon was noticed to the instant the upper ball faded amounted to two minutes; the movement of the lower ball close above the ground lasted about 3/4 minute.... (Meteorologische Zeitschrift, 29:384, 1912)

# MINIATURE BALL LIGHTNING

## GLOBULAR LIGHTNING DISCHARGE
**Chattock, A. P.; *Nature,* 109:106, 1922.**

The two ladies were sitting at table about 8 p.m., with the window open. It was raining heavily at the time, and there was no wind. Stormy clouds were about, but it was not unusually hot. Thunder and lightning at the same time were afterwards reported from London---a distance of, say, 50 miles---but there was no thunderstorm at Eastbourne. There had been no rain during the few preceding days. As one of the ladies took up a knife to cut bread the ball of light was seen to flash past the knife (without touching it) on to the table, travelling a distance of about 9 in. at an average height of about 3 in. from the table, but moving towards the latter.

When the ball touched the tablecloth it "went out with a spitting sound," leaving no mark or trace of any sort. Until it touched the cloth there was no sound, and the whole thing was over in such a "flash of time" that it was impossible to say how fast the ball travelled. There seems to have been an impression that the ball came from the direction of the open window, but it was only under dependable observation during its 9-in. path from the bread-knife to the tablecloth.

As to the appearance of the ball itself, it was "about the size of a pea, the light encircling it being about the size of a golf ball. The light was white and intensely bright, like electricity." "Too dazzling to see through." (Nature, 109:106-1922)

# TRANSPARENT BALL LIGHTNING

## TRANSPARENT BALL LIGHTNING
**Anonymous; *Das Wetter,* 6:68, 1889.**

In August (?) of the year 1868 I was engaged in the construction of a tunnel on the Upper Ruhr Valley railroad in the vicinity of Arnsberg. One day during a thunderstorm we took shelter under the roof of the nearby blacksmith and wheelwright shop; the shelter provided by the roof against the downpouring rain was not too good, since the roof had leaks at various places.

The sun was already shining again for some time and only a few, fairly transparent clouds could be seen in the sky. We had already forgotten the storm and were even less occupied with the thought of another thunderstorm. The four of us were busy on the trestle and were just about to lift a stone plate approximately 80 cm square onto the wall; we were standing in a circle, suspecting nothing, when suddenly a lightning flashed and a round, yellowish, transparent ball of about 20 cm diameter appeared in our midst at approximately 90 cm above the stone we were about to lift onto the wall; the ball oscillated continually about 4 cm up and down above the plate. At the center of this ball there was a blue flame, which was pear-shaped with the tapered portion slanted downward and had a length of 4 cm. This flame rapidly rotated around a vertical circle of about 7 cm diameter inside the large ball. Anyone can imagine how startled we must have been. My eyes were fixed upon the frightful intruder and my only thoughts were how to get rid of it. However, I shall not waste time elaborating about this. After about 3-4 seconds there followed a loud bang, such as I had never heard before, and the ball vanished, none of us knowing where it had gone. However, we breathed more freely now and I felt as if the full load of the heavy stone which lay before our feet had been taken off my shoulders. When we recovered from the shock and lifted the stone onto the wall our limbs were almost paralyzed. (Das Wetter, 6:68, 1889)

# FRAGMENTING BALL LIGHTNING

## IN SUPPORT OF A PHYSICAL EXPLANATION OF BALL LIGHTNING
**Wittmann, A.; *Nature,* 232:625, 1971.**

During a dinner on the evening of July 6, 1971, at Orselina, near Locarno, Switzerland, I observed a flash of lightning which struck an unidentified target near the roof of a building about 200 m away. An intense point discharge immediately followed the flash and was visible clearly at the target for 1 or 2 s. I immediately told the other people at the dinner (Professor E. H. Schroter and Dr. E. Wiehr from Gottingen and Mr. C. Kuhne from Oberkochen) of my observation because they were not looking out of the window at the time.

This occurrence was unmistakably different from another lightning event, which I observed several years ago and which I consider to have been a rare event of actual ball lightning. At that time a thunderstorm accompanied by heavy rain took place in the area of Neustadt near Coburg, Germany. I was sitting with some other people (among them Mr. C. Forster from Berlin) right behind a ground floor window so that I could look at the weather; the field of view out to the street was limited by the frame of the window. Suddenly, at a height of about 16 m above the ground and at a short range of about 24 m, I saw a spherical plasma ball coloured bright yellow-white. This object appeared to have a diameter of 50 to 100 cm. It moved vertically downwards with a speed of about 4 m s, and its path ended in the top branches of a tree, at a height of approximately 9 m. On touching the tree the ball instantly disintegrated into eight to twelve smaller spheres. These were the same colour as the large one and each had a diameter to 12 to 15 cm. They fell to the ground, guided by the outer contour of the tree, and moved vertically during the last few metres in the absence of branches. On reaching the ground (an asphalt roadway and a neighbouring footpath) the spheres instantly disappeared. There was no noise apart from that of the rain and no lightning associated with the primary plasma-like sphere. Three to five minutes afterwards, the same phenomenon occurred again in precisely the same way as before, indicating that the conditions needed to produce and guide the primary sphere were still maintained or re-established during the time that had elapsed. Immediately after the rain had stopped, I went out into the street to look for further evidence. There were circular patches of melted asphalt on the wet asphalt cover of the roadway which showed the interference colours of thin layers. Their diameters were each 12 to 15 cm and they obviously marked the impact areas of the smaller spheres. This event was described similarly by all the witnesses.

Because roadway asphalt in general contains B-80 bitumen---a thermoplast from which liquid components disintegrate at about 170°C---one may very roughly calculate the minimum amount of energy needed to produce the observed patches. Assuming that a water layer of thickness 0.5 mm at 20°C was evaporated and that an asphalt layer of thickness 1 mm was then heated to 170°C, and assuming a mean density of 1.0 g $cm^{-3}$ and a mean specific heat $c_p$= 0.46 calorie $g^{-1}$ for B-80, one may easily calculate the energy density of the plasma spheres to be at least 1.9 x $10^7$ J $m^{-3}$. (Nature, 232:625, 1971)

# BALL LIGHTNING IN ENCLOSURES

## A TRULY REMARKABLE FLY

**Mohr, Frederick B.; *Science,* 151:34-, 1966. (Copyright 1966 by the American Association for the Advancement of Science)**

On 25 August 1965, I was editing an article entitled "Soviet research on ball lightning" prepared by Arsen Iwanovsky of this division for publication in the September issue of the Foreign Science Bulletin. We discussed at some length the unusual behavior of ball lightning and the fact that the very few eyewitness reports available contained conflicting statements.

On the same day my uncle and aunt, Mr. and Mrs. Robert B. Greenlee, were relaxing on their fibreglass-screened, roofed patio in Dunnellon, Florida. The temperature was in the 90's, the sky was overcast, and there was a slight drizzle;

Ball lightning materializes inside screened patio.

the Greenlees had heard thunder some distance to the west of their immediate vicinity. Mrs. Greenlee and a neighbor, Mrs. Riggs, were seated a few feet apart in aluminum chairs, and Mr. Greenlee was standing about three feet from Mrs. Greenlee. Mrs. Greenlee had just swatted a fly when a ball of lightning the size of a basketball appeared immediately in front of her. The ball was later described as being of a color and brightness comparable to the flash seen in arc welding, with a fuzzy appearance around the edges. Mrs. Riggs did not see the ball itself, but saw the flyswatter "edged in fire" dropping on the floor. The movement of the ball to the floor was accompanied by a report "like a shotgun blast." The entire incident was over in seconds.

None of the witnesses felt any heat from the ball, and Mrs. Greenlee showed no signs of external injuries, although she complained of pain in the back of her neck and has had occasional headaches since. The explosion was heard by a neighbor about 150 feet away, and it was subsequently learned that another neighbor's electric range had been shorted out at the same time. There was no damage of any sort at the Greenlees, nor were there any marks on the patio floor where the flyswatter had fallen.

With regard to the fly, Mrs. Riggs commented, "You sure got him that time."

## BALL LIGHTNING
**Jennison, R. C.; *Nature,* 224:895, 1969.**

I was seated near the front of the passenger cabin of an all-metal airliner (Eastern Airlines Flight EA 539) on a late night flight from New York to Washington. The aircraft encountered an electrical storm during which it was enveloped in a sudden bright and loud electrical discharge (0005 h EST, March 19, 1963). Some seconds after this a glowing sphere a little more than 20 cm in diameter emerged from the pilot's cabin and passed down the aisle of the aircraft approximately 50 cm from me, maintaining the same height and course for the whole distance over which it could be observed.

The observation was remarkable for the following reasons. (i) The appearance of the phenomenon in an almost totally screened environment; (ii) the relative velocity of the ball to that of the containing aircraft was $1.5 \pm 0.5$ m $s^{-1}$, typical of most ground observations; (iii) the object seemed perfectly symmetrical in all three dimensions and had no polar or torroidal structure; (iv) it was slightly limb darkened having an almost solid appearance and indicating that it was optically thick; (v) the object did not seem to radiate heat; (vi) the optical output could be assessed as approximately 5 to 10 W and its colour was blue-white; (vii) the diameter was $22 \pm 2$ cm, assessed by eye relative to the surroundings; (viii) the height above the floor was approximately 75 cm; (ix) the course was straight down the whole central aisle of the aircraft; (x) the object seemed to be in perfect equilibrium; (xi) the symmetry of the object was such that it was not possible to assess whether or not it was spinning. (Nature, 224:895, 1969)

# UNUSUAL FORMS OF LIGHTNING

Lightning is simply the rapid, concentrated discharge of electricity through the atmosphere. The adjectives "rapid" and "concentrated" must be specified to distinguish lightning and the "slow" and "diffuse" discharges of electricity described in the section on aurora-like phenomena. Actually, some lightning phenomena, such as the so-called "thunderless lightning" described below, may be a rare form of diffuse glow discharge rather than conventional lightning. Whatever else thunderless lightning may be, the reports below prove that it is weird and awe-inspiring.

The overwhelming majority of lightning strokes are from cloud to earth. A small fraction of strokes, however, reach upwards from clouds toward the ionosphere and outer space---and vice versa. Somewhere high above the earth exists an electrical terminal for both conventional lightning and slow, aurora-like discharges. Does the effectiveness of this high altitude terminal vary with solar activity and possibly meteor influx rates?

Horizontal lightning poses a problem, for there seems no obvious reason for lightning to digress five miles to strike a house when the cloud base is only a thousand feet above the ground. Do certain areas have a special attraction for lightning; that is, do local geological and meteorological conditions become more important than the length of the stroke?

Rarely, lightning will descend with no clouds in sight. In such cases, electric charge must accumulate in the atmosphere without the creation of a visible cloud. Next to nothing is known about such invisible charge concentrations, assuming they do exist.

Pinched and bead lightning may be related to ball lightning. Terrestrial experiments with high-power current flows in gases show that a "pinch effect" exists in which a column of electrical current can be self-constricting; that is, it "strangles" itself. Bead lightning may have its origin in this effect, but it is hard to explain dozens of pinches forming simultaneously along a lightning channel many feet long. Terrestrial laboratories cannot duplicate nature in all of lightning's idiosyncracies; just as terrestrial theorists cannot explain all of lightning's quirks.

## ROCKET LIGHTNING

### ROCKET LIGHTNING
**Everett, W. H.; *Nature*, 68:599, 1903.**

We saw some strange lightning yesterday evening at about 9 p.m. [July 22, 1926, at Calcutta] It was a clear, moonless night, with just a bank of cloud very low in the S.S.W., with a well-marked edge, height say from horizon (flat) to 5° up. There was a misty cloud above this. These clouds we could only see properly when the flashes came. Stars were visible at about 10° above the horizon at this point, and the sky was quite clear all over elsewhere. Now and then flashes showed from behind the lower cloud (the flashes themselves were mostly hidden, and thunder was not audible). The flashes were not so frequent as usual, say one per minute or so. Generally here they are almost incessant during thunderstorms.

At intervals of three minutes or so, immediately after a flash---which, as common here, was mostly multiple, lasting a second or so altogether---a luminous trail shot straight up to 15° or so, about as fast as, or rather faster than, a rocket, and of very similar appearance, but with minute waves, like ribbon lightning. It was hardly as bright as most lightning. S. and I saw it repeated seven times, and Prof. Bruhl (physics) three or four times after we directed his attention to it. He was equally surprised at the novelty, and he has been out here some eighteen years. One of the trails turned off, as shown; the others were about vertical as seen from here. Each grew up steadily from below, and then disappeared at once. The upper end was definite, and did not branch or spread.

In each case it followed immediately on a vivid flash or set of flashes. It was certainly not fireworks of any kind. It terminated in apparently clear sky. Its appearance as a uniformly and very bright ribbon was different from any fireworks. It was somewhat yellowish, not purple as lightning often is. It was much too far off for fireworks to be so high and bright. No thunder was audible. (Nature, 68:599, 1903)

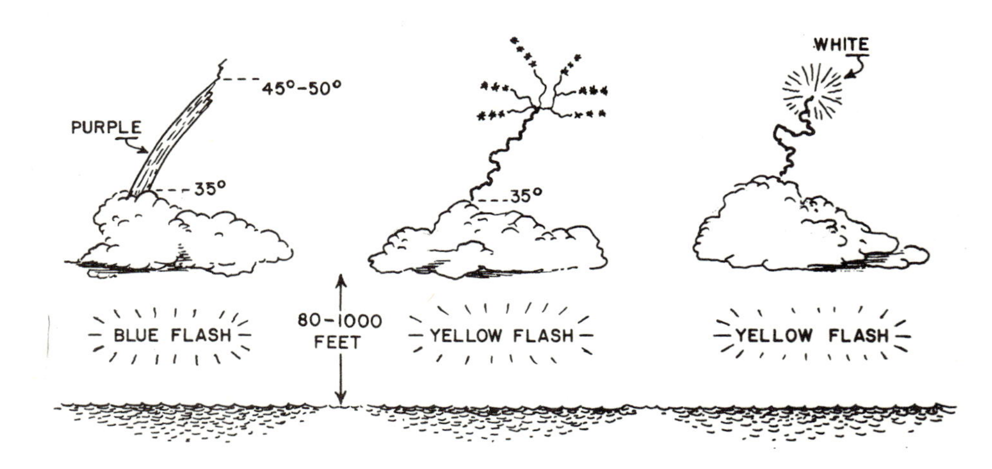


Displays of rocket lightning at sea

## LIGHTNING
**Powell, G.; *Marine Observer,* 38:173, 1968.**

m.v. Geesthaven. Captain G. Powell. South Wales to West Indies. Observers, the Master, Mr. I. McLachlan, 3rd Officer and the Radio Officer.

13th October 1967. At 0215 GMT lightning of unusual character was seen in the south-east. Instead of travelling downwards towards the sea it began at an elevation of about 35° and travelled upwards to an altitude of 45° to 50°. In some respects it resembled a firework display. No thunder was heard, nor was there any rain in the vicinity of the ship, but the radar indicated heavy rain clouds about 15 miles away and moving in a NW'ly direction. This unusual type of lightning was seen over a period of 3 hours. The general appearance is shown in the accompanying sketches. Air temp. 79°F, wet bulb 76.8°, sea 83.3°. Wind light and variable. (Marine Observer, 38:173, 1968)

## LIGHTNING
**Taylor, T. J.; *Marine Observer*, 42:12, 1972.**

s. s. British Bombardier. Captain T. J. Taylor. Tripoli to Genoa. Observers, Mr. P. N. Duffield, Extra 2nd Officer and Mr. D. Houston, A. B.

27th March 1971. At 0120 GMT, during a period of thundery showers, lightning was observed to the NW striking upwards from the clouds which, when illuminated, seemed to be Cb. Several strokes of lightning were seen to discharge upwards simultaneously, radiating from a point below the top of the clouds. There seemed to be no simultaneous downward strokes, nor were there any flashes visible from that direction for a few minutes previous to and after the described flash. No thunder was heard. Afterwards, numerous flashes were seen in the same direction but at no time were the upward strokes again observed.

Position of ship: 37° 40'N, 16° 26'E.

Note. Single discharges from Cb cloud tops to space have occasionally been observed, but the observation of several simultaneous discharges is extremely rare. (Marine Observer, 42:12, 1972)

## UNUSUAL LIGHTNING
**Wood, C. A.; *Weather*, 6:64, 1951.**

Your correspondent J. B. Wright, in the overseas section of Weather, June 1950, gives an account of an unusual lightning display observed above the anvil of a cumulonimbus cloud in the vicinity of Fiji. The captain of the aircraft from which the phenomenon was noted asked if such 'firework displays' are common in the tropics, and will therefore be interested to learn that an almost similar display was seen on 7 February 1945 at Broome, Australia. A report by W/O N. J. Keay in the RAAF Tropical Weather Research Bulletin for April 1945, runs as follows:

'At 2040 WST on 7 February 1945 lightning flashes of a very unusual nature were observed. At the time there was a trace of cumulonimbus and some anvil cirrus. There was considerable intense sheet-lightning (almost continuous) and some normal fork-lightning at the time. The unusual flashes were observed during a lull in the sheet lightning.

The flashes commenced as an intense point of light in the tops of the cumulonimbus. A normal lightning-flash extended upwards from this point for a short distance, after which the discharge assumed a shape similar to roots of a tree in an inverted position. This "beam" extended upwards for some distance into the cloudless part of the sky---above the cirrus.

The "beam" was purple in colour. The colour was similar to that of an electrical discharge when it passes through air at a reduced pressure. The flashes were of quite long duration, from approximately half to one second, and appeared to develop and diminish gradually. About six such flashes were observed and from two separate points in the sky'. (Weather, 6:64, 1951)

# PINCHED OR BEAD LIGHTNING

## BEAD LIGHTNING
**Steward, J. F.; *Popular Astronomy,* 2:89, 1894.**

Having given some attention to photographing lightning and studying the records made upon the sensitive plate, I have been more observant than ever.

To-day while driving over the country midst a terrific thunderstorm, I observed many fine flashes, one instance of which deserves particular mention. Two substantially parallel simultaneous flashes, somewhat wavy, seemed to pass toward the Earth. Length about 40°, distance apart about 7°, both brilliant, but not blindingly so. Both as lines of light, vanished instantly, but the one to the right broke into a multiplicity of what seemed to be spheres. This string of beads, as it seemed, appeared to be formed of material furnished by the original flash.

In the accompanying sketch the two full lines represent the original flashes, while the line of dots represents the after effect of the flash nearer it. The phenomenon was observed by Mr. C. A. Rand who was riding with me. That there was no optical illusion is evidenced by the fact that only one of the flashes made the showing referred to. (Popular Astronomy, 2:89, 1894)

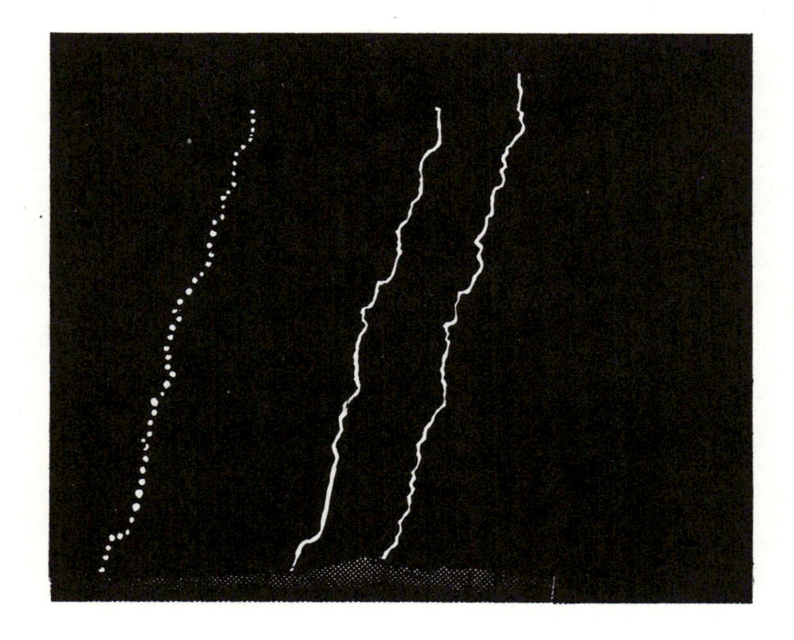

Lightning flash breaks up into beads

## A CURIOUS AFTER-EFFECT OF LIGHTNING
**Beadle, D. G.; *Nature,* 137:112, 1936.**

An interesting observation was made during a thunderstorm which took place over Johannesburg on December 5. The storm, which was accompanied by rain, was the severest experienced for some years.

During the height of the storm, a particularly bright flash struck the ground about a hundred yards away; the flash appeared to be approximately a foot wide and to last for at least a second. After the flash had died away, there remained a string of bright luminous beads in the path of the flash. The beads, of which there were twenty or thirty, appeared to be about a quarter of the width of the flash, that is, say, three inches in diameter. The distance between the beads, which appeared to be nearly constant, seemed about two feet. They remained visible for approximately half a second; during this time they gave no indication of any movement.

Measurements were afterwards made to endeavour to check the above dimensions; they appear to be of the right order. (Nature, 137:112, 1936)

# COLORED LIGHTNING

## RED LIGHTNING
**Boys, H. A.; *Meteorological Magazine,* 60:192, 1925.**

On Sunday, May 13th, 1906, there was a notable thunderstorm in Somerset. At Melles and also at Downside 3 in. of rain fell. At North Cadbury the fall was 1.19 in. between 4.30 p.m. and 11 p.m. Thunder and lightning were incessant from 3.30 till 11; the bulk of the rain fell between 4.30 and 6. Evensong began in the church at 6 with a sadly depleted congregation, and from the pulpit at 6.45 I had my third experience of red lightning, for I could see through the windows dull red horizontal flashes which seemed to be darting about the churchyard; and later till it was dusk the whole country-side seemed to be full of this dull red lightning, apparently independent of the thunder. From time to time there were bright flashes of ordinary colour almost simultaneous with the thunder, but the non-luminous character of the red lightning was very noticeable. (Meteorological Magazine, 60:192, 1925)

## GREEN LIGHTNING
**Stephenson, H. H.; *Nature,* 120:695, 1927.**

During a thunderstorm last night [Ontario, October 7, 1927] a flash of lightning started from the top white, and about half-way down turned to a vivid green. Possibly I am very unobservant, but I do not remember ever to have seen green lightning before. (Nature, 120:695, 1927)

# LIGHTNING WITHOUT THUNDER

## THE STRANGEST LIGHTNING
**Talman, C. F.; *American Meteorological Society, Bulletin,* 12:130, 1931.**

Probably the queerest display of lightning ever reported by a scientific observer was one encountered on October 3, 1927, by Dr. Walter Knoche, a well-known German meteorologist, during a steamboat journey down the Rio Paraguay in South America. A severe drought had prevailed in the surrounding country for months and no rain fell before or during the electrical storm, which, without warning, suddenly began in all parts of the sky at 7 p.m. "It would have been hopeless," says Dr. Knoche, "to attempt even an approximate count of the flashes." Some were ordinary streak lightning, but reddish in color. Almost equally numerous, however, were the flashes of dazzling white "beaded lightning," which looks like a string of glowing pearls in the sky. Such lightning has occasionally been reported in connection with other storms, but never as occurring with such frequency.

Dr. Knoche tells also of various anomalous phenomena seen during this display, including curious, rapidly-moving orange-colored discharges, which he says resembled cylindrical masses of glowing gas; flashes that revolved like pinwheels; and, at one period of the storm, hundreds of luminous arcs crowded together near the zenith, so dazzling in their brilliancy that he had to close his eyes.

Strange to say, this spectacular display went on for hours without thunder. A ghastly quiet prevailed; not a breath of air stirred. Thunder began abruptly at 1.30 a.m., and was then also continuous until 4 a.m. The last lightning was seen astern of the steamer about 8 a.m. (American Meteorological Society, Bulletin, 12:130, 1931)

## SILENT LIGHTNING
**Collingwood, C.; *Knowledge,* 4:381, 1883.**

I have met with two remarkable instances of long-continued lightning, unaccompanied by thunder. Once, at Shanghai, in July, the sky was illuminated with one incessant unintermittent glare, lasting several hours, but no thunder was heard. But the most remarkable instance occurred in May, off the south-coast of Madagascar, when, from 7 to 11 p.m., I watched a glare, which left the sky dark only for a second once in an hour or so, the central point of which seemed elevated 10° or 15° above the horizon. As the nearer clouds cleared away, I watched for hours the unceasing flashes, tongues of fire darting out round the distant clouds, radiating in five or six distinct streams of flame from a given point, like the thunderbolt in the hand of Jove, coursing along the sky, or dashing down into the sea at the horizon like liquid fire; yet all the while not a sound was heard. After I had retired, at 11, I still saw, as long as I lay awake, the reflection playing upon the walls of the cabin as from a flickering lamp. The day following was marked by a brilliancy of atmosphere and freshness of temperature not before experienced. On this, and another occasion at Sarawak, the lightning was unusually vivid, but the flashes were not simple instantaneous sparks, but had the appearance of liquid fire poured from a vessel in a continuous stream, and lasting a definite time, during which the lightning vibrated upon the retina, the zigzag form of the flash, however, being meanwhile perfectly retained. (Knowledge, 4:381, 1883)

## LIGHTNING WITHOUT SOUND
**Anonymous; *Nature,* 60:511, 1899.**

Mr. T. Kingsmill sends us two cuttings from the Shanghai Mercury referring to two electric displays of an unusual character observed at Shanghai on July 19 and August 10. On the former occasion it is stated:---"The northern sky was in an almost constant blaze of light. Flashes came sometimes from two centres, as though there were an elliptical area of disturbance from whose foci were sent forth the shafts of lightning. At times these flashes would take the opposite course, and starting from the circumference make their way to the foci. Though the lightning flashes reached within twenty-five degrees of the zenith, and were vigorous enough in all conscience, yet nothing but the faintest distant rumbling could be heard." On August 10 "the reflection of lightning was seen from the S. W. and gradually increased in brightness until at about 7. 50 it had reached the zenith." The report states that "the lightning played over nearly the whole of the exposed sky, sometimes six and seven streamers at a time lighting up the sky. They were different in appearance from ordinary forked lightning, having rather the appearance of a network of ribbons crossing the exposed sky in all directions, like the discharges in a vacuum tube. The most unusual circumstance was that these discharges, though most vivid, were almost noiseless, and could scarcely be heard above the ordinary jinrickshaw traffic of the street, the only accompanying sound to the brightest display even in the zenith being a low rumbling, as of ordinary very distant thunder." Mr. Kingsmill remarks that both displays were synchronous with distant typhoons. (Nature, 60:511, 1899)

# HORIZONTAL LIGHTNING

## EXTRAORDINARY FLASH OF LIGHTNING
**Ley, W. Clement; *Symons's Monthly Meteorological Magazine,* 8:106-107, 1873.**

Sir,---A very remarkable flash of lightning occurred here [Hereford, England] on July 16th. The morning was showery, and thunder-clouds formed in different directions at 9 a. m. About 10 a large cumulus rose in W. S. W., and broke rain about 5 miles from this place, its summit at the same time, as is usual in incipient storms, assuming the cirriform appearance. The first thunder-clap soon followed, and was succeeded by seven or eight others as the storm travelled to N. W. The sky overhead was quite clear, and also over Hereford, 3 miles east of this place, and the storm-cloud had a very isolated appearance, though there were other distant clouds on the horizon. I was standing in my garden watching the distant lightning, when a flash left the cloud 2 or 3 miles to W. N. W. of this place, passed almost directly overhead, but a little to N. E., and descended upon Hereford, traversing in a horizontal direction a space of about five miles of clear blue sky, devoid of cloud. The clap commenced nearly in the zenith, the time interval showing the part of the flash nearest to this place to be about 1-1/2 miles; it then became loud both over Hereford, in the east, and in the storm in the west. Hereford lies comparatively low,

Five-mile horizontal lightning stroke

and the electric fluid travelled near the earth over high ground covered with trees, buildings, &c; avoided all Saints' and St. Peter's spires, very near which it must have passed, and singled out a house in the more eastern part of the city. The house struck is lower than others in the same row, and the adjoining house on the west has much higher chimneys. Not much injury was done, though two persons suffered from a temporary paralysis. The inhabitants of Hereford, few of whom had noticed the distant storm, were greatly startled by the terrific flash and clap under a clear sky and brilliant sun.

As observed here, the flash appeared straight, almost resembling a rocket fired horizontally, but Mr. Isbell informs me that as seen in Hereford it was very zigzag, and seemed to come along near the ground.

I sent you last summer an account of several similar flashes, which passed in succession out of a thunder-cloud west of this place. Once before, in April, 1865, I was watching from Sellack Vicarage, near Ross, a violent storm fully six miles distant, the sky overhead being almost perfectly clear, when I was astonished by a tremendous flash of lightning at a distance of about 500 yards. These are the only examples of lightning striking the earth at a distance from a storm which I have myself observed in the course of about twenty years, during which I have minutely watched every thunder-storm in my neighbourhood; but there are records of occasional accidents from lightning at the distance of one or two miles from a thunder-storm.

The ancients, as school boys know, regarded such occurrences as portents of evil omen. (Symons's Monthly Meteorological Magazine, 8:106-107, 1873)

## LIGHTNING
## Cook, J. W.; *Marine Observer,* 36:53, 1966.

s.s. Caltex Edinburgh, Captain J. W. Cook. Bahrain to Manila.
14th May 1965. At 2040 GMT, a single jagged flash of lightning was seen coming from a well developed Cb cloud with anvil, several miles to the southwards. Instead of following the normal cloud-to-earth direction, this flash appeared to begin in a horizontal direction, curving to earth a considerable distance to the east and well clear of the cloud itself. Although there was another similar cloud over Sumatra the sky to the east was devoid of any type of cloud. Air temp. 82.5°F, wet bulb 78.7°, sea 84.3°.
Position of ship: 3° 05'N, 100° 38'E. (Marine Observer, 36:53, 1966)

## A LIGHTNING STROKE FAR FROM THE THUNDERSTORM CLOUD
## V., B. M.; *Monthly Weather Review,* 54:344, 1926.

Mr. J. H. Armington sends us from Indianapolis the report printed below, which was sent to him by Prof. Z. A. McCaughan, of Bloomington, Ind. The stroke occurred about 1 p.m. on July 23, 1926, in Monroe County; it killed two children.

"I drove to the place referred to and made personal inquiry of people who were within 100 yards of the place. The sun was shining, the nearest cloud seemed to the witnesses 2-1/2 or 3 miles north (toward Clear Creek and Bloomington). They had heard no thunder previous to this stroke and heard only two or three of distant thunder afterward. Their sky stayed clear for two hours afterwards. At the time of this stroke we were having frequent strokes of lightning and thunder here at Bloomington and we had 0.23 inch of rain. Three miles southeast of Bloomington there was a small tornado that broke limbs of trees and carried away anything small that was loose. The lightning was severe. Witnesses near where the children were killed say the lightning travelled horizontally from north to south. It passed three buildings, missing them by about 100 feet, and struck this little house just above the top of the corner foundation post."

The striking of lightning, through clear sky, at points somewhat distant from the region immediately beneath the storm cloud, while relatively rare, occurs probably more frequently than is realized.

During a three-year residence in east-central Florida I observed the phenomenon at least three times. The typical local thunderstorm cloud of the Florida summer grows with great rapidity, and is usually an entity quite unconnected with storms of the same kind that may be developing elsewhere within the observer's field of view. Opportunities for watching the lightning strokes from individual clouds are therefore excellent.

It is my recollection that the distance along the ground between a vertical dropped from the edge of the cloud and the striking point of the bolt was of the order of the height of the cloud base above ground. How foreshortening affected this estimate is of course impossible to say. But it is probably true that the distances were never of the order of 2 to 3 miles, as in the extraordinary case described by Professor McCaughan. In one instance (and I think this was true of all these far flung bolts) the spark seemed to leave the cloud from a point at least halfway up from cloud base to summit, and in this one instance which I recall especially vividly it was about three quarters of the way. (Monthly Weather Review, 54:344, 1926)

# LIGHTNING FROM A CLEAR SKY

## ABNORMAL LIGHTNING
**Smith, N. W.; *Marine Observer,* 27:208-209, 1957.**

S. S. Oronsay. Captain N. W. Smith. Las Palmas to Cape Town. Observers, Mr. E. Pickles, Senior 2nd Officer, and Mr. J. M. V. Boyde, Junior 3rd Officer.

17th October, 1956. At 0330 G. M. T. there was about 3/8 cloud, consisting of patches of tangled Ci which in the opinion of the observers was associated with Cu, there were also three small fair weather Cu almost directly beneath the Ci. In the light of the full moon the height of the Ci was estimated at 25,000-30,000 ft and that of the Cu at about 2,500 ft. Lightning started to flash from a point slightly behind the Ci and travelled in a path about 30 from the vertical in an absolutely straight line, at which seemed to be a comparatively slow speed, towards the small Cu. It appeared to enter the Cu and then reappear, moving downwards in a different direction at normal speed. The lightning track from the Cu, which had a greenish glow

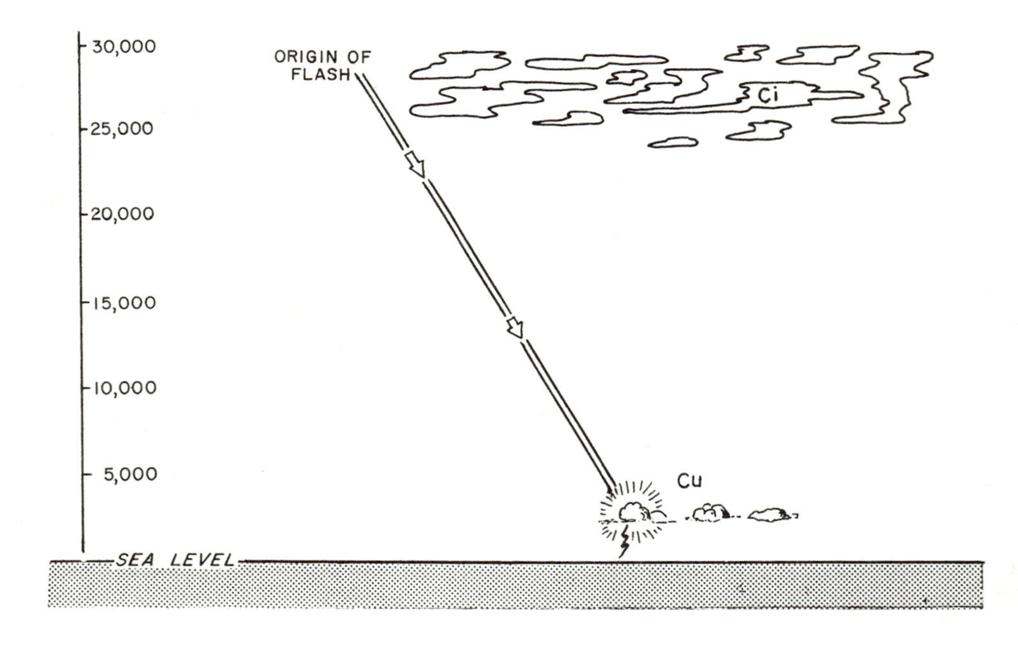


Lightning from a clear sky

when "hit", did not seem to be so wide as the original one. Similar flashes occurred at regular intervals of between 2-3 min and each followed the same path to the Cu. The watch was relieved at 0400 but it is believed the lightning continued regularly for another 15-20 min. Neither of the observers had ever seen lightning before seeming to emerge from an almost cloudless part of the sky, and which travelled so straight and apparently slowly. Air temp. 79°F, wet bulb 75°, sea 84°.

Note. This is an extremely interesting observation of an abnormal form of lightning, differing entirely from other forms of unusual character, a few observations of which have been published in this journal in recent years. The origin of the flashes in a region of clear sky is very remarkable.

Dr. B. V. J. Schonland, who is an international authority on lightning, makes the following comments:

"The observation described by Captain Smith in which a lightning flash travelled downwards to a small cloud and then emerged in a different direction has been made from time to time in the past. The first person to report it was Benjamin Franklin, who carried out model experiments in which the small cloud was replaced by tufts of cotton wool. It would seem that the effect is due to the positive charging of the small cloud by ions rising upwards in the negative field due to the cloud above. These ions are captured on the water drops of the small cloud and the result is to direct the field from the main cloud to the small one below, from which the flash travels on to earth. That the subsequent portion of the discharge was weaker could be accounted for by the neutralisation process involved in the first portion.

"One would expect this effect to occur several times if conditions remain the same, though I do not recall any incidents of the regularity reported by Captain Smith.

"The other point which is interesting from the description is the apparent slowness of the discharge. This could be explained if its lower portion did not reach the sea, for flashes which do not involve a return stroke from the ground are composed of comparatively slow leader discharges. If, however, the lightning passed to the sea I cannot offer any explanation of its apparently slow speed." (Marine Observer, 27:208-209, 1957)

## THUNDERSTORM IN A CLEAR SKY
**Anonymous; *Scientific American,* 54:325, 1886.**

Capt. Anderson, of the British bark Siddartha, which lately arrived in New York, reported a peculiar thunderstorm on April 27, while on the northern edge of the Gulf Stream. The sky was quite clear at the time and the sun shining brightly, although there appeared to be a thin mist about the ship. Suddenly there appeared a vivid flash of lightning, accompanied by violent thunder. The compass was caused to vibrate perceptibly for a period of 15 minutes. (Scientific American, 54:325, 1886)

## LIGHTNING IN BLUE SKY
**Williams, W.; *Royal Meteorological Society, Quarterly Journal,* 36:126, 1910.**

I hope you will pardon my troubling you with the record of what to me appears an unusual meteorological occurrence, and I should be much obliged if you would let me know whether it has often occurred before. This phenomenon was a vivid flash of forked lightning in blue sky. I was riding along the foothills to the south of Mount Troodos at an elevation of about 3000 feet about 10.30 a.m. on September 10, 1909, when straight in front of me I distinctly saw this flash. It was exactly similar to the ordinary flash usually seen against a dark thundercloud. It was followed by a very sharp crash of thunder, so that there can be no doubt it was an electric discharge. Heavy thunderclouds were then forming on the high mountains to the side, and low thunder was rumbling continuously, but the flash was quite away from these clouds and where the sky was entirely blue. It may, of course, be a fairly common occurrence, but as I have never seen it before it may be worth recording. (Royal Meteorological Society, Quarterly Journal, 36:126, 1910)

# LIGHTNING IN SNOW STORMS

## SNOWSTORM LIGHTNING
**Anonymous; *Monthly Weather Review,* 31:534, 1903.**

The following from the Cleveland Leader is kindly communicated by Father Odenbach of Ignatius College in that city:

Geneva, Ohio, November 19.---A phenomenon was seen in Unionville between 5 and 6 o'clock yesterday afternoon, during the snowstorm. There was a flash of lightning, seeming to emanate from the snow itself and illuminating surrounding buildings and objects quite brightly. It consisted of two almost simultaneous, flashes, one stronger than the other, and of a purple and milky-white color. They were followed by a faint roll of thunder like the approach of a distant storm. Such a freak of nature was known to occur during a snowstorm twenty years or more ago. (Monthly Weather Reveiw, 31:534, 1903)

# CROWN FLASH

## POSSIBLE NEWLY RECOGNIZED METEOROLOGICAL PHENOMENON CALLED CROWN FLASH
**Gall, John C., Jun., and Graves, Maurice E.; *Nature,* 229:185-186, 1971.**

On July 2, 1970, at about 1945 h EST a thunderstorm cell passed a few miles north of Ann Arbor, Michigan. The column of cumulus cloud towered in the light of the setting Sun, far above the dark mass below, which occasionally flickered with lightning. Thin lamellae of cirriform cloud began to form above the peak of the cumulus column and streamed off to the north-east, in advance of the storm cloud. The cumulus column impinged on these lamellae as it boiled upward, causing the horizontal layers of thin cloud to become somewhat dome-shaped in the region of the rising column. At and just above the peak of the storm cell the cloud mass seemed to be undergoing sudden changes in brightness lasting for several seconds at a time. J. C. G.'s wife and two children also noticed the effect when it was called to their attention. M. E. G., a meteorologist, also verified the observation. The phenomenon continued to occur repeatedly at intervals of 30-60 s during the next 15 or 20 min, providing the basis for the following description.

The sudden brightening effect began concurrently with lightning strokes in the main cloud mass, but continued after the lightning flash was over. It had the appearance of a ripple-like upward and outward spread of radiance from the region just west of the peak of the cumulus column, resembling somewhat a fan-like display of aurora borealis. It lasted a substantial fraction of a second with each lightning stroke. On one or two occasions it had the appearance of a bright ring moving rapidly outward and upward above the cumulus peak. On these occasions it was clearly observed to extend beyond the cloud and into blue sky. A linear shadow, apparently cast by one of the cumulus masses, appeared to shift its position suddenly up or down with each occurrence of the event.

While the effect was observable, the portion of the cloud mass most immediately involved seemed considerably brighter than the remainder of the cumulus peak, which was still in sunlight. This area of the cloud seemed to glow from within; the individual puffs of cumulus cloud did not seem to cast shadows but glowed with equal intensity in all portions; the zones between puffs were not dark as in the remainder of the cloud, but bright as if light were escaping through them. The colour of the entire phenomenon was white.

Hourly radar log reports from Detroit Metropolitan Airport, 24 miles ESE of Ann Arbor, for the evening of July 2, 1970, were later consulted. At 1900 h radar echoes were reported 117 miles due west of the airport, moving from the WNW ($290^{o}$) at 25 knots, with prominent peaks at 50,000 feet and containing 0.75 inch hail. At 2000 h, echoes with peaks at 45,000 feat were recorded 90 miles due west of the airport. At 2100 h, large storm cells with peaks at 40,000 feet were recorded in the Ann Arbor area. Official sunset occurred at 1941 h EST on that day.

We speculated that the phenomenon observed represented some sort of ionization discharge, perhaps occurring at the top of a column of ionized air associated with the thunderhead. Conceivably, the effect tends to occur in favourable conditions of rapid updraught with strong winds aloft, associated with a relatively energetic thunderstorm. If so, it may represent one early stage in the formation of a vortex that could eventuate in production of a tornado.

We have been unable to find published reports describing a similar phenomenon. We invite readers who may know of such published reports, or who may have observed similar effects themselves to communicate their findings. We have tentatively designated this phenomenon the "crown flash". (Nature, 229:185-186, 1971)

# VAGARIES AND ODD EFFECTS OF LIGHTNING

Lightning kills animals, rends trees, fuses sand into fulgerites, and fires houses and barns, as befits a powerful manifestation of electricity. But lightning can also be choosey and fastidious in what it does. Indeed, the tales of lightning's pranks verge on the unbelievable. It is not too hard to understand that certain high conductivity trees are favored by lightning, but why does it spare a child and a fraction of a second later strike down cows and pigs in a barnyard? The peculiar itinerary of rampaging lightning can often be explained in terms of paths of lowest electrical conductivity, but anomalies always crop up, as described below.

The sounds and magnetic effects of lightning are not particularly mysterious either---merely strange! As terrestrial electrical current rushes in the direction of a lightning strike, "brush" discharge from trees, rocks, and even people is to be expected. The "zit" sounds and swishes occasionally heard between the flash and thunder probably owe their origins to these discharges, although there is also evidence that some individuals can detect strong electromagnetic waves (such as radar transmissions) directly as sound. The passage of strong currents also creates the ozone smell typical of electric discharge as well as magnetic perturbations. All the foregoing is within the realm of conventional explanation. But just as lightning sometimes reaches out miles from a cloud to smash some object under a clear sky, it can also be perversely gentle and particular as it picks and chooses its way through a home.

## LIGHTNING'S PREFERENCE FOR CERTAIN TREES

### WHICH TREES ATTRACT LIGHTNING?
**Anonymous; *Monthly Weather R*eview, 26:257-258, 1898.**

I have the honor to invite attention to the need of an authoritative answer to the question "Why some trees are more frequently struck by lightning than others?" At first glance the subject may seem to be foreign to the work of the Weather Bureau; but investigation will show some phases of the question are germane to the work of this Bureau and proper subjects for study.

Primarily it is a matter of saving human life; and in that direction the Weather Bureau has always put forward its best efforts. Many people, particularly farmers and those who work in the fields exposed to thunderstorms, will work until the storm is almost upon them, and then run to the nearest tree for shelter. If the tree is an oak and the charged thunder clouds are moving toward it, with high electrical potential, the person or persons under the tree are in the line of strain, and all unconsciously are contributing to the establishment of a path for the lightning discharge through themselves. Records show how frequently death results, and how dangerous it is to stand under certain trees during thunderstorms. On the other hand, if it had happened to be a beech tree, there is some reason to believe that it will afford safety as well as shelter, though the reason why is not at present known. It is known that the oak is relatively most frequently and the beech least frequently struck. If the relative frequency of the beech is represented by 1, that for the pine is 15, trees collectively about 40, and oaks, 54. Trees struck are not necessarily the highest and most prominent. Oak trees have been struck twice in the same place on successive days. Trees have been struck before rain began and split; and trees have been struck during rain and only scorched. It is plain then that before any statement is made as to the danger of standing under certain kinds of trees during thunderstorms, the more general questions of the effects of lightning upon trees should be gone into. Such a study would be best undertaken by cooperated effort of statistician, physicist, and vegetable pathologist. (Monthly Weather Review, 26:257, 1898)

## TREES AND LIGHTNING
**Anonymous; *Royal Meteorological Society, Quarterly Journal,* 33:255, 1907.**

Mr. W. R. Fisher in his recently published book on Forest Protection gives some particulars about the liability of trees to be struck by lightning. As very little reliable information is available on the subject, the following extract from Mr. Fisher's book will be of interest:---

All species of trees are liable to be struck by lightning, but oaks and other species with deep roots appear to be most exposed to this danger, perhaps on account of their roots forming better conductors to the moist subsoil than those of shallow-rooted species.

According to the valuable observations made annually by Dr. Hess from 1874 to 1890 in the forests of Lippe-Detmold, among broad-leaved trees the oak suffers most, among conifers the Scots pine. Then follow spruce and beech. The birch, poplars, ash, alder, willows, larch, and other trees suffer only exceptionally.

Trees Struck in Lippe-Detmold, 1874-1890.

| Broad-leaved trees | | Conifers | |
|---|---|---|---|
| Oaks | 310 | Scots pine | 108 |
| Beech | 33 | Spruce | 39 |
| Birch | 10 | Larch | 11 |
| Poplars | 6 | Austrian pine | 1 |
| Ash | 4 | Weymouth pine | 1 |
| Willows | 2 | Others | 5 |
| Other trees | 8 | Total | 165 |
| Total | 373 | | |

The forest was stocked as follows:---

| | Per cent. |
|---|---|
| Oak | 11 |
| Beech | 70 |
| Spruce | 13 |
| Scots pine | 6 |
| | 100 |

The danger, therefore, considering the beech as 1, was 6 for a spruce, 37 for a Scots pine, and 60 for an oak. (Royal Meteorological Society, Quarterly Journal, 33:255, 1907)

# LIGHTNING FIGURES

## DEATH BY LIGHTNING
**Anonymous; *Monthly Weather Review,* 47:729, 1919.**

A. G. Newell describes an interesting phenomenon in connection with the burns received by a man who was killed by a lightning stroke in London. The man and his wife were walking in the open, near a row of elm trees when he was struck. The wife was momentarily stunned, but upon recovering she saw her husband standing erect with a blazing line up and down the back of his coat. He died immediately, and fell backwards against an iron fence. Severe burns on his back indicated that he was struck from the right side of his back.

From the right shoulder and across the chest and down to the lower of the front of the abdomen impressions of branches and leaves were clearly imprinted on the skin, showing like an X-ray plate how certain rays of light were impeded by the branches and foliage, whilst others made the contours of these. There were two distinct branches with leaves, one occupying the space between the right iliac crest to near the ensiform cartilage and the other proceeding down on the left side from the stomach to the left iliac crest. To my mind it would appear as if these were implanted while the man was falling back with a flash coming over the right shoulder. (Monthly Weather Review, 47:729, 1919)

## A "LIGHTNING FIGURE" PHOTOGRAPHED
**Anonymous; *Symons's Monthly Meteorological Magazine,* 18:81-82, 1883.**

At a meeting of the Meteorological Society, held March 24th, 1857, Prof. Poey read a paper entitled "On the connection of photographic impressions with those of the images which lightning produces on the surface of the human body." A short abstract of the paper is printed in the Annual Report of the Meteorological Society for 1857, and from it I quote the first two cases of the seven there given.

1. The first authentic, but not the earliest, mention of this singular phenomenon, was made by Benjamin Franklin in 1786, who himself stated to a member of the French Academy, M. Leroy, that, twenty years previous, a man, who was standing opposite to a tree which had just been struck by lightning, was very much surprised to perceive on his breast a fac-simile of that tree.

A similar case was pointed out in the "Journal of Commerce" of New York, on the 26th of August, 1853, which gives the following account:---"A writer in the 'Intelligencer' communicates a curious incident. A little girl was standing at a window, before which was a young maple tree. After a brilliant flash of lightning, a complete image of the tree was found imprinted on her body. This is not the first instance of the kind."

The Report of the Society goes on to state:---

Shortly after the reading of M. Poey's paper before the Society, the following communication was addressed to C. V. Walker, Esq., Member of the Council.

"On the Photographic Effects of Lightning," by James B. Shaw, Esq.

"I beg to give you the particulars connected with the photographic effects of lightning on six sheep, which were killed by a stroke of electricity about the year 1812.

"About four miles from the city of Bath, near the village of Coombe Hay, was at that time an extensive wood of hazel and detached oak-trees. In the centre of this wood was a field of about 50 yards square, in which were six sheep, all of which were struck dead by lightning. When the skins were taken from the animals, a fac-simile of a portion of the surrounding scenery was visible on the inner sur-surface of each skin. I do not remember whether these six beautiful photographic images were precisely similar, but I think they were.

"I remember that it caused great local sensation at the time, and the skins were exhibited in Bath, and I have no doubt there are many living now in that neighbourhood who could vouch for the correctness of this statement. I may add, that the small field and its surrounding wood were so familiar to me and my

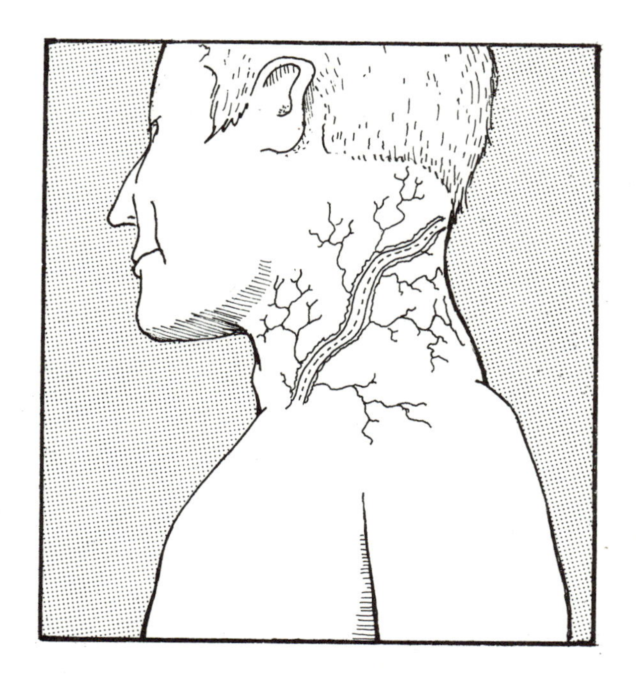

Dendritic "breath figure" of an electric discharge

schoolfellows, that when the skins were shown to us, we at once identified the local scenery so wonderfully represented. I was much pleased to read the several cases mentioned before the Meteorological Society, proving the photographic effects of lightning on animals, as my statement of the circumstances herein described, frequently detailed by me, has always been received with credulity." [?incredulity---Ed.]

In 1861, Prof. Poey published an extremely fascinating memoir upon the subject.

In the same year, and in 1862 Prof. Tomlinson, F. R. S., of King's College, published two papers in the Edinburgh Philosophical Journal on "Lightning Figures," in which he maintained that there was no photographic action whatever, but that the ramifications of the electric fluid, such as occur on deflagrating a sheet of gold leaf, had been erroneously identified with the branches of trees. (Symons's Monthly Meteorological Magazine, 18:81-82, 1883)

## ON LIGHTNING FIGURES

**Tomlinson, Charles; *Symons's Monthly Meteorological Magazine*, 18:104-105, 1883.**

Sir,---I am glad to see that you and Mr. Bruce have adopted my views as to the formation of ramified figures, which have frequently been noticed on human beings and animals that have been struck by lightning. A ramified figure, produced under such circumstances, would naturally lead a non-scientific observer to associate such a figure with that of a neighbouring tree, and in reporting the occurrence, he would be likely to call it "an exact representation of the tree;" while any little blotch he would represent as "a bird and nest on one of its branches." I am not aware that any scientific man has ever actually witnessed a case of this kind, but when the patient is taken to a medical man, he is apt to fall back upon his knowledge of anatomy for an explanation, and speak of "the eruption of blood in the vessels of the skin, producing an effect like that of an injection." In a case reported by a medical man in the Lancet, for July, 1864, Dr. Mackintosh, referring to a boy who took refuge under a straw stack, says, "The figures on either hip were so exceedingly alike, and so striking, that an observer could not but be impressed with the idea that they were formed in obedience to some prevailing law." In this case, and many others, no tree was near. One case recorded in Nature, for March 25th, 1875, by Mr. Pidgeon, at Torbay, close to the sea, is very well reported, because he relates the facts as they occurred, without any attempt to explain them. Mr. Pidgeon, his wife, and his son, were standing under a flag-staff, looking out upon the bay, when the vane was struck by lightning, and shivered. Mrs. Pidgeon was struck, and fell to the ground, and her lower limbs and left hand became rigid. On examining the body from the feet to the knees she was splashed with rose-coloured tree-like marks, branching upwards; while a large tree-like mark, with six principal branches diverging from a common centre, 13 in. in its largest diameter, and bright rose-red, covered the body." On recovering, Mrs. Pidgeon declared that she saw nothing of the flash, but described her feeling as that of "dying away gently into darkness, and being roused by a tremendous blow on the body, where the chief mark was afterwards found."

When my paper was read before the Physical Section of the British Association at Manchester, in 1861, the astronomer royal was in the chair, and on showing him some of my electrical figures after the sitting, he patted me on the back, and said, "You have settled that matter!" I also produced the figures in the presence of my friend, the late Professor Miller, of King's College; and on breathing on one of the plates, he exclaimed, "There's the tree, capitally made, bird's nest, and all!" I may remark that persons have complained to me of their inability to produce these figures. The following directions will make their production easy. Squares of common crown window glass, about 4 m. to the side should be steeped in a strong solution of soap and water; and before an experiment, a plate should be taken out, and wiped dry with a duster. This will leave an exceedingly thin film of soap on the surface of the glass. A Leyden jar of about a pint capacity should be charged, and the plate being held by one corner, he brought up to the knob of the jar, while one knob of the discharging rod should be placed on the outer coating of the jar, and the other knob be brought opposite the knob of the jar, in contact with the glass. A discharge will now take place over the surface of the plate; and passing over its edge reach the upper knob of the discharging rod. If we now breathe on the surface of the plate, a tree-like figure, consisting of trunk, branches, and spray, will be beautifully made out, because wherever the electricity touched the plate, the soapy film is burnt off, and the plate rendered chemically clean, so that the breath condenses in watery lines on those parts, and in minute globules of dew on those parts where the film still remains. On the other side of the plate, there is also a figure, but it is marred by contact with the knob of the discharging rod. (Symons's Monthly Meteorological Magazine, 18:104-105, 1883)

## THE PRODUCTION BY LIGHTNING OF A SHADOW-PICTURE ON BARE BOARDS

**Anonymous; *Meteorological Magazine,* 58:166, 1923.**

During the storm of July 9th-10th a remarkable "photograph" of a waste-paper basket was made by lightning on the floor of a Mincing Lane office, that of Messrs. Thompson Bros. & Co.

The impression had made on bare, unpolished, ordinary flooring boards, and consists of a bleached image of the basket, which was apparently lying across the boards with its open end roughly towards south. The direction of the length of the boards is about perpendicular to Mincing Lane and Mark Lane.

The discovery was made next morning by the staff as they entered the door of the office. The cleaners had been in previously and had disturbed everything but had apparently not noticed the impression. They could not recollect anything definite about the position of the basket when they first entered the office. So far as could be seen, no other similar image was obtained of any other object. The waste-paper basket is of the ordinary wicker type, very old and begrimed, otherwise calling for no remark.

The roof is of clear glass, some 14 or 15 feet high. On each side running parallel to Mark Lane and Mincing Lane are very tall blocks of buildings, and the room in which the impression was made is on the ground floor, in fact, it is merely part of the space between the buildings roofed over. The firm being rice merchants and samples being handled continually, no doubt much rice is trampled on the floor.

The impression had the appearance of a photographic print of the waste-paper basket lying on its side on a piece of photographic paper, as might be obtained by means of a point source of light nearly directly overhead, but perhaps slightly towards the south. It was so clear and unblurred that it is not likely that the same effect would be obtained if the "printing" were repeated, unless the source of light were in substantially the same position every time. The boards carrying the "photograph" have been cut out and are now in the Science Museum. Unfortunately the image seems to be fading. Our illustration is from a photograph taken at the Museum and reproduced by the courtesy of the Director. Photographs taken in situ show the image rather more distinctly but with exaggerated perspective. As will be noticed in the illustration opposite images of the ribs (apparently the ribs nearest the ground) can be seen as well as the complete images of the rim, the girdle and the bottom of the basket.

The image is seen most distinctly with oblique lighting. It can not be detected, however, when the wood is examined at close quarters with a lens. (Meteorological Magazine, 58:166, 1923)

# LIGHTNING SOUNDS (OTHER THAN THUNDER)

## LIGHTNING
**Cave, C. J. P.; *Nature,* 115:801, 1925.**

Ever since I was a child I have heard of the idea that lightning makes a "swishing" noise when one is quite close to it, but I have looked on this as a popular superstition. I have recently, however, had occasion to wonder whether there may not be some foundation for the idea. On April 24 there was a very severe thunderstorm here, quite a number of flashes having been within a kilometre and a half of this house; a barn was struck one kilometre away, and probably also a cottage 450 metres in another direction. During the storm three men were working in a field; two of them were together close to a holly tree in a hedge; there was a very bright flash of lightning, with a just perceptible interval between the flash and the thunder. At the moment of the lightning there was quite a loud swishing sound in the holly tree, as though, they said, a sudden blast of air went through the tree; the sound occurred definitely before the thunder.

At about the time of the occurrence the wind rose to 30 miles per hour and gradually fell off to about 13 miles per hour; both men, however, are positive that there was no wind at the time, but that it got up shortly afterwards when the rain began. The flash must have been very close as they both smelt "sulphur"---nitrogen peroxide; and they could scarcely see anything for some moments. The third man was about 230 metres away and was close to an oak tree to which he had his back; he says that when the flash came there was a noise in the tree as though it were "on fire." He turned round expecting to see that it had been struck, but neither oak nor holly showed any signs of having been struck. Is it possible that in the neighbourhood of a flash, brush discharges may take place from trees and other points? (Nature, 115:801, 1925)

## A PHENOMENON ACCOMPANYING LIGHTNING
**Belasco, J. E.; *Meteorological Magazine,* 69:91, 1934.**

This afternoon this locality [Blenheim Gardens, London] was on the edge of a thunderstorm which at one time appeared to spread from north to west and later to withdraw from the west and spread from north-eastwards. In spite of this there was some electrical display close at hand. My wife and I were sitting by the open window looking west and the circumstances of one flash of lightning are, I think, of some interest. We did not see the actual discharge as it was on the other side of the house. The illumination from it was, however, brilliantly white (no rain was falling) and was accompanied by a peculiar sound as if of the noise of something swiftly rushing through the air across the garden from north to south or the swishing of a long whip. This sound was also heard by my wife, who describes it like that of a sudden rush of wind, although there was almost a calm. This was followed almost immediately by a terrific crash of thunder. I learned later that the Methodist Church in Walm Lane, about 100 yards away from here was struck, although happily without serious consequences. (Meteorological Magazine, 69: 91, 1934)

## THE SOUND OF LIGHTNING
**Cave, C. J. P.; *Nature,* 116:98, 1925.**

Since my letter on the above subject in Nature of May 23, several other instances have been brought to my notice. Mr. W. H. Dines has heard the sound six times certainly, and probably more; Mr. J. S. Dines has heard it once, as has also my brother, Capt. A. L. Cave, in London, when he was indoors; two other correspondents also write to say that they have heard the sound, one of them three times. But perhaps the most remarkable case is that given in the Marine Observer for July (page 112); Capt. J. Burton Davies of s. s. Hurunui reports that from 10 p. m. on July 30, 1921, to 3.45 a. m. on July 31, when in about lat. 38 N. and long. 71 W., "a terrific electric storm was playing about the ship. . . . On three occasions the officer of the watch and myself were momentarily completely dazzled by flashes, and it appeared that immediately before the flash we heard a tearing noise as of canvas being ripped violently; in fact, after the first of these flashes I caused the quartermaster to inspect the boat covers on boat deck to see if any were torn. This noise interested me very much." The fact that the noise was heard before the flash seems to indicate that it may have been caused by a brush discharge. In any event, it proves that the noise must be real, and not an illusion like the rushing noise that some have imagined they have heard when watching a bright meteor, or the rustling sometimes attributed to the aurora. (Nature, 116:98, 1925)

# MISCELLANEOUS PRANKS OF LIGHTNING

## LIGHTNING BREAKS EVERY OTHER PLATE
**Anonymous; *Nature,* 66:158, 1902.**

A curious effect produced by lightning is described to us by Dr. Enfield, writing from Jefferson, Iowa, U. S. A house which he visited was struck by lightning so that much damage was done. After the occurrence, a pile of dinner plates, twelve in number, was found to have every other plate broken. It would seem as if the plates constituted a condenser under the intensely electrified condition of the atmosphere. The particulars are, however, so meagre that it is difficult to decide whether the phenomenon was electrical or merely mechanical. (Nature, 66:158, 1902)

## A LIGHTNING WELL BORER
**Anonymous; *Scientific American,*11:344, 1856.**

During a recent thunder storm at Kensington, N. H., the lightning descended perpendicularly in an intense discharge into a pasture field, and made a hole about a foot in diameter and 30 feet deep, forming a well which soon filled up with good water. (Scientific American, 11:344, 1856)

## ACTION OF LIGHTNING IN FORFARSHIRE
**Anonymous; *Symons's Monthly Meteorological Magazine,* 2:105, 1867.**

Professor Kelland read the following paper of Sir David Brewster's: "In the summer of 1827 a hay stack in the parish of Dun, in Forfar, was struck by lightning. The stack was on fire, but before much of the hay was consumed the fire was extinguished by the farm servants. Upon examining the hay-stack, a circular passage was observed in the middle of it, as if it had been cut out with a sharp instrument. This circular passage extended to the bottom of the stack, and terminated in a hole in the ground. Captain Thomson, of Montrose, who had a farm in the neighbourhood, examined the stack, and found in the hole in the hay-stack, a substance which he described as resembling lava. A portion of this substance was sent by Captain Thomson to my brother, Dr. Brewster of Craig, who forwarded it to me with the preceding statement. The substance found was a mass of silex obviously formed by the fusion of the silex in the hay. It had a highly greenish tinge, and contained burnt portions of the hay. I presented the specimen to the museum of St. Andrews." (Symons's Monthly Meteorological Magazine, 2:105, 1867)

## VAGARIES OF LIGHTNING
**Anonymous; *Nature,* 37:64, 1887.**

During a hailstorm at Mors, in Denmark, a few days ago, a flash of forked lightning ---the only one occurring---struck a farm, and, having demolished the chimney-stack and made a wreck of the loft, descended into the living-rooms on the ground floor below. Here its career appears to have been most extraordinary; all the plaster around doors and windows having been torn down, and the bed-curtains in the bed-rooms rent to pieces. An old Dutch clock was smashed into atoms, but a canary and cage hanging a few inches from it were quite uninjured. The lightning also broke sixty windows and all the mirrors in the house. On leaving the rooms it passed clean through the door into the yard, where it killed a cat, two fowls, and a pig, and then buried itself in the earth. In one of the rooms were two women, both of whom were struck to the ground, but neither was injured. (Nature, 37:64, 1887)

## NOTE ON A LIGHTNING STROKE PRESENTING SOME FEATURES OF INTEREST
**Scott, Robert H.; *Royal Meteorological Society, Quarterly Journal,* 17:18-19, 189 .**

On my recent visit to the West of Ireland, I learnt that a remarkably violent lightning stroke had occurred within the peninsula of the Mullet, in the barony of Erris, Co. Mayo, and I went to the spot to collect what information I could.

The occurrence took place January 5th, 1890, in a violent storm of wind. I did not visit the place until October 12th, so that I have for the most part only hear-say evidence to report. The locality is the fishing village of Tip, about one mile from Ballyglass Coastguard Station, on Broad Haven. The house struck belonged to Michael Moran, who with his wife were the sole inmates. It is a thatched cottage, one storey high.

At 10.30 p.m. January 5th, 1890, Moran was in bed, his wife was standing up. She heard a very loud clap of thunder, and called to her husband to get up. Immediately after she was struck down and stunned for a short time, a few minutes. The bedroom showed marks of the lightning on the walls in various places.

The next morning it was found that the kitchen adjoining the bedroom was not much damaged, but that the sitting room on the other side of the kitchen had been completely wrecked.

The grate was forced out of its setting and thrown across the room. Various marks were seen on the walls, and a hole was pierced through the wall opposite the head of a crowbar which was leaning against the wall in the adjacent workshop. The hangings on the mantelpiece, &c., were much torn; there were no signs of fire or fusion, except that two corners of a photograph on a metal plate standing on the chimney piece were fused; but the photograph itself was otherwise uninjured. All objects of glass or china in the room were upset, but only a few of them broken; a corner was cut clean off a glass ink bottle, without spilling ink, &c. The most extraordinary occurrence was what happened to a basket of eggs lying on the floor of the room. The shells were shattered, so that they fell off when the eggs were put in boiling water, but the inner membrane was not broken. The eggs tasted quite sound. Moran's account is that he boiled a few eggs from the top of the basket, the rest were "made into a mummy, the lower ones all flattened, but not broken."

I have referred this latter statement to my friend, Mr. C. W. Boys, F.R.S., whose reply is as follows:---

"I can only conclude from your account of the destruction of the egg shells by the lightning, that the action was purely mechanical, the result of a detonation among the eggs, rather than electrical. I do not think the eggs within would have remained undamaged if the lightning had really struck them.

"It seems very similar to the result of an experiment which I have seen in which an egg thrown up in the air and falling on hard turf, was not broken, though no doubt the shell at the point of impact was cracked up a good deal.

"The egg was not boiled, for on being thrown up a second time, which was not fair, it spread itself about in a way which is not possible with a boiled egg." (Royal Meteorological Society, Quarterly Journal, 17:18-19, 1891.

## LIGHTNING: MAGNETIC COMPASSES AFFECTED
**Sutcliffe, H.; *Marine Observer*, 38:63 1968.**

s.s. Crofter. Captain R. Sutcliffe, Pensacola to Newport News. Observer, Mr. R.A. Francis, 3rd Officer.

24th April 1967. At 0200 GMT the vessel entered a severe thunderstorm which radar indicated was 16 miles in radius. The forked lightning in the storm centre frequently affected the off-course alarm and the two magnetic compasses showed signs of disturbance. In the course of an hour the air temp. fell from 76°F to 70° and the wet bulb from 70° to 69°.

Position of ship: 30° 35'N, 78° 20'W. (Marine Observer, 38:63, 1968)

# METEOR-LIKE PHENOMENA

The usual meteors are streaks of light in the sky popularly called "shooting stars." Some meteors are not completely burned up by atmospheric friction and impact the earth's surface as meteorites. Most, however, detonate into pieces at high altitudes or are silently burnt to nothingness. Thus are the fates of most "conventional" meteors.

Some meteors do not follow the pattern. The really great meteoric displays are worldwide in scope, as was the case with the 1908 and 1913 events described below. The reader will discover that the phenomena produced in such cases are decidedly nonconventional. Then, there are those meteors that cruise along with difficult-to-explain slowness and those that deviate repeatedly from "normal" trajectories. Most awe-inspiring are meteor-like luminous phenomena close to the earth's surface that emulate ball lightning and electric discharge effects. The dividing lines between these phenomena becomes distressingly vague. If something palpable remains---a meteorite---we class these phenomena as meteoritic and hope for the best.

Other strange phenomena are superficially meteoric but may turn out to be something quite different. In this category are "telescopic meteors" (possibly just seeds and other debris seen floating high in the air through the telescope), peculiar darting gleams of light (electrical phenomena?), and the famous "fireflies" seen from some spacecraft (supposedly material flaking off the spacecraft itself).

## THE GREAT SIBERIAN METEOR OR TUNGUSKA EVENT

### POSSIBLE ANTI-MATTER CONTENT OF THE TUNGUSKA METEOR OF 1908
**Cowan, Clyde, et al; *Nature,* 206:861, 1965.**

Perhaps the most spectacular meteor fall to be observed in modern times occurred on June 30, 1908 0h. 17m. 11s. U. T. in the basin of the Podkanemaia Tunguska River, Siberia (60$^{o}$ 55' N, 101$^{o}$ 57' E), some 500 miles to the north of Lake Baykal. It was seen in a sunlit cloudless sky over an area of about 1, 500 km in diameter and was described as the flight and explosion of a blindingly bright bolide which "made even the light of the sun appear dark". The fall was accompanied by exceptionally violent radiation and shock phenomena. Although seismic, meteorological and geomagnetic field disturbances were registered at points around the world at the time, and descriptive accounts of the phenomena accompanying the fall were collected from witnesses during the years following, the first inspection of the place of fall was not made until 1927. No trace of a crater was found, though great damage of the forest was still evident due to thermal and blast effects.

Various hypotheses have been advanced to explain the massive phenomena as having been caused by a large meteor, a small comet or a nuclear explosion. All are argued cogently, for and against. Estimates of the total yield of energy, made from the records of the disturbances already mentioned and from deductions based on blast and thermal damage at the site of fall, agree with one another quite well and place the yield at something in excess of $10^{23}$ ergs, probably about $10^{24}$ ergs. If this were the result of a nuclear explosion of some sort, then the yield of neutrons into the atmosphere with the consequent formation of carbon-14 from atmospheric nitrogen should be detectable by analysis of plant material which was growing at that time. It seemed worth while, therefore, to make such an analysis, and the growth rings of a tree were chosen for this purpose.

Before reporting the present work it may be of interest to repeat some of the accounts of the effects of the Tunguska meteor. The general area of the fall is composed of tiaga with peat bogs and forest and is (fortunately) very sparsely populated. One eye-witness, S. B. Semenov, a farmer at Vanovara some 60 km away, told L. A. Kulik, who investigated the meteor first in 1927, that he was sitting on the steps outside his house around 8 a.m., facing north, when a fiery explosion occurred which emitted so much heat that he could not stand it: "My shirt was almost burnt on my body". However, the fireball did not last long. He just managed to lower his eyes. When he looked again, the fireball had disappeared. At the same time, an explosion threw him off the steps for several feet, leaving him briefly unconscious. After regaining his senses, a tremendous sound occurred, shaking all the houses, breaking the glass in the windows, and damaging his barn considerably.

Another observer, P. O. Kosolopov, a farmer and neighbor of S. B. Semenov, was working on the outside of his house, when suddenly he felt his ears being burnt. He covered them with his hands and ran into his house after asking Semenov if he had seen anything, on which Semenov answered that he too had been burnt. Inside the house, suddenly earth started falling from the ceiling and a piece from his large stove flew out. The windows broke and he heard thunder disappearing to the north. Then he ran outside, but could not see anything.

A Tungus, Liuchetken, told Kulik on April 16, 1927, that his relative, Vassili Ilich, had some 500 reindeer in the area of the fall and many "storage places". With the exception of several dozen tame deer, the rest were grazing in that area. "The fire came by and destroyed the forest, the reindeer and all other animals". Then several Tungus went to investigate and found the burnt remains of several deer: the rest had completely disappeared. Everything was burnt in Vassili Ilich's storage including his clothing. His silverware and samovars (tin?) were molten. Only some large buckets were left intact.

According to Krinov the dazzling fireball moved within a few seconds from the south-east to north-west leaving a trail of dust. Flames and a cloud of smoke were seen over the area of the fall. Visible phenomena were observed from a distance as great as 700 km, and loud explosions were heard after the passage of the fireball at distances up to 1,000 km. (Nature, 206:861, 1965)

## THE QUESTION OF THE METEORITE OF JUNE 30, 1908, IN CENTRAL SIBERIA

**Kulik, L. A.; *Popular Astronomy*, 20:559-560, 1937.**

On June 30, 1908, at 7:00 a.m. in mild weather and a cloudless sky, a large bolide or fireball rushed along in the general direction from south to north over the basin of the River Yenissei in central Siberia. The fall of the mass onto the earth was completed by the appearance of a column of fire raised over the taiga [virgin forest] ---and observed from a distance of 400 km. at Kirensk---by three or four powerful thunderclaps and a crash, recorded at a distance of more than a thousand kilometers from the center of the fall, and by powerful air waves. To distances of some scores of kilometers around the center of the fall, these waves blew down the trees of the taiga in an eccentric-radial direction, tops outward, and were so powerful that they set into motion the baromicrographs of western Europe as well as those of North America and other countries. These waves went around the world and were recorded a second time in Potsdam, Germany. Seismographs all over the world (at such widely separated stations as, e.g., Irkutsk, Tashkent, Tbilissi, Jena, Washington, and Java) recorded powerful earth waves, which, as just stated, twice encircled the globe, as a result of the explosive action of the meteorite on the earth's crust. The epicenter of this disturbance was located behind [the] Podkamennaya Tunguska, in the region of the factory [of] Vanovara. Huge masses of the finest substance, sprayed by the meteorite in its flight through the atmosphere and raised by the explosion in the earth's crust (due to the cosmic speed of the impact of the meteorite), caused the heavy blanket of dust in the upper layers of the atmosphere, and the formation, at a height of from 83 to 85 km., of "silvery clouds" (light clouds), and dust screens on the "ceiling" and in the lower layers of the stratosphere. Thus were produced those remarkable phenomena called "night dawns," which were of incomparable beauty. These were observed on the day of the fall, from the place of its occurrence as far as Spain, and from Fenno-Scandia to the Black Sea. During the night of June 30 to July 1 and the next few days, these "dawns" became gradually weaker.

The first information concerning the fall of the meteorite was gathered by the writer's expedition, which had been equipped for work in Siberia in 1921 by A. V. Loonatcharsky, People's Commissary for Public Instruction. The official scientific circles at that time did not, however, regard the accounts of this phenomenon as sufficiently trustworthy to warrant an investigation of it. In 1925-26, the writer's opinion was confirmed by new data about the fall; first, from information concerning the seismic waves, furnished by A. V. Vosnessensky, Director of the Irkutsk Observatory, and, second, from other evidence supplied by S. V. Obrootcheff, a geologist, and I. M. Soussloff, an enthnographer. Finally, in 1927, the writer had the opportunity to go to [the] Podkamennaya Tunguska to examine the place of this sensational fall. The investigation brought to light a continuous, eccentric, radial "windfall" in a huge area of radius about 30 km., which extended on some of the hills to a distance of 60 km. and even farther. Observers relate that individual trees were thrown down on the hills in the neighborhood even of Vanovara. The exceptional power of these air waves is evidenced further by the fact that in Kirensk (distant 400 km.), fences were overturned; in Kejma, people carrying grain-sacks were knocked down; and in Kansk (600 km. distant), workers were thrown from rafts into the river, while at a distance of even 700 km. to the south of Kansk, horses could not stand on their feet!

The probable figure of this "windfall," in consequence of the presence of many hills around the center of the fall, in plan reminds one of the meteorological graph called the "wind-rose." It is difficult now, of course, to estimate the dimensions of the whole area of this "windfall" without a survey. The outer fringe of the "windfall" (about 20 km. from the center) bears traces of a continuous burn from above. Moreover, the branches of the fallen trees, as well as those which still remain standing, are, as a rule, broken and destroyed. Especially characteristic is the fact that every surface of a break bears a little bit of charcoal, i.e. a trace of the burn. There is "no break without a burn." The central area of the "windfall" lies on permanently frozen hilly peat mosses, which alternate with swamps among the hills of the mountain watershed. This central area is surrounded by burned trees, still standing, but totally devoid of branches. The continuous "windfall" begins a little farther from the center, at a distance of one-half to one kilometer. (Popular Astronomy, 20:559-560, 1937)

# THE METEORIC PROCESSION OF FEBRUARY 9, 1913

## A CURIOUS METEORIC DISPLAY
**Anonymous; *Nature*, 92:87-88, 1913.**

Our interest is, however, again wakened by what is described as "an extraordinary meteoric display" which was seen over a very extensive area in the United States of America and Canada on the evening of February 9 this year. The magnitude of the display was such that a very great number of people distributed in the path of observation had their attention drawn to it, and its peculiar nature was so marked that nearly every observer remarked similarly of the extraordinary feature of the event.

Fortunately Prof. C. A. Chant, of Toronto, although not an eye-witness of the phenomenon, undertook to collect all the available information of this very exceptional, if not unique, occurrence. In the May-June number of the Journal of the Royal Astronomical Society of Canada he presents a very judicious summary of the observations made, and accompanies this with extracts from letters received from observers.

The sum total of the discussion of the data is to show that the apparition took the following form:---

As seen from western Ontario there suddenly appeared in the north-western sky a fiery red or golden-yellow body, which quickly grew larger as it approached, and had attached to it a long tail; observers vary in their descriptions as to whether the body was single or composed of three or four parts with a tail to each part.

This body or group of bodies moved forward on an apparently perfectly horizontal path "with peculiar, majestic, dignified deliberation; and continuing its course without the least apparent sinking towards the earth, it moved on to the south-east, where it simply disappeared in the distance." After this group of bodies had vanished, another group emerged from precisely the same region. "Onward they moved, at the same deliberate pace, in twos or threes or fours, with tails streaming behind them, though not so long or bright as in the first case." This group disappeared in the same direction. A third group followed with less luminosity and shorter tails.

In reading some of the communications from the numerous observers, the extraordinary feature of the phenomenon seems to have been the regular order and movement of the groups. Thus some compared them to a fleet of airships with lights on either side and fore and aft; others to a number of battleships attended by cruisers and destroyers; others again to a brilliantly lighted passenger train travelling in sections and seen from a distance of several miles.

Such descriptions indicate that the display was of a very unusual kind, very different from the usual quick-moving and scattered bodies. It may be of interest to reprint here in full one of the accounts. Mr. J. G. MacArthur writes:---

"There were probably thirty or thirty-two bodies, and the peculiar thing about them was their moving in fours, threes, and twos, abreast of one another, and so perfect was the lining-up you would have thought it was an aerial fleet manoeuvring after rigid drilling. About half of them had passed when an unusually large one hove in sight, fully ten times as large as the others. Five or six would appear in two detachments, probably five seconds apart; then another wait of five or ten seconds, and another detachment would come into view. We could see each detachment for probably twenty or twenty-five seconds. The display lasted about three minutes. As the last detachment vanished, the booming as of thunder was heard---about five or six very pronounced reports. It sounded in the valley as if some of the balls of fire had dashed into Humber Bay. The bodies vanished in the south-east, but the booming appeared to come from the west or north-west, and the time it was heard was close to 9.12 p.m." (Nature, 92:87-88, 1913)

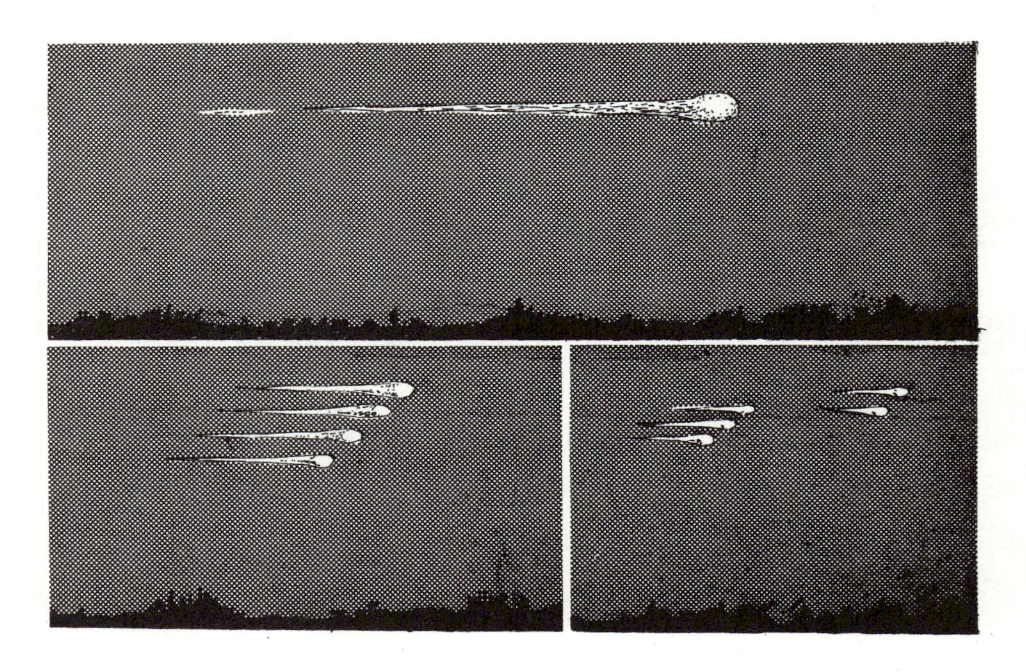

Stately progress of the 1913 meteors

## THE METEORIC PROCESSION OF FEBRUARY 9, 1913
**Pickering, William H.; *Popular Astronomy,* 30:632-633, 1922.**

This remarkable phenomenon was in no sense a meteoric shower. It was a different kind of event altogether, and while undoubtedly much less spectacular than a great shower, such as that of Nov. 1833, was in some respects more interesting and instructive. It consisted of a procession of fire balls and meteors all moving very slowly in practically the same path across the sky, from horizon to horizon. It was first seen near Mortlach, 65 miles west of Regina, Saskatchewan, lat. 50°.5 N., long. 106° W. It traversed successively Manitoba, Minnesota, Michigan, Ontario, New York, Pennsylvania, and New Jersey a few miles to the south of New York City. It then went out to sea, and appeared next in Bermuda, was seen from the steamship Bellusia, lat. 17° 35' N., long. 51° 11'W., and last from the steamer Newlands, lat. 3° 20'S., long. 32° 30'W. The distance between the first and last stations is 5659 miles. Computation indicates that the meteors traversed several thousand miles more, while still within the limits of the earth's atmosphere, thus forming prior to their final destruction a series of minute temporary terrestrial satellites.

There appear, when first and last seen, to have been some 40 to 60 fire balls, arranged at first in 4 or 5 separate groups. Several of these had long tails. In part of the route they were accompanied by innumerable finer particles that were swept off of them in their rush through our atmosphere. The interval during which the larger fire balls were visible under favorable circumstances seems to have been from 30 to 40 seconds, and the duration of the display about 5 minutes. In Canada their color was usually described as yellow or reddish, but not white. Generally they were compared to bright stars, sometimes to Venus, but in Bermuda the two leaders looked like large arc lights, slightly violet in color and of diameter equal to the moon. Their true angular diameter was of course much less than this. An incandescent steel globule 0.2 inches in diameter, at a distance of one half mile was found by Professor I. L. Smith in 1857 to appear of the moon's diameter (Jour. R. A. S. Canada, 1914, 8, 109). This would correspond to a diameter of 24 feet. It therefore appeared 1400 times its true size. If the meteors were at a distance of 70 miles from Bermuda, this would indicate that the diameter of the two leaders was about 28 inches.

A very full description of them, with a computation of their path, and estimates of their height, speed, and size is given by Professor Chant in the Journ. R. A. S. Canada 1913, 7, 145 (See also 404, 438; 8, 108, 112; 9, 287, and 10, 294). The last two papers are by Mr. Denning, and are particularly valuable as giving the data of the two observations made at sea. These were not available when Professor Chant made his calculations, and they not only permit us to correct them, but they serve to give us much more definite information regarding the height, speed, and size of the meteors, regarding which his paper left us in considerable doubt.

The path of the meteors across the United States was about as long as it was in Canada, and they must have been visible over an area of rather more than a quarter of a million square miles, including all the large cities and observatories of the important states of Michigan, New York, Pennsylvania, New Jersey, Delaware, Maryland, the District of Columbia, and southern New England. In fact if the skies were clear they might have been seen by over 30,000,000 people. It is regrettable that it did not occur to any one in this area that this very unusual phenomenon was worth investigating. A collection of observations might at that time have been made, culled from newspapers and by private correspondence, similar to that secured by Professor Chant. His collection is most valuable and unique, and it is proper to say that all that we know about the meteors is really due to him, since it was his work which called Mr. Denning's attention to the matter. My own first knowledge of the phenomenon was due to reading his papers. The total number of reports on which our knowledge is based is 141. Of these 135 came from Canada, 1 from southern Michigan, 1 from a lady in New Jersey, 2 from Bermuda, and 2 at sea. It is gratifying to find that America is not left out entirely in the investigation.

It is certainly unfortunate for our scientific reputation however that the most important and interesting meteoric event of the past ninety years, visible in the early evening, and to more advantage in our country than in any other, should be described by us in only two very brief reports. From these two the most important information that we gather is that the sky was clear in those two regions. (Popular Astronomy, 30:632-633, 1922)

## THE REMARKABLE METEORS OF FEBRUARY 9, 1913
**Denning, W. F.; *Nature,* 97:181, 1916.**

The large meteors which passed over Northern America on February 9, 1913, presented some unique features. The length of their observed flight was about 2600 miles, and they must have been moving in paths concentric, or nearly concentric, with the earth's surface, so that they temporarily formed new terrestrial satellites. Their height was about 42 miles, and in the Journal of the R. A. S. of Canada there are 70 pages occupied with the observations and deductions made from them by Prof. C. A. Chant.

The meteors were last seen from the Bermuda Islands, according to the descriptions in the journal named (May-June, 1913).

I have since made efforts to obtain further observations from seafaring men through the medium of the Nautical Magazine, and have succeeded in procuring data which prove that the meteors were observed during a course of 5500 miles from about lat. 51°N., long. 107°W., to lat. 5-1/2°S., Long. 32-1/2°W.

Mr. W. W. Waddell, first mate of the s. s. Newlands, writes me that at 12.13 p. m., February 9, 1913, he saw a brilliant stream of meteors passing from the N. W. to the S. E. during a period of six minutes. The ship was in lat. 3° 20'S. and long. 32° 30'W. at the time. He says the meteors disappeared in the region of Argo to the south, and I have assumed they were over about lat. 5-1/2°S. and long. 32-1/2°W. when he lost sight of them.

Such an extended trajectory is without parallel in this branch of astronomy. Further reports from navigators in the South Atlantic Ocean might show that the observed flight was even greater than 5500 miles. (Nature, 97:181, 1916)

# METEORS WITH ERRATIC COURSES

## REMARKABLE METEOR
**Anonymous; *English Mechanic,* 7:351, 1868.**

A remarkable meteor was seen at the Radcliffe Observatory, Oxford, on the 8th ult., at 9h. 50m. When first seen by Mr. Lucas it had the appearance of a fine white cloud about 5° in length and 1° in breadth, a little to the west of Polaris. As the observer was pointing out its comet-like appearance to some persons who happened to be with him in the lower meadow of the Observatory, it appeared to start into motion, taking a course directly west, and passing just below $\alpha$ and $\beta$ Ursae Majoris, and leaving a train behind of a greater breadth than itself, which remained visible through its whole course after it had disappeared below the N. W. horizon. When it approached Leo it deviated from the straight line which it had previously taken, and turned somewhat towards the south, passing near Regulus, and then bent northwards again. The time that it was visible must have been nearly four minutes. Its appearance at one time was very like that of the flame and smoke combined which sometimes issue from a railway engine, only very faint on account of the brightness of the still remaining twilight. There was a thick haze all the night. A parselenes, or mock moon, was seen on the same night at 13h. 40m. (English Mechanic, 7:351, 1868)

## UNDULATING METEOR
**Hanney, R. A.; *Marine Observer*, 31:189, 1961.**

s. s. Hector, Captain R. A. Hanney, Adelaide to Aden. Observer, Mr. J. H. Watterson, 2nd Officer.

14th November 1960. At 2046 GMT a meteor of unusual brilliance was observed ahead, and moving in azimuth from 310° to 295° approx. at a mean altitude of 4°. Its flight was slower than is usual, though it was in view for only 2 sec., but it was nevertheless remarkable in that its track showed a dip and a climb, as shown in the accompanying sketch. Only 1/8 $C_L8$ was present. There was a slight haze, making the horizon indistinct. Air temp. 80°F, wet bulb 75°, sea 83°.

Position of ship: 8° 49'N, 56° 00'E. (Marine Observer, 31:189, 1961)

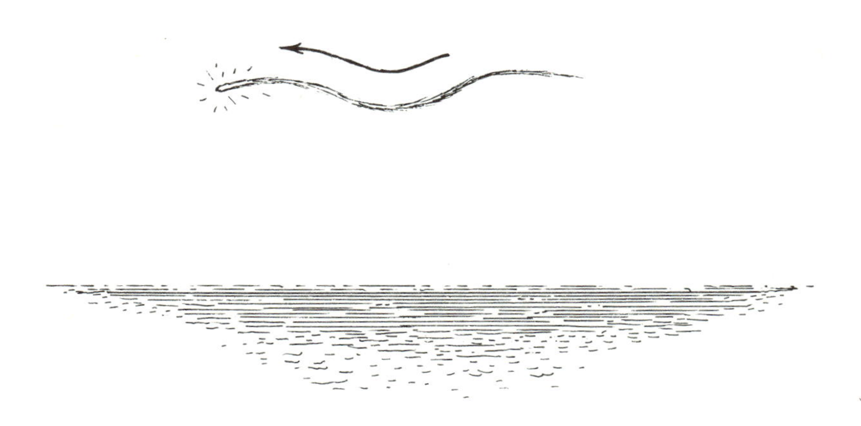

Meteor dips and then climbs

## PHENOMENON OBSERVED AT SEA
**Knevitt, H. P.; *Royal Astronomical Society, Monthly Notices*, 33:411-412, 1873.**

A phenomenon having been seen from this ship when on the passage from Manzanilla to this place, I think it advisable to make it known at once, instead of waiting the transmission of my remark-book at the end of the year.

On the 16th May, 1872, at 2h 45m a.m. (the weather having been squally since midnight) a phenomenon was seen in the heavens at an altitude of about 50° and bearing east of compass; the ship at the time being in lat. 14° 55'N. and Long. 99° 58'W.

I did not see it myself, but the following is a description given of it by Lieutenant Cecil G. Horne, who was officer of the watch:---

"Attention was first drawn by a very bright flash, resembling a small flash of vivid lightning, but being much more solid, and lasting 4 to 5 seconds; the passage of the luminous body was towards the horizon for a short distance (say $3^{o}$ or $4^{o}$), in a zig-zag course; it then appeared to burst and throw off a tail such as a comet has; the tail forming a ring and spreading itself round the body, till the whole had very much the appearance of a large Catherine wheel; it then gradually faded out of sight, having been visible from first to last, about 10 to 15 minutes." (Royal Astronomical Society, Monthly Notices, 33:411-412, 1873)

## DISGRESSING METEOR
**Anonymous; *Nature,* 27:423, 1883.**

On February 5, at 6.45 p.m., a meteor of unusual size and appearance was observed near Arvika, in Sweden. An observer who happened at the time to be passing a lake---Glasfjorden---states that he first observed the meteor high on the horizon, going from south-east to north-west, when, after about eighteen seconds, it suddenly changed its course to south-east. During its progress to north-west, calculated at eighteen seconds, the meteor made several digressions from its plane, while its size varied from that of an ordinary star to that of the sun, sometimes emitting a white, at others a yellow light, and at times discharging showers of sparks. At the point of changing its direction, when it was so near the surface of the lake that its path was reflected therein, it possessed a distinct tail, and with this adjunct it passed out of the range of sight in a south-easterly direction, after being observed for nearly fifty seconds. (Nature, 27:423, 1883)

# REPORTS OF VERY-LOW-LEVEL METEORS

## A STRANGE PHENOMENON, EVIDENTLY A DAYLIGHT METEOR
**Keppler, F. W.; *Popular Astronomy,* 44:569, 1936.**

While duck hunting on Lake Butte des Mortes, a few miles northwest of Oshkosh, Wisconsin, my two companions, Arthur Nelson of Oshkosh, and F. E. Winans of Wauwatosa, and I were witnesses to a strange phenomenon, which occurred sometime between the hours of eight and ten in the morning, on October 29, 1936.

The point at which we were situated was on the south shore of the lake. We were facing the north watching a large raft of duck, when from the east, seemingly low over the water, came a rocket black formation throwing sparks of various colors, which shot across our front and seemed to disintegrate in the north-west. Had any of us been alone we would have thought we were seeing things, but all three of us witnessed the incident.

Both Mr. Arthur Nelson who lives at Oshkosh, Wisconsin, and Mr. F. E. Winans, 1868 N. 70th Street, Wauwatosa, Wisconsin, will verify this report. (Popular Astronomy, 44:569, 1936)

## UFO IN 1800: METEOR?
**Anonymous; *American Philosophical Society, Transactions,* 6:25, 1804, as reprinted in *Science,* 160:1260, 1968.**

A phenomenon was seen to pass Baton Rouge on the night of the 5th April 1800, of which the following is the best description I have been able to obtain.

It was first seen in the South West, and moved so rapidly, passing over the heads of the spectators, as to disappear in the North East in about a quarter of a minute.

It appeared to be of the size of a large house, 70 to 80 feet long and of a form nearly resembling Fig. 5 in Plate, IV. [Not reproduced]

It appeared to be about 200 yards above the surface of the earth, wholly luminous, but not emitting sparks; of a colour resembling the sun near the horizon in a cold frosty evening, which may be called a crimson red. When passing right over the heads of the spectators, the light on the surface of the earth, was little short of the effect of sun-beams, though at the same time, looking another way, the stars were visible, which appears to be a confirmation of the opinion formed of its moderate elevation. In passing, a considerable degree of heat was felt but no electric sensation. Immediately after it disappeared in the North East, a violent rushing noise was heard, as if the phenomenon was bearing down the forest before it, and in a few seconds a tremendous crash was heard similar to that of the largest piece of ordnance, causing a very sensible earthquake.

I have been informed, that search has been made in the place where the burning body fell, and that a considerable portion of the surface of the earth was found broken up, and every vegetable body burned or greatly scorched. I have not yet received answers to a number of queries I have sent on, which may perhaps bring to light more particulars. (American Philosophical Society, Transactions, 6:25, 1804, and Science, 160:1260, 1968)

## FALL OF AN AEROLITE AT FOREST HILL, ARKANSAS
**Anonymous; *American Journal of Science,* 2:5:293-294, 1848.**

A letter from Henry Hicks, P. M. dated Dec. 12, 1847, to the editor of the Philadelphia Courier, states, that on the 8th of December the sky was perfectly clear, and had been so through the day; but at quarter past 3 p. m., it became suddenly very dark. The clouds, or what appeared to be such, whirled in the strangest contortions, and appeared like a solid black fleece lighted from above by a red glare as of many torches, but scarcely penetrating the masses of clouds below. In the presence of hundreds of spectators there was a deafening explosion, and the concussion shook the houses and caused the bell of the village church to toll several times. At the moment of the explosion an ignited body---compared for size to a hogshead---"descended like lightning," and struck the earth about twenty feet east of a cotton gin which stands near the village. Within twenty minutes, the sky was as clear and the sun shone as brightly as ever. The hole made by the aerolite was eight feet deep and two feet two and a half inches in diameter; at the bottom of it was a black mass, which was still so hot that water thrown upon it hissed as it would do on hot iron. The exhumed mass, recovered with considerable labor, was covered by a black incrustation; it smelled of sulphur, and was, for weight, compared to common rock. It is not stated that it was weighed. [We look for specimens confirming this event.] (American Journal of Science, 2:5:293-294, 1848)

## A PILLAR OF FIRE (?)
**Anonymous; *Symons's Monthly Meteorological Magazine,* 4:37, 1869.**

On Saturday morning, between four and five o'clock, a ball---or more properly speaking, a pillar of fire---passed from east to west over the City of Carlisle, and was plainly visible for fully a score of miles around. It resembled an ordinary gate-post in size and shape, and seemed as though it were prevented from falling by some invisible connecting cord. It travelled in a westerly direction, and was plainly visible at Cummerdale, and at Glasson, 10 miles distant from Carlisle. At Glasson, a respectable yeoman watched the pillar intently, and he says that it caused a great heat, while another person in the same locality says that the fiery substance seemed to pass within a score of yards of him, and that the heat was almost overpowering. It exploded in the air, and immediately after the report, resembling the sound of the discharge of cannon, a singular brightness lit up the heavens. (Symons's Monthly Meteorological Magazine, 4:37, 1869)

# SLOW OR "LAZY" METEORS

## THE "LAZY METEOR" OF APRIL 17, 1934
**Pruett, J. Hugh; *Popular Astronomy,* 45:277, 1937.**

A beautiful reddish meteor, followed by a long trail, floated in from the ocean and over the mid-Oregon coastline at 9:35 p.m., P.S.T., April 17, 1934. Its seeming leisure caused a newspaper to dub it the "Lazy Meteor." Travelling almost east-south-east, it finally disappeared over a point well in the State of Nevada. Many reports from widely separated observers were received. All indicated that it seemed almost to crawl and to bore its way across the sky.

The writer observed this meteor from Eugene. Its slowness bewildered him. It was first seen low in the west. It picked up some speed as it neared the meridian to the south. It was hardly losing any altitude at this point. It disappeared behind nearby tall trees when only slightly east of south. From words spoken while it was visible, it is certain that 10 seconds elapsed. Another Eugene observer saw the entire flight. He reported that the meteor remained visible fully twice as long after it crossed the meridian as before. He saw it finally, barely moving, low in the southeast. Many reported that the meteor was in sight fully a minute. Unreasonable as it may seem, the writer is certain that it lasted 25 or 30 seconds. The calculated projected path was over 300 miles in length. This fact would indicate a reasonable speed for a fireball overtaking the earth.

At Roseburg, 60 miles almost south of Eugene, the meridian passage appeared north of the zenith. Two observers at Roseburg were furnished with a clinometer and measured the estimated meridian altitude as $75^{\circ}$. The Eugene altitude was carefully measured as $59^{\circ}$. The height when passing between the two towns was thus almost 70 miles. (Popular Astronomy, 45:277, 1937)

## A SUPPOSED METEOR
**Drayton, N. S.; *Scientific American,* 47:53, 1882.**

On the evening of the 6th, while engaged in "sweeping" the vicinity of Ursa Minor for double stars, my attention was drawn to a bright object about the size of a star of the second magnitude moving slowly from west to east. It passed within a degree of Polaris and continued steadily in its course eastward, disappearing from view in the neighborhood of Capricornus. In color this object, a meteor doubtless, was deep red, without scintillations or train of any kind, and its slow movement was in marked contrast with the rapid flashing of the common "shooting star." It was visible to me fully three-fourths of a minute, varying but slightly in brightness during that time. In the closeness of my attention to its movement I neglected to note the time of its appearance, but judge it to have been near half past ten. Perhaps there were others of your readers who observed the phenomenon, and can add more specifically to my testimony. (Scientific American, 47:53, 1882)

## UNIDENTIFIED PHENOMENON
**Cutler, R. J. H.; *Marine Observer,* 41:170-171, 1971.**

s.s. Oriana. Captain R. J. H. Cutler. Honolulu to San Francisco. Observers, Mr. M. S. Cavaghan, 1st Officer, Captain P. W. Love, Staff Captain and Mr. T. P. Watkins, Navigator.

24th December 1970. At 0158 GMT---evening twilight---an object was first observed, bearing 075°, altitude 15°. It appeared similar to a planet, about the same size as Jupiter, and was travelling fast. A big delta-shaped glow emanated from it and it was leaving a vast trail which glowed green/white. The object stopped leaving a trail when it was bearing approx. 150°, thereafter appearing only as a bright point, still moving at great speed. It finally disappeared from view 4 min after first sighting at an altitude of approx. 60°, bearing 175°. The trail still remained visible for about 3 min after the object had disappeared, glowing a bright luminous green towards the first point of observation, fading to a faint whitish glow towards the point where it vanished. There was no sound from the object throughout the observation. Air temp. 12.2°C, wet bulb 9.4°. Pressure 1017.5 mb. Wind N'ly, force 1. Cloud 3/10 Sc, 4/10 Cs almost too thin to be visible. Course 068° at 25 kt.

Position of ship: 36° 03'N, 127° 53'W.

Note. When forwarding the above report, the Administrator of the Canadian Meteorological Service enclosed the following comments from the Astronomer Dr. P. M. Millman, Head of Upper Atmospheric Research in the National Research Council of Canada:

"The sighting has all the earmarks of a slow bright fireball entering the atmosphere at near-orbital velocity, except for the unusually long time of flight. This is near the extreme possible but since the timing is only to the nearest minute there is some leeway in the interpretation. We have no records of a satellite decay on this date, and the description of the luminous train agrees very well with what one would expect from the typical but uncommon brilliant evening fireball overtaking the earth in its orbit around the sun. With only one observation we cannot, of course, identify this with 100 per cent certainty." (Marine Observer, 41:170-171, 1971)

# SO-CALLED "TELESCOPIC" METEORS

## STREAM OF LUMINOUS BODIES
Powell, Baden; *Reports of the British Association*, 235-237, 1852.

"I was then residing at the Vicarage, South Mimms, Middlesex, in a situation peculiarly favourable for astronomical observation.

"I had been engaged for several consecutive days in observing the planet Mercury during his approach to the sun; partly to test the accuracy of my power of observation by the calculations of the Nautical Almanack, but chiefly to remark how nearly I could trace the planet in his course to the sun, before he should be wholly lost in his rays.

"For this purpose I used the most careful adjustments my instrument was capable of, and continued my observations without noticing anything peculiar.

"When, however, on the morning of the 4th of September I was preparing my equatoreal before it was fixed on the planet, I observed, passing through the field of view, in a continuous stream, a great number of luminous bodies and I cannot more correctly describe the whole appearance, than by employing the same language which I used when I communicated the circumstance to the Royal Astronomical Society, in the Monthly Notices of Dec. 13, 1850, and Dec. 12th, 1851.

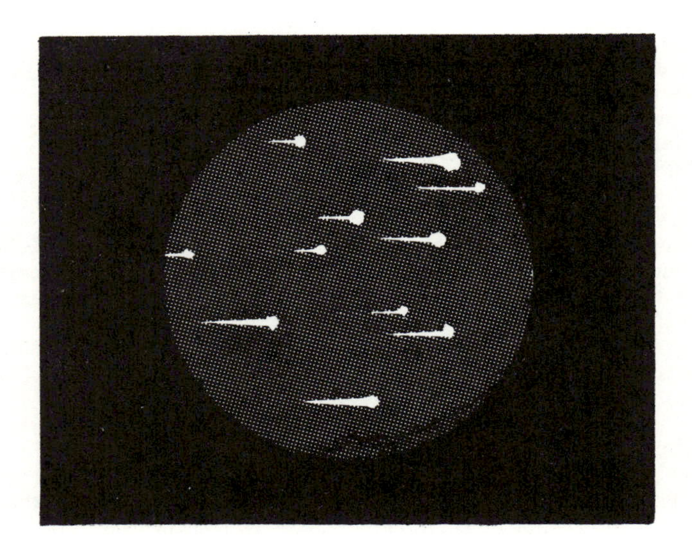

Stream of luminous bodies seen through telescope

"When I first saw them I was filled with surprise, and endeavoured to account for the strange appearance by supposing that they were bodies floating in the atmosphere, such as the seeds of plants, as we are accustomed to witness them in the open country about this season; but nothing was visible to the naked eye.

"The sky was perfectly cloudless; and so serene was the atmosphere, that there was not a breath of wind through the day, even so much as to cause any perceptible tremor of the instrument; and I subjected the luminous bodies to examination by all the eye-pieces and coloured glasses that were needful; but they bore every such examination just as the planets Mercury and Venus did, both of which were frequently looked at by me, for the purpose of comparison, during the day; so that it was impossible I could resist the conclusion (much as I was early disposed to hesitate) that they were real celestial bodies moving in an orbit of their own, and far removed beyond the limits of our atmosphere.

"They continued passing, often in inconceivable numbers, from 1/2 past 9 a. m., when I first saw them, almost without intermission, till about 1/2 past 3 p. m., when they became fewer, passed at longer intervals, and then finally ceased.

"The bodies were all perfectly round, with about the brightness of Venus, as seen in the same field of view with them; and their light was white, or with a slight tinge of blue; and they appeared self-luminous, as though they did not cross the sun's disc; yet when seen near him they did not change their shape, or diminish in brightness.

"They passed with different velocities, some slowly, and others with great rapidity; and they were very various in size, some having a diameter, as nearly as I could estimate, about 2", while others were approaching to 20".

"I tried various powers upon them, and used both direct and diagonal eyepieces; but with every one I employed they showed the same appearance, being as sharply defined as the planet Jupiter, without haze or spot, or inequality of brightness.

"I naturally anticipated some such appearance at night, but after 1/2 past 3 I saw nothing peculiar, though I waited till 11 p. m.; but have since been informed that at 1/2 past 11 (it is believed on the same night) a meteor of amazing brilliance and size, and passing in the same direction and about the same altitude, was observed by Mr. Ballan of Wrotham Park, in the immediate neighbourhood of South Mimms.

"I repeated my observations the following morning, and then saw one such single body pass in the same direction as those of the preceding day.

"They occupied a tolerably well-defined zone of about 18$^{o}$ in breadth; and, though with some exceptions, their direction was due east and west. Their motion was perfectly uniform, so far as I was able to follow them with the instrument at liberty; and they were observed continuously by myself and members of my family, accustomed to the use of instruments, both by day and night.

"The telescope I employed on this occasion is one of 3-1/2 feet focal length, and 2-3/4 inches aperture, by Mr. Dollond, of faultless performance and mounted equatoreally by Mr. Jones of Charing Cross, the circles divided by Mr. Rothwell of London, and reading off to 5".

"I understand that a similar phaenomenon has been witnessed by Mr. Cooper of Markree Castle, County of Sligo, though I have not communicated with that gentleman on the subject; but I take the opportunity of subjoining a portion of the contents of a letter to me from Charles B. Chalmers, Esq., F. R. A. S., now residing at Jugon, Cotes du Nord, France.

"He thus writes:---'About the latter end of the year 1849, I witnessed a phaenomenon similar to that which you saw in September 1850, in every respect, excepting that I thought some of the bodies were elongated, though certainly the majority were globular; and their brightness appeared to me about equal to that of Venus, as seen at the same time.

"'I was then residing at Weston-Super-Mare, in Somersetshire; and the instrument with which I saw them was a 5-feet telescope, equatoreally mounted, in a fixed observatory.

"'I was engaged similarly to yourself in observing the planet Mercury; about 1/2 past 10 a. m. I was at first inclined to believe it must be the seed of some plants of the thistle nature floating in the air, but from my position that could not have been the case.

"'The wind on the day I observed the phaenomenon was very slight; but such as it was it came from the sea. The bodies all appeared sharply defined, no feathery appearances that I could detect; and I did not observe any difference in their brightness during the time I observed them. (Reports of the British Association, 235-237, 1852)

# DARTING GLEAMS OF LIGHT

## A CURIOUS VARIETY OF METEOR
**Denning, W. F.; *Popular Astronomy*, 22:404-405, 1914.**

In Monthly Notices of the R. A. S. for June 1885 I wrote a short paper on a special class of meteors which I had occasionally observed. They exhibit exceedingly swift motion, give an impression of extraordinary nearness and show no heads, trains or streaks. In fact they are mere gleams of white or bluish-white light, seemingly quite close, and very dissimilar to the ordinary meteors which are probably what their appearances suggest, viz., solid bodies in a state of combustion and dissipation. The special kind, of great velocity, to which I have alluded is somewhat rare and perhaps I have not seen more than one or two in every 500 meteors of normal type and they certainly suggest electrical origin.

Were it not for the fact that in recent years several ladies in the United Kingdom have taken up observing meteors with an enthusiasm equal to their ability I am afraid that this branch of science would have made little progress here on its practical side. I mention this matter because two of the excellent observers referred to have succeeded in recording specimens of the rather exceptional phenomena sometimes seen here.

On March 19, 1914 Mrs. Fiammetta Wilson recorded two of them as "appearing quite near, moving terribly rapid and suggesting little lines of light rushing close to one like a lighted wire." Miss A. Grace Cook has also witnessed the flight of several similar objects, viz., on 1913 September 13, November 1 and December 1 and 1914 February 1. She remarks, "This class of meteors is so different and distinct from ordinary ones that they are unmistakable when seen. They give one a little shock of almost fear, certainly surprise, at their seeming nearness and tremendous rapidity. They show no heads and I never saw one with either a tail or an after-streak."

In late years I have not been able to pursue meteoric work to the same extent as formerly but I have occasionally caught one of these transient flashes. The last one seen here was on 1913 August 23 near the star Polaris and the meteor had a duration of less than the tenth of a second. It looked like a mere gleam of light; not a burning missile.

It is a noteworthy point that I have witnessed a few of these very rapid meteors directed from the western sky and from a region where we should expect the apparent motion to be very slow.

Colonel Tupman saw one of these rapid objects in 1870 January 9 and he says of it that though it had a path of 31° its duration was only 0.1 second. "An instantaneous flash seemed to be in the air, quite near."

There are many varieties of meteor visible in the heavens and it is very desirable that observations of curious objects should be fully recorded whenever they are seen. (Popular Astronomy, 22:404-405, 1914)

# JOHN GLENN'S SPACE FIREFLIES

## IF YOU'RE SHOOK UP, YOU SHOULDN'T BE THERE
**Glenn, John; *Life,* 52:29, March 9 1962.**

The strangest sight of all came with the very first ray of sunrise as I was crossing the Pacific toward the U. S. I was checking the instrument panel and when I looked back out the window I thought for a minute that I must have tumbled upside-down and was looking up at a new field of stars. I checked my instruments to make sure I was right-side-up. Then I looked again. There, spread as far as I could see, were literally thousands of tiny luminous objects that glowed in the black sky like fireflies. I was riding slowly through them, and the sensation was like walking backwards through a pasture where someone had waved a wand and made all the fireflies stop right where they were and glow steadily. They were greenish yellow in color, and they appeared to be about six to 10 feet apart. I seemed to be passing through them at a speed of from three to five miles an hour. They were all around me, and those nearest the capsule would occasionally move across the window as if I had slightly interrupted their flow. On the next pass I turned the capsule around so that I was looking right into the flow, and though I could see far fewer of them in the light of the rising sun, they were still there. Watching them come toward me, I felt certain they were not caused by anything emanating from the capsule. I thought perhaps I'd stumbled into the lost batch of needles the Air Force had tried to set up in orbit for communications purposes. But I could think of no reason why needles should glow like fireflies, nor did they look like needles. As far as I know, the true identity of these particles is still a mystery. (Life, 52:29, March 9, 1962)

# CURIOUS MANIFESTATIONS OF ASCENDING ROCKETS

## UNIDENTIFIED PHENOMENON
**Browne, A. C., et al; *Marine Observer,* 34:181-183, 1964.**

m.v. British Oak. Teneriffe to Monrovia. Captain A. C. Browne. Observers, the Master and Mr. P. M. Edge, Chief Officer.

27th November 1963. A point of light of about 2nd magnitude with an elliptical glow of approx. $3^{o}$ diameter and concentric circles of light was observed through binoculars at 1925 GMT. It was first seen bearing $230^{o}$ at $18^{o}$ altitude and disappeared 4 min. later bearing $190^{o}$, altitude $8^{o}$. The sky was cloudless and the atmosphere clear.

Position of ship: $24^{o}$ 27'N, $17^{o}$ 14'W.

m.v. Ripon. Captain Smith. On passage to Freetown. Observers, the Master and Mr. G. W. Brown, Chief Officer.

27th November 1963. At 1926 GMT an illuminated body was observed bearing $270^{o}$, altitude $30^{o}$. It appeared at first to have a suffused glow around it, but as the object moved parallel with the ship's course, the glow assumed the definite form of a tight spiral of blue-white light. The spiral expanded to a maximum radius of about 5° with about 12 turns visible at one time when bearing 200°, altitude 20°. The size afterwards diminished until the body faded from sight bearing 155°, altitude 12°, at 1931. As the object moved in azimuth, it also appeared to be gyrating about a centre in an anticlockwise direction and to vary in brilliance. At its brightest the object had a brilliance less than Venus and greater than Altair; its track passed between these two bodies.

The whole phenomenon gave the impression of looking into a conically formed spring and was indeed a most sensational sight. We can only conjecture that it was an artificial satellite 'gone wrong' or passing through a cloud of meteoric dust.

The accompanying sketches show how the phenomenon appeared to the observers. There was a cloudless sky and bright moonlight at the time.

Position of vessel: 10° 5'N, 15° 59'W.

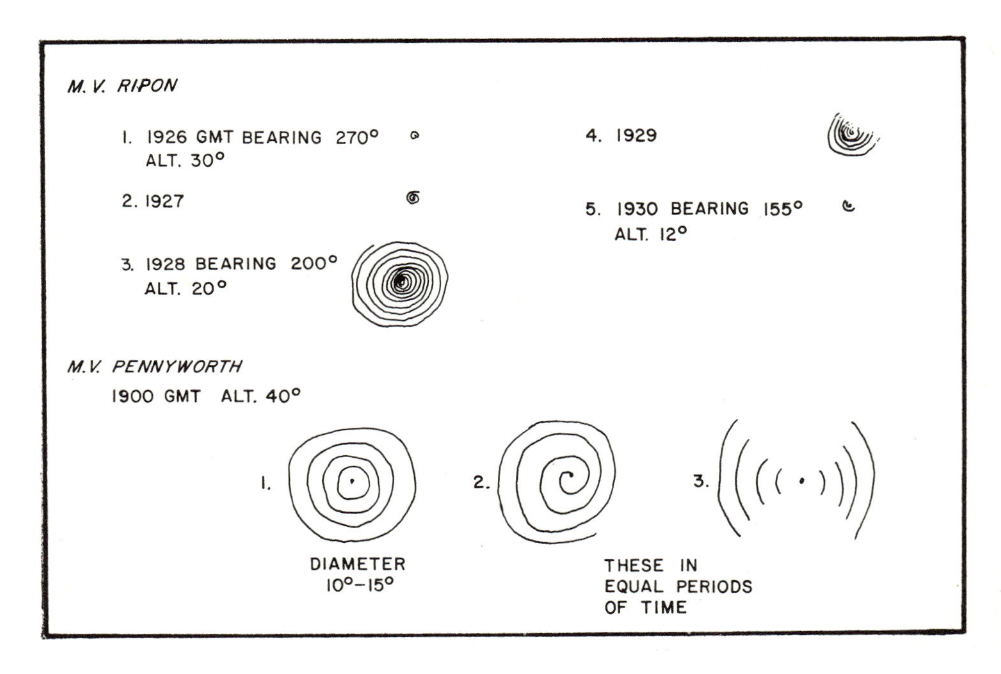


Spiral trails of ascending rockets seen at sea

m.v. Pennyworth. Captain I. Gault. Middlesbrough to Monrovia. Observers, Mr. J. H. Edwards, 2nd Radio Officer, the Master, Mr. J. Nielsen, Chief Officer, Mr. J. MacKenzie, 3rd Officer, Mr. T. Walker, 1st Radio Officer, and the Chief Engineer.

27th November 1963. At 1900 GMT for approx. 5 min. a bright object having a magnitude greater than any other star or planet was seen in the sky. It appeared to be stationary in the west at an elevation of 40°, for about 2 min. It then moved off rapidly in a SE'ly direction, disappearing about 2 min. later. The bright light from the object radiated outwards, like the ripples from a pebble thrown into a pond; at first in concentric circles, then in a spiral and finally in concentric half-circles. The general impressions of the phenomena seen are shown in the accompanying sketches. The object was definitely not a meteorite, and the course was too erratic for an earth satellite.

Position of ship: 7° 39'N, 14° 13'W.

Note 1. What was seen by the ships was undoubtedly an American rocket, Centaur 2, launched from Cape Kennedy at 1900 GMT on 27th November 1963. The times and positions indicated by the three ships agree very closely with calculated values. The odd appearance cannot be explained precisely but it is no doubt associated with the fact that the rocket when seen was still under power or had very recently been so. (Marine Observer, 34:181-183, 1964)

# SOME REPORTED STRANGE EFFECTS OF METEORS

Meteors reaching the ground are often associated with meteorological events, especially those thunderstorms in which the meteorite itself survives, giving rise to the "thunderstone" superstition. Despite the relegation of such tales to the nonsense category, the thunderstone phenomenon seems to persist, and we should inquire why. For further exploration of thunderstorm/meteorite coincidences, see the chapter on falling material.

Another old wive's tale relates how meteors sometimes swish or hum as they streak through the upper atmosphere 50 miles up. Obviously they are too high up for real sound to reach the ground so quickly. The possibility is now recognized that electromagnetic waves created by the meteor's passage, travelling at the speed of light, may be detected by some observers as sounds. In this sense, meteor sounds other than the usual long-delayed rumblings and detonations are in the same category as auroral sounds.

Terrestrial electrical effects due to meteors entering the atmosphere are rarely reported but do exist. No obvious mechanism exists for the production of electrical shocks on the ground unless, perhaps, the meteor is highly charged electrically, perhaps after the fashion of ball lightning. Once again, the phenomena of ball lightning, meteors, and electrical discharges overlap.

## METEORS ASSOCIATED WITH METEOROLOGICAL EVENTS

### METEOR AND WHIRLWIND
**Anonymous; *Nature*, 29:15, 1883.**

On Monday, September 24, about 9 p.m., a remarkable phenomenon occurred at Karingon, in the province of Bohus, Sweden. During a perfect calm a violent whirlwind suddenly arose from the south-east, carrying with it a quantity of sand, earth, and straw, when suddenly a bright light lit up every object and made the night as clear as day. This was caused by a magnificent meteor, egg-shaped in form, which appeared in the zenith, and which at first seemed to consist of myriads of large sparks, gradually changing into a star shining with a blinding lustre, and which burst, with all the colours of the rainbow, in the north-west, four to five metres above the horizon. When the meteor had disappeared the wind suddenly fell, and it was again perfectly calm. The phenomenon lasted about sixty seconds. The wind had throughout the day been south and very slight. (Nature, 29:15, 1883)

## METEORS AND STORMS IN INDIA
**Bruit, Dr.; *Reports of the British Association,* 239, 1852.**

"We have had three instances this season of what seems to have been the fall of an aerolite during thunder-storms. On the 25th of September a violent explosion occurred in the air at Bombay, followed by a wild rushing sound overhead, heard at various points over an area of thirty miles in length and eight in breadth, followed by a severe concussion, as if a heavy body had fallen, just before the occurrence of which a large fire-ball was seen plunging into the sea. On the 18th of March, during a violent thunder-storm near Dhutmah in the north-west provinces, at seven p. m., a thunderbolt, as it was called, was seen to fall and strike the ground, giving out in the course of the concussion a clear ringing sound like the crack of a rifle; there was no echo or reverberation at all like thunder. It appeared 150 yards from Choki, and resembled in its descent a huge ball of red-hot iron with a band of fire estimated at about thirty feet in length. On the 30th of April, about midnight, a violent explosion was heard during a storm of wind and rain at Kurrachee, resembling the discharge of a vast artillery battery, and about half a minute afterwards a meteor, partially obscured by the rain, but still distinct and visible, was seen descending into the sea. It is now well-established that in India at all events earthquakes are almost always accompanied by furious storms of thunder, lighting, wind and rain: it is difficult to trace the cause of coincidences so remarkable in the commotions of the earth and air, still more so to imagine any connection whatever betwixt the perturbations within the limits of our atmosphere and the movements of solid bodies entering it from regions beyond its boundaries; yet it is surely possible to suppose a thunderstorm propitious to the precipitation on the surface of the ground of bodies which might otherwise have passed on in their career." (Reports of the British Association, 1852, 239)

# ANOMALOUS SOUNDS FROM METEORS

## AN EXCEPTIONAL AND BRILLIANT METEOR
**Zoller, Harper F.; *Popular Astronomy,* 21:62, 1913.**

On Saturday evening, November 23, 7:35 p. m., while walking home from my daily tasks, I witnessed one of the most beautiful sights I have ever experienced. A large meteor of exceeding brilliancy lighted up the south-western portion of the sky. First seen from a point sixty degrees above the horizon, it described an arc of about seventy-five degrees in length, approached diagonally to within five degrees of the earth, when it suddenly flew into many, revolving, illuminated pieces. No apparent report accompanied the rupturing of the object; however, a dull swish was audible as it passed swiftly through the atmosphere. A major portion of the object, cleaved by the quickly produced heat and contact with the denser strata of atmosphere, was diverted from its original course and proceeded in an entirely new path. The trail left in the atmosphere much resembled a cissoid curve. (Popular Astronomy, 21:62, 1913)

## METEOR OF SEPTEMBER 15, 1902
**Moseley, E. L.; *Popular Astronomy,* 12:191-192, 1904.**

The meteor passed over eastern Ohio and southwestern Ontario at about 5:42 a.m. Washington time [September 15, 1902]. According to most observers it continued visible between ten and thirty seconds. It was egg-shaped or pear-shaped with the large end in front. To many it appeared to have about half the diameter of the full Moon, but was much brighter, giving probably several times as much light as the full Moon. The color was like that of an arc light, white or with a slightly bluish or possibly purplish tinge.

At Waverly in southern Ohio a rumbling sound was heard, but correspondents in southeastern Ohio do not report any sound. At many places in northern Ohio sounds were heard, but not very loud. At Defiance, from which the meteor when nearest was 175 miles distant, four observers interviewed by Dr. C. E. Slocum heard "a hissing noise." An observer near Sandusky heard "a slight hissing noise about as loud as a bee." M. F. Roberts, directly under the meteor at Mentor, heard "a rushing sound" that attracted his wife's attention.

In Michigan E. J. Smith and W. Kearns at Detroit heard "a loud sizzling noise." Near Port Huron an observer reported by C. K. Dodge heard "a great crackling and hissing, supposing at first it was his stove."

In Ontario, Andrew Smale, of Union, compared the noise to "that of an electric car running." The noise was heard by a number of persons in and near London, Ontario. J. B. McMurphy says of the sounds: "first like the swish of a falling tree, then changing to a noise similar to the striking of a parlor match on some hard surface with not quite sufficient force to ignite it but enough to make it snap. It was something like this---Bir-rup-bir-rup-bir-rup, then changing to a sound like distant cannon. There were three such sounds as those. All those sounds were as if they had been produced from an echo and reproduced several times, each time growing fainter." (Popular Astronomy, 12:191-192, 1904)

## COMPANION METEORS
**Hopkins, Warren; *American Meteorological Journal,* 6:33-34, 1890.**

In the year 1838 when a small boy only nine years old I saw what I now know were two falling aerolites. As 51 years have elapsed since that time I can not remember the exact day or even month, but only that it was sometime during the summer or early autumn of the year above named. It would seem strange if others did not see the same, yet I have no recollection of hearing them spoken of by others, nor of myself at the time telling any one what I saw. My parents were then living in the northwestern part of Monroe county, N. Y.; to be exact, about 1-1/2 miles east of the west line of the county and about the same distance south of the lake. Early in the evening, sometime between sundown and dark, I was alone in the yard on the south side of the house, when suddenly I heard a humming or whizzing noise as of some solid body moving with great velocity through the air. Quickly looking up in the direction from which the sound came I saw directly south of me and at an elevation of about 45 degrees two meteors, the larger one ahead and the smaller following directly in the larger one's path and so close that they apparently touched each other. Their apparent distance from me was about 125 or 150 feet, but I have no doubt their real distance was very much greater. They were, however, so near that I could clearly see they were solid bodies and could plainly hear the humming noise (it sounded much like, and about as loud as, the droning of a large bee) that they made in rushing through the atmosphere. They were moving with great velocity in a direction a very little north of west. In a very few seconds they had passed out of my sight. In regard to their magnitude, it is almost impossible to form any idea of even the apparent magnitude of a body at a considerable, but entirely unknown, distance, and at the same time moving with great velocity. But perhaps I can convey to the reader some idea of how large the meteors looked to me by saying that had they passed directly between my eye and the full moon, I think their entire length would have extended across the moon's disc. They were shaped alike and each was about one-third as thick as it was long. They were at, or nearly at, a white heat, and from the front end of the larger one sparks of fire were thickly flying that resembled the sparks from a large piece of iron heated to a welding heat. Fifty-one years have elapsed since that time and I have written from memory, but have been very careful to state nothing but what I distinctly remember. (American Meteorological Journal, 6:33-34, 1890)

# ELECTRICAL EFFECTS APPARENTLY CAUSED BY METEORS

## MASS OF FIRE FALLS INTO SEA
**Anonymous; *Science,* 5:242, 1885.**

The following account of unusual phenomena was received March 10, at the Hydrographic office, Washington, from the branch office in San Francisco. The bark Innerwich, Capt. Waters, has just arrived at Victoria from Yokohama. At midnight of Feb. 24, in latitude 37° north, longitude 170° 15' east, the captain was aroused by the mate, and went on deck to find the sky changing to a fiery red. All at once a large mass of fire appeared over the vessel, completely blinding the spectators; and, as it fell into the sea some fifty yards to leeward, it caused a hissing sound, which was heard above the blast, and made the vessel quiver from stem to stern. Hardly had this disappeared, when a lowering mass of white foam was seen rapidly approaching the vessel. The noise from the advancing volume of water is described as deafening. The bark was struck flat aback; but, before there was time to touch a brace, the sails had filled again, and the roaring white sea had passed ahead. To increase the horror of the situation, another 'vast sheet of flame' ran down the mizzen-mast, and 'poured in myriads of sparks' from the rigging. The strange redness of the sky remained for twenty minutes. The master, an old and experienced mariner, declares that the awfulness of the sight was beyond description, and considers that the ship had a narrow escape from destruction. (Science, 5:242, 1885)

## DESCRIPTION OF A NEW METEORIC IRON FROM CHILI...
**Greg, R. P.; *American Journal of Science,* 2:23:119-120, 1857.**

Having been deeply engaged in Argentine politics and wars in 1843 to 1844, I accompanied the Corrientine army in its invasion of the province of Entre Rios. This army returned from that expedition in January, 1844. Our rear, in which I marched, was so continually harrassed by Entrerian skirmishers, that for ten days before we had gained the Corrientine frontier we had no time to sleep or change clothes; but soon after passing this, in Carritas Paso, on the river Mocorita, we placed a guard in the pass, and deeming ourselves secure, the whole division abandoned itself to the profoundest sleep.

From this sleep we were all simultaneously awakened at about two o'clock in the morning; and as if actuated by electricity, each individual of our division (about 1400 men) sprung on his feet at the same moment. An aerolite was falling. The light that accompanied it was intense beyond description. It fell in an oblique direction, probably at an angle of about $60^{o}$ with the earth, and its course was from east to west.

Its appearance was that of an oblongated sphere of fire, and its tract from the sky was marked by a fiery streak, gradually fading in proportion to the distance from the mass, but as intensely luminous as itself in its immediate vicinity. The noise that accompanied it, though unlike thunder, or anything else that I have heard, was unbroken, exceedingly loud and terrific. Its fall was accompanied by a most sensible movement of the atmosphere, which I thought at first repellent from the falling body, and afterwards it became something of a short whirlwind. At the same time I and my companions all agreed that we had experienced a violent electric shock; but probably this sensation may have been but the effect on our drowsy senses of the indescribably intense light and noise. The spot where it fell was about one hundred yards from the extreme right of our division, and perhaps four hundred from the place where I had been sleeping. Accompanied by our general (Dr. Joaquin Madauaga), I went within ten or twelve yards from it, which was as near as its heat allowed us to approach.

The mass appeared to be considerably imbedded in the earth, which was so heated that it was quite bubbling around it. Its size above the earth was perhaps a cubic yard, and its shape was somewhat spherical; it was intensely ignited and radiantly light, and in this state it continued until early dawn, when the enemy, having brought his artillery to the pass, forced us to abandon it to continue our march. I may mention, that, at the time of its fall, the sky above us was beautifully clear, and the stars were perhaps more than usually bright; there had been sheet lightning the previous evening. (American Journal of Science, 2:23:118-120, 1857)

# POSSIBLE EFFECT OF THE MOON ON METEORS

## A LUNAR EFFECT ON THE INCOMING METEOR RATES
**Bowen, E. G.; *Journal of Geophysical Research,* 68:1401, 1963.**

Following the recent discovery of a lunar influence on rainfall and on the freezing nucleus count, it is now shown that the radar meteor rate varies in a similar way with lunar phase. The physical mechanism is uncertain, but an electrostatic charge on the moon could produce an effect on micrometeorites of the right order of magnitude. (Journal of Geophysical Research, 68: 1401, 1963)

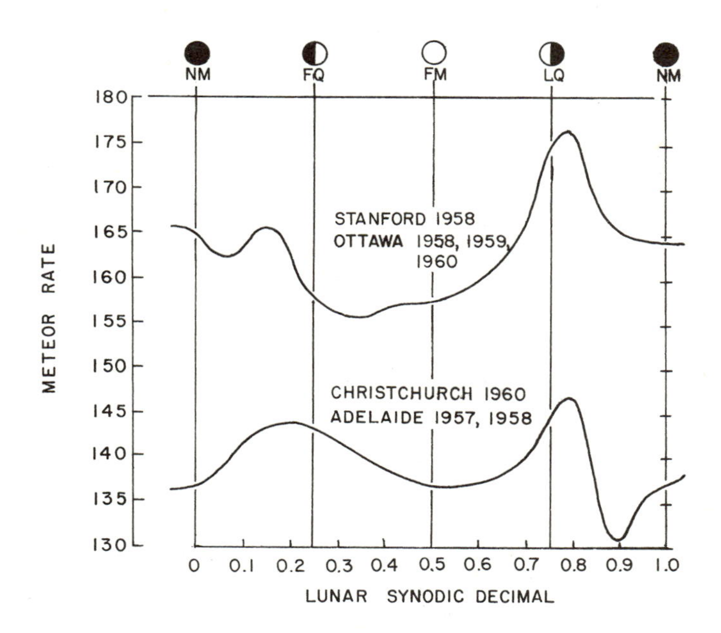


Radar meteor rates

# NOCTURNAL LIGHTS

Nocturnal lights are not associated in any obvious way with meteors, electric storms, or other weather activity. The name "nocturnal light" had its origin in the UFO literature, but many nocturnal lights, especially will o' the wisps, probably have natural origins. The characteristics of the class are: (1) small size (a few feet in diameter on the average); (2) strange movements that might be categorized as "playful"; and (3) their soft, eerie light. Of all the luminous phenomena reported in this chapter, nocturnal lights are most closely related to psychic phenomena and, in some instances, may have no objective existence.

Nocturnal lights are divided into three groups by virtue of the circumstances of their appearance:

| | |
|---|---|
| Will o' the wisps | Usually associated with swamps and decaying organic matter |
| Ghost lights | Reappearing repeatedly in the same place, acquiring local names as a consequence |
| Psychic lights | Said to occur in connection with religious revivals and other intense emotional activity. |
| Moving and flashing lights | Anomalous lights appearing more or less at random. Often called UFOs. |

## WILL O' THE WISPS

### THE "IGNIS FATUUS," OR WILL O' THE WISP
**Anonymous; *Royal Meteorological Magazine, Quarterly Journal,* 17:260, 1891.**

For some years now, since I have been living at the Warren, Crowborough, in Sussex, I have, in common with many others, been struck by seeing Will o' the Wisp, or Ignis Fatuus, playing about in the evening at various times, sometimes in one place, sometimes in another, and sometimes two at once, but not in the same place. Now I have observed that these appearances always coincide with unsettled weather, that is, that bad weather has invariably followed their appearance ---in fact, it is quite a storm warning to us, as bad weather is sure to follow sooner or later, generally sooner.

On Saturday, September 26th, 1891, one of my brothers and myself watched one for some time, and last night (October 1st) it appeared in the same place and was visible quite half-an-hour. It does not dance about, but now and then takes a graceful sweep, now to quite a height, and then makes a gentle curve downwards, after sparkling and scintillating away for ten minutes or more.

In the winter there is generally a very remarkable one which appears to rise in Ashdown Forest beyond this estate, which I have known to keep steadily in the air for half-an-hour, and then sail away a long space and stop again. Our keeper has seen it dozens of times, and in every case bad weather followed. (Royal Meteorological Society, Quarterly Journal, 17:260, 1891)

## WILL-WI'-THE WISP
**Elder, David; *Symons's Monthly Meteorological Magazine,* 15:156, 1880.**

I think you said in the Magazine some time ago that you had never met with, in the flesh, the man who had seen the ignis fatuus, or Will-wi'-the-Wisp. I beg to enclose a paragraph which I cut out of the Fife News newspaper some weeks since, and you will see from it that it has been seen there. I can also tell you that in 1832, when a boy herding cows on the farm of Crosshouses, near Kingskettle, Fifeshire, that I had frequent opportunities of seeing Will-wi'-the-Wisp, or "Spunkie," as it is called in the old Doric.

On the farm of Crosshouses was a bog, about 300 yards from the door of the dwelling-house, and it was in the bog---or rather over the bog---that I saw the light moving about. The best description I can give of it is that it resembled that of a lantern being carried about by a person for some twenty or thirty yards. Sometimes I had seen it move zig-zagly about these distances, and sometimes in a straight line, and always at the same rate of speed---that of a person's usual speed when walking.

My grandfather frequently saw it. Sometimes, when coming in from suppering the horses, he would cry in haste to come out and see "Spunkie" in Pilkim Moss, when I, and the others at the fireside sitting, would run to the door to see it. When all of us were either less or more afraid of the wandering light, many times I have gone to bed at once---after prayers though---and buried my head among the bed-clothes that "Spunkie" might not get hold of me; for every person regarded the phenomenon with dread.

I saw it also in a bog called Powglass, on the farm of Dawnfield, close by Crosshouses, several times in the same year, 1832. I have lived since, as a gardener, in four counties in England, and four in Scotland, but never saw "Spunkie" in either of them. I must say, however, that no bots---or rather quagmires, for Powglass bog partook more of the quagmire---were in close enough proximity to where I lived in those eight counties, to give me a chance of seeing Will-wi'-the-Wisp again.

Powglass bog and Pilkim bog have been both drained, and "Spunkie" is no longer seen there. (Symons's Monthly Meteorological Magazine, 15:156, 1880)

## OBSERVATIONS ON IGNIS FATUUS
**Mitchell, John; *American Journal of Science,* 1:16:246-249, 1829.**

Those luminous appearances, which are popularly called "Will-o'-the-wisp" and "Jack-a-lantern," have been alike the object of vulgar superstition and philosophical curiosity; and notwithstanding all attempts to apprehend and subject them to examination, they are not much more the subjects of knowledge now than they were centuries ago. They are still but an ignis fatuus to the philosopher, and a thing of mystery to the credulous.

I was myself, formerly, familiar with these appearances; they were of frequent occurrence near my father's residence, owing, probably, to the proximity of extensive wet grounds, over which they are usually seen. The house stood upon a ridge, which sloped down on three sides to the beautiful meadows which form the margin of the Connecticut, and of its tributary creeks, and which, owing to their own luxuriance and the deposits of the vernal freshets, are covered with rich and constantly decaying vegetable matter. From the circumstance, also, that we had no neighbors in the direction of these grounds, a light could not be seen over them without attracting our notice. I mention this by way of suggesting, that probably the ignis fatuus, in consequence of its not being always distinguished from the lights of surrounding houses, and therefore exciting no curiosity, is oftener seen than it is supposed to be.

These mysterious luminaries used often to be seen by the fishermen; who plied their nets by night as well as by day. They commonly reported that they saw them a little above the surface of the meadow, dancing up and down, or gliding quietly along in a horizontal line. Sometimes two, or even three, would be seen together, skipping and dancing or sailing away in concert, as if rejoicing in their mutual companionship. I might entertain you with abundance of fabulous accounts of them---the offspring of imaginations tinctured with superstition, and of minds credulous from a natural love of the marvellous. Fables, however, are of little value for the purposes of science: if the following account of some of the phenomena of the ignis fatuus, shall, with the observations of others, contribute towards a true theory of its nature, you will think them worthy of a place in your Journal.

A friend of mine, returning from abroad late in the evening, had to cross a strip of marsh. As he approached the causeway, he noticed a light towards the opposite end, which he supposed to be a lantern in the hand of some person whom he was about to meet. It proved, however, to be a solitary flame, a few inches above the marsh, at the distance of a few feet from the edge of the causeway. He stopped some time to look at it; and was strongly tempted, notwithstanding the miriness of the place, to get nearer to it, for the purpose of closer examination. It was evidently a vapor, [phosphuretted hydrogen?] issuing from the mud, and becoming ignited, or at least luminous, in contact with the air. It exhibited a flickering appearance, like that of a candle expiring in its socket; alternately burning with a large flame and then sinking to a small taper; and occasionally, for a moment, becoming quite extinct. It constantly appeared over the same spot.

Typical will o' the wisp

With the phenomena exhibited in this instance, I have been accustomed to compare those exhibited in other instances, whether observed by myself or others; and generally, making due allowance for the illusion of the senses and the credulity of the imagination in a dark and misty night, (for it is on such nights that they usually appear,) I have found these phenomena sufficient for the explanation of all the fantastic tricks which are reported of these phantoms.

They are supposed to be endowed with a locomotive power. They appear to recede from the spectator, or to advance towards him. But this may be explained without locomotion---by their variation in respect to quantity of flame. As the light dwindles away, it will seem to move from you, and with a velocity proportioned to the rapidity of its diminution. Again as it grows larger, it will appear to approach you. If it expires, by several flickerings or flashes, it will seem to skip from you, and when it reappears you will easily imagine that it has assumed a new position. This reasoning accounts for their apparent motion, either to or from the spectator; and I never could ascertain that they moved in any other direction, that is, in a line oblique or perpendicular to that in which they first appeared. In one instance, indeed, I thought this was the fact, and what struck me as more singular, the light appeared to move, with great rapidity, directly against a very strong wind. But after looking some time, I reflected that I had not changed the direction of my eye at all, whereas if the apparent motion had been real, I ought to have turned half round. The deception was occasioned by the motion of the wind itself---as a stake standing in a rapid stream will appear to move against the current.

It is a common notion that the ignis fatuus cannot be approached, but will move off as rapidly as you advance. This characteristic is mentioned in the Edinburgh Encyclopoedia. It is doubtless a mistake. Persons attempting to approach them, have been deceived perhaps as to their distance, and finding them farther off than they imagined, have proceeded a little way and given over, under the impression that pursuit was vain. An acquaintance of mine, a plain man, told me he actually stole up close to one, and caught it in his hat, as he thought;---"and what was it?" I asked. "It was'nt nothin." On looking into his hat for the "shining jelly," it had wholly disappeared. His motions had dissipated the vapor, or perhaps his foot had closed the orifice from which it issued. To this instance another may be added. A young man and woman, walking home from an evening visit, approached a light which they took for a lantern carried by some neighbor, but which on actually passing it, they found to be borne by no visible being; and taking themselves to flight, burst into the nearest house, with such precipitation as to overturn the furniture, and impart no small share of their fright to the family.

The circumstance that these lights usually appear over marshy grounds, explains another popular notion respecting them; namely, that they possess the power of beguiling persons into swamps and fens. To this superstition Parnell alludes in his Fairy Tale, in which he makes Will-o'-the-wisp one of his dancing fairies;

> "Then Will who bears the wispy fire,
> To trail the swains among the mire," &c.

In a misty night, they are easily mistaken for the light of a neighboring house, and the deceived traveller, directing his course towards it, meets with fences, ditches, and other obstacles, and by perseverance, lands at length, quite bewildered, in the swamp itself. By this time, he perceives that the false lamp is only a mischievous jack-a-lantern. An adventure of this kind I remember to have occurred in my own neighbourhood. A man left his neighbor's house late in the evening, and at daylight had not reached his own, a quarter of a mile distant; at which his family being concerned, a number of persons went out to search for him. We found him near a swamp, with soiled clothes and a thoughtful countenance, reclining by a fence. The account he gave was, that he had been led into the swamp by a jack-a-lantern. His story was no doubt true, and yet had little of the marvellous in it. The night being dark, and the man's senses a little disordered withal, by a glass too much of his neighbor's sherry, on approaching his house, he saw a light, and not suspecting that it was not upon his own mantel, made towards it. A bush or a bog, might have led to the same place, if he had happened to take it for his chimney top. (American Journal of Science, 1:10:246-249, 1829)

## A BLAZING BEACH
**Penhallow, D. P.; *Science*, 22:794-796, 1905.**

In the early part of September the papers throughout the country gave wide publicity to the occurrence of a phenomenon at Kittery Point, Me., which attracted much local consideration because of its sensational aspects, and which might be correctly described as a 'blazing beach.' On the evening of Friday, September 1, the guests at the Hotel Parkfield were startled by the appearance of flames rising from the beach and from the surface of the water, an event of so remarkable and unusual a character as to excite great curiosity and some alarm. The conflagration occurred between seven and eight o'clock in the evening, and lasted for upwards of forty-five minutes. The flames were about one foot in neight. They were accompanied by a loud and continuous crackling noise which could be distinctly heard one hundred yards away, while at the same time there was a very strong liberation of sulphurous acid fumes which penetrated the hotel, drove the proprietor and his staff from the office and filled the other rooms to such an extent as to cause great inconvenience to the guests. One guest of an investigating turn of mind secured some ot the sand in his hand, but was obliged to drop it on account of the heat. When some of the sand was taken into the hotel and stirred in water, bubbles of gas were liberated and produced flame as they broke at the surface in contact with the air.

Some of the attempts at explanation were of a remarkable character and illustrate how far one may be carried when the imagination is not controlled by an adequate knowledge of facts. One observer stated that some vessel in the harbor had thrown overboard a quantity of calcium carbide which had washed ashore and caught fire. The most popular explanations referred the phenomenon to the effects of the blast at Henderson's Point, some six weeks before, the theory being that the explosion of fifty tons of dynamite had opened up rock fissures to such an extent as to liberate volcanic gases; while a somewhat similar theory attributed it to the earthquake of the day before, and the consequent opening up of rock fissures. With respect to the latter it may suffice to state that the earthquake may have been a contributary factor in so far as it served to give just that shaking at a critical moment which would suffice to liberate gases stored under slight pressure. The most sensational explanation was that of a resident of the town who refused to accept the explanation I offered as altogether too commonplace, and who had 'always told the people of Kittery Point that the town was built on the edge of hell, the proof of which had now been given.'

Divesting the phenomenon of its sensational aspects, it was not difficult to reach a satisfactory explanation of all the features presented, and to eliminate explanations which had some semblance of reasonableness. The beach at the point where the fire occurred is composed of a beach ridge at its upper margin, made up of pebbles of varying size. From this ridge, a somewhat sharp slope continues the same formation outward and downward for perhaps seventy-five feet, where the pebbles are replaced by sand. This latter begins at about the half-tide mark and extends outward with a very gentle slope beyond lowwater mark, so that during even the lowest tides a portion is covered by very shallow water. This sand beach extends laterally for a distance of about 175 to 200 feet, being limited in each direction by solid ledge, which forms the general construction of the shore all along the river. Over the outer portion of the sand, as also for great distances beyond, wherever there is a muddy bottom, there is an abundant growth of eel grass (Zostera marina) which, together with other debris of a similar nature, is continually washed upon the beach, broken up by the action of the waves and gradually buried, so that each year the deposit is increased by definite though rather slight increments. One of the well-defined features of the fire was, that it was limited to that area which is occupied by the sand. It occurred over that portion of the sand which was exposed by the falling tide, but it was also observed to extend out over the water for a distance of thirty or forty feet. Gas was found to be liberated from the exposed sand when stirred in water, and similarly gas was seen to rise from that portion of the sand covered with water. Such facts showed conclusively that the evolution of the gas was immediately connected with the sand itself and not with the adjacent rock formation, hence the theory that rock fissures had been opened could not be regarded as resting upon a valid basis.

Observation has shown that in the salt marsh lands of the coast the underlying portions of the sod are continually undergoing decay with the formation of large quantities of sulphuretted hydrogen, with which there must also be associated certain amounts of the light carburetted and possibly also of the phosphuretted hydrogen. Personal experience has shown that such gases are stored in the decaying turf in large quantities, being often held in pockets, so that when the turf is cut they may escape in such volume as to drive one away for the time. It is also known that any decaying vegetation will produce similar results, and two explanations were, therefore, suggested as offering a solution of the problem: (1) that there was an area of buried marsh such as is known to exist in places along the coast, and that its decay had given rise to combustible gases; (2) that the accumulations of organic debris in the formation of the beach had been productive of the results observed.

That one or both of these causes would offer an adequate explanation was adopted as a tentative hypothesis, and an examination of the beach was proceeded with. It was found that the superficial layer to a depth of about one inch, consisted of freshly washed sand with which there were mingled fragments of marine plants and even fragments of land plants. Successive accumulations are thus transferred from the superficial layer to that below, which was found to be about six inches in thickness, and to consist of sand filled with all sorts of organic debris, including marine plants, fragments of wood and bones. Moreover, this layer was perfectly black, and when washed it exhibited very small, carbonized fragments of zostera and other marine plants, fragments of wood with a distinct surface charring, and bones of animals, the surface of which was like ebony. Below this layer there was a deposit of beach pebbles mingled with sand, and as this formation continued to the limits which it was possible to reach with the implements at hand---about two feet---the conclusion was reached that such was the lower construction of the beach and that no buried marsh was present. This naturally led to the final conclusion that the six-inch layer, rich in organic matter, was entirely responsible for the production of inflammable gases which had been accumulated there until favorable conditions for their release were presented.

An explanation of the spontaneous combustion of these gases is not difficult. The light carburetted and the phosphuretted hydrogen are well known to ignite spontaneously wherever produced in marsh lands, thus giving rise to the well-known 'will-o'-the-wisp,' 'Jack-o'-lantern' and the ignis fatuus, 'corpse candle,' etc., which are well known to the folk-lore of England. That sulphuretted hydrogen was also present has been abundantly shown, and since this would naturally be set on fire by the other gases, it is possible to reach a complete explanation of a phenomenon which must have occurred at more or less frequent intervals in the past, though escaping observation through lack of combination in those circumstances which would bring it under direct notice. It would seem, however, that the possibility of such combustion on a rather large scale offers a most reasonable explanation of many forest fires, the origin of which it has hitherto been impossible to account for in a satisfactory manner. (Science, 22:794-796, 1905)

# GHOST LIGHTS

## AN ATMOSPHERIC PHENOMENON IN THE NORTH CHINA SEA
**Norcock, Chas. J.; *Nature,* 48:76-77, 1893.**

During a recent wintry cruise in H. M. S. Caroline in the North China Sea, a curious phenomenon was seen which may be of interest to your readers. The ship was on passage between Shanghai and the western entrance of the famous inland sea of Japan. On 24th February, at 10 p. m., when in latitude 32° 58'N., longitude 126° 33'E., which, on reference to the map, will be seen to be sixteen to seventeen miles south of Quelpart island (south of the Korean peninsula) some unusual lights were reported by the officer of the watch between the ship and Mount Auckland, a mountain 6,000 feet high. It was a windy, cold, moonlight night. My first impres-

Nocturnal lights on the North China Sea

sion was that they were either some fires on shore, apparently higher from the horizon than a ship's masthead, or some junk's "flare up" lights raised by mirage. To the naked eye they appeared sometimes as a mass; at others, spread out in an irregular line, and, being globular in form, they resembled Chinese lanterns festooned between the masts of a lofty vessel. They bore north (magnetic), and remained on that bearing until lost sight of about midnight. As the ship was passing the land to the eastward at the rate of seven knots an hour, it seen became obvious that the lights were not on the land, though observed with the mountain behind them.

On the following night, February 25th, about the same time, 10 p. m., the ship having cleared Port Hamilton, was steering east, on the parallel of 34°, when these curious lights were again observed on the same bearing, at an altitude of 3° or 4° above the horizon. It was a clear, still, moonlight night, and cold. On this occasion there was no land in sight on a north bearing when the lights were first observed, but soon afterwards a small islet was passed, which for the time eclipsed the lights. As the ship steamed on at a rate of seven knots an hour, the lights maintained a constant bearing (magnetic) of N. 2°W., as if carried by some vessel travelling in the same direction and at the same speed. The globes of fire altered in their formation as on the previous night, now in a massed group, with an outlying light away to the right, then the isolated one would disappear, and the others would take the form of a crescent or diamond, or hang festoon-fashion in a curved line. A clear reflection or glare could be seen on the horizon beneath the lights. Through a telescope the globes appeared to be of a reddish colour, and to emit a thin smoke.

I watched them for several hours, and could distinguish no perceptible alteration in their bearing or altitude, the changes occurring only in their relative formation, but each light maintained its oval, globular form.

They remained in sight from 10 p. m. until daylight (about 5. 30 a. m.). When lost sight of the bearing was one or two points to the westward of north. At daylight land 1300 feet high was seen to the north and north-north-west, distant fifty miles, the mirage being extraordinary.

Thus, these lights were seen first in longitude 126° 33'E., and last in longitude 128° 29'E. At first the land was behind them, but during the greater part of the distance run it was forty-five or fifty miles away to the north; and the bearing of the lights for at least three-fourths of the distance did not change.

On arrival at Kobe I read in a daily paper that the "Unknown light of Japan" had, as was customary at this season of the year when the weather is very cold, stormy, and clear, been observed by fishermen in the Shimbara Gulf and Japanese waters. The article went on to say that these lights were referred to in native school-books, and attributed to electrical phenomena. On mentioning the matter, however, to the leading Europeans in Yokohama and Tokio, they appeared to have no knowledge of the matter.

Captain Castle, of H. M. S. Leander, informed me that, not long ago, the officers of his ship saw lights in the same locality which they thought at first were caused by a ship on fire. The course of the vessel was altered at once with a view of rendering assistance, but finding that the lights increased their altitude as he approached, he attributed them to some volcanic disturbance, and being pressed for time, resumed his course.

The background of high land seen on the first night dispels all idea of these extraordinary lights being due to a distant volcano. The uniformity of the bearing renders the theory of their being fires on the shore most improbable. I am inclined to the belief that they were something in the nature of St. Elmo's fires. It is probable that there are travellers among the readers of your interesting journal who have seen or heard of this phenomenon, and will be able to describe its origin and the atmospheric conditions necessary for its appearance. (Nature, 48:76-77, 1893)

## THE QUEER LIGHTS ON BROWN MOUNTAIN
**Anonymous; *Literary Digest*, 87:48-49, 1925.**

What would you think or say, if you were standing on Rattlesnake Knob, on the Morganton, North Carolina, road about 7:30 p. m. and saw in a southeasterly direction a curious light, about the size of a toy balloon, smaller than the full moon, and very red, rise mysteriously over Brown Mountain, proceed into the air a short distance, waver as if it were palsied, and then in less than a minute disappear?

The lights, however, do not appear regularly at the same spot. Brown Mountain has a table-shaped top at 2,600 feet elevation. Small tributaries of the John River flowing south make many indentations in its surface. Geologists say that there is nothing unusual in the formation of Brown Mountain---that it is simply a pile of Cranberry granite.

In 1913, at the request of one of North Carolina's representatives in Congress, the United States Geological Survey sent a man down there to study the origin of the strange lights. He did not tarry long before he branded them as coming from the headlights of locomotives flashing up over the mountain. The people laughed at his simple explanation. They continued to let the strange lights amuse and entertain them, and work upon their imaginations.

The descriptions of the strange light made by various observers do not agree. One person says that it is pale white, as is ordinarily observed through a ground-glass globe, with a faint, irregular halo encircling it. He claims that it is restricted to a prescribed circle, and appears from three to four times in rapid succession, then conceals itself for twenty minutes, when it reappears within the same circle. Another observer, who was standing about eight miles from Brown Mountain, says that suddenly after sunset there blazed into the sky above the mountain a steady glowing ball of light. To him, the light appeared yellowish, and it lasted about half a minute, when it disappeared rather abruptly. It appeared to him like a star from a bursting skyrocket, but much brighter.

To some people it appears stationary; to others, it moves sometimes upward, downward, or horizontally. A minister says that it appeared like a ball of incandescent light in which he could observe a seething motion.

Some persons have suggested the will-o'-the-wisp as an explanation, but since Brown Mountain bears the reputation of owning no marshy ground, it could hardly be such a phenomenon. Then some one said it might be phosphorus, but phosphorus is oxidized so readily that it is never found in a free state. Of course a number of people thought it might be fox-fire, but fox-fire makes such a very weak, pale light that their guesses were obviously erroneous.

..........

And of course, some person mentioned 'St. Elmo's fire'---an electrical discharge from sharp objects during a thunder-storm. St. Elmo's fire is brushlike in appearance, and since the strange lights could not be it, the amateur scientist's next assertion was that probably the phenomenon known as 'Andes light' was the guilty chap! The Weather Bureau says that the Andes light is a striking luminous discharge of electricity observed over the crest of the Andes in Chile, where thunder-storms are almost unknown. Silent discharges of electricity take place between the peaks and the clouds. The Andes light is said often to produce glimmering lights with circular borders, sometimes observed for more than 300 miles.

The citizens of the mountainous part of southwestern Virginia, where the ridges run parallel with each other, separated by deep, narrow valleys, report that they sometimes observe a noiseless electrical discharge from one summit to the air. However, the contour of Brown Mountain is said to be unfavorable for any phenomenon like Andes light.

..........

The first geologist who visited Brown Mountain in 1913, after checking up the train schedule, and watching the lights, let the schedule and lights convince him, Mr. Walker tells us, that the headlights from locomotives were responsible for it all, and in short order left the mountain. Objections followed, for the simple reason that headlights make a beam in the air like a search-light. The suggestion of automobile lights was objected to for the same reason. In 1922, when the lights of Brown Mountain had persisted in arousing the curiosity of the people, another representative of the people asked the Department to send a second geologist to make a more complete investigation. This was made in the same year. Says the writer:

..........

With maps spread out on the table, and directions properly noted and landmarks sighted and marked, the investigator proceeded to make observations. When the observation of the lights was made, the time of arrival and disappearance of the strange lights was checked with the train schedule, and after remaining there a fortnight, the investigator came to the conclusion that he had observed a fair representation of the 'strange lights' of Brown Mountain, and that 47 per cent of them originated from automobile headlights; 33 per cent from locomotive headlights; 10 per cent from fixt lights, and 10 per cent brush fires.

It was agreed that while the queer lights appeared above Brown Mountain, none of them originated there, but in a broad valley several miles beyond.

The atmospheric conditions in the valley beyond are said to be ideal for juggling distant lights that penetrate the atmosphere. The air varies in density, which makes it very refractive. The dust and mist are credited with giving the lights the various colors and tints.

So much for the conclusion of the geological investigator. Last year when I heard that intelligent people living in that part of North Carolina refused to accept the report of the investigator, I sent a letter of inquiry to a few prominent business men there, those who had been observing the lights for many years. One of these men exprest the opinion that the geologist had tried hard to find out the source of the strange lights, but had failed. That the lights had been observed for the last sixty years, and long before railroads were constructed in that part of the country, and before the advent of the automobile; that the flood of 1916 put the railroads out of business for a solid week, as well as the automobiles, yet the strange lights continued to put in their regular appearance. As conclusive proof that the lights do not originate from headlights, he states that the mysterious lights fail to appear following a long drouth; that after the report was made, he went to a certain site, turned his back to the railroad and highway, and witnessed the strange light suddenly come up from the mountain-top, and keep moving until it was lost in the sky. (Literary Digest, 87:48-49, 1925)

## DANCING LIGHTS OF ADA, OKLA.
## Fuller, Curtis; *Fate,* 15:16-18, November 1962.

Recently crowds of folks have been lining up at the Busby Ranch near Ada to watch the fox fire dance. Finally John Bennett of the Ada Evening News went out to watch the goings on. Here is part of his report.

"...and there it was. It looked orange, like light filtering through trees from a window of a house. 'That's just a house light,' I scoffed. The boys and girls (watching the nightly show) answered in a chorus that if that was a house light, it was the first one they ever had seen that danced and changed colors. And, by golly, it did.

It started glowing bigger and bigger and giving off a diffused orange, then red, then yellow light. For a full five minutes the light glowed like a dying ember. Then things started happening.

"The lights began to dance. They flickered eerily up and down like a bouncing luminous ball, then darted sideways. The single ball of light appeared about three feet in diameter.... During its fantastic flight back and forth it changed colors: first orange, then yellow, then red. But it stayed in one general area, behind what looked like a sparse growth of trees....

"In the tree line the light had changed colors again and was beginning to get more active. Suddenly a piece of the glow broke away and started a rapid bouncy course across the field in front of us. It looked like a luminous basketball, and about the same size. It danced before our eyes about 100 yards out front. We traced the glowing trajectory that appeared like a giant lightning bug, until it went out....

"The big show ended there. Shortly after the field orbit by the satellite the glow resumed but it didn't dance around much. In about an hour I left. My suspicion is that it is a hoax but I can't explain the dancing lights. It would take a clever manipulation of light. And it's beyond me how anyone could make the darn things go so fast."

During his visit Bennett talked with one of the boys, Ronnie Black, who said that a week before a piece of a ball of fire had bounded away from the main body and came right up to the fence.

"It just bounced across the field and came right up a few feet from me," Ronnie said. "I didn't move and it was like it was looking right at me. It wasn't any higher than the top strand of the fence here and you could see the tops of the weeds under its glow. Then it darted off fast as lightning." (Fate, 15:16-18, 1962)

## PHANTOM LIGHTS AT SEA
**Anonymous; *Scientific American,* 47:56, 1882.**

A Fulton Market fish dealer gives the following explanation of some of the strange lights, phantom vessels, and other mysterious appearances that puzzle seamen:

"Two years ago I went menhaden fishing, and one day as we were going up the Sound one of the hands said he hoped we were not going off the Point, meaning Montauk. I asked him why. He seemed kind of offish, but at last let out that he had seen ships sailing about in the dead of night in a dead calm. I laughed at him, but two nights later we came to anchor at Gardiner's Bay, and as it was a hot night we stretched out on deck. In the middle of the night I was awakened by some one giving me a tremendous jerk, and when I found myself on my feet my mate, shaking like a leaf, was pointing over the rail. I looked, and, sure enough, there was a big schooner about an eighth of a mile away, bearing down on us. There wasn't a breath of wind in the bay, but on she came at a ten-knot rate, headed right for us. 'Sing out to the skipper,' I said, 'It's no use,' said my mate, hanging on to me, 'It's no vessel.' But there she was, within a hundred yards of us. Shaking him off, I swung into the rigging, and yelled 'Schooner ahoy!' and shouted to her to bear away, but in a second the white sails were right aboard of us. I yelled to the hands, and made ready to jump, when, like a flash, she disappeared, and the skipper came on deck with all hands and wanted to know if we had the jimjams. I'd have sworn that I had seen the Flying Dutchman but for one thing. We saw the same thing about a week afterward. The light passed around us and went up the bay. I got out the men and seine and followed in the path of the phantom schooner, and as sure as you are alive, we made the biggest single haul of menhaden on record. The light, to my mind, was nothing more or less than the phosphorescence that hovered over the big shoal. The oil from so many millions of fish moving along was enough to produce a light; but you will find men all along the shores of Long Island that believe there is a regular phantom craft that comes in on and off---sort of a coaster in the spirit trade. I saw an account of something like this in the Portland papers some time after, and they thought it was very remarkable; but wherever you find menhaden you may look out for queer lights on the water---phantom ships and the like." (Scientific American, 47:56, 1882)

## THE FIRE-SHIP OF BAY CHALEUR
**Anonymous; *Science,* 24:501, 1906.**

In his 'Notes on the Natural History and Physiography of New Brunswick' (Bull. Nat. Hist. Soc. New Brunswick, xxiv, Vol. V., 1905) Professor W. F. Ganong has a short paper, 'On the Fact Basis of the Fire (or Phantom) Ship of Bay Chaleur.' After an examination of all the evidence it appears to the author plain (1) that a physical light is frequently seen over the waters of Bay Chaleur and its vicinity; (2) that it occurs at all seasons, or at least in winter and summer; (3) that it usually precedes a storm; (4) "that its usual form is roughly hemispherical with the flat side to the water, and that at times it simply glows without much change of form, but that at other times it rises into slender moving columns, giving rise to an appearance capable of interpretation as the flaming rigging of a ship, its vibrating and dancing movements increasing the illusion." This is doubtless a manifestation of St. Elmo's Fire, but the compiler of these notes is not aware of any reports of similar phenomena, of such frequency in one locality, and of such considerable development. Professor Ganong cites the case of some lights reported around Tremadoc Bay in Wales, but notes that they in all probability had only a subjective basis. Lights of unexplained origin, the author notes, were reported as common off the Welsh coast two hundred years ago, and mention is made of St. Elmo's Fire observed at Anticosti. The phenomenon described by Professor Ganong is an interesting one, well worthy of careful study. (Science, 24:501, 1906)

# PSYCHIC LIGHTS

## PSYCHOLOGICAL ASPECTS OF THE WELSH REVIVAL
**Fryer, A. T.; *Society for Psychical Research, Proceedings,* 19:148-149, 1905.**

[Wales, during the great religious revival of 1905] With regard to the lights which appear in this neighbourhood, perhaps one instance of my experience will suffice. I have been an eye-witness of them on more than one occasion. I happened to be with Mr. Beriah Evans, Carnarvon, on that night, the report of which has been given to the world by Mr. Evans himself. I can testify to the truth of the report.

The night which I am going to relate you my experience was Saturday evening, March 25, 1905, when Mrs. Jones, the evangelist, of Egryn, was conducting a service in the Calvinistic Methodist Chapel at Llanfair, a place about a mile and half from Harlech on the main road between Harlech and Barmouth.

My wife and myself went down that night specially to see if the light accompanied Mrs. Jones from outside Egryn. We happened to reach Llanfair about 9.15 p.m. It was a rather dark and damp evening. In nearing the chapel, which can be seen from a distance, we saw balls of light, deep red, ascending from one side of the chapel, the side which is in a field. There was nothing in this field to cause this phenomenon---i.e. no houses, etc. After that we walked to and fro on the main road for nearly two hours without seeing any light except from a distance in the direction of Llanbedr. This time it appeared brilliant, ascending high into the sky from amongst the trees where lives the well-known Rev. C. E. The distance between us and the light which appeared this time was about a mile. Then about eleven o'clock, when the service which Mrs. Jones conducted was brought to a close, two balls of light ascended from the same place and of similar appearance to those we saw first. In a few minutes afterwards Mrs. Jones was passing us home in her carriage, and in a few seconds after she passed, on the main road, and within a yard of us, there appeared a brilliant light twice, tinged with blue. In two or three seconds after this disappeared, on our right hand, within 150 or 200 yards, there appeared twice very huge balls of similar appearance as that which appeared on the road. It was so brilliant and powerful this time that we were dazed for a second or two. Then immediately there appeared a brilliant light ascending from the woods where the Rev. C. E. lives. It appeared twice this time. On the other side of the main road, close by, there appeared, ascending from a field high into the sky, three balls of light, deep red. Two of these appeared to split up, whilst the middle one remained unchanged. Then we left for home, having been watching these last phenomena for a quarter of an hour.

Perhaps I ought to say that I had an intense desire to see the light this night for a special purpose; in fact, I prayed for it, not as a mere curiosity, but for a higher object, which I need not mention. Some will ridicule this idea, but I have a great faith in prayer.... (Society for Psychical Research, Proceedings, 19:148-149, 1905)

# MOVING AND FLASHING LIGHTS

## LIGHT FLASHES IN THE SKY
**Rutledge, Harley D.; *Physics Today*, 27:11-12, September 1974.**

Jack Epstein suggested (March, page 15) that "some of the phenomena that have been designated as 'flying saucers' could possibly be sightings of small specks of antimatter in the process of annihilation in the earth's atmosphere." On two occasions the past year, I have observed a series of random but localized light flashes that one might attribute to annihilation process. The observations of what I call the "random flash-bulb effect" were made by continuously scanning the twilight sky with 10 x 35 binoculars. The color and duration of each flash was similar to that of a xenon bulb.

The explanation given by Epstein had occurred to me because no material objects could be discerned as the sources of these lights. The first observation was a series of localized light flashes in a clear, blue sky, estimated to be less than 10 miles away. Following this 3-4 second display, an offwhite light "switched on" and proceeded slowly across the sky. Apparently it traveled a straight path at an altitude less than 10,000 feet. When viewed through binoculars, the latter light appeared as a single, spherical shape unattached to a material object. Assuming that the angular resolution of the eye is one second of arc, a solid object greater than about two feet in its largest dimensions would be resolved in the binocular field, assuming a distance of 10 miles between the observer and the object.

The second observation was of an off-white light that "switched on" above a long, narrow cloud. The intensity of the light varied at a rate of about one "blink" per second. Within a few seconds, this was accompanied by the transient "random flash-bulb effect," all visible in the seven-degree binocular field. Upon depressing the microphone switch of a 37.1 MHz radio, the blinking light extinguished. No other strange events occurred during the evening, although three viewing stations, two of which were separated by 22 miles, were in operation until 2:00 a.m.

The fact that the off-white lights were observed in conjunction with the flash-bulb effect and the fact that the phenomena was observed at relatively low altitude seem to preclude annihilation as a plausible explanation. Two other members of the research team have witnessed the "random flash-bulb effect" independently and on more than one occasion. After personally recording some 100 anomalous events, it is my contention that there is no simple, physical explanation for the variety of phenomena observed. (Physics Today, 27:11-12, 1974)

## MORE LIGHTS IN THE SKY
**Maccabee, Bruce S.; *Physics Today,* 29:90, March 1976.**

Over the past two years several letters have appeared in Physics Today in which the authors have discussed the possible origins of some randomly occuring lights in the sky (Epstein, March 1974, page 15; Rutledge, September 1974, page 11; Heaton and Epstein, February 1975, page 11). In view of these letters, and especially in view of Epstein's suggestion (February) that antimatter meteorites could conceivably last up to several minutes and describe somewhat complex (Brownian) motion in Earth's atmosphere, I would like to recount a recent observation made by three witnesses, one of whom is a technically competent civilian employee of a military installation near Washington, D.C.

The witnesses reported seeing two bright, apparently self-luminous, circular objects at midday when the sky was cloudless, empty of aircraft, balloons, and so on, and the visibility was about twenty miles according to the weather records of the date and time. The objects were observed to descend "from the blue" one after the other, and remain at a fixed angular altitude for a time estimated to be a minute or more. During this time the observers noted faint dark rings about the central bright regions. The second appearing object then executed a left-right, zig-zag motion and then rose rapidly "straight up." Moments later the first appearing object also began a rapid uniform, apparently vertical, ascent. The object shrank visibly in apparent size as it ascended and was lost to sight nearly directly overhead. The brightness of the objects were reported to be constant throughout the sighting.

The angular subtense of each object when at its lowest angular elevation of $25^{o}$ has been estimated to be at least 1 milliradian but no greater than 10 milliradians. Unfortunately, it was not possible to estimate the distance to the objects.

It is difficult to decide what to make of a report such as this. If the objects had been observed to disappear after their descent, one could argue that the observation is roughly consistent with the antimatter meteor hypothesis of Epstein. However, the final ascent of the objects seems to conflict with this suggestion. In any case, the observational data are quite good and provide useful information with which to test any alternative hypothesis. A more complete report is available from the author. (Physics Today, 29:90, March 1976)

## UNUSUAL AERIAL PHENOMENA
**Hynek, J. A.; *Optical Society of America, Journal,* 43:312-313, 1953.**

One of these patterns might be called "Nocturnal Meandering Lights." Reports falling into this category are characterized by the sighting of a bright star-like light, perhaps of -2 or -3 stellar magnitude which floats along without sound, frequently hovers, reverses its field without appearing to turn, and often abruptly speeds up. The light is most frequently described as a yellow amber or orange, changing to blue or red occasionally, and changing in brightness markedly. Sometimes the description states that the light went out as if someone had pushed a button; at other times the light is reported only as variable. A very characteristic statement by those making the reports is: "I have never seen anything like this in my whole life." The desire to identify these sightings as balloons is thwarted by the tactics observed.

As an example of a report of this kind, let us take one that came in from Florida this past July. On one night several airmen independently observed a light approach at a very slow speed, come to a halt nearly overhead, then reverse direction with no apparent turn. On two other nights, three other lights appeared in other sections of the sky, of similar appearance, but maneuvering more rapidly. They were observed for some 10 minutes by 9 airmen, including a control tower operator, an aircraft dispatcher, and two pilots from Wright Field.

In the words of one of the men, "For the next fifteen minutes we watched this light and speculated on what it might be. It was not a sharp light like a bare bulb but more like a light shining through frosted glass. No shape of any kind was discernible. It appeared to blink, but with no regularity whatever."

Also this past July at an air base in New Mexico, a similar sighting was made. Paraphrasing from sworn statements made by observers, "Our station was notified that an unconventional aircraft had been picked up with both electronic and visual contact. Our station made electronic contact with the object and two of our men and I had gone outside the building and saw it hovering under a cloud layer to the east of us. It appeared as a large light, at an uncertain distance, and was hovering at the time. A minute or so later, it moved rapidly toward the north for a short distance and stopped as suddenly as it had begun to move."

And from another statement, "Our scope operator at that time reported a strange target about thirty miles east of our station. Two of us went outside and sighted a very bright light traveling at what we estimated to be around 200 miles an hour. The light went out at least two times but did not stay out more than two or three minutes. The light seemed to have a floating effect and made no sound. At one time around seven or eight smaller lights could be seen. The object seemed to drop to about 10 or 12 thousand feet and then climbed to about 25,000 taking a northern course."

Radar observations as well as visual observations are involved in this problem. Early last month shortly before dawn colored lights were observed in the sky southeast of the radar station. At the same time and the same azimuth, unidentified targets appeared on the scope. Only a very slight temperature inversion was present, $1^{o}$ at 25,000 feet. No more than two lights appeared at one time. They were observed to be moving in a rather erratic pattern and changing colors occasionally. The last thirty minutes of observation revealed the lights remaining yellow---prior to that they were red, green, and blue. They moved in no apparent formation but mostly appeared in one area and disappeared in another, when either the light went out or the objects dived behind clouds. They were starlike objects and appeared to develop long, white vapor trails, when they dived. They were motionless at times and moved rapidly at other times. This corresponded to similar movements observed on the radar scope.

One white light went out as it changed direction and continued as a black silhouette against the dawn sky. Observation was for a period of about an hour and was made by two airmen and a radar operator---all three observers were experienced aircraft control and warning operators. Objects were observed 20 to $40^{o}$ above horizon. Radar gave distances of 50 to 80 miles. This implies a height of about 40 miles. There was no air traffic on radar within 100 miles.

Quoting from the observer's statement, "receiving a call concerning a strange light in the sky, I went out and scanned the sky in several directions before I saw a light. My first glimpse was a very bright blue light, but it lasted only about a minute, then it faded into a light green. It moved in a slow orbit.

I was startled at first so I closed my eyes and opened them again. The light was still there. I stared at it a few minutes and now the light seemed more yellow than before.

I did not think anyone would believe me, so I went inside the building and relieved the radar scope operator. I found a target at $123^{o}$, 53 miles. After that it appeared as a permanent echo. In about two minutes, it disappeared and almost immediately another pip appeared, at $134^{o}$, 73 miles. It also seemed like a permanent echo. It stayed on the scope for 1-1/2 minutes. These pips were at no time caused by malfunction of the radar set.

It was daylight when it (the object) seemed to fade both visually outdoors and electronically indoors."

And another sighting---in Northern Michigan---on July 29 of last year, a pilot chased a brilliant multicolored object close to the horizon, and due north. He flew at 21,000 feet, followed the object for over a half-hour but could not gain on it. Radar operator reported contact with the object for about thirty seconds. And ground control interceptor station reported blips too. In this case, it seems certain that our harried pilot was pursuing Capella! Reference to a star map will show that at his latitude, at the time of his sighting, Capella was at lower culmination, that is, at the lowest point of its swing around the pole just skirting the horizon. I have seen it at that position myself in Canada, and, can vouch for the fact that its blue, yellow, and red twinkling can be spectacular.

Unfortunately, neither Capella nor any other star can explain many other nocturnal meandering lights. But there is no question in my mind, just to make this point exceedingly clear, that there exists a relatively simple, natural explanation for them, perhaps even ordinary aircraft under special test conditions. The chief point here, is to suggest that nothing constructive is accomplished for the public at large---and therefore for science in the long run---by mere ridicule and the implication that sightings are the products of "birdbrains" and "intellectual flyweights." In short, it would appear that the flying saucer situation has always been a problem in science-public-relations, and that fine chance has consistently been missed to demonstrate on a national scale how scientists can go about analyzing a problem.

A lot is said about the proper interpretation of science to the public, but the only answer they receive to a question about which they are more widely concerned than perhaps any other in this century, is ridicule. Ridicule is not a part of the scientific method and the public should not be taught that it is. (Optical Society of America, Journal, 43:312-313, 1953)

# OCEANIC PHOSPHORESCENT DISPLAYS

Ships that ply the Indian Ocean, particularly the waters leading to the oil-sodden lands around the Persian Gulf, frequently encounter dazzling phosphorescent seas. As Kipling described it, the ship's wake is "a welt of light that holds the hot sky tame." Huge globs of light rise from the depths and explode on the surface. Wavetops sparkle; and broad, geometrically precise corridors of bioluminescence stretch from horizon to horizon. Buckets lowered into these glowing seas prove that luminescent marine organisms are the cause of these displays.

Bursts of phosphorescence are mundane compared to the vast, rotating wheels of light and other fantastic luminescent displays seen in these same seas. Ridiculed as wild sailor's tales for centuries, ships have reported scores of bona fide sightings of geometric displays since World War II. Mariners tell of great, spoke-like bands of light seeming to rotate about some distant hub. Some rotate clockwise; others counterclockwise. Occasionally, two wheels will be rotating in opposite directions around separate hubs. Great waves of light and whirling crescents of luminescence also engulf ships in the Persian Gulf area. Crews that see these fantastic apparitions do not soon forget them, and their testimonies recorded below betray their awe and bewilderment.

An immediate reaction is to explain the wheels of light and their kin in terms of marine bioluminescence stimulated by natural forces that, like the wake of a ship, leave behind glowing evidence of their passage. Sound waves emanating from submarine earthquakes have been the most popular explanation. But what combination of seismic waves can create overlapping, counterrotating spirals? Furthermore, there are a few well-attested cases where the luminescence is seen in the air above the water. This fact, the persistence of the phenomenon (half an hour), and the complex nature of the displays, suggest that these apparitions may not even be connected with bioluminescence, although bioluminescence seem the most probable explanation at the moment.

Another possible answer is the electromagnetic stimulation of bioluminescence, perhaps something connected with terrestrial discharges of electricity to outer space, like the mountain-top glows seen along the Andes. To illustrate a possibility, a luminous "mist" is sometimes observed during very low auroral displays and marine phosphorescent activity. Also, the collective behavior of marine organisms should not be ruled out. (Travelers in the tropics tell amazing accounts of synchronized flashing of huge masses of fireflies.)

Before we can really understand the geometrical phosphorescent displays, we must answer the following questions: Why are they concentrated in the Indian Ocean to the near-exclusion of other seas where bioluminescence is common? Do they occur only at certain times of the year? Are they correlated with solar activity? Could they be earthquake-stimulated? What are the effects of ship radar and underwater explosions? Are the wheels related to the more common "white" or "milky" seas?

# LONG, PARALLEL PHOSPHORESCENT BANDS

## BRILLIANT GULF WATERS
**Anonymous; *Monthly Weather Review*, 36:371, 1908.**

A remarkable marine phenomenon was observed by the steamship Dover, Capt. Yon A. Carlson, as that vessel steamed to Tampa from Mobile. When at a point 35 miles from Mobile light, at 7 o'clock in the evening of the 24th, the ship ran suddenly in a streak of light coming from the water which alternated blue and green, the colors being so brilliant that the vessel was lighted up as if she were covered with arc lights with colored globes.

A half mile streak of dark water, and a blackness that settled like a pall over the ship followed, and a second streak of the same brilliant-hued waters was encountered. The second streak was about as wide as the first one, and when the ship ran out of it the same black waters and a night of exceptional blackness were all encountered. * * *

"I have sailed the high seas for twenty years," declared Captain Carlson, "and have seen interesting phenomena, both meteorological and otherwise, in the waters of every known ocean, but I never saw anything that approached this blue and green light from the water phenomena. The night was dark, but clear, and we ran into the streaks without any seeming warning. I was in the pilot house when we struck it, and I ran on deck, thinking that something was on fire.

"The crew tumbled out to witness it also, and it was magnificent. It was so light that it was remarked by the chief engineer that it could be read by, and to make sure I grabbed a paper, and the finest print that I could find was easily discernible. We ran out of the streak into a streak of black water, and the darkness of the night seemed to increase as we did so. From the streak of blackness we ran into the second streak of lighted waters. Each of the streaks and the intermediate streak of black water was about half a mile wide. The wind at the time was a light northwest. The sea was smooth and we were bearing southeast by east half east, 35 miles from Mobile light." (Monthly Weather Review, 36:371, 1908)

Brilliant parallel bands off Mobile Light

# MOVING PHOSPHORESCENT BANDS

## PHOSPHORESCENT WHEEL
**Rutherford, W.; *Marine Observer,* 30:128-129, 1960.**

S. S. Stanvac Bangkok. Captain W. Rutherford. Lautoka (Fiji) to Tandjong Uban (Bintan Island, Indonesia). Observer, the Master.

Between 2350 ship's time on 27th September and 0010 on 28th September, 1959. Light easterly wind, slight sea, very dark and clear. Course 290°, speed 15 kt.

As we were passing through a fleet of fishing vessels, I and the Officer on watch were keeping a sharp look out through binoculars and thus observed the phenomenon from start to finish. The first indication of anything unusual was the appearance of white caps on the sea here and there, which made me think that the wind had freshened, but I could feel that this was not so. Then flashing beams appeared over the water, which made the Officer on watch think that the fishing boats were using powerful flashlights. These beams of light became more intense and appeared absolutely parallel, about 8 ft wide, and could be seen coming from right ahead at about 1/2 sec intervals. At the time, I thought I could hear a swish as they passed, but

Moving phosphorescent bands in East Indian seas

decided that this was imagination. They did not appear like rings or arcs of a circle, unless it was a circle so big as to make them appear as straight lines. It was like the pedestrian's angle of a huge zebra crossing passing under him whilst he is standing still. When this part of the phenomenon was at its height it looked as if huge seas were dashing towards the vessel, and the sea surface appeared to be boiling, but it was more or less normal around a fishing vessel which we passed fairly close. The lights of various fishing vessels were visible through the beams of light, though dimmed by the brightness of the latter. The character of the flashes changed and took on the appearance of beams from a lighthouse situated about two miles on the starboard bow, or as if the centre of a giant wheel was somewhere on the starboard bow with the beams as its spokes. As the beams from the wheel on the starboard bow weakened, the same pattern appeared on the port bow at the same distance and regularity. The wheel on the starboard bow revolved anti-clockwise and the one on the port bow revolved clockwise, i.e. both wheels were revolving towards the ship. The wheel on the starboard bow diminished as the one on the port bow increased: when the latter was at its peak the one on the starboard bow had disappeared.

The next change was that the beams appeared to be travelling in the exact course of the ship, like a following sea, i.e. the beams now seen were a reversal of those seen at first. At the time they appeared I asked the Officer on watch what he saw, in case through blinking my eyes I was not seeing these correctly. The Officer on watch agreed that the beams were now 'chasing' us. I cannot tell from how far astern these were forming, as I could not actually see them further aft than the funnel (which is right aft, this ship being a tanker), but I could clearly see them passing from aft along the ship's side. I was not looking aft very much because I was picking my way through these small fishing vessels.

Presently all the beams gradually ceased and the surface of the sea could be seen again. At that time for about 2 min, as far as the eye could see, there were rings of light about 2 ft in diameter and 6 ft apart in the sea, flashing in and out with rhythm. The flashing reminded me of a treeful of glow-worms. Although the flashes or beams of light appeared to be above the surface of the water, I think this was an illusion and that the actual light was in the water, flashing in and out at regular intervals. During the wheel effect, I was reminded of the apparent motion of these electric signs which give the appearance of objects moving by the flashing on and off of various lights. The ship seemed to be in the centre of the disturbance and at one time I had the feeling that she was actually causing it, and that if I reduced speed or altered course, I would alter the pattern accordingly. I could not try this, however, as I had to get clear of the fishing vessels as soon as possible. (Marine Observer, 30:128-129, 1960)

## PHOSPHORESCENCE OF THE SEA
**Hatfield, J.; *Marine Observer,* 8:230-231, 1931.**

S. S. Laomedon, Captain J. Hatfield, bound Suez to Penang, records that on "July 31st, 1929, 20.00 A. T. S. in Latitude 12° 22' N., Longitude 56° 28' E., Barometer 29.77 in., Wind S. S. W. force 6, with rough sea and swell; at 19.00 when darkness set in, the ship appeared to be surrounded by a luminous halo reflected from the water. By 20.00 the water had turned a greyish white colour which, however, did not appear to be caused by surface phosphorescence as there was little or no sparkle from the bow wave. This discolouration extended to the horizon in all directions giving it a misty appearance except between S. S. E. and W. S. W. where the horizon remained clearly defined. This phenomenon continued until 21.50 when it gradually closed in round the ship again and finally disappeared leaving the sea a normal colour with no phosphorescence. While the ship was in this discoloured water, the sea appeared to calm down considerably though the wind remained force 6. The air temperature remained steady at 79° F. but the sea temperature fell from 78° F. to 76° F. but had risen to 79° by 23.00 hrs.".

The apparent calming of a rough sea during extensive phosphorescence is a feature frequently mentioned in accounts of such displays and possibly the explanation is that the presence of myriads of these animalculae, many of which secrete mucous to adjust their specific gravity in order to keep afloat, has a similar effect to pouring oil on the sea. On the other hand the glare of the light might cause the sea to appear calm when actually there is no change, but marine observers are quite definite that there is actually less sea disturbance. (Marine Observer, 8:230-231, 1931)

## PHOSPHORESCENCE OF THE SEA
**Mordue, J. A.; *Marine Observer,* 8:231, 1931.**

The following is an extract from the Meteorological Report of S. S. Koranna, Captain J. A. Mordue, Port Sudan to Bushire.

"On May 30th, 1926, about 1 a.m. when standing to N'd. between Quoin Is. and Larak Is. in about Latitude 26° 40' N., Longitude 56° 33' E., passed through a phenomenal, scintillating, phosphorescent belt of water. It was first sighted as a line of phosphorescent water stretching across the horizon ahead from East to West. As the ship approached the area, it presented a curious, scintillating effect. On passing through it, it was found to be a belt about 1/2 a mile in width extending to the horizon in an East and West direction. The effect at close quarters was as though thousands of powerful beams of light directed upwards from under water, each illuminating a patch of some twenty to thirty square yards of sea surface, were being switched on and off alternately, independently of each other. If any one of these patches were watched, the intervals of light and darkness were found to be of surprising regularity about 1 to 1-1/2 seconds.

The belt gradually receded astern and the display continued until it was lost over the horizon to the Southward. There was a moderate, Westerly breeze blowing at the time, with a clear sky, whilst the gentle ripples and small wavelets gave the surface of the sea sufficient movement to enhance the most startlingly bizarre effect of phosphorescence I have ever witnessed." (Marine Observer, 8:231, 1931)

# PHOSPHORESCENT WAVES IN THE AIR

## PHOSPHORESCENCE SEEN IN THE AIR
**Austen, G. A.; *Marine Observer,* 26:78, 1956.**

S. S. Dunkery Beacon. Captain G. A. Austen, Mombasa to Glasgow. Observers, the Master, Mr. J. McDonald 1st officer, Mr. D. Watson 2nd officer, Mr. D. Mac Farlane 3rd officer, Mr. N. Wyatt, Radio Officer.

13th May 1955, 2106-2110 G. M. T. Shortly after leaving Aden for Suez bright pulsating waves of light having a period of 1.5 sec were observed emanating from the sea in close proximity to the ship. They were traveling in a direction of 070°

for some distance and extended to a height of about 12 ft above the sea. This phenomenon was definitely not caused by the proximity of shore lights; there was no sign of phosphorescence in the sea at the time of observation. Sky cloudless; wind 100°, force 2; sea slight; visibility excellent; moon's alt. 7° 30' approx. (Marine Observer, 26:78, 1956)

# PHOSPHORESCENT WHEELS

## REPORT OF AN UNUSUAL PHENOMENON OBSERVED AT SEA
**Pringle, J. Eliot; *Nature*, 20:291, 1889.**

The following Report to the Admiralty has been communicated to us for publication by Capt. Evans, C.B., F.R.S., the Hydrographer to the Navy:---

H.M.S. Vulture, Bahrein, May 17, 1879

Typical Indian Ocean light wheel

Sir,---I have the honour to inform you that at about 9.40 p.m. on May 15 when in lat. 26° 26' N. and long. 53° 11' E., [the Persian Gulf] a clear unclouded, star-light night, Arcturus being within some 7° of zenith, and Venus about to set; wind north-west, force 3, sea smooth, with slight swell from the same direction; ship on starboard tack, heading west-south-west and going three knots, an unusual phenomenon was seen from the vessel.

I noticed luminous waves or pulsations in the water, moving at great speed and passing under the ship from the south-south-west. On looking towards the east, the appearance was that of a revolving wheel with centre on that bearing, and whose spokes were illuminated, and looking towards the west a similar wheel appeared to be revolving, but in the opposite direction. I then went to the mizen top (fifty feet above water) with the first lieutenant, and saw that the luminous waves or pulsations were really travelling parallel to each other, and that their apparently rotatory motion, as seen from the deck, was caused by their high speed and the greater angular motion of the nearer than the more remote part of the waves. The light of these waves looked homogeneous, and lighter, but not so sparkling, as phosphorescent appearances at sea usually are, and extended from the surface well under water; they lit up the white bottoms of the quarter-boats in passing. I judged them to be twenty-five feet broad, with dark intervals of about seventy-five between each, or 100 from crest to crest, and their period was seventy-four to seventy-five per minute, giving a speed roughly of eighty-four English miles an hour.

From this height of fifty feet, looking with or against their direction, I could only distinguish six or seven waves; but, looking along them as they passed under the ship, the luminosity showed much further.

The phenomenon was beautiful and striking, commencing at about 6h. 3m. Greenwich mean time, and lasting some thirty-five minutes. The direction from which the luminous waves travelled changed from south-south-west by degrees to south-east and to east. During the last five minutes concentric waves appeared to emanate from a spot about 200 yards east, and these meeting the parallel waves from south-east did not cross, but appeared to obliterate each other at the moving point of contact, and approached the ship, inclosing an angle about 90°. Soundings were taken in twenty-nine fathoms; Stiffe's Bank, with fifteen to twenty fathoms, being west about one mile. The barometer was already at 29.25 from 8 to 12 p.m.

| | At 8 p.m. | 10.15 p.m. | Midnight |
|---|---|---|---|
| Temperature of air | 84 | 83 | 83 |
| Temperature of sea-water | 84 | 82 | 82 |

I observed no kind of change in the wind, the swell, or in any part of the heavens, nor were the compasses disturbed. A bucket of water was drawn, but was unfortunately capsized before daylight. The ship passed through oily-looking fish spawn on the evening of the 15th and morning of the 16th inst. (Nature, 20:291, 1889)

## PHOSPHORESCENT WHEELS
**Barth, H.; *Marine Observer*, 24:74-75, 1954.**

M.V. British Empress. Captain A. Henney, O.B.E. Port Okha to Persian Gulf. Observer, Mr. P. M. Alderton, 3rd Officer.

5th April, 1953, 2125 Indian Standard Time. Commencing from about NNW, shafts of pale white diffused light appeared, apparently travelling on the surface of the water at a great speed. Each shaft was several feet wide and they stretched as far as the eye could see. At first they appeared in perfectly parallel lines, equally spaced, passing the ship at about one every second, but after five minutes they wheeled round in perfect formation and approached the ship from all points of the compass. They came from only one compass point at a time and each change of direction was swift and definite, though not abrupt. The most frequent directions were from NNW and SSE.

After about 15 minutes the shafts occasionally formed into a rotating radial movement in which they retained their equal geometrical precision and the frequency of about one per second. At this time the pattern was continually changing about every 20-30 seconds from the parallel lines to the wheel. The periods of transition were hardly noticeable, but they were not abrupt. Each time the wheel appeared it was in a different place. On one occasion there were two distinct wheels visible at the same time. Throughout the period the wheels appeared they varied in direction of rotation, some clockwise and some anticlockwise. Five minutes later the pattern became still more complicated but remained perfectly regular and at 2150 the light faded out over a period of 30 seconds.

Although the light appeared to be on the surface of the water it was unaffected by the wind and no disburbance of the water was produced. The most notable feature of the phenomenon was the effortless speed and mathematical precision of movement. The only near analogy I think of is that of being placed in the middle of a large A scan when a large variable AC current is supplied. The whole effect was one of great weirdness and eerieness, so much so that the look-out man came on to the bridge quite scared, believing that he was suffering from hallucinations.

The ship's course was 290° (T), speed 10 kt and no alteration of either took place during the observation. The sea was that corresponding to Beaufort Scale wind force 2-3, swell negligible, sea temperature 78° F. The sky was cloudless, with perfect visibility, wind NW force 2-3, air temperature 80° F.

Position of ship: 22° 42'N, 68° 08'E. (Marine Observer, 24: 74-75, 1954)

M. V. Rafaela. Pladju to Bangkok. Observer, Mr. H. Barth, 2nd Officer.

24th April, 1953, 0215 S. A. T. (1915 G. M. T.). Faint flashes of light with oscillating movements were observed on the sea. The flashes gradually increased in strength until at 0230 they suddenly changed into rather intensive rays of light moving around centres lying near the horizon. Three groups of rays were present, as shown in the sketch.

(a) One on the port bow having a bearing of about 300° with the rays rotating anticlockwise.
(b) One on the port bow having a bearing of about 230°, rotating clockwise.
(c) One on the starboard bow having a bearing of about 95°, rotating anti-clockwise.

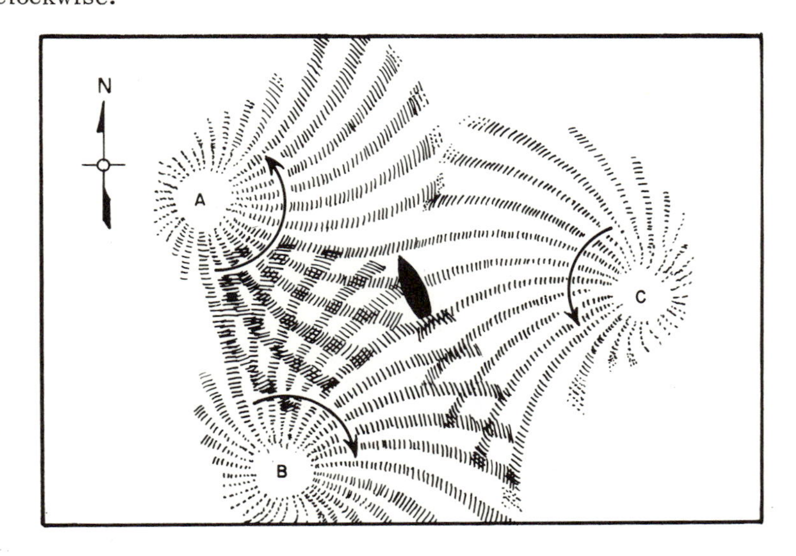


System of three wheels observed on the Gulf of Thailand

The beams were curved with the concave side in the direction of the movement, and were passing the ship continuously with a frequency of about three a second; they looked more like glowing shafts than beams of light. Reflections on the ship were clearly visible. The Chinese quartermaster became panic-stricken, left the wheel and did not return until he had been called three times.

The phenomenon lasted till 0250, and it had been clear by the increasing strength of the group ahead and the decreasing strength of the groups astern, that the ship was advancing through the area of phosphorescence. Soon only the oscillating flashes could be seen and they also disappeared shortly afterwards. At 0300 the situation was normal again. The third engineer also saw the phenomenon clearly, as described above. (Marine Observer, 24:74-75, 1954)

## PHOSPHORESCENT WHEEL
**Hocking, R. W.; *Marine Observer,* 7:240, 1930.**

The following is an extract from the Meteorological Report of S. S. Talma, Captain R. W. Hocking, R. D., R. N. R., Calcutta to Far East. Observers, Messrs. L. T. Carter and H. F. Wright.

At 1845 G. M. T., 28th December, 1929, when in Latitude 14° 15' N., Longitude 96° 41' E., the vessel entered what appeared to be an area of unusual phosphorescent disturbance.

At first what appeared like small globules of phosphorescence rising from below and breaking at the surface were observed, later these gradually assumed an appearance almost like flashes of lightning under the water, which rapidly formed into regular beams, curved as curved spokes of wheel might be and of a width at the ship of about 30 feet, and revolving rapidly from right to left, at the rate of two a second, timed as the beams passed the bridge, around a distant centre which could not actually be seen clearly but appeared to be about five miles off. This centre passed ahead of the ship being first observed on the port beam and from there drawing slowly ahead of and across the bows of the ship fading gradually till on the starboard bow when the whole phenomenon finally disappeared. For a short period when the centre was on the port bow the beams appeared revolving in the opposite direction, this latter phase was not clearly marked as the beams had already begun to fade at that time. The beams could clearly be followed on both sides of the ship though the illumination was much greater on the side nearest the centre of revolution (port), their brilliance on that side being dazzling; the whole phenomenon lasting 15 minutes.

The vessel at the time was in 50 fathoms of water over a bottom composed mostly of fine sand and mud with shingle here and there, the compass remained entirely unaffected and no difference was noticed in the steering. The weather at the time being cloudless and calm with smooth sea and steady barometer.

It was later reported from the engine room that at this time the revolutions dropped considerably and the main engines were straining. As this straining of the engines appeared to me to point to the possibility of marine volcanic disturbance, I considered it advisable to send out a wireless warning to all ships and stations. (Marine Observer, 7:240, 1930)

# EXPANDING PHOSPHORESCENT RINGS

## CIRCLES OF LIGHT RADIATE FROM POINTS ON SEA SURFACE
**Whyborn, J. A.; *Marine Observer*, 27:92, 1957.**

S. S. San Leopoldo, Captain J. A. Whyborn. Lyttleton to Mena al Ahmad. Observer, Mr. R. G. Broderick, 3rd Officer.

17th April, 1956, at about 2030 S. M. T. When in 26° 23' N., 54° 38' E., about 8-1/2 miles NW. of Jazireh-e Farur, a strange effect was noticed on the sea surface about 2 miles away on the port bow. On approaching the area large bands or ripples of faint light appeared to be travelling in a confused way across the surface. Further inspection showed that these bands were spreading out from several central points distributed fairly evenly over an area about 2 miles across, in the same way as the circular ripples produced by raindrops in a puddle of water. Each band was about 70 yd wide and gave a faint white light. They spread, in unison, from the various central points, at intervals of a little more than 1 sec. They travelled at a considerable speed, so that by the time one circular band was formed the previous one was vanishing, at a radius of about 1/2 mile. When first seen it was supposed that the effect was being produced by moon shadows, as the moon was at an altitude of about 35° in the west, and there was a large number of small misty clouds. The clouds were, however, all moving in the same direction and could not have produced the effect of localised circular ripples. There was some general phosphorescence in the water but the presence of moonlight made observation of this difficult. The depth of water was about 24 fm according to the chart. The phenomenon was also seen by the helmsman and the lookout. (Marine Observer, 27:92, 1957)

# PHOSPHORESCENT PATCHES MOVING IN CIRCULAR ARCS

## ROTATING PATCHES OF LIGHT
**Arthur, D. E. O.; *Marine Observer*, 27:92-93. 1957.**

M. V. British Caution. Captain D. E. O. Arthur. Fao, Iraq, to Venice, Observer Mr. Scott, 3rd Officer.

30th May, 1956, about 2245 S. M. T. When in approximate position 26° 30'N., 54° 41'E., about 14 miles E. of Kais (Qais) Island light-vessel, lights appeared on the horizon resembling those of a town bearing roughly SW. There was, however, no town in this direction. When examined with binoculars the appearance of a town still persisted, with the addition of a shimmering effect, probably due to abnormal refraction. Some minutes later the lights appeared to merge together and come nearer the ship, and then resembled a luminiferous island parallel to which the vessel was steaming at a distance of about a mile. Patches of luminosity then began to appear, each about 6 ft across, arranged in lines which were spaced a cable apart; they crossed the bow from port to starboard, extending towards the main body of light. The lines then became wider and more numerous, but the patches composing them did not increase in brightness. The patches seemed to be on or above the sea surface, but were not affected by the waves.

After a further few minutes a change occurred. The parallel line formation disappeared and the patches were seen to be moving fairly slowly in an anticlockwise direction round circles of from 100 ft to 300 ft in diameter. When the vessel passed through a circle the patches of light disappeared on reaching the ship's port side and reappeared in the same formation on the starboard side. After about 20 min the whole effect ceased near the ship but was still seen in the distance on the starboard quarter. The vessel continued on the same course, 113°, and at the same speed. After a further 15 or 20 min more circles appeared, of varying sizes. The patches on the starboard side were small, about 1-2 ft in diameter, and were rotating in circles of about 20 ft diameter. After 10 min these dropped astern and no further phosphorescence was seen. (Marine Observer, 27:92-93, 1957)

# RADAR-STIMULATED SPINNING CRESCENTS

## PHOSPHORESCENCE
**Baker, F. G.; *Marine Observer*, 22:190-191, 1952.**

M. V. British Premier. Captain F. G. Baker. Observer, the Master.

30th November, 1951, 1800 G. M. T. The ship's radar apparatus had been switched on with a view to checking her position, when, in the same instant that this gear became operative, most brilliant boomerang-shaped arcs of phosphorescent light appeared in the sea, gyrating in a clockwise direction to starboard and anticlockwise to port, but all sweeping inwards towards the ship from points situated from five to six points on either bow and some two miles distant, and conveying the impression that they ricocheted from each other on meeting at the ship's bows and then turned and travelled away astern to similar points which were equidistant on either side and about four points on each quarter.

Having read with considerable interest the paragraph "Luminosity of the sea" ---which is given on page 18 of the West Coast of India Pilot---I do not suggest there is any definite connection between this phenomenon and the operation of radar, but it certainly appears peculiar that it should have been coincident with the switching on of the gear and only lasting during the period of its operation; this I suggest gives room for the thought that the ultra-high frequency radiation from the ship might possibly have had some effect upon some submarine substance that led to this result, especially as the whole aspect can best be described as flashes of fluorescent green light of such brilliance that the whole ship was illuminated as though by deliberate floodlighting, each one being separate in itself and having the size and shape of a "Cossor Radar Scanner" and revolving in the water as though synchronized with that unit, and all travelling to or from the ship as described and as I have attempted to portray in the accompanying sketch. (Marine Observer, 22:190-192, 1952)

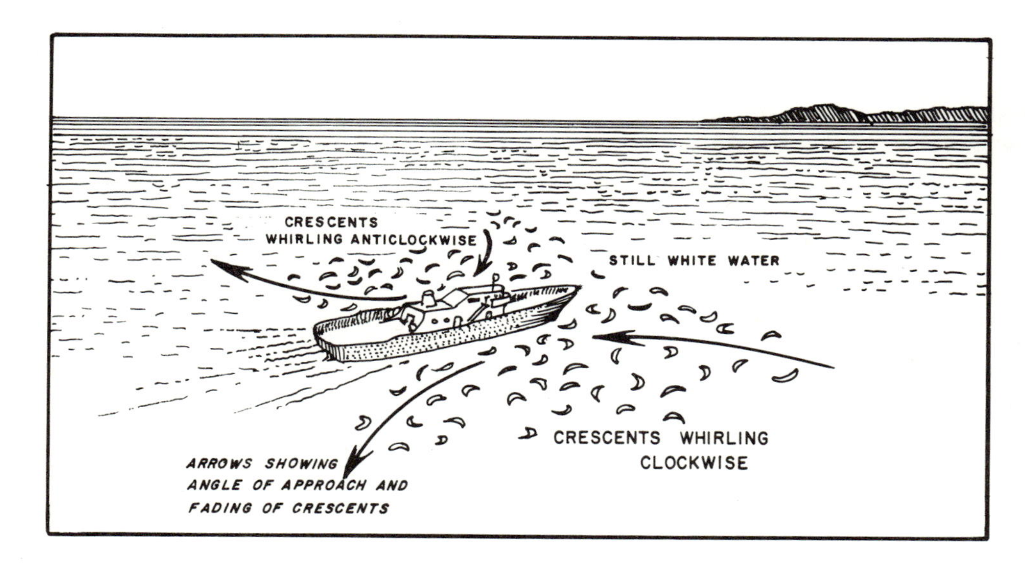


Radar-stimulated crescents of light on sea surface

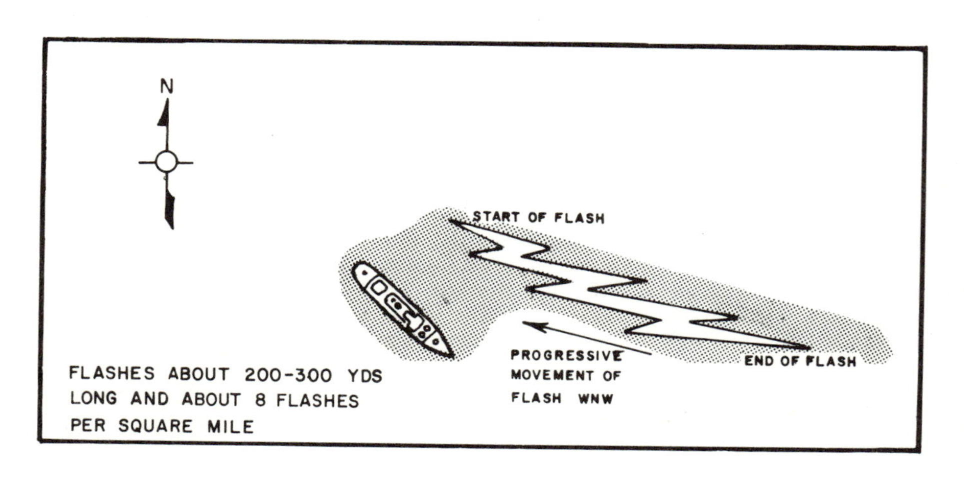


Zigzag flashes of phosphorescence

# FLASHING ZIGZAG PHOSPHORESCENT STREAKS

## PHOSPHORESCENCE
**Spencely, T. A.; *Marine Observer,* 20:139, 1950.**

S. S. Dara. Captain T. A. Spencely. Bombay to Karachi. Observer, Mr. G. R. McLutosh, 3rd Officer.

30th August, 1949, 2000-2200 G. M. T. In vicinity 22° 30'N., 68° 10'E., at mouth of Gulf of Cutch, brilliant phosphorescence in lightning-like zig-zag form was observed in patches of about a square mile each. These patches seemed to have a slight directional movement WNW, but the zig-zag phosphorescence flashed from left to right in a WNW-ESE direction. (Marine Observer, 20:139, 1950)

# MILKY SEAS OR WHITE WATER

## LUMINOUS SEA
**Ressorp; *Knowledge,* 2:438-439, 1882.**

In 1871 I was ordered to South America for the benefit of my health, and with a view to having as long a sea trip as possible, I took passage in a sailing-vessel. While lying becalmed a couple of degrees south of the equator I had the opportunity of observing the grand phenomenon of a luminous sea, and I must say that of all the phenomena I have witnessed, none gave me greater pleasure than that alluded to. About 10 p. m. the luminosity commenced, and the whole of the sea, from the centre we occupied to the horizon all around, appeared to be one mass of phosphorescent light. Even the stars above seemed dimmed, and the sails of the ship were brightly illuminated, while every cord in the rigging was clearly distinguishable. I drew up a bucketful of the water from alongside, and found that directly the bucket touched the water the phosphorescence suddenly ceased for a few minutes in the immediate vicinity of the ship. For about ten minutes the water in the bucket was as dark as ink, but after that time again shone out, though immediately a finger was dipped in it again became dark. I have no doubt that the effect was due to certain marine animalculae, the name of which I cannot recall to mind, and the light appeared to me to have a bluer tinge in it than given off when a vessel drives through the water. With regard to the depth to which this phosphorescence occurred, I believe that it was confined to the surface in this case, because when I dipped a lead line some 8 or 10 in. below the surface I failed to make out the lead. At the time several dolphins were swimming about, and I noticed that at whatever point one of these fish made its appearance, the inky darkness contrasted strangely with the illuminated portion of the sea. At midnight the glare gradually decreased, and about one o'clock had entirely disappeared. (Knowledge, 2:438-439, 1882)

## ILLUMINATION OF THE SEA
**Edgell, J. A.; *Marine Observer,* 3:132-133, 1926.**

The following is an extract from the Meteorological Log of H. M. A. S. Moresby, Captain J. A. Edgell, O. B. E., R. N., Singapore to Thursday Island. Observer, Lieutenant J. Donovan, R. A. N. :---

"At 0230, August 27th, 1925, when in estimated position 8° 29' S., 128° 10' E., the following unusual phenomenon was observed: The sea suddenly appeared to be illuminated by a soft lambent light as far as the horizon on all sides. At and above the horizon to an elevation of 2° there was a dark band of very deep blue, which abruptly faded to an exceptionally light and clear sky.

"At this time the sky was clear, with brilliant stars, and the moon had set. There was no phosphorescence. Sea surface temperature was 77° and Specific Gravity 1011.00. Wind E. S. E., force 3. Sea surface: waves E. S. E., disturbance 4, with a slight easterly swell.

"The phenomenon lasted until 0305, fading away quickly." (Marine Observer, 3:132-133, 1926)

## PHOSPHORESCENCE AND ELECTRICAL STORM
**Taylor, F. C.; *Marine Observer,* 4:190, 1927.**

The following is an extract from the Meteorological Report of S. S. Socrates, Captain F. C. Taylor, Norfolk, Va., to Brazil, Observer, Mr. W. E. Jordan, 2nd Officer.

"October 4th, 1926, bound to Rio Grande do Sul from Santos. The weather during the day was normal, wind S. E., force 4. Swell S. S. E. 3. Barometer steady around 1025.7 mb., mean temperature of air 62°F., water 58°F., sky cloudless and exceptionally clear. At 6.30 p. m., being abeam of Mostardos Lighthouse, in Latitude 31° 20' S., Longitude 50° 45' W., light wisps of Cirrus were observed in the zenith moving from W. by S. These, however, showed no signs of increasing, but apparently disappeared with coming of dark. At the same time a heavy bank of Cu-Nb and Nb was seen to be banking up from W. S. W. accompanied by vivid fork and sheet lightning. By 8.00 p. m., the sky was completely overcast with heavy Cu-Nb, heavy drops of rain began to fall but ceased after a few minutes. The barometer remained steady at 1025.7 mbs., temperature of air 59°. Winds, variable, force 0-3. This prevailed until 10.00 p. m. when the sea became absolutely white with phosphorescence so that the vessel seemed to be moving in a sea of milk, at the same time a curious phenomenon was observed, the sky which was composed of very heavy Cu-Nb appeared to break up and turn absolutely white, looking like white Cumulus clouds in the sunlight and had the general appearance of a gigantic honeycomb extending over the whole vault of the heavens. No stars were visible. At 11.00 p. m., the phosphorescence subsided to isolated patches on breaking waves. Co-incident with its subsidence the sky again became pitch black, vivid lightning now playing across the heavens from all directions, thunder being heard in the S. W. At midnight the conditions were the same, the barometer steady at 1026.1 mb., air 50°, wind N. E., force 0-3, swell S. S. E. 3. Between midnight and 1.00 a. m. a few drops of heavy rain fell and the barometer fell 3.4 mb., in the hour (1026.1 mb. to 1022.7 mb.). At 1.30 a. m. the sea again became milky white with phosphorescence, so as to render the shore lights invisible, and the same phenomenon was observed in the sky although not in such a marked degree as previously, rather the sky had a curious dappled white appearance and seemed to be composed of all sorts of erratic shapes, diamonds, circles, &c.; this lasted for fifteen minutes, the sea again losing its phosphorescent appearance, the sky becoming black at the same time. A small break now appeared to the westward through which stars were visible, this however disappeared in a few minutes, the heavy thunder clouds rolling up as black if not blacker than before. The lightning flashes became incessant, roll after roll of thunder pealing, but very little, if any, rain. During this period we arrived off the port and anchored at 2.00 a.m. The barometer now commenced to rise slowly, the storm appeared to die away to isolated flashes and peals. At 4.00 a.m. heavy rain set in, a steady downpour; the wind remaining at N.E., force 3, the barometer continued rising until 5.00 a.m. when it commenced to fall again. The wind now shifted to N.N.E., freshening, force 6-7. Heavy rain still falling. This type of weather prevailed until 7.30 a.m., the rain then ceased and the sky clearing slightly; there was however no change in the wind, the thunderclouds having also passed on. It is almost impossible to describe the awe-inspiring grandeur of that sky, it seemed to hint at the supernatural, this I know sounds ridiculous, but any others who witnessed it will, I am sure, agree with me. Other outstanding points are the absence of rain and wind until after the passing of the electric storm." (Marine Observer, 4:190, 1927)

# COLORED RAYS EMANATING FROM SHIP STERNS

## COLORED RAYS EMANATE FROM SHIP'S STERN
**Rattray, J.; *Marine Observer,* 30:64-65, 1960.**

S.S. Esso Manchester. Captain J. Rattray. Suez to Fao. Observers, Mr. P. O'Connor, 2nd officer and Able Seaman W. Burke.

1st June, 1959, 2315 G.M.T. Brief quick flashes of white light were seen on the surface of the sea ahead and on each side of the bow. They appeared to be elliptical in shape rather than circular and varied in size from 10-40 ft along the major

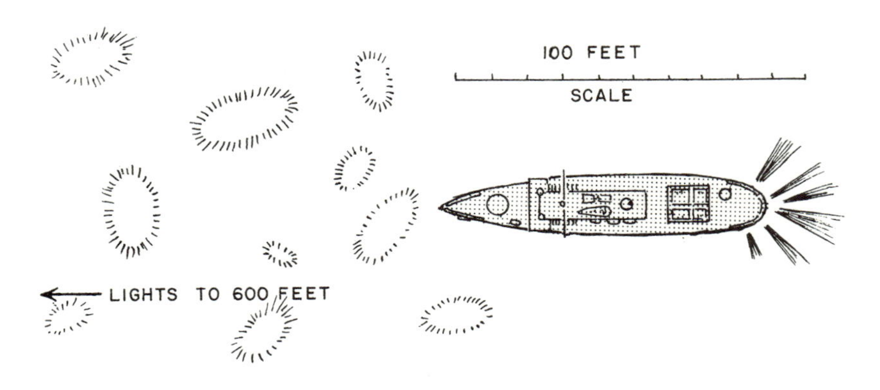


Rays of light emanating from ship's stern

axis and 5-15 ft along the minor axis. The flashes were seen up to a distance of 600 ft from the vessel. At the same time the look-out on the forecastle reported coloured rays of light radiating from the stem of the ship, over the surface of the water to a distance of about 20 ft. Some of the colours noted and remembered by the look-out were red, purple, blue and green. The whole phenomenon lasted about 3/4 hour. (Marine Observer, 30:64-65, 1960)

# Chapter 2

# OPTICAL AND RADIO ANOMALIES IN THE ATMOSPHERE

## INTRODUCTION

The atmosphere plays many optical tricks, from mirages to rainbows to the crepuscular rays at sunset. The laws of reflection, refraction, and color dispersion are sufficient to explain the great majority of these often impressive displays of colors, shadows, and distorted images. Undoubtedly, the same laws can explain most of the rarer and more spectacular phenomena described below, but the theoretical attempts made so far have not been completely satisfying. How does one account for elliptical haloes, kaleidoscopic suns, long-delayed radio echoes, and all the trickery of mirages?

Actually, few scientists take much interest in atmospheric optical phenomena any more. They are for the most part content that atmospheric anomalies can be brought into the fold of the scientifically explained with little effort; that we already know all that is really important about light, color, and radio waves in the atmosphere. This attitude is, of course, smug and dangerous in science. The ephemeral, frequently beautiful phenomena of this chapter pinpoint some areas of genuine scientific ignorance---not so much ignorance of the basic laws of optics as appreciation of the fine structure of our atmosphere and its interfaces with outer space.

It is regretable that so many of the irregular atmospheric phenomena occur at sea, on mountain tops, in the reaches of the Arctic, and in other remote places. Few have really observed all the natural nuances of sunlight and moonlight. By necessity, this chapter is the testimony of mariners, explorers, and those who frequent the open places of the world.

# COLOR FLASHES FROM RISING AND SETTING ASTRONOMICAL BODIES

The "green flash" or "green ray" is an elusive and long-debated phenomenon of the rising and setting sun. The general consensus today is that the green flash is physically real and caused by atmospheric dispersion (i.e., the vertical fanning out of the solar spectrum). As the sun sinks below the horizon, the blue-green end of the spectrum is the last to disappear to the observer. Why then are not purple or blue flashes seen rather than green? The fact is that these colors are seen, but only very rarely. Why does green predominate? Another puzzle is why the green flash is not seen at all when conditions seem ideal. Finally, some observations of the green flash may be physiological in nature; that is, the green flash may be the green afterimage of the red tip of the setting sun. Indeed, not too long ago, the green flash was believed to be entirely illusory, but photographs have proved this to be incorrect. The green flash is real---some of the time, at least---but there are some things about it we do not understand.

## THE GREEN FLASH
**Porter, Alfred W.; *Nature,* 94:672, 1915.**

So much has been written about the green-ray, at sunset that I am somewhat diffident about adding anything. But as I find myself unable to accept the orthodox explanation of the phenomenon usually seen I write this note. This phenomenon, as seen by me on several occasions during the last summer on my way to Australia, always consisted in the last segment of the red sun before disappearance becoming a bright green (without any transition through intermediate tints); this green was as nearly as could be judged the complementary to the red of the sun itself. On one occasion I shut my eyes immediately after the green tint appeared, and it remained visible. There could be no doubt that what I saw was the purely subjective after-image of the disappearing segment of the sun. (Nature, 94:672, 1915)

## THE GREEN FLASH
**Whitmell, C. T.; *Nature,* 95:35-36, 1915.**

Prof. Porter's interesting letter (Nature, February 18, p. 672) on this subject must be my excuse for sending a summary of my own experience during the last eighteen years in which I have observed the flash more than a hundred times, and in no single case did I find anything not explainable by atmospheric dispersion, nor anything that could be put down as a subjective or complementary after-image.

I may add that I have observed with the naked eye, with an opera-glass (power 3), with binoculars (power 9), and with a telescope (power 100).

Whenever on a clear day a low sun is observed through a telescope, the upper limb appears bordered with a marine, i. e., blue-green, fringe, the lower with an orange-red fringe, the side-limbs are unaltered. (The telescope should have a solar diagonal and other means of reducing the brightness of the sun.)

The marine upper fringe develops ultimately into the green flash, the blue element weakening as the sun descends. I have watched this change with the telescope, and it is perfectly continuous.

Again, if the sun descends behind a low cloud, parallel to the horizon, but with a clear space between, the base of the sun, just as it becomes visible, shows the red flash. I have seen this only thrice, as the necessary conditions are obviously seldom satisfied. The red flash seems inexplicable save by dispersion.

Under favourable conditions at sunset, as the upper segment of a yellow sun gradually diminishes, the right and left corners of the segment become green; this colour gradually spreads inwards, becoming marine, until finally the last tip of the sun has sunk, a very faint wisp of blue light is glimpsed directly above the point of disappearance. One friend even records a violet wisp.

But when the sun is orange the blue is replaced by green, and when the sun is really red no green flash at all is seen, the atmosphere cutting off the green as well as the blue rays. To see these changes it is desirable to use a power of 8 or 9.

Prof. Barnard, writing to me some years ago, said he preferred the title, "blue flash," as in sunsets seen over the Pacific from the Lick Observatory the final flash was usually blue. Doubtless this is due to clear atmosphere.

It is well known that at sunrise, when no exciting colour can be present, the flash has been seen, sometimes green, sometimes blue.

In the 1906 volume of Symons's Meteorological Magazine appears correspondence on this subject by Dr. Rambaut, myself, and others, and the editor, Dr. Mill, in summing up the matter, decided strongly in favour of the dispersion theory. Capt. Carpenter, R. N., whose numerous observations appear in the British Astron. Assoc. Journal, holds the same view.

It is quite true that, if I look steadily at a bright red sun, and then close my eyes, I see a green after-image, but this is just what the observer of the green flash should not do. He should avoid looking at the sun when it is bright, and should wait until it is so low that the eye can easily bear the light---should wait, in fact, until only a very small segment is visible. As before stated, with a really red sun the flash fails to appear, so far as my experience goes. If proper precautions are taken I do not think any appreciable after-image will be present.

I may point out that the nature of the horizon, provided it is clean-cut and low down, makes but little difference. It may be of cloud, land, or water, and, of course, the last is the best. My experience relates to all three.

But few persons appear to have used a telescope for observing the flash. If those who have not done so would observe a low sun with a power of, say, 100, I think that they would be convinced that the true cause was atmospheric dispersion.

The real mystery about the flash is that it so often fails to appear, when apparently all conditions seem favourable. I have not yet found any explanation of this, but I am inclined to the opinion that at a clear sunset the flash could always be seen if a telescope were available, though it might be too feeble for the naked eye, for a telescope invariably reveals the upper green fringe when the sun is low. The telescope, of course, was achromatic, and showed no colour with a high sun. (Nature, 95:35-36, 1915)

## THE GREEN FLASH
**Starr, C. Vaughan; *Meteorological Magazine,* 64:290-291, 1930.**

Many of the people one has met in Jaffa are familiar with the "green flash," and I too have seen it on several occasions when on the beach near Jaffa. As in the letter referred to, the sun right down on the horizon retains its brilliant orange-yellow colour until just as the last portion disappears a bright green "spot" is seen, which, to me, has always appeared as a small disc of green light and lasts about a quarter of a second.

One evening when expecting to see this flash, a party of four of us, including two doctors, had discussed the possibility of this phenomenon being due to some effect in one's eyes and not to refraction; to try this out two of us watched the sun carefully until it had set, while the other two looked eastwards and on a signal from the sun watchers turned quickly just as the sun was disappearing. The two who had watched continuously saw the green flash, while to those who did not look till the very last moment, the sun went down orange-yellow, and no green was seen. (Meteorological Magazine, 64:290-291, 1930)

## GREEN AND RED FLASHES
**Thompson, H. M.; *Marine Observer,* 23:144-145, 1953.**

M. V. Coptic. Captain H. M. Thompson. Aruba to London. Observer, Mr. B. J. Pratt, 4th Officer.

21st August, 1952, sunset. As the sun's lower limb, altitude 35', appeared to emerge at 2208 G. M. T. from the base of a large Cu (estimated at 2, 500 ft), a brilliant red flash was observed. As the sun's disc sank below the horizon (2215) a brilliant bluish-green flash was seen from its upper limb.

Position of ship: 26° 20'N, 55° 40'W.

Note. This observation is of very great interest, observations of the red flash being extremely rare, though it might be seen more often if it were definitely looked for in suitable conditions. In conditions of normal refraction the sun's upper limb is bordered by a narrow rim of bluish-green or green colour and the lower limb by a similar rim of red colour, when the sun is near the horizon. These rims are wider and therefore more likely to be observed in conditions of abnormal refraction. The upper green rim gives the well-known green flash when the sun sinks below the horizon; almost all the sun is hidden before the green coloration can be seen. The green flash has also been seen when the sun goes down behind a cloud with a sufficiently well-defined upper edge. The red flash can never be seen at the horizon since the red rim would be the last part of the sun to rise and it would be lost in the sun's general light. It can therefore only be seen in the circumstances of the present observation, when almost all the sun is hidden by a cloud with a well-defined lower edge at which the emerging red rim is seen. The only previous observation of which we can find a record is that made by Mr. Whitmell in 1906, on the sun's emergence below a horizontal bank of cloud. The phenomenon has been artificially produced by inserting a metal diaphragm to block out part of the field of view of a telescope and watching the sun's image emerge from behind this. (Marine Observer, 23:144-145, 1953)

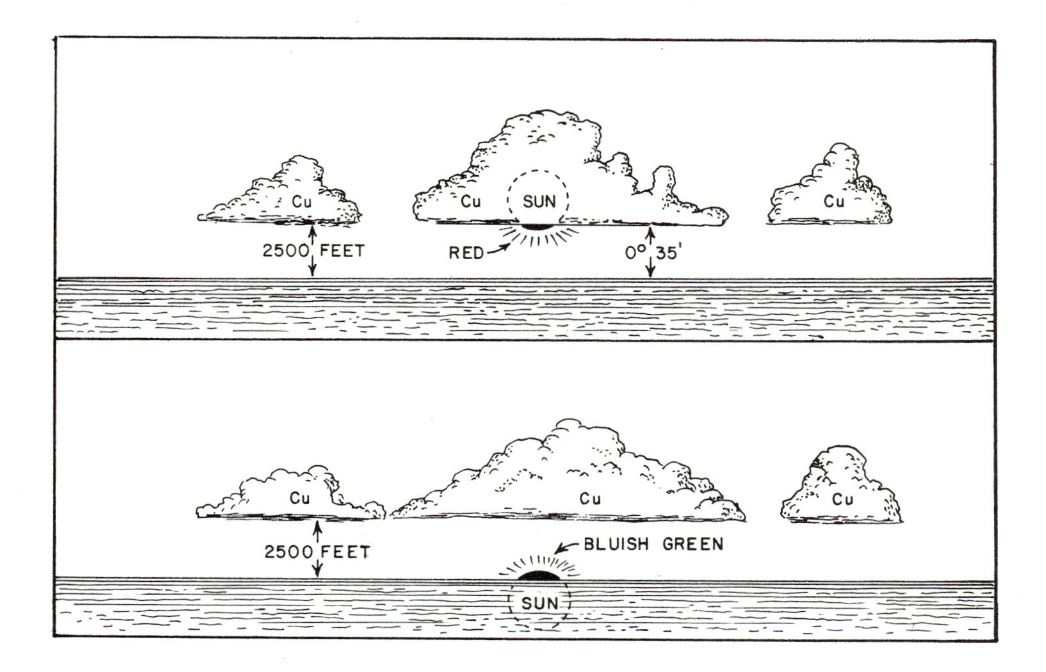


Red and green flashes from setting sun

## GREEN-TO-PURPLE FLASH
**Bennett, J. D.; *Marine Observer*, 28:76-77, 1958.**

M. V. Otaki. Captain J. D. Bennett. Napier to Balboa. Observer, Mr. E. Norman, 3rd Officer.

12th April 1957. At 0053 G. M. T., the green flash was observed under almost perfect conditions. There was no evidence of any abnormal refraction, the sun sinking uniformly below the horizon without distortion. There was, however, a more striking series of colour changes than is normally seen. Just before the top edge of the sun's disc finally disappeared, the green coloration quickly changed to blue, from pale blue to intense blue and finally to a purple, almost violet, shade as the last rays of light vanished.

Position of ship: 15° 52's., 104° 50'w.

Note. Observations of the violet flash are rare because light of such short wavelength is almost always absorbed in passing horizontally from the setting sun through the lower atmosphere. The blue light is usually absorbed also, although observations of the blue flash are not so rare. A number of observations of the blue flash have been published in this journal, also a few of the violet flash. In some of these cases the change of colour from green to blue or from blue to violet was seen, but as far as we can remember we have never had an observation before in which the whole change from green, through shades of blue, to violet was seen, and Mr. Norman is to be congratulated on having observed what must have been a very beautiful spectacle. There is evidence that the rate of colour change in this phenomenon varies on different occasions, for in an observation by S. S. Asia, published on page 211 of the

October 1954 number, the whole change seems to have been telescoped into a split second, the flash being described as "greenish-purple". (Marine Observer, 28:76-77 1958)

## DOUBLE GREEN FLASH
**Tanner, W. B.; *Marine Observer*, 29:57, 1959.**

S. S. Arabia. Captain W. B. Tanner, R. D. Montreal to Liverpool. Observer, Mr. I. K. Grindrod, 3rd Officer.

11th June, 1958. At 2300 G. M. T. the sun was on the point of setting on the NW. horizon. As the upper limb sank below the horizon there was a distinct green flash lasting 1/2 sec: the top of the sun's disc then reappeared for approximately 6 sec and set again, when there was a second well-defined green flash. At 2400: vis. over 10 miles, no cloud. Very heavy SW'ly swell.

Position of ship: $55^{o}$ 46'N., $32^{o}$ 56'W. (Marine Observer, 29:57, 1959)

## THE GREEN FLASH IN ANTARCTICA
**Anonymous; *Nature*, 135:992, 1935.**

Dr. Raymond M. Bell, of the Pennsylvania State College, directs attention to Admiral Byrd's account of the phenomenon in the National Geographic Magazine (58, 186; 1930). Admiral Byrd observed the green colour for as long as thirty-five minutes in Antarctica as the sun rolled along the horizon. (Nature, 135: 992, 1935)

## THE GREEN FLASH—AN UNUSUAL FEATURE
**Barlow, E. W.; *Meteorological Magazine*, 57:246, 1922.**

On September 20th I observed the "green flash" at sunset from my house at Wadhurst, Sussex, and noted a feature which I have not seen described in previous accounts of the phenomenon.

The sun set over low hills in the Ashdown Forest region and at a distance estimated at eight miles from the point of observation and several isolated rounded trees were projected on the disc for some time before the sun's disappearance.

The largest of these trees subtended an angle estimated at about two minutes of arc. The observation was made with a telescope of three inches aperture magnifying eighteen diameters.

The phenomenon occurred in two stages: (a) Every tree projected on the disc became suddenly outlined with a narrow hazy border of green light, the rest of the sun retaining its normal colour; (b) after a perceptible interval of time the whole of the remaining segment of the sun turned a brilliant clear emerald green and almost simultaneously disappeared below the horizon.

The sky was entirely clear at the time, with a little haze at low altitudes, so that while the setting sun was a golden colour only slightly tinged with red, the light was sufficiently reduced to enable it to be viewed directly without any dazzling or retinal fatigue effects. (Meteorological Magazine, 57:246, 1922)

## BLUE FLASH AT MOONSET
**Lawson, A.; *Marine Observer,* 36:182, 1966.**

s. s. British Destiny. Captain A. Lawson. Kharg Island to Suez. Observer, Mr. D. Roberts, 3rd Officer.

1st December 1965. Through binoculars, a blue flash was observed as the moon set behind the mountains at Muscat. The atmosphere was exceptionally clear and dust free. Sky cloudless. Air temp. 75.7°F.

Position of ship at 1800: 24° 12N, 58° 24'E.

Note. The blue flash phenomenon seen at sunset or moonset is essentially the same as the green flash, but in the case of the moon it has been less often observed. Due to the long passage of the white rays of the sun or moon through the atmosphere, almost the only rays which escape the effects of absorption or scattering are the red and blue-green. Refraction causes the blue-green rays to appear at the top of the disc, while the red occur at the lower edge. The blue-green rays are thus the last to be seen at the moment of setting. (Marine Observer, 36:182, 1966)

## RED FLASH AT MOONRISE
**Thomas, W. S.; *Marine Observer,* 31:68, 1961.**

m.v. Deseado. Captain W. S. Thomas. Santos to Las Palmas. Observer, Mr. J. McCaughrean, 2nd Officer.

19th May 1960. The moon on rising at 0339 GMT was seen, when examined through binoculars, to have assumed the distorted appearance shown in the sketch. As the lowest part of the limb cleared the horizon a brilliant red flash was observed which lasted for 1-1/2 sec. approx. Cloud present was 1/8 $C_I$ 1 and 2/8 $C_I$ 8.3. Air temp. 79°F, wet bulb 71°, sea 81°. Wind E'S, force 3. Sea slight, no swell.

Position of ship: 14° 25'S, 36° 30'W.

Note. The red flash is a similar phenomenon to the green flash, and the abnormal vertical distribution of temperature and water vapour content in the atmosphere that produced the abnormal refraction intensified the red flash. (Marine Observer, 31:68, 1961)

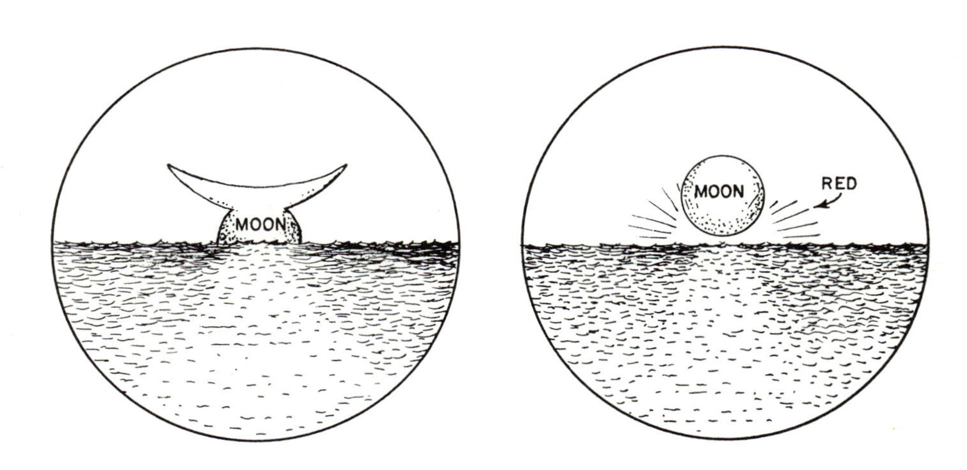


Red flash and moon image distorted by refraction

## MOON'S LIMB SCINTILLATES IN BRILLIANT SCARLET
**Brittain, D. B.; *Marine Observer*, 30:135, 1960.**

S. S. Devon. Captain D. B. Brittain. Fremantle to Melbourne. Observers, Mr. A. E. Robinson, 3rd officer and the Quartermaster.

24th August, 1959. At 2200 S. M. T. the side light and poop lights of a vessel were seen plainly, its distance away being estimated at about 5 miles: radar, however, gave the distance as 12-1/2 miles. An unusual degree of scintillation was observed in the case of stars, which became visible at 1/2$^{o}$ altitude. In these conditions of abnormal refraction, it was hoped to see the green flash at moonrise, but a cloud bank low on the horizon prevented an observation being made. However, when the moon had risen to 15 min altitude and was clear of the cloud, its lower limb was seen to scintillate with a brilliant scarlet colour. The coloured rim gradually contracted in length and the last of the scarlet disappeared when the moon was at an altitude of of about 1$^{o}$. Air temp. 53$^{o}$F, sea 56$^{o}$; wind, light and variable; low swell.

Position of ship: 37$^{o}$ 22'S., 132$^{o}$ 32'E.

Note. Phenomena similar to this have been reported by aircraft crews, usually at sunrise, and have been found to be associated with low level temperature inversions, probably at about 1,000 ft. (Marine Observer, 30:135, 1960)

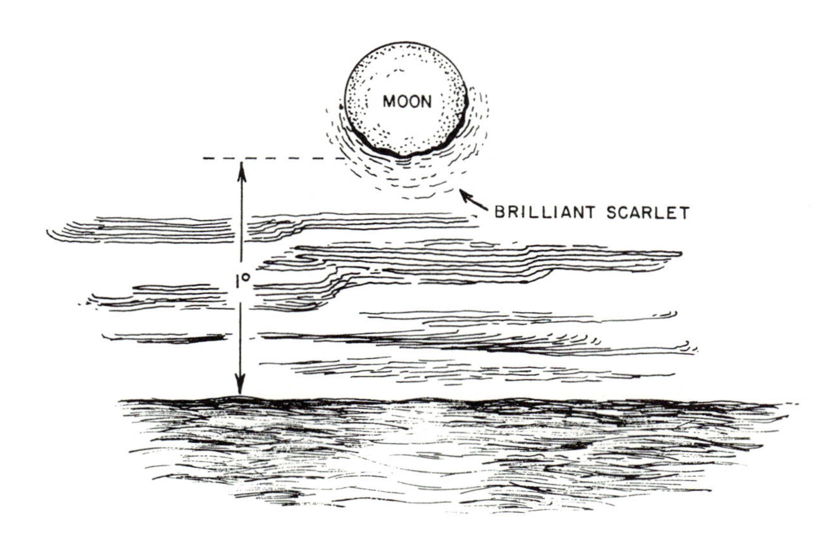


Lower edge of moon scintillates in bright scarlet

## VENUS—ABNORMAL REFRACTION
**Borthwick, C. A.; *Marine Observer*, 36:183, 1966.**

m. v. Delphic. Captain C. A. Borthwick. Aden to Suez. Observer, Mr. M. H. Murray, 3rd Officer.

11th November 1965. At 1744 GMT when Venus was about to set, bearing 237°, it changed from yellowish white to red. Shortly afterwards the planet appeared double, i. e. with a red image a short distance beneath the true disc. The image was below the horizon line and it gradually rose to meet the horizon as Venus decreased in altitude. At the moment of disappearing the two red discs coalesced and for an instant changed to a light green colour. The various phases of the phenomenon are shown in the accompanying sketch. Air temp. 79°F, wet bulb 75°, sea 79°. Sky cloudless. Wind NE'E, force 3.

Position of ship: 12° 35'N, 44° 37'E

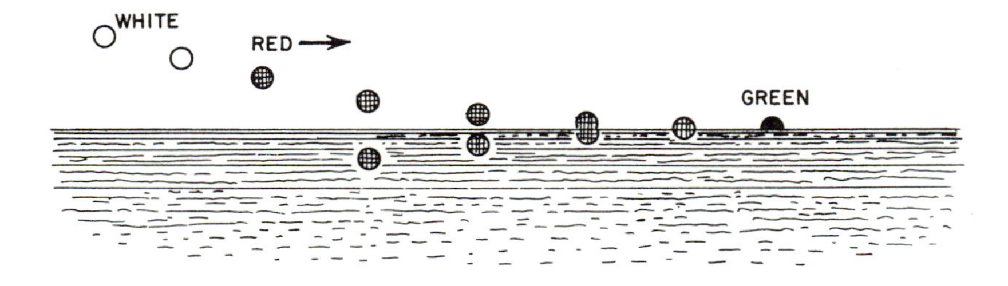


Double Venus and green flash

Note. The phenomenon described by Mr. Murray seems to be due to a combination of the effects of mirage and the well-known green flash. Venus is seen directly, while underneath it is its reflected image, with the horizon lying between. The yellow-white rays from the planet have an increasingly long distance to travel through the atmosphere as it nears the horizon and, due to absorption and scattering, only the red and blue green rays are left, the red being visible at the bottom of the disc and the green at the top. As the planet disappeared below the horizon the last colour to be seen would be the green at the top of its disc. (Marine Observer, 36:183, 1966)

## VENUS GREEN AND BLUE FLASHES
**Pilcher, C.R., and Bankier, A.; *Marine Observer*, 25:153-154, 1955.**

S.S. Captain Cook. Captain A. Bankier. Balboa to Wellington. Observer, Mr. A. Maclean, 2nd Officer.

29th July, 1954, 2120 L.T. The planet Venus was observed setting on bearing 275°T and altitude 3°. By the unaided eye and later by binoculars (magnification 7x) the planet's colouring changed alternately from orange through brilliant red to light and bright green. The shades remained virtually the same, but a "shimmer" of light appeared round the body just prior to its setting. At the instant the planet set a green flash was observed, and a small arc of whitish light over the point of setting gradually faded as the planet set below the horizon.

Position of ship: 6° 50'N, 80° 30'W.

Note. The arc of whitish light in the sky over the point of setting of Venus is very interesting as it seems likely to have been an observation of twilight due to the planet. This would naturally be of small extent and duration. (Marine Observer, 25:153-154, 1955)

# CURIOUS SUNSET PHENOMENA

When the sun sinks below the horizon and twilight falls, the rays of the vanished sun can still illuminate objects at high altitudes. High-flying aircraft may become bright points of light and generate UFO reports from inexperienced observers. When sunlight streams up from below the horizon through breaks in the clouds, it creates the well-known fan-like crepuscular rays. How though, are the similar and much rarer antisolar rays formed in the east when the sun as set in the west? If atmospheric conditions are favorable, peculiar bands of light and shadow span the entire heavens from the sunset point to the eastern horizon, where the dark blue earth's shadow is cast against the sky. Such are some of the puzzling phenomena seemingly caused by a sun already gone from the western sky.

## ANTISOLAR RAYS

### SUNSET PHENOMENON
**Wambaugh, W. J.; *American Meteorological Journal*, 8:229-230, 1891.**

On the evening of the 18th inst. my attention was attracted to a peculiar phenomenon, for which for a time I was unable to account.

On the eastern sky there were what appeared to be radiant beams of light emanating and diverging from a certain point upon the horizon. The appearance presented was quite similar to the shadows of clouds as are frequently seen projected upon the sky shortly after sunset, and before sunrise. In this instance, however, the sun was at an opposite point of the compass from where the phenomenon appeared, hence it was apparently not the source from which the light emanated. The sky at the time, from the point of observation westward was covered with cumulo-stratus clouds and a heavy rain was passing to the northwestward, while near the zenith it was quite clear and in the east was overcast with cirro-stratus clouds. There were no shadows visible in the west, for the entire sky, in that direction and the sun were obscured. Evidently clouds in the west which were not visible from the point of observation, cast shadows which, with alternate beams of light passed overhead in parallel lines, but which as a result of perspective were rendered invisible near the zenith and then converged and made visible near the eastern horizon. This view is sustained by the fact that the beams had a lateral movement toward the north,---first appearing at a point south of their apparent center and then gradually moving northward and disappearing at a point north of that center. This movement corresponded with the directions in which the clouds in the west were moving. (American Meteorological Journal, 8:229-230, 1891)

## CONVERGENT SUNBEAMS
**Hopfield, John J.; *Nature,* 141:333, 1938.**

The phenomenon of 'divergent' beams of sunlight is known to everyone. The illusion of divergence of the beams is so good that one has forcibly to remind himself that they are really parallel.

One day last summer, I observed 'convergent' beams and this observation was corroborated by others with me. It was late afternoon in a flat country. Clouds hid the sun, which, perhaps, was nearly eclipsed by the horizon. It had been raining in the surrounding country and cumulus clouds were still prominent in the west. The rays of the setting sun gave the familiar fan-like pattern in the west, passed almost invisibly overhead and then 'converged' in the east as if they emanated from a sun on the eastern horizon. These rays were of lower intensity than those in the west but were plainly visible.

The explanation of the phenomenon is obvious; one is under the parallel beams, and looking along them in opposite directions. The spectacle was new to me. (Nature, 141:333, 1938)

## CONVERGENT SUNBEAMS
**Trotter, A. P.; *Nature,* 141:558, 1938.**

The necessary conditions for the phenomenon of convergent sunbeams are a slight haze spread evenly overhead, and a few small cumulus clouds on the eastern horizon. The best example that I have seen was at sea while travelling from Cape Town to Durban at Easter 1899. By good fortune, Mr. R. T. A. Innes, then secretary of the Cape Observatory, and others were on board, and there was opportunity for careful observation and comparison of notes.

There was a beautiful sunset; cloud shadows radiated from the sun, converging toward a point as far below the horizon as the sun was above it. As the sun set this point rose and the rays met at the eastern horizon. Long bands of pink and blue arches over our heads and over most of the sky like marking of a huge melon. One could have sworn that they were curved---obviously curved---but they were really straight and parallel. A drawing or photograph of any small portion of the sky would have shown straight lines. To the north, south or overhead they would have been parallel, to the east or west they would have been straight but converging. After the sun had set, the converging point rose and a solid looking blue band touching it, rose too. This was the shadow of the earth, which may often be seen in the east after sunset. (Nature, 141:558, 1938)

## ANTI-SOLAR RAYS
**Thomas, W. S.; *Marine Observer,* 30:19, 1960.**

S. S. Loch Ryan. Captain W. S. Thomas. Cristobal to Rotterdam. Observers, Mr. F. G. Nickson, 3rd officer and the watch.

29th January, 1959, at 2055 G. M. T. Before sunset, when the sun was 7° above the horizon, bearing 246°, several bands showing a darker colour than the sky, which was rose pink, were observed to be radiating from a point just below the horizon diametrically opposite to the sun (see sketch). As the sun set, the pink colouring gradually faded and the bands disappeared. Air temp. 73°F, sea 73°. Visibility excellent. 2/8 Cu.

Position of ship: 28° 40'N., 47° 51'W.

Note. The darker sections of the sky are probably associated with the shadow of the earth thrown by the sun on to the earth's atmosphere. (See Marine Observer's Handbook, page 77.) (Marine Observer, 30:19, 1960)

Pink anti-solar rays

## UNUSUAL TWILIGHT PHENOMENA OVER ETHIOPIA
**Gouin, Pierre; *Weather,* 23:70-71, 1968.**

In Ethiopia, at sunset, the lower part of the eastern horizon is normally pink. This pink layer moves up, becomes better delineated as a dark segment develops underneath it; it seen becomes an anti-twilight arc with a horizontal extent of some 180 deg.

Three to five minutes after the upper limb of the sun has disappeared below the western horizon, from a point 4 to 5 deg above the eastern horizon and in the centre of the anti-twilight arc, dark blue bands suddenly radiate in all directions through the pink-coloured sky, progressively cover the whole hemisphere and converge towards a focus apparently lower under the western horizon than the actual position of the sun. Then the blue bands slowly vanish from the east, covered by the earth's shadow advancing westward.

The number of dark blue bands may vary from one to seven or nine. When unique, the preferred zenithal angle seems to be 45 north or south. By the time of their disappearance in the earth's cone of shadow at the upper limit of the twilight are (on 16 December 1966, about 10 deg above the western horizon 33 min after apparent sunset) the height of the dark blue bands was estimated to be 30-35 km.

The same phenomenon is also seen before sunrise, but I never observed one in the morning as geometrically perfect and complete as in the evening. Moreover, the time of disappearance before sunrise seems shorter (20 min, 30 December 1966) than at sunset. This difference in time of disappearance at dusk and dawn could probably be interpreted as daily variation in height of the same atmospheric layer, provided the elevation of the point of disappearance be the same in both cases. The pink or reddish colour of the evening may be attributed to volcanic dust in the stratosphere.

The pattern is usually observed to be more or less geometrically perfect, from October to the end of December in the Addis Ababa region. Although it is more spectacular when the sky is clear, the clearness of the sky is not a guarantee that, on a particular day during that period, the phenomenon will take place. The azimuth of the solar point at sunset is not an indication either; the solid angle swept by the solar point on the horizon from January to 15 March is the same as that covered from October to January, and no complete display has been reported after 1 January. After January, dark rays can be seen in the west, but they are not of constant width. (Weather, 23:70-71, 1968)

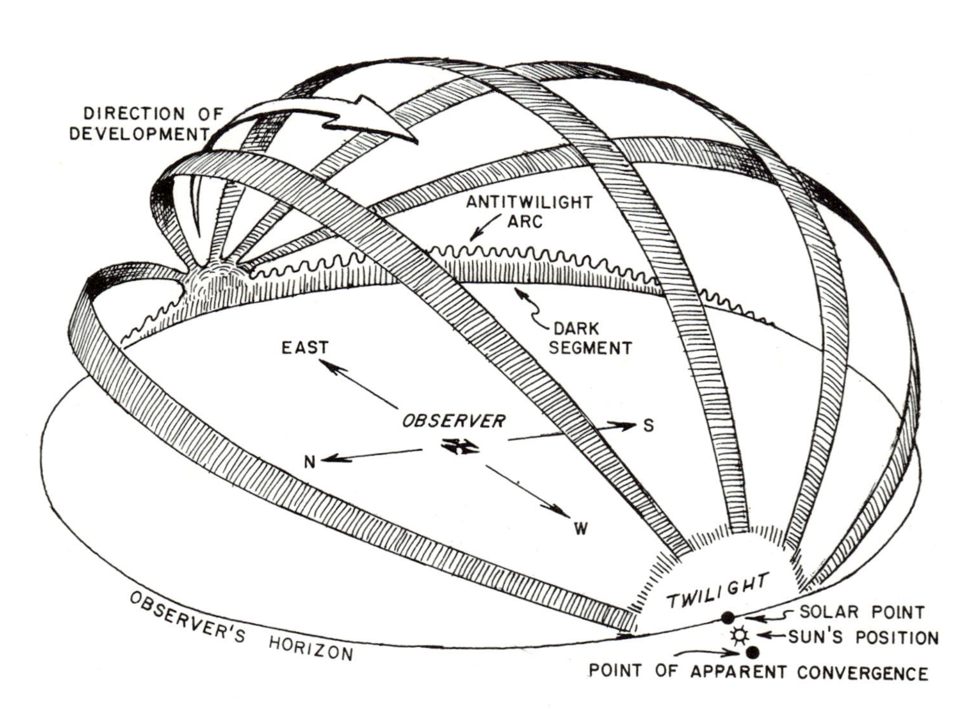


Geometry of dark-blue twilight bands seen in Ethiopia

# THE EARTH'S SHADOW CAST ON THE SKY

## THE EARTH'S SHADOW
**Anonymous; *Nature*, 173:475, 1954.**

An article in l'Astronomie (November 1953), entitled "Sur l'Ombre de la Terre", by J. Dubois, of the Bordeaux Observatory, gives the results of some recent research, and more especially of his own work carried out at the Pic-du-Midi Observatory and at Bordeaux, on the phenomenon of the earth's shadow or the anticrepuscular arch. This phenomenon, which occurs a short time before sunrise and a short time after sunset, is briefly described for sunset; but the same description in inverse chronological order applies also at sunrise. As soon as the apparent altitude of the sun is less than about $2^{o}$, a greyish-blue zone rises above the eastern horizon, slowly ascending as the sun sinks, changing in colour from blue to an increasing dark hue, and ending as a vast surbased arc, forming what is known as "the earth's shadow". Some minutes after sunset a whitish zone, somewhat bright, called "Albe" by M. Durand-Greville, is clearly distinguished, and to it is due the reillumination of clouds or of the tops of mountains---phenomena which have puzzled many observers and for which various theories have been proposed. In a discussion of the history of the observations and aim of the researches, reference is made by M. Dubois to the observations of Gruner and R. P. Combier, the latter having published the results of his work during the years 1938-40. However, it is to M. Dubois that the most important work is due, and all who are interested in the phenomenon would be well advised to read the description of his equipment and his conclusions. In this short note, attention can merely be directed to a few of his deductions, from the time when he started work in 1945 until he completed it four years later. He points out that the conclusions deduced by observations with the spectrophotometer are consistent with his theories, and that the main factor in promoting the anticrepuscular arch is the ozone in the atmosphere localized between heights of 20 and 30 km. The blue tint observed must be attributed to the feeble intensity of the yellow and red radiations which are rather strongly absorbed by the ozone. It appears that absorption by aqueous vapour and oxygen has very little effect on the intensity and colour of the arch. (Nature, 173:475, 1954)

## SUNSET PHENOMENON
**Rees, R. G.; *Marine Observer*, 18:193-194, 1948.**

The following is an extract from the Meteorological Record of S. S. Pipiriki. Captain R. G. Rees. Aden to Suez. Observer, Mr. J. Laidlow, 4th Officer.

11th November, 1947, 1440 G. M. T. (sunset). At about 1450 the sunset glow broke up, leaving blue sky from horizon to zenith in the sector between W x S and SW. On the W x S bearing, the division between bright orange glow and blue sky was at right angles to the horizon, whereas on the SW bearing the orange glow was at an angle of $20^{o}$ to the horizon. This condition prevailed for about 15 minutes, when the orange glow in the W x S direction gradually covered the blue sky and joined the orange light in the SW, which had remained stationary. When the phenomenon was at its best, the summit of Jebel Aduali could be seen between the two sections of orange light.

Position of Ship: Latitude 13° 20'N., Longitude 43° 06'E.

Note.---The most probable explanation of this phenomenon is that the blue sector was formed by the shadow of some inland mountain or plateau below which the sun had sunk. (Marine Observer, 18:193-194, 1948)

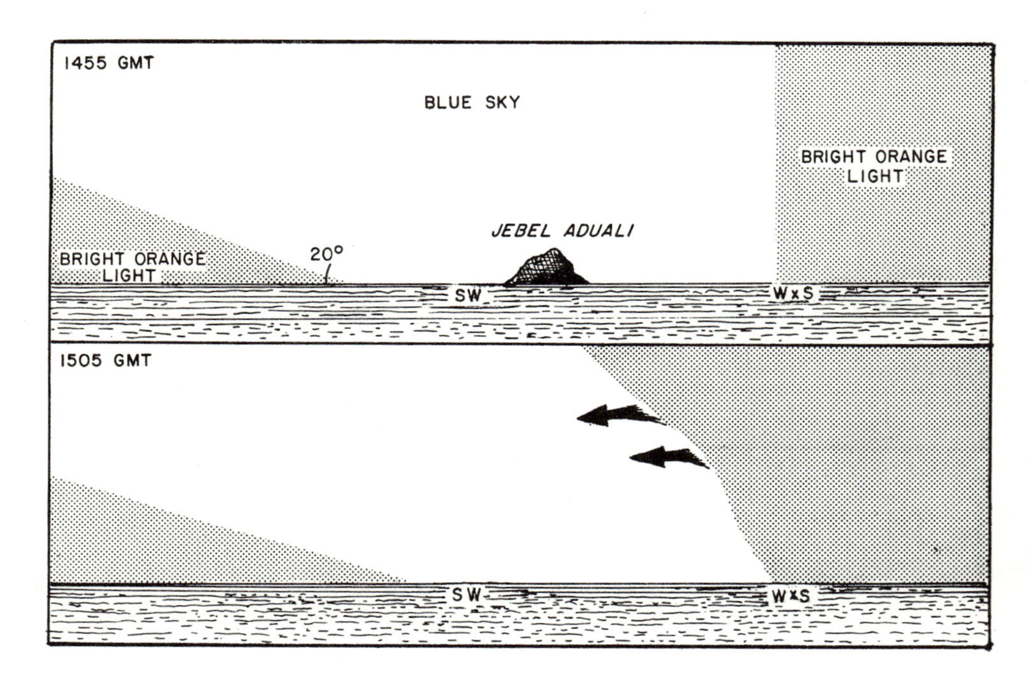


Strange sunset sky colors

## BLUE RAYS AFTER SUNSET
**Pattison, P. V.; *Marine Observer*, 29:14-15, 1959.**

M. V. Angelina. Captain P. V. Pattison. Observer, the Master.

28th February, 1958. Just after sunset when off the coast of Viet Nam, a peculiar blue light was observed in the western sky in the form of a fan-shaped shaft of Prussian blue extending from approximately 4° to 40° above the horizon. The base of the shaft bore from between 264° and 269° and it inclined to the N. at an angle of approximately 15°. The sun had set behind the land and there were low layers of brownish-red colouring in the sky, which faded with altitude to a pale pink. Some small low Cu type clouds were also present.

The blue ray was very clearly defined and there were four other pale thin indistinct rays lower down to the N. The phenomenon was visible for some time, the southern edge of the principal ray remaining well defined until the sky darkened.

Position of ship: 12° 18'N., 109° 36'E.

Note. This vessel is one of the voluntary observing ships on the Hong Kong Fleet List and the observation has been sent by the Director of the Royal Observatory, Hong Kong, who remarks that both the 0900 and 1200 G. M. T. charts on 28th February show isolated Cb clouds over Viet Nam which might have caused the phenomenon.

Deep blue fans or rays seen in the sky after sunset are shadow phenomena produced by the obstruction of the sun's rays by some object of relatively high altitude below the observer's horizon. The clearly defined shaft seen may well have been the shadow of a Cb cloud. An alternative explanation is that it was the shadow of a high hill or mountain, as the hinterland in this region rises steeply in height. The indistinct rays seen further to the N. were similar phenomena, where the obstructing objects were further away and also further, angularly, from the line from the observer to the sun's position below the horizon. (Marine Observer, 29:14-15, 1959)

## SKY COLORATION
**Jackson, I. W.; *Marine Observer*, 27:86, 1957.**

M. V. Wanstead. Captain I. W. Jackson. Auckland to Panama. Observers, Mr. R. Hall-Soloman, Chief Officer, and the Radio Officer.

6th April, 1956, 0315 G. M. T. An interesting example of sky coloration was observed at sunset. A wedge-shaped piece of sky, in the E., was coloured deep blue, almost purple, and the sky on either side of the wedge was a very pale blue.

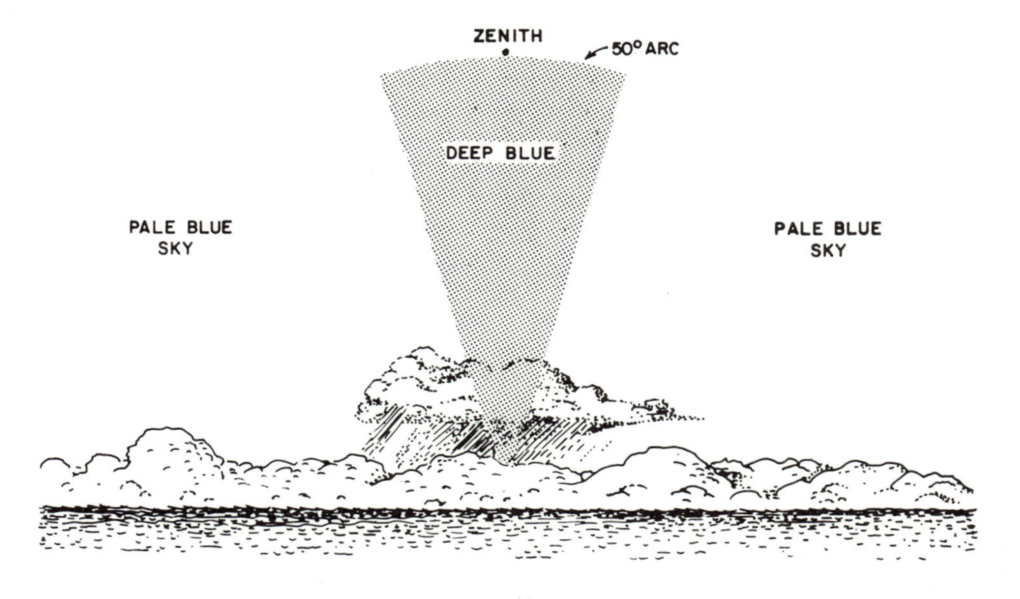


Shadow of earth thrown on sky

The wedge was extremely well defined and covered an arc of from approximately 10° above the horizon, where it disappeared into the clouds, to the observers zenith. The thick end of the wedge, directly overhead, was covering an arc in the sky of about 50°. A particularly heavy shower of rain was directly in line between the sun and the observed phenomenon. The outstanding feature of this peculiarity was the extremely fine definition of the sides of the wedge and the sudden change from purple to pale blue. Air temp. 79°F, barometer 1011.5 mb.

Position of ship: 27° 05'S., 133° 38'W.

Note. This is an interesting observation since the colour effect described is not what is usually seen. The normal aspect of the clear eastern sky after sunset is as follows. At sunset a steely blue segment, darker than the rest of the sky, begins to rise up from the horizon opposite the place of sunset. This is the shadow of the earth thrown by the sun on to the earth's atmosphere. The earth-shadow has a narrow band of rose or purple colour, known as the counterglow, along its curved upper boundary. The whole rises fairly quickly in altitude, with the shadow encroaching on the eventually obliterating the counterglow. The shadow rises to the zenith, but before this it generally becomes invisible owing to the progressive darkening of the whole sky. (Marine Observer, 27:86, 1957)

# MISCELLANEOUS SUNSET PHENOMENA

## PINK SEMICIRCLE OPPOSITE THE SUN

**Backhouse, Thos. W.; *Symons's Monthly Meteorological Magazine,* 18:187, 1884.**

There is one striking phenomenon of the remarkable sunrises and sunsets which I have been very much surprised not to see mentioned, and yet it is a phenomenon I never saw before they began. I allude to a great pink semi-circle opposite the sun, which is at its best about twenty minutes before sunrise and after sunset. It does not last many minutes at each time, which may be the reason it has not attracted so much notice as the wide pink ring round the sun or moon, which is much less conspicuous, and doubtless arises from the same cause, for it, also, I never saw before. It has been more or less visible each morning that the sky has been clear enough, ever since the strange sunsets began. I have not seen it so often in the evening. I first noticed it on the morning of November 27th. It consists of the upper half of a very broad bright ring, whose outer radius is perhaps 25°, and inner about half that; it is more or less pink, sometimes a deep pink, but sometimes the inner half is white. Its edges are indefinite. The space within is darker, and always more or less blue. The centre seems just opposite the sun.

I would ask your readers whether this phenomenon is of frequent occurrence, for though I never saw it before, I am aware that this is no actual proof that it does not often occur; for it is much easier to see a phenomenon again, than to discover it. But if, as I believe, it is a phenomenon connected with the sunrises and sunsets, it is to my mind more striking and characteristic than any other of their features, which makes it the more strange that it has been so little observed. I suppose, also, that if any one has sufficient optical knowledge, it should throw much light on the cause of the fine sunrises, &c.---Yours truly, (Thos. Wm. Backhouse)

P. S.---I found yesterday afternoon, and again this morning and afternoon, that the semicircle is visible, though very faintly, when the sun is above the horizon; but there appears to be an interval, about sunrise and sunset, when it disappears entirely. This morning it was distinctly visible from 8.48 to 8.58, and had a slight pink tinge, the interior being blue as usual. It then gradually faded, but at 9.45 was still faintly visible, and nearly white as compared with the neighbouring sky. (Symons's Monthly Meteorological Magazine, 18:187, 1884)

## A MOCK SUN AND RAINBOW COLORS AFTER SUNSET

**Nichols, A. E.; *Meteorological Magazine*, 55:149, 1920.**

On July 11th, shortly after sunset (20 h. 13 m. to 20 h. 35 m. G. M. T.), a remarkable colouration of the sky occurred. The clouds at the time, which covered about nine-tenths of the sky, were all Fr. St. Cu., and they did not change materially for two hours. Just as the green light---the greenish sky often seen at sunset, but more brilliant---was disappearing, an image of the sun, yellow surrounded by red, appeared on the clouds in the eastern horizon. The cliffs of France then drew attention by changing from white to pink and to a brilliant luminous carmine. The sun image faded and the colouration of the clouds became a brilliant pink, and about 50° above the horizon, in the same azimuth as the image which had gone, there appeared a portion of an arc of rainbow colours. The surrounding clouds were of different shades of purple and paynes-grey, and there was a large V-shaped cloud a pure white in the north-east. On the south horizon a replica of the irisation occurred, including the rainbow colouring. The cliffs of France changed to a thin purple line and the irisation gradually faded, and nearer the zenith a light brown colouring of the clouds took place. The phenomena lasted 20 minutes.

The weather following was overcast to continuous thunder, with lightning and rain. The rain reading on Monday 12th was 0.45 inches. (Meteorological Magazine, 55:149, 1920)

# UNUSUAL HALOES AND MOCK SUNS

When the sourcebooks were first planned, haloes, mock suns, sun pillars, and allied phenomena were considered too well understood to warrant space in these collections of curiosa and rarities. Research soon demonstrated that many of these optical apparitions remain rather mysterious and not fully explainable in the usual terms of sunlight reflected from ice crystals in the atmosphere. For example, some haloes are not centered on the sun or moon, and their origins are thus obscure. Mock suns are also asymmetric on rare occasions. While the overwhelming majority of haloes are circular, a few are elliptical, posing a theoretical puzzle. In sum, to quote several of the notes following observations from the Marine Observer, a goodly residuum of solar and lunar haloes and arcs are "unrecognized".

## PRIMER ON THE COMMON HALOES

### HALOS AND CORONAE
Botley, Cicely M.; *Weather,* 18:5-7, 1963.

Halo phenomena are due to refraction and reflection of the light of the sun or moon by ice crystals; they are often very complicated and diverse owing to the variety of crystal forms and the large possible number of refractions and reflections. Refraction phenomena, especially those due to sunlight are sometimes beautifully coloured.

Figure 1 gives a schematic drawing of some of the most common halo phenomena. It is well to know these so as to be able to recognize unusual forms when they occur.

A. The common halo of 22°radius, formed by refraction in hexagonal crystals at 60° angle, and which may be seen, entire or in part, on about 200 days of the year. This halo enjoys an enormous reputation as a sign of bad weather, which, as it is often seen in the cirrostratus in front of a depression, is not unfounded; it also occurs, often in fragments, in fine-weather cirrus. It is usually not well coloured, appearing as a narrow whitish ring, reddish on the inside.

B. The circumscribed halo, oval in shape, which may surround the 22° ring when the sun is higher than 29°. At lower altitudes it breaks up into upper and lower arcs of contact. Uncommon, but to be looked for, is a curved bar across the upper contact arc---the Parry arc. Anomalous contact arcs are sometimes to be seen.

C. The mock suns or parhelia of 22°, which are two bright patches, often brilliant and beautifully coloured and sometimes with long white 'tails', on a level with the sun. They often appear by themselves and arise when the axes of the ice-crystals in large number are vertical. They are on the halo at low altitudes, but as the sun rises they recede from it and sometimes become connected with it by the arcs of Lowitz.

D. The halo of 46°, due to refraction at 90° in hexagonal ice crystals oriented at random. It is not common and is faint, though the colours are better than in the common halo. It has parhelia and certain contact arcs, but these are rare.

E. What has been described as the 'most beautiful of all halo phenomena', the circumzenithal arc, which appears as a miniature rainbow, red towards the sun and parallel to the horizon, a few degrees above the summit of the $46^{\circ}$ halo when this is visible. Often this arc appears alone. It is not infrequent but somewhat transient, so it is not well known. Three or four times it has been seen as a complete circle. It arises from refraction through crystals, shaped like plates or umbrellas, with vertical axes floating in a stable position.

F. The parhelic circle, a white ring passing through the sun parallel with the horizon and due, as its lack of colour shows, to reflection from the side faces of crystals floating with axes vertical. On it are the parhelia, those already mentioned and others, notably those at $90^{\circ}$ and $120^{\circ}$. Other odd ones also occur and, if possible, the distances of these from the sun or moon should be carefully measured with instruments. Opposite the sun, on the ring, is the very uncommon anthelion, G, through which may pass certain arcs. The parhelic circle is often fragmentary. (Weather, 18:5-6, 1963)

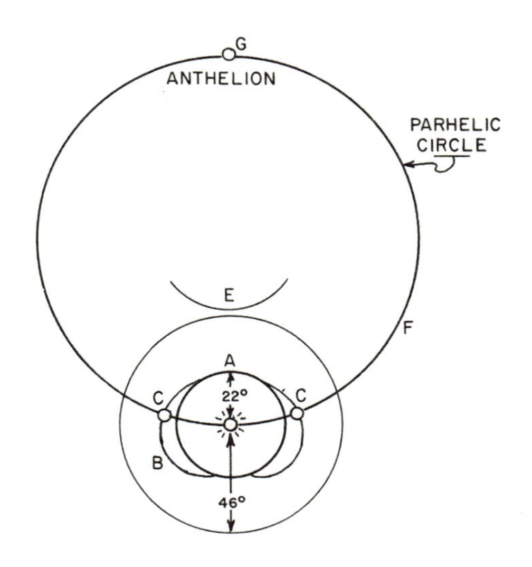


The common haloes. See text for explanation.

# RARE AND DIFFICULT-TO-EXPLAIN HALOES

## RARE SOLAR HALOES
**Crabbe, J. A.; *Weather*, 10:359, 1955.**

I hereby pass on to you some details of a sky phenomenon which I observed last year, in the hope it may be of some value to you. My host at the time, a local Italian chemical engineer, had never seen such a sight before.

Place: N. Italy: Sistiana (on the coast between Trieste and Monfalcone).

Date: 13 June 1954.

Time: 1145 - 1315.

Description: we first noticed a circle of darker sky, concentric with the sun, at 11.45. At noon we noticed the second circle (same diameter as the first): it was paler than the first, and its circumference cut the centre of the sun. The peripherae of both circles were concentric rings of spectrum colours (unfortunately I did not note the positional order of the colours). At 1210 we noticed the horizontal bar of dark grey "cloud" in front of us, between the almost overhead sun and the horizon: it was composed of horizontal layers of the spectrum colours (again, unfortunately, I did not note the positional order of the colours). Meanwhile the two circles overhead maintained their position in relation to the sun, but were gradually becoming fainter.

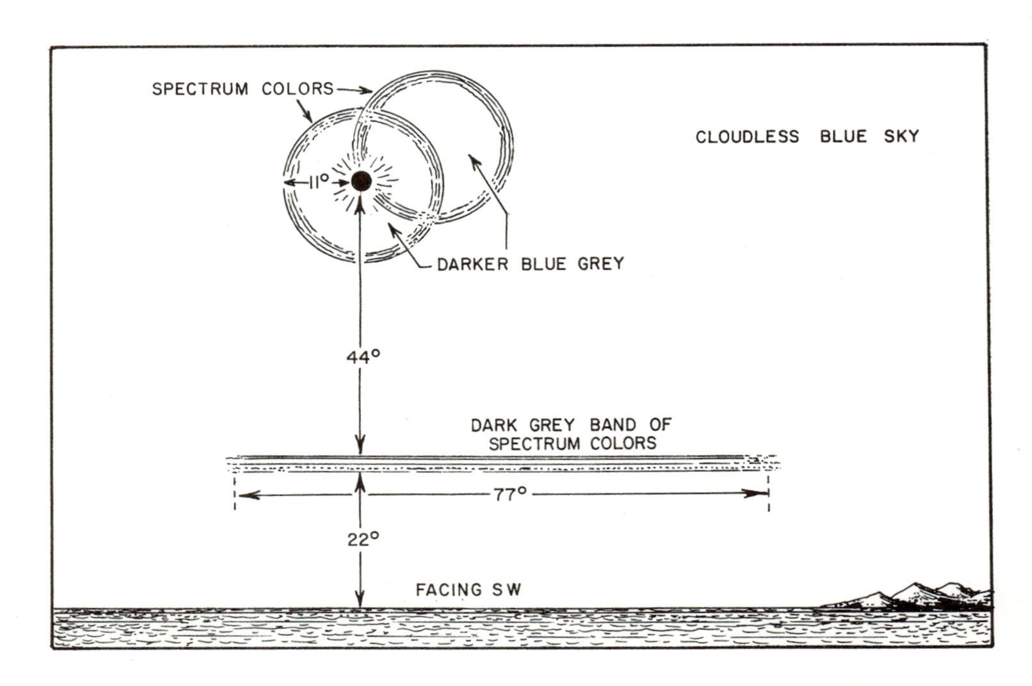


Offset solar halo

By 1315, when we left the beach everything had practically disappeared. It was a fine day, with a cloudless blue sky and blazing sunshine.

Measurements: the only source I could contrive was to use the four knuckles of the fist on an outstretched arm. If this be A, then the two circles were both of radius A. The distance from the sun to the horizontal bar was 4A. The length of the horizontal bar was 7A, and it was 2A from the horizon.

I enclose a copy of a sketch I made at the time, the rough measurements being given here in degrees. (*Weather*, 10:359, 1955)

## A DESCRIPTION OF A HALO OR CORONA OF GREAT SPLENDOR

**King, Alfred T.; *American Journal of Science*, 1:40:25-27, 1841.**

If you consider the subjoined description of one of those meteorological phenomena, usually denominated by philosophers coronas or halos, which was observed in this town about eleven o'clock, A. M. on the 28th of August last, and which excited considerable interest among the intelligent portion of the community, and apprehension and alarm in the minds of the uninformed, worthy of a place in your excellent Journal, it is much at your service.

This phenomenon consisted of from three to five circular belts or zones of light, one of which emulated, in appearance, the splendor and magnificence of the most gorgeous rainbow. The arrangement of these rings was somewhat singular; the first or inner one, which had the sun in its center, was truly brilliant, exhibiting all the prismatic hues of the rainbow, the colors of which were so dazzling that the unprotected eye could scarcely rest upon it a moment. This, I presume, was occasioned by the sun being near the meridian, and consequently many of his rays would impinge upon the halo, without passing through the mass of vapor, to the existence of which I attributed the formation of the halo. The outer circles, however, one only of which appeared to be perfect, were composed of pure white light, and had for their centres the circumference, or a point near it, of the inner ring. Consequently, their circumferences, if all the circles had been perfect, would necessarily have passed through the apparent situation of the sun. I mentioned, however, that one only of these rings was perfect, the others were concentric arcs of circles which crossed one another, as seen in the accompanying diagram.

In the centre of the inner circle and bounded by it, a bluish mass of dense vapor was perceptible, which gave to the whole an embossed appearance, and added much to the beauty and brilliancy of the scene. Around and within the exterior circles there were also perceptible masses of vapor, though obviously much less dense than the mass which was nearer the sun. With the exception of these masses of vapor, and a large cumulus which lay to the south of us, and here and there a few scattered cirri, the sky was cloudless and the atmosphere calm and serene. The mercury in the thermometer stood at 86°. The weather continued thus for thirty six hours, when we had a smart fall of rain, and a descent of the mercury in the thermometer to 36°, at which point or near this, it has remained until about three days since, when it rose to 66°.

Coronas and parhelia have frequently been observed and accurately and glowingly described, by many scientific gentlemen, and various and conflicting opinions have been entertained respecting their causes, some attributing them to the peculiar state of the air consequent upon intense cold, while others, probably more correctly,

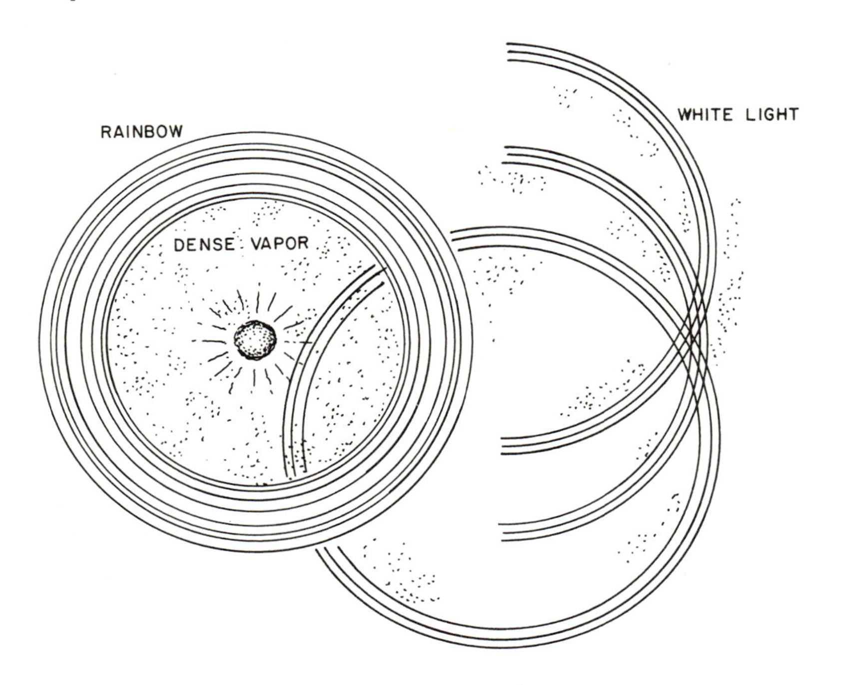


Rainbow-colored sun-centered halo with white offset haloes

attribute them to the refraction and reflection of the rays of light through masses of vapor which are formed in such aggregations as are not heavy enough to fall in the form of drops. Descartes remarks, that halos never appear when it rains. Coronas have frequently been observed around the moon, and even around Sirius and Jupiter, but, as far as my information extends, they have been but seldom variegated, even when they have encircled the sun.

I know not to what cause this phenomenon can be attributed, unless it be to the refraction and reflection of the sun's rays through the masses of vapor. Doubtless the first circle was thus formed, and if we suppose the rays of light from the circumference of this circle to be again refracted and reflected through another mass of vapor, an outer ring would evidently result. Again, if we suppose the same to take place from another point of this circle, a second ring would be formed which would cross the other in some point of its circumference, and in like manner, I presume, any number of rings may be formed. I offer this explanation, however, with much diffidence. (American Journal of Science, 1:40:25-27, 1841)

## LUNAR HALO COMPLEX
**Davison, I. C.; *Marine Observer*, 34:73, 1964.**

s. s. Devon. Captain I. C. Davison. Adelaide to Aden. Observer, Mr. D. Southworth, Jnr. 3rd Officer.

29th May 1963. Between 1515 and 1520 GMT when the moon was at an altitude of 60° and bearing 302°, the halo complex shown in the sketch was observed. The halo (a) and the arc (b) formed first; (x) and (y) appeared at the time when (a) and (b) touched. The arc (z) formed when (x) and (y) moved outwards and joined: the whole system faded when (z) lay along the top of (a). The amount of Cs. increased during the afternoon, being 8/8 at 1800.

Position of ship: 2° 21'S, 62° 33'E.

Note. This very complex system cannot be explained by any simple theory of haloes. (Marine Observer, 34:73, 1964)

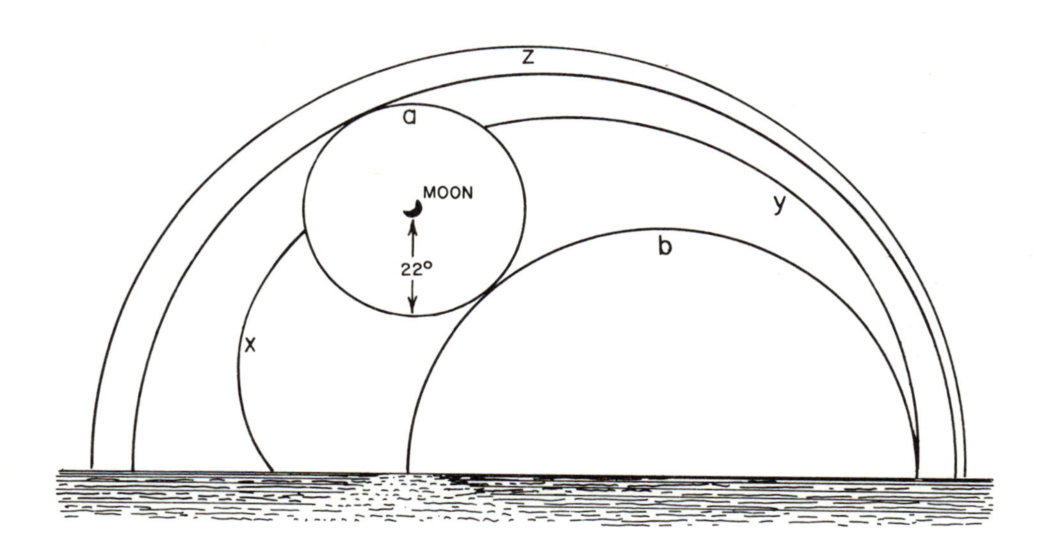


Curious complex of lunar haloes

## UNUSUAL LUNAR HALO
**Hazlewood, H. W.; *Marine Observer,* 21:16-17, 1951.**

M. V. Port Chalmers. Captain H. W. Hazlewood. Curacao to Colon. Observer, Mr. D. J. A. Pritchard, 3rd Officer.

7th March, 1950, 0630 G. M. T. A distinct lunar halo of radius 22° was observed. The halo was obscured from time to time by Cu. At 0640 an arc of a second halo, of apparently slightly larger radius, appeared. It intersected the primary and the outer perimeter passed to within about 2° of the moon, as shown in the sketch. The secondary was visible for 5 minutes, after which the area became obscured by Cu. After a few minutes the primary was again observed, but the secondary had vanished. The moon, which was 20 days old, had an altitude of 65°. The cloud formation consisted of 4/8 Cu. and 3/8 Cs.

Mean position of ship: Latitude 12° 40'N., Longitude 71° 19'W.

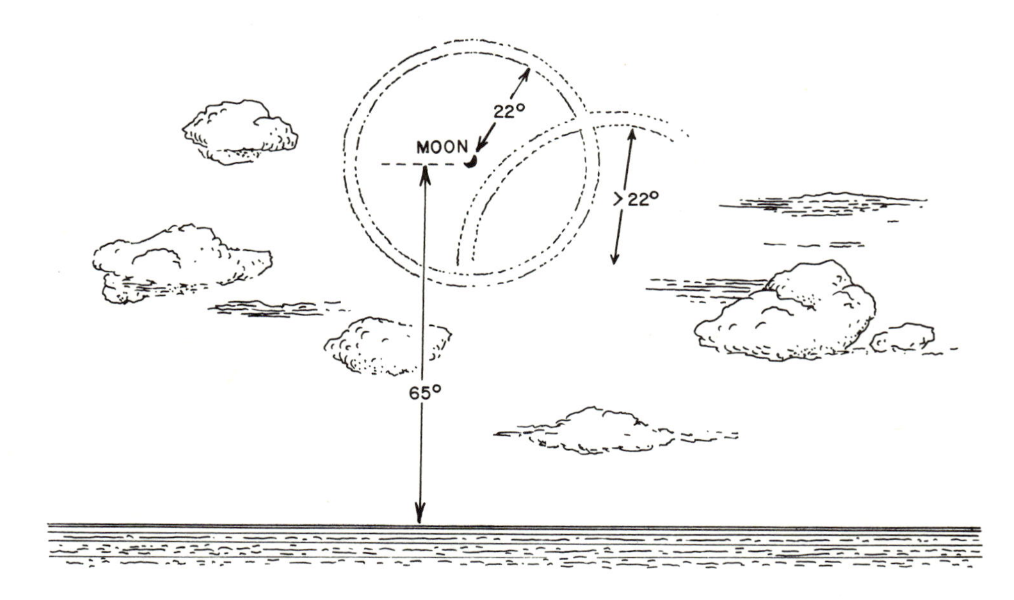


Offset lunar halo

Note. This is another interesting halo observation which cannot be fully explained. At first sight the arc of the halo intersecting the ordinary halo appears to be a secondary halo, such as was observed by O. W. S. Weather Explorer, see page 14. There are, however, two reasons why this cannot be the case: (i) that the halo is stated to be of slightly larger radius; (ii) the altitude of the moon is 65°. A secondary halo cannot be formed with an altitude of the moon (or sun) greater than 60° 45'. (Marine Observer, 21:16-17, 1951)

## HALO AROUND THE MOON
**Pryer, H.; *Knowledge,* 1:521, 1882.**

On January 29, between 6.20 and 6.40 p.m., I saw a very peculiar halo round the moon. I was in Tokiyo (Yedo) at the time, and the same appearance was seen by many of my friends at Yokokama, 18 miles distant. It was like the figure herewith. At first I was under the impression that the figure was slightly elongated on the outer circle, that is to say, the distance from the centre of the moon to the edge of the first halo was less than that from the edge of the first halo to the farthest point in the second; but on drawing it I am convinced that this was an optical illusion. The lower part of the second halo interesected the lower horn of the moon in a peculiar way, as given in the drawing. The moon was nearly half full. (Knowledge, 1:521, 1882)

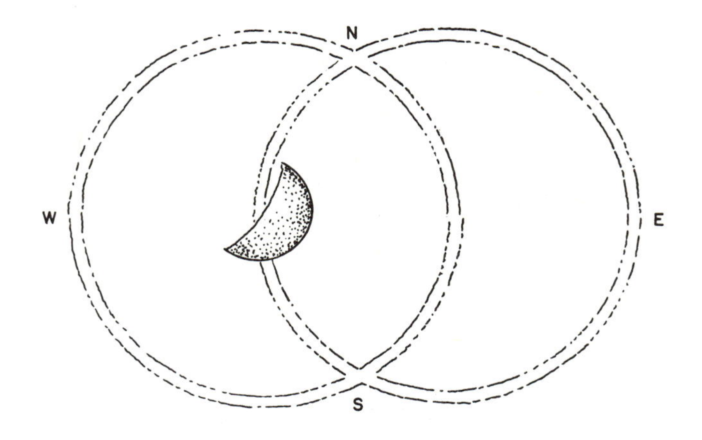


Intersecting lunar haloes

## UNRECOGNIZED HALO PHENOMENON
**Dick, W.; *Marine Observer*, 24:143, 1954.**

S. S. City of Khios. Captain W. Dick. Bombay to Vizagapatam. Observer, Mr. J. A. Parsons, 3rd Officer.

19th September, 1953, 1530 G. M. T. A lunar halo was observed with two arcs of contact. The inner radius of the halo was 20° 46'. The edges of the halo and the contact arcs were soft and merged gradually into the night sky. The arcs were slightly fainter than the halo. The phenomenon was observed for one hour, when it was obscured by Ac. Altitude of moon, 54° 35'.

Position of ship: 14° 32'N, 83° 06'E.

Note. This is an interesting observation, as the two arcs intersecting the halo, which seem to have been quite clearly seen, are not included in the recognised halo phenomena. (Marine Observer, 24:143, 1954)

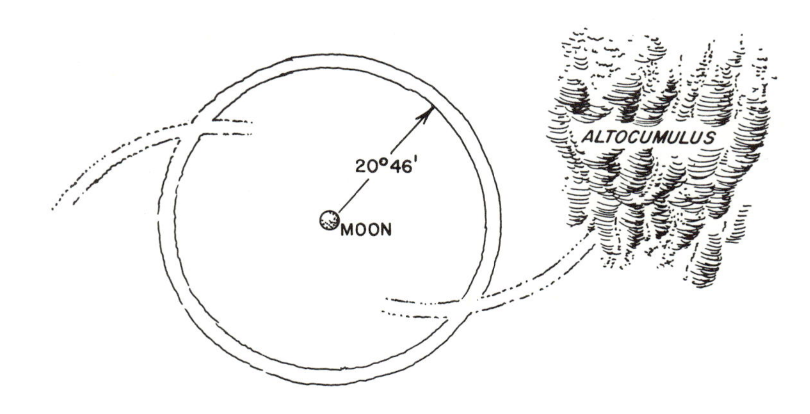


Lunar halo with intersecting arcs

## PECULIAR ARCS AND LUNAR HALO
**Robertson, A. I.; *Marine Observer*, 21:217, 1951.**

M. V. Rakaia. Captain A. I. Robertson, R. D., A. D. C., R. N. R. Auckland to Balbao. Observer, Mr. D. E. Moran, 3rd Officer.

20th October, 1950, 2005 A. T. S. (Zone +8). A lunar halo of radius 21-3/4° appeared as a plain white ring. At 2040, two arcs, meeting as in the sketch, appeared and were visible for about 12 minutes. At 2220 an area around the moon appeared green, tinged with orange red. The halo was visible until 2235. Cloud 4/8 Ci.

Position of ship: 24° 12'S, 122° 21'W.

Note. This is an interesting observation; the two arcs shown outside the ordinary lunar halo do not form part of halo phenomena as at present known. (Marine Observer, 21:217, 1951)

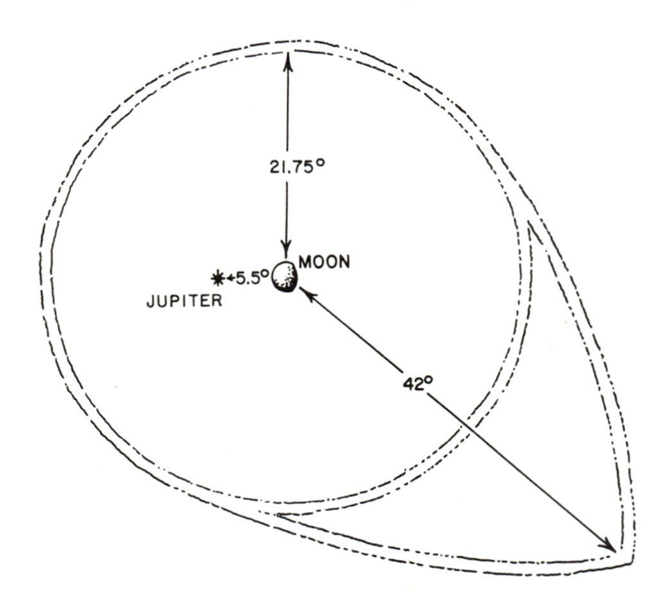


Lunar halo with unexplained arcs

# NONCIRCULAR HALOES

## ELLIPTICAL LUNAR HALO
**Goble, C. J.; *Marine Observer*, 6:221, 1929.**

The following is an extract from the Meteorological Log of S. S. Culebra, Commander C. J. Goble, R. D., R. N. R., London to Bermuda, Observer Mr. W. S. Thomas, 3rd Officer.

"25th October, 1928, 9.50 p.m. A.T.S. (0313 G.M.T.), Latitude D.R. 34° 08' N., Longitude D.R. 56° 33' W. A lunar halo was formed by the passage of a patch of Cirrus clouds across the moon. This phenomenon was specially interesting in that the halo took the distinct form of an ellipse. Careful measurements were taken and found to be as follows, viz.: Moon 3/4 full. Altitude 48° bearing 8.1° W. Halo major axis 25°, minor axis 20°. Temperature, dry bulb 75.4° F., wet bulb 72.4°F. Sea 76° F. Wind W.S.W., force 2. The halo maintained this elliptical form for about ten minutes."

Note.---Elliptical halos have occasionally been observed, but are among the rarest of halo phenomena. (Marine Observer, 6:221, 1929)

## ELLIPTICAL HALO
**Chapman, E. G.; *Marine Observer*, 29:116, 1959.**

S.S. City of Brisbane. Captain E. G. Chapman. London to Corner Brook (Newfoundland). Observer, Mr.R. E. W. Butcher, Junior 3rd Officer.

8th July, 1958. The halo shown in the sketch was seen from 1200 to 1230 G.M.T. It was silvery in appearance, against the 7/8 layer of white cs cloud present.

Position of ship: 45° 45'N., 45° 40'W.

Note. This halo is of special interest because it is elliptical. Almost all such luminous rings are circular. (Marine Observer, 29:116, 1959)

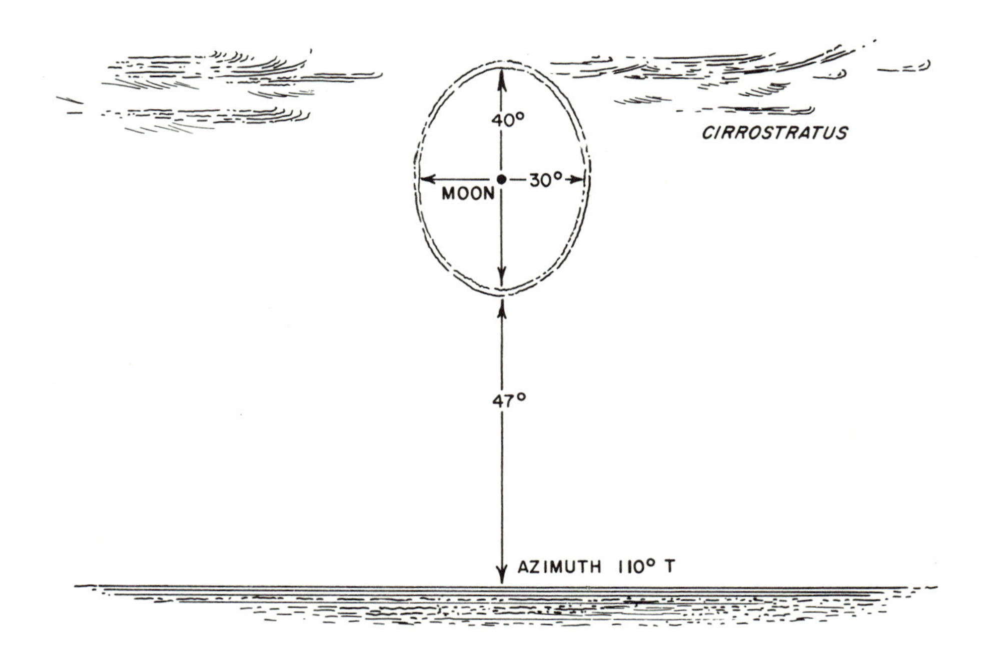


Elliptical solar halo

## REMARKABLE LUNAR HALO

**Evans, Lewis, *Royal Meteorological Society, Quarterly Journal*, 39:154, 1913.**

I send an account of an interesting lunar halo which I saw from the Union-Castle Line R. M. S. Balmoral Castle on January 21, 1913. I first noticed it about 6.15 p.m., ship's time; which would be about 6.30 local time, as the ship's time changes 12 to 15 minutes at midnight.

The halo was square with quite straight sides and the angles quite sharp, and I estimated the length of each side of the square at 3 diameters of the moon. The halo did not seem to vary in size as the moon rose higher, though I think at times it became less distinct, or possibly became rather less bright as the moon got higher, but that was hard to judge.

I pointed out the halo to some of the ship's officers, none of whom seems to have seen one like it before, so it may be very uncommon.

The following is an extract from the ship's log:

"Lat. $4^{o}$ S. Long. $8^{o}$ S. January 21, 1913, 8 p.m. The moon (aged 14 days) is framed in a dull, yellowish halo of square shape, which is about 3 lunar diameters in extent and is placed with a corner towards the horizon after this manner.

"January 22. 5 a.m. The halo is still visible." (Royal Meteorological Society, Quarterly Journal, 39:154, 1913)

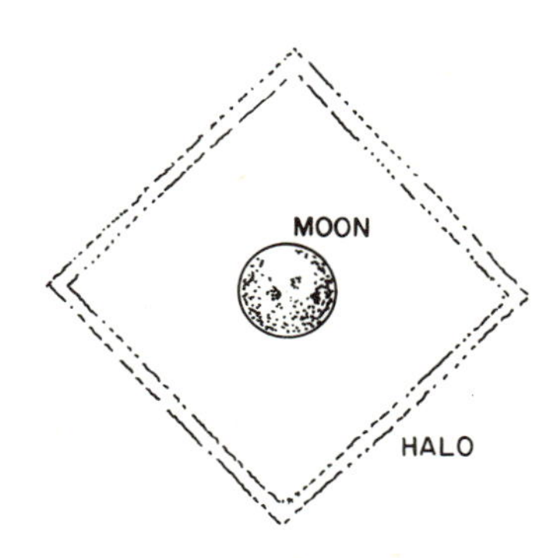


Square lunar halo

# ONE-SIDED MOCK SUNS

## AN EXCEPTIONAL PHENOMENON OF ATMOSPHERIC OPTICS: A ONE-SIDED MOCKSUN AT $3^{o}$ 30'

**Minnaert, M.; *Weather*, 21:250-251, 1966.**

Although many phenomena of atmospheric optics have been described, it now and then occurs that a completely new and astonishing observation is reported, as yet unexplained, and which one would unhesitatingly reject, were it not that very strong testimonies are presented. Several times such an observation, once published and brought to the attention, was found to be less unique than at first supposed. Sometimes even an explanation was found. These considerations may perhaps justify the publication of the following observation.

On 15 June 1965 the motorship Fairstar navigated in the Sea of Timor. The sea was very quiet, the sky almost perfectly clear. A few passengers were on deck, in the open air, looking at the sea, no glass pane or any other object being interposed between them and the prospect. About 10 minutes before the real one, somewhat to the left, but at the same height above the horizon. There was nothing at the symmetrical point to the right. She pointed this out to two other passengers, who then also noticed the phenomenon. She ran to her cabin, grasped her camera and made a colour photograph, which by a piece of good luck had almost the right exposure. After deveopment, it was found to confirm fully the visual observation. A reproduction of this slide is of course much less clear than the original (Fig. 1), which shows the sun and the mocksun as nicely round discs. [Photo not reproduced]

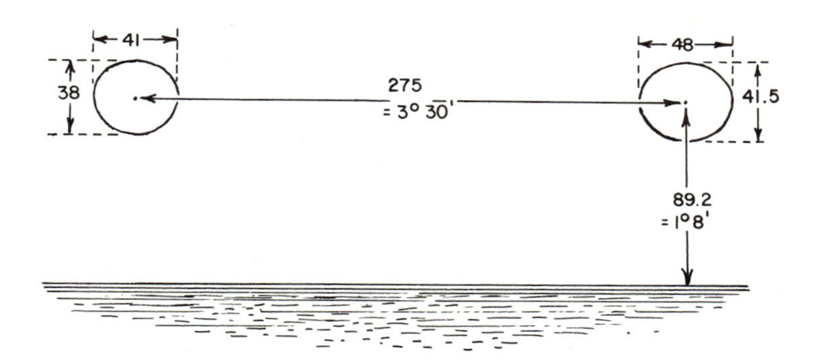


One-sided mock sun

The original was measured with a plate measuring machine and gave results, accurate to about 0.01 mm and well reproducible. The focal distance of the camera objective amounted to 45 mm. so that $1^{o}$ corresponds to 0.786 mm and 1 minute of arc to 0.0131 mm. The results are summarized in the diagram Fig. 2, where all dimensions are given in units of 0.01 mm. The following conclusions may be drawn.

(a) The sun and the mocksun are indeed precisely at the same height above the horizon: $1^{o}8'$.

(b) The horizontal diameter of the sun of 31'32' (after the ephemeris) should correspond to 41.2 of our units on the photograph; the mocksun is found to have precisely the geometrical diameter, the real sun is slightly bigger (irradiation?).

(c) The vertical diameter of the mocksun is 38, that of the real sun is 42 (irradiation); for a solar altitude of $1^{o}8'$ the observed ratio between the vertical and the horizontal diameter corresponds to the normal flattening by atmospheric retraction.

(d) The mocksun was at a distance of $3^{o}36'$ to the left of the sun.

(e) Below the mocksun there is a slight brightening of the sky, just as it is seen below the real sun, only much weaker. It looks as if the mocksun was simply a faint duplicate of the real sun.

The phenomenon remained visible till the sun was setting. It seems impossible to explain the observation by any reflexion or by instrumental effects inside the camera, since it was first noticed visually by three persons. Mrs. N. S., who reported the observation, makes a very reliable impression; however she has no special scientific education. (Weather, 21:250-251, 1966)

## AN UNUSUAL MOCKSUN

**Vigurs, J. T. C.; *Meteorological Magazine*, 61:91, 1926.**

Dr. C. C. Vigurs, of Newquay, sends us the following note which he has received from his son. No explanation of this curious phenomenon has been suggested.

Tuesday, Feb. 16th, 1926. S. S. "Manchester Civilian" on passage from U. S. A. to Avonlouth, in Lat. 51° 32'N Long. 6° 34'W. Hook Point, Waterford bearing 339° (clockwise from North) and distant 38 miles, and St. Ann's Head bearing 80-1/2° and distant 54 miles. Barometer 29.54 in. falling slowly. Dry bulb 47.2° F. Wet bulb 46.0° F. Wind W'S, force 5, Cloud 7, cirro-stratus around the sun and to the south and cumulo-nimbus below the sun and to the west; clear blue arc of sky to the north-northwest.

The phenomenon was first seen at 15h. 46m. G. M. T. when the true sun, with an altitude of 14° and bearing 227° was shining brightly with good limb in medium sextant shade, and a heavy squall was moving from west to south-southeast with the upper edge of the cumulo-nimbus cloud at a point vertically below the sun and having an

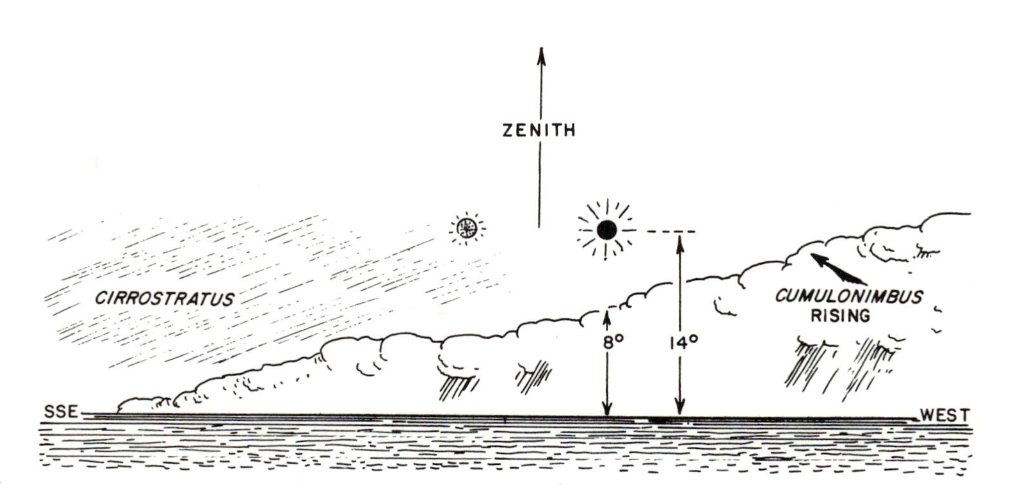


Mock sun 10° left of true sun.
Most mock suns are separated from true sun by 22°.

altitude of 8°. The phenomenon was an uncoloured image of the true sun at the same altitude and at an angular distance of 10° to the left, certainly not 22°. It was fairly well defined, sufficiently so for an azimuth bearing, but not enough for anything but a doubtful sextant altitude. The image had the appearance of the true sun shining through denser layers of cloud, although the cirro-stratus was fairly uniform in density; but with a slightly mottled appearance. No image was seen to the right of the true sun, and probably, though not certainly, the corresponding position to the right of the sun was already occupied by the moving cumulo-nimbus. No halo or mock-sun-ring was observed.

The image dimmed as the squall approached the true sun and disappeared when the true sun became covered. (Meteorological Magazine, 61:91, 1966)

# RARE RAINBOWS AND ALLIED SPECTRAL PHENOMENA

Everyone knows the rainbow, and just about everyone knows that it is caused by sunlight falling on a veil of raindrops. In more scientific terms, the sunlight (or moonlight) is dispersed into a spectrum as it is reflected back by spherical drops of water. Such commonplace rainbows are excluded here as unmysterious, but this still leaves us with red rainbows, displaced and intersecting rainbows, fogbows, spurious rainbows, and a host of colorful, often remarkable spectral phenomena.

In addition to rainbow displays, which are usually of circular configuration, one must recognize that linear cloud bows, broadcast rainbows, and isolated patches of colors often appear in the sky. Some of these phenomena are closely related to haloes; others have more obscure origins.

## EXTRAORDINARY RAINBOWS AND SPURIOUS BOWS

### SIX RAINBOWS AT ONCE

**Anonymous; *Royal Meteorological Society, Quarterly Journal,* 40:75, 1914.**

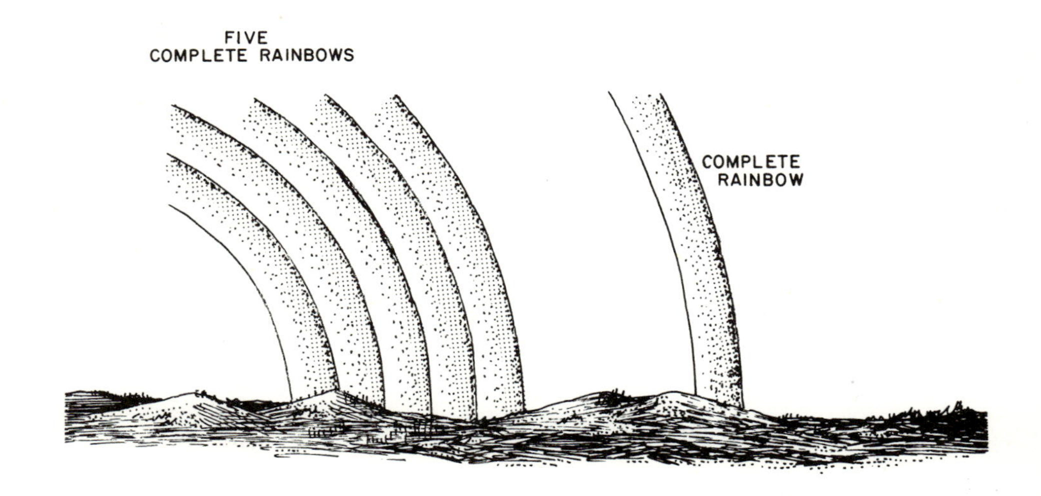


Six rainbows at once

Yesterday, November 5, at 5.40 p.m., a little previous to the sun disappearing behind the mountains, I noticed in the east six brilliant rainbows, extending for about one-quarter of the semicircle. Five were quite close together, and the sixth some little distance away as per accompanying rough sketch. (Royal Meteorological Society, Quarterly Journal, 40:75, 1914)

## FOG BOW
## Gardner, G. F.; *Marine Observer*, 6:11, 1929.

The following is an extract from the Meteorological Report of S. S. Saxon, Captain G. F. Gardner, O. B. E., Madeira to Cape Town. Observer, Mr. G. H. Pickering, 4th Officer:

"January 7th, 1928, between 4 a.m. and 7 a.m. Weather, passing fog banks and low Cumulus cloud. Visibility poor. Air 54°, sea 54°. Barometer 30.34 in. A single halo was observed at intervals, opposite the moon, i.e., 180°. The angle between the horizon and upper limb varied in size from about 6° to 10°, the lower limb appearing to reach to side of the ship. Position of ship at 8 a.m., Latitude 42° 08' N., Longitude 10° 09' W."

Note.---A fog bow is formed in the same manner as an ordinary rainbow by refraction and reflection of light from the sun or moon within drops of water. In the case of fog the phenomenon is produced by the very small drops of water suspended in the air. It can be proved that when the drops of water are very small, either in the form of fog or very fine rain, the resulting bow is white and not coloured as in the ordinary rainbow, because of the feebleness and overlapping of the colours. When the moon is the source of light the separate colours are rarely seen in a rainbow on account of its lack of brilliancy. (Marine Observer, 6:11, 1929)

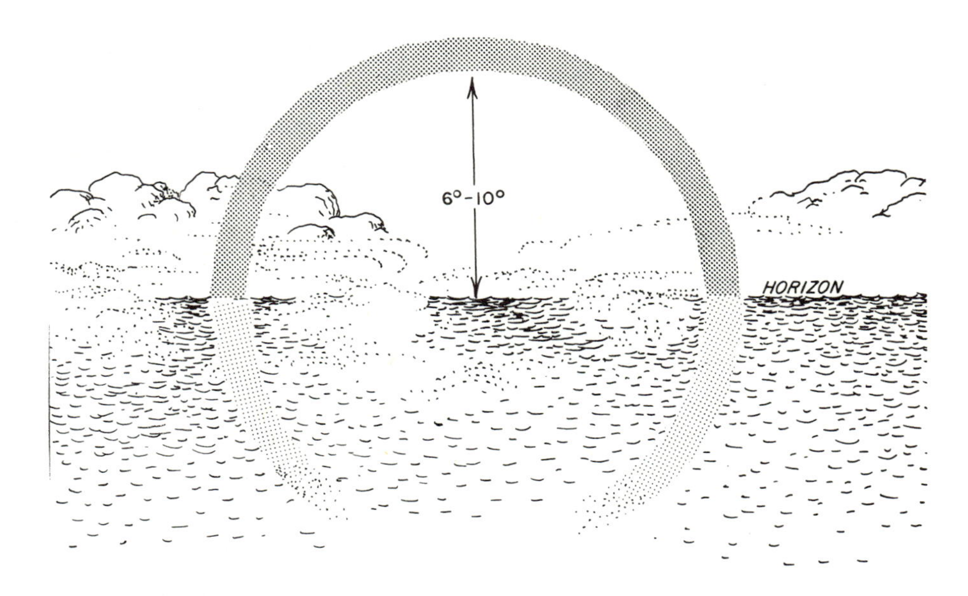


Typical fog bow

## A RED RAINBOW
**Harries, H.; *Meteorological Magazine,* 57:246-247, 1922.**

At 19h. 40m. on Sunday, July 30th, 1922, from the Western Explanade, Hove, I observed over the sea about 20° of the southern end of a very remarkable rainbow---a pillar of a dark red brick colour, the red so intense that none of the other prismatic colours could be detected. The western sky at this time was gorgeous in brilliant yellow gold, the sun itself being hidden by a cloud-bank. Gradually the red pillar stretched up and curved over until by 19h. 47m. it reached to a point above the South Downs and about 150° from the sea. The intense redness of the bow was maintained, and continued to swamp other colours, except for a faint indication of green in one or two short spaces immediately above the sea, near which was a faint secondary of about 15°, the same red colour only. Sunset was at 19h. 51m., but the bow, fading away from the northern end, did not vanish until 19h. 56m. The clouds overhead then assumed a soft rose tint, and the sky to west and north-west became a brilliant red. The automatic gauge in the Old Steine, Brighton, registered rain from 19h. to 23h., total .15 in.; the ordinary gauge, .11 in.

In my long life I have seen many rainbows, but never before have I beheld a red one---it even imparted a tinge of its redness to the atmosphere! A singular feature deserving of special mention was the appreciably greater breadth of the bow in the section above Brighton, a fact which attracted the attention of others as well as myself. (Meteorological Magazine, 57:246-247, 1922)

## THE EFFECT OF LIGHTNING ON A RAINBOW
**Minnaert, M.; *The Nature of Light and Color in the Open Air,* 182, 1954.**

A striking observation was carried out by J. W. Laine. Each time it thundered he noticed that the boundaries of the colours in the rainbow became obliterated. The change was particularly noticeable in the spurious (or supernumerary) bows: the space between the violet and the first spurious bow disappeared entirely and the yellow grew brighter. It was as if the whole rainbow vibrated. According to the table in Section 123, these alterations indicate an increase in the size of the drops.

This optical effect did not occur simultaneously with the lightning, but several seconds later. together with the sound of the thunder. One might imagine that owing to the vibration of the air, the drops tended to merge into each other; but this tendency is so slight that a really perceptible effect seems improbable. It is also possible that the electric discharge brings about a change in the surface-tension of the drops, so that they merge together more easily; but then it would be a mere coincidence if the space of time required for this change should equal that between the lightning and the thunder. (The Nature of Light & Color in the Open Air, 182, 1954)

## LUNAR RAINBOW
**Owen, G.; *Marine Observer,* 18:145-146,1948.**

The following is an extract from the Meteorological Record of M. V. New Zealand Star. Captain G. Owen, O. B. E. London to Australasia. Observer, Mr. D. J. Thomas, 3rd Officer.

29th July, 1947, 0230 G.M.T. The moon three-quarters full, with an approximate altitude of 25°, was observed dead ahead. Astern of the ship and covering some 3/10 of the sky was a mass of white cumulus. Ahead, was about 2/10 stratocumulus, through a gap in which the moon shone. The remainder of the sky was clear. Immediately astern of the ship was observed a lunar rainbow of a startlingly bright milky hue and, just ahead, another, larger but not so vividly bright as the one astern. These bows appeared to be not much more than the ship's length apart (600 ft.) and both were completely visible throughout their entire arcs. There had been no rain or drizzle for the previous eight hours, the sky being mainly clear of cloud. Neither had there been any suspicion of fog or haze, although these had been experienced 12 hours earlier. These phenomena were also seen by Cadet M. J. Slessor. Course 210°. Speed 16 knots.

Position of Ship: Latitude 47° 45'N., Longitude 06° 00'W.

Note.---This observation of the second bow, that ahead of the ship and thus between the moon and the observer, is a very remarkable one, but in the absence of angular measurements it cannot be fully explained. Two points of interest arise: (i) a rainbow is known to be formed in the position stated but it is so faint that observations of it are extremely rare---furthermore, according to optical theory it should be of approximately the same size as the primary rainbow, whereas the bow observed was larger; (ii) if conditions were so exceptional that this tertiary bow was fairly bright, one would have expected that the ordinary secondary bow, seen above and outside the primary bow, would have been readily visible. It does not, however, appear to have been seen. (Marine Observer, 18:145-146, 1948)

## DOUBLE LUNAR RAINBOW AND OFFSET BOW

**Robertson, G. M.; *Marine Observer*, 19:188, 1949.**

S. S. Corinthic. Captain G. M. Robertson, D. S. C. Curacao to London. Observer, Mr. S. H. Wilde, 3rd Officer.

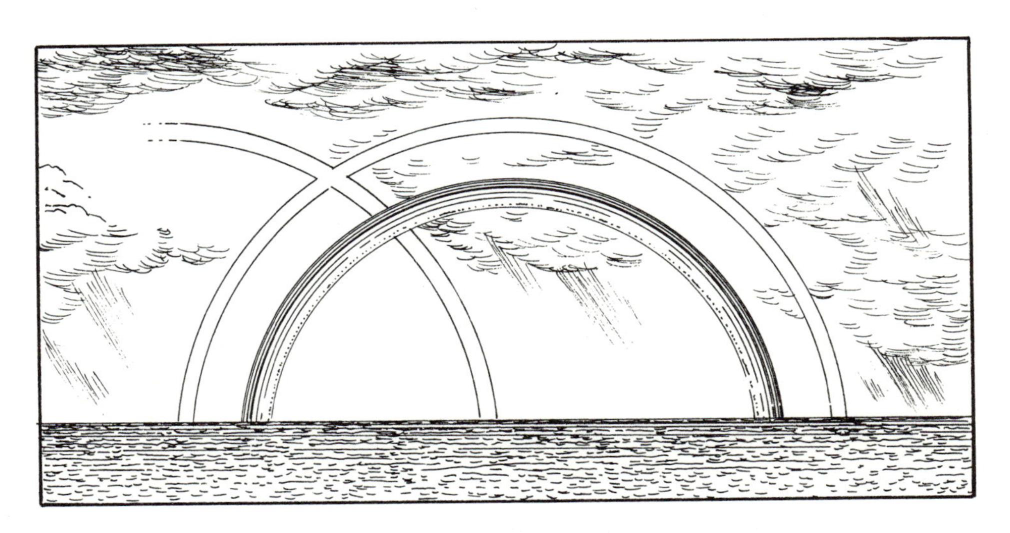

Lunar rainbows; primary, secondary, and intersecting third bow

10th November, 1948, 2200 G. M. T. Two complete lunar rainbows were observed and one arc of a third bow. The lower complete bow showed the spectrum colours very distinctly, but the upper bow and remaining arc were white. Bearing NNE, height of centre 28°. Passing rain squalls with Cb. 6/10 and As. 3/10, wind NW, 8. Course 065°. Speed 14-1/2 knots.

Position of Ship: Latitude 39° 01'N., Longitude 39° 06'W.

Note. This is a very interesting observation. The secondary lunar rainbow is not very often seen, as it is usually too faint to be visible. The most remarkable part of the observation is the arc of the third bow, which is an example of an abnormal bow. Such bows are very rarely seen and cannot be explained on the basis of the ordinary optical theory of rainbows. (Marine Observer, 19:188, 1949)

## RAINBOW AND WHITE ARC
**Eckford, R. D. S.; *Marine Observer,* 21:215, 1951.**

M. V. Laguna. Captain R. D. S. Eckford. Nassau to Havana. Observer, Mr. P. A. A. James, 3rd Officer.

31st October, 1950, 1330 G. M. T. Off Mollasses Reef an arc of a rainbow was

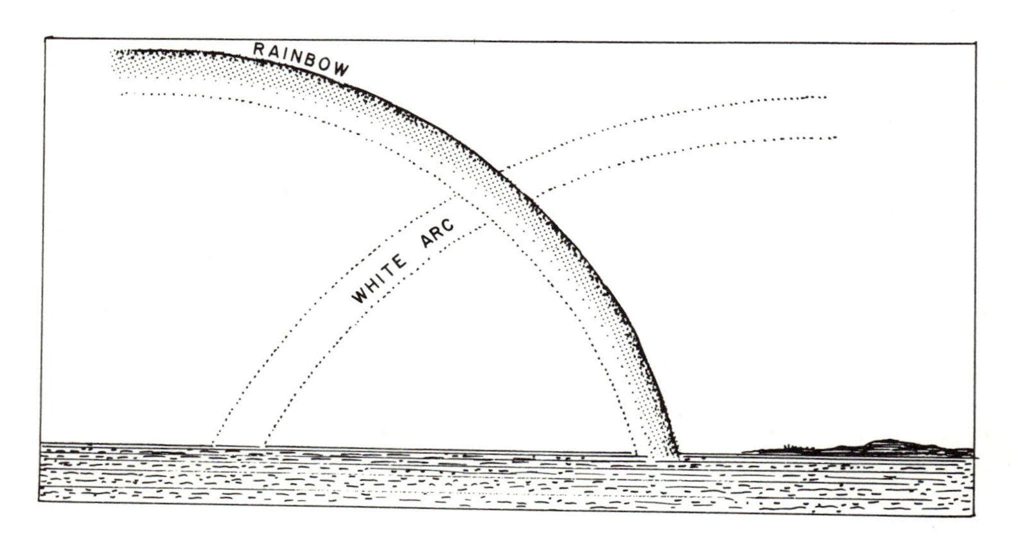


Rainbow intersected by white arc

observed extending from right to left for approximately a quarter of a semi-circle. At the same time a white arc intersected the other as shown in the accompanying sketch.

Note. This is an unusual and interesting observation and no explanation can be given of the cause of the white bow intersecting the primary rainbow. (Marine Observer, 21:215, 1951)

## INTERSECTING RAINBOWS
**Crichton, J.; *Meteorological Magazine,* 65:213-214, 1930.**

On the evening of August 12th, whilst camping on the foreshore, on the Scottish side of the Solway Firth near Brow Houses, towards 7 p.m. G.M.T. during a local thunderstorm which was moving southwestwards across the Firth, I observed an unusually brilliant primary rainbow together with a secondary with numerous spurious bows on the under side of the southernmost portion of the primary. The bows were not complete as there was an interrupting cloud layer at approximately 2,000 ft. level. The important part, however, of the observation was the appearance of an intersecting bow between the southern extremity of the primary and the secondary bow, the intersection at the primary being sharp at the extremity whilst the intersection and ending of the secondary bow was less sharp, inclining towards a fade out, the intersection here taking place near the cloud level. The colours of the intersecting bow were brilliant and of the same order as those of the primary. The whole phenomena on the southern portion of the rainbow resembled the letter N. but with the connecting stroke in the reverse direction, as shown in the sketch.

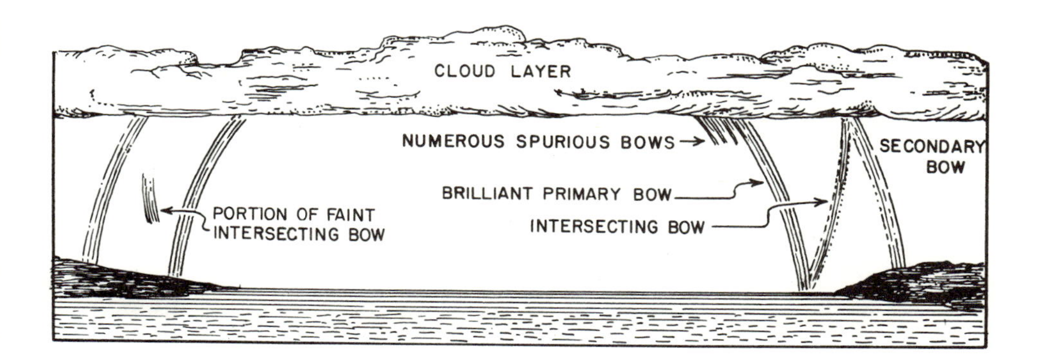


intersecting rainbows

Such an intersecting bow arises from the presence of two sources of parallel rays. In the present instance one set was that coming directly from the sun giving the primary and the secondary bows and the other set from the image of the sun in the bay to the westward of the point of observation. The image being as much below the horizon as the sun was above it, the centre of the intersecting bow was thus as much above the horizon as the centre of the primary was below it. The display lasted for nearly half an hour and was watched by a number of people. The water in the Solway was at low tide and almost perfectly calm. To the north side there were at times only faint traces of the intersecting bow. (Meteorological Magazine, 65:213-214, 1930)

## PECULIAR RAINBOWS
**Arthur, R. B.; *Marine Observer*, 27:210, 1957.**

M. V. Arabistan. Captain R. B. Arthur. Cape Town to Las Palmas. Observer, Mr. H. Sennett, 3rd Mate, Cadets G. Bashford and R. Tusler.

22nd October, 1956. At 0802 G. M. T. during generally cloudy conditions giving light showers of rain near the vessel, the end of a rainbow of large radius was seen on the horizon, the rest of the bow being hidden by low Sc (sketch 1). The band was 4° wide and fairly bright. Reading from the inside, the colours were red, yellow, green, blue, yellow, white, pink and red. The sun was then clear of cloud.

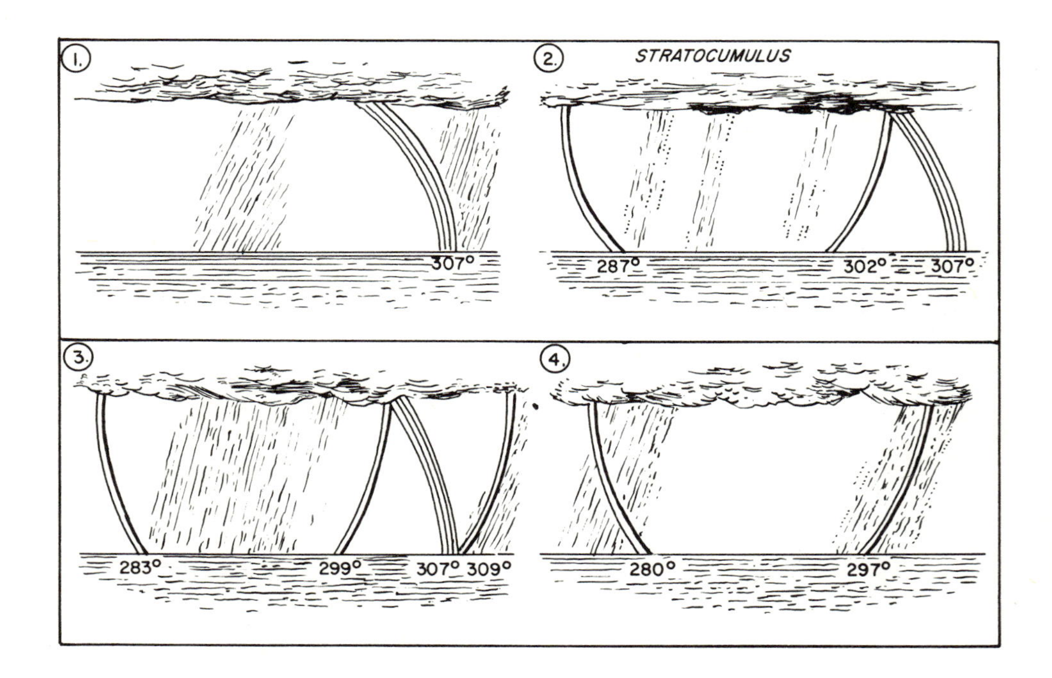


Spurious rainbows and white arcs

At 0805, the colour sequence from the inside was yellow, purple, blue, green, yellow, orange and red. By now the sun was shining through thin As.

At 0806, during showers in the area, the bright, white ends of a second bow became visible inside the rainbow (sketch 2). The ends of these bands were 1° wide. At 0813 yet another fainter white bow, 2° wide, appeared outside the others. The bearings then were as shown in sketch 3. When the rain ceased at 0820, the original rainbow disappeared, as did the second white arc, leaving the first two faintly visible (sketch 4). At 0833 these also faded out.

Position of ship: 09° 47'S., 04° 08'W.

Note. This is a very interesting observation. The colours of the main rainbow seem to be somewhat abnormal even if allowance is made for a possible supernumerary bow inside and in contact with it. The white rainbows are abnormal phenomena which cannot be explained on the ordinary geometrical theory which accounts for the primary, secondary and supernumerary bows. An observation in which a white bow was seen crossing a rainbow was made by M. V. Laguna (see page 215 of the October 1951 number). (Marine Observer, 27:210, 1957)

## RAINBOW PHENOMENA
**Watson, L. E.; *Meteorological Magazine,* 82:218, 1953.**

The rare phenomenon mentioned in Mr. Pilsbury's letter published in the Meteorological Magazine for February 1953 was observed at Benbecula on December 16, 1952. The arcs observed were travelling slowly through the straits separating Benbecula from North Uist.

The weather had been cloudy with continuous moderate rain from 8 oktas of nimbostratus at 1, 500 ft. with patches of fractostratus at 1, 000 ft., the wind at the time being 180° 15 kt. At 1055 G. M. T. a break in the cloud became visible in the south-west, and at 1115 the wind veered to 280° 12-15 kt. indicating the passage of a trough. After this, the wind dropped slightly and the rain became intermittent, and during this period a faint but complete rainbow was observed at 1125 between Benbecula and North Uist. At 1135 an arc formed in the opposite direction to the original bow which had faded and now had only an arc composed of the western half of the bow, red being uppermost in both arcs. Subsequent developments are shown in Fig. 1, red being uppermost in all the arcs shown. In this diagram the lengths of the arcs and the angles between the bows are only approximations. (Meteorological Magazine, 82:218, 1953)

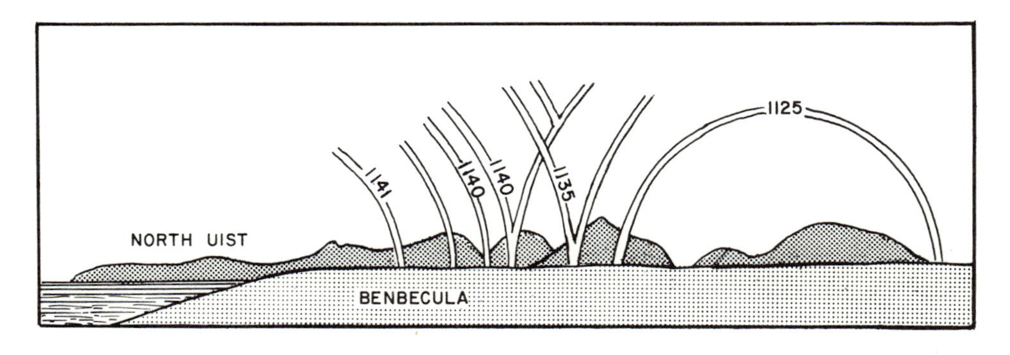


Moving rainbow arcs

# CLOUDBOWS, IRRIDESCENCE, AND OTHER RAINBOW-LIKE OPTICAL EFFECTS

## A BROADCAST RAINBOW
**McLean, R. C.; *Nature,* 110:605, 1922.**

On Friday, September 16, I witnessed an atmospheric phenomenon sufficiently unusual, I believe, to merit a record. Standing on Ogmore Down near Bridgend in this county (Glamorgan) at 2.30 p.m. and looking northwards across the board vale towards the Maesteg hills, there appeared to me a broadcast rainbow colouring, stretching east and west for several miles along the vale. The day was exceptionally fine, with brilliant visibility and no trace of mist. The clouds were small and scattered, with a distant bank of cumulus beyond the hills, while the colours were clear and unmistakable, covering, from red in the west to blue in the east, an angle of about fifty degrees. The height of my point of view was about 300 ft. above the sea, and the whole apparition hung, like a veil of pure, immaterial colour, at about the level of my eyes, covering the distant hills but without screening their smallest particular. (Nature, 110:605, 1922)

## A GIGANTIC HORIZONTAL CLOUDBOW
**McDonald, James E.; *Weather,* 17:243-245, 1962.**

We normally perceive rainbows as arcs on roughly vertical sheets of raindrops; but the same optical effects can be produced by horizontal arrays of drops. Thus dewdrops on a lawn or uncoalesced raindrops on a calm soot-covered pond can form horizontal bows of the rainbow or cloudbow type, as Minnaert (1954) points out. Clerk Maxwell described seeing such a horizontal bow produced by many droplets resting unfrozen on an ice surface! For all such cases, the dimensions of the luminous arc must measure only tens of feet at most. Here I wish to report an observation of the same phenomenon in which the luminous are spanned many tens of miles!

Shortly after taking off from Honolulu en route to Los Angeles in a Pan American 707 jet on 19 September 1961, we attained the chosen cruising altitude of 29,000 ft. At about 10 o'clock Honolulu time, I noted from my port window that we were accompanied by a luminous arc below the aircraft. My first, wrong, impression was that I was viewing some kind of ice-crystal halo involving both reflection and refraction since it had the almost colourless luminosity characteristic of the majority of haloes. However, in a moment of further scrutiny, I noted that it seemed to originate on the cloud deck whose tops we had earlier penetrated at about 8,000 ft altitude. Its true nature was then unambiguously revealed when I noted further that here and there along its course it assumed a most vivid rainbow banding where I was looking down through breaks in the 8,000-ft deck to where the trade cumuli were raining. The near edges of the rain columns constituted for me small elements in a huge and essentially horizontal field of drops of two quite distinct size-classes:

first, the tiny cloud-drops in the broken deck whose tops lay near 8,000 ft; second, the scattered patches of relatively large raindrops visible through the breaks. Having recognized its real nature, I scanned along the expected arc far to the north of our flight path, and was pleased to find that the luminosity, though very pale in the distance, constituted well over half of a full bow from my vantage point.

Subsequent check on the solar altitude for that time and place permitted assigning the numberical values indicated in Fig. 1, which is drawn in a vertical plane through sun and aircraft. The sun S at altitude very near $50^{o}$ sent down rays of which the one

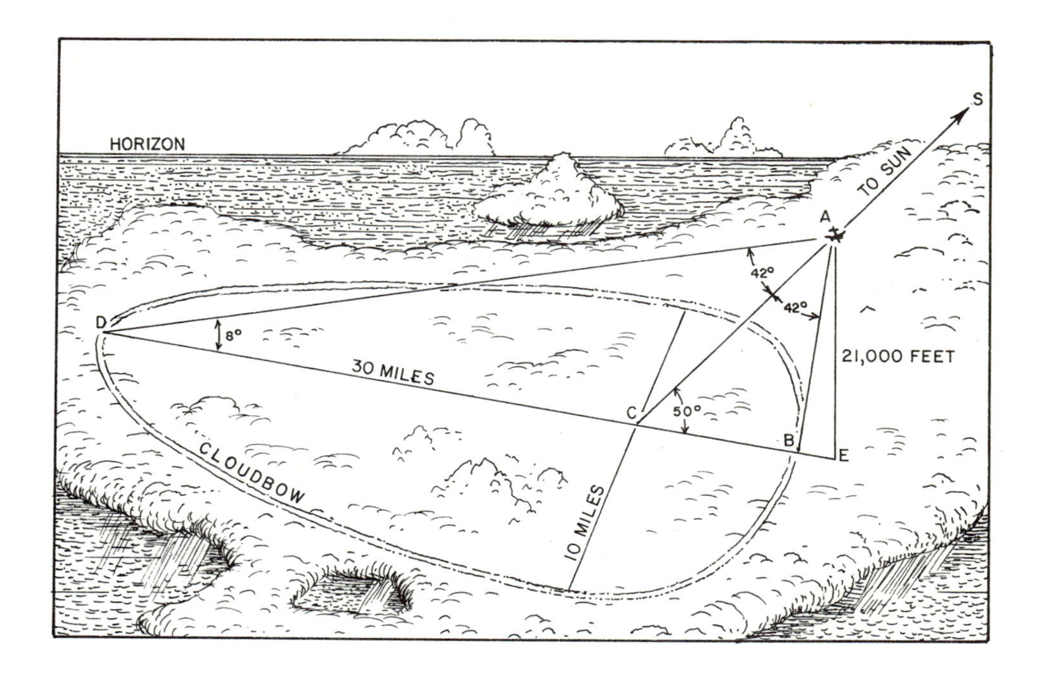


Formation of mixed rainbow-cloudbow

passing near an observer in the aircraft at A strikes the top of the 8,000-ft cloud deck at C. A cone of semi-vertex angle $42^{o}$ revolved about that ray SC as axis intersects the deck in the rainbow-(more precisely, in the cloudbow-) locus, whose extrema D and B are indicated. Since the top of the cloud deck lay about 21,000 ft below the aircraft, the indicated angles imply that the 'summit' of the bow at D lay some 30 miles away, measured from the point E directly under the aircraft. The bow is here not the hyperbola which must almost always be expected with dew-bows (whose occurrence can only be expected in the morning hours of low sun), but rather an ellipse. Further calculation shows that the minor axis running perpendicular to the plane of the figure slightly to the left of C is here about 10 mi. The geometry of the small windows in the 707 precluded my seeing the full ellipse (indeed I could barely see the weak glory at C on half-standing to peer downwards), a limitation arguing the need for glass-bottomed jet transports for meteorologically inclined passengers.

Where the bow changed from cloudbow to rainbow at a break in the underdeck near a rain area, the difference in angular spread of the two parts was easily perceived. I made rough attempts to time the passage of the cloudbow over distinctive cloud turrets, estimating 3 to 4 seconds between transit of leading and trailing edge of the bow where the latter lay normal to our flight path. Assuming a ground speed of 600 mph, this gave an actual width of the cloudbow luminosity of somewhat under a half-mile, implying an angular width of about 3 to 4°. The angular width of a typical rainbow is about 1.7°, so the cloudbow was about twice as wide as a rainbow, in good enough agreement with theory (Minnaert).

This phenomenon seems noteworthy on a number of grounds: here was a bow spanning a space of 10 by 30 miles, large beyond all ordinary circumstances; its locus of luminosity was elliptical rather than hyperbolic; and, finally, it was actually a mixed rainbow-cloudbow, chiefly the latter. Searching correspondence in Weather over the past dozen years disclosed no similar observation. With the rising volume of jet traffic, surely many opportunities to see similar gigantic cloudbows should arise. A little thought will reveal that the most favourable circumstances will be with maximum flight-altitude and minimum underdeck-altitude, for then the 'graininess' of the cloud deck is favourably suppressed, giving a bow a better chance to become perceptible. Also, I noted that the cloudbow was most readily perceived where the deck was thinnest (within limits). An optically thick underdeck will send back so much multiply-scattered light not satisfying cloudbow geometry that the cloudbow luminosity will be washed out. Perhaps the same kind of trade cumulus situation as that here described will prove nearly optimal for seeing such cloudbows. Surely it is optimal for the added treat of intermittently visible rainbow patches. (Weather, 17:243-245, 1962)

## A 'SAND BOW'—AN UNUSUAL OPTICAL PHENOMENON
**Talmage, James E.; *Science*, 13:992, 1901.**

The following description, based on personal observation is presented without discussion of the optical principles involved.

On the evening of May 16, the writer was crossing the main ridge of Antelope Island---the largest land body within the area of the Great Salt Lake. As he began the descent on the eastern slope, there appeared between the island and the mainland what seemed at first glance to be segment of a brilliant rainbow of unusual width. It was evident, however, that no rain was falling in that direction. Clouds were gathering in the south and west, but the sun was yet unobscured. A wind setting toward the mainland had lifted from the dry flats large quantities of the 'oolitic sand,' with which the lake bottom and the recently dried patches on this side of the island are covered to a depth varying from a few inches to several feet. This so-called 'sand' consists of calcareous spherules, fairly uniform in size between the limits of No. 8 and No. 10 shot. The oolitic bodies are polished and exhibit a pearly luster.

It would seem that the outer spherical surfaces reflected the light in such a manner as to produce the bow. The colored column appeared almost to touch the lake bed, and its ends subtended with the observer an angle of about 40°. The prismatic colors were distinct, the red being outside, i.e., away from the sun. In apparent width the column was fully double that of the ordinary rainbow. A fainter secondary bow was plainly visible beyond the primary, with the colors in reverse order. The phenomenon was so brilliant as to attract the attention of all members of the party, and it remained visible for over five minutes, then, as the sun sank lower, it rapidly died away.

The production of a color bow by reflection from the outer surfaces of opaque spherules is a new phenomenon to the writer. It is inexplicable on the principle of refraction and total reflection from the interior of transparent spheroids, according to which the rainbow is generally explained. (Science, 13:992, 1901)

## AN OPTICAL PHENOMENON ON A FLORIDA LAKE
**Donnelly, William T.; *Natural History,* 22:372, 1922.**

My yachts, "Dawn" and "New Era," are anchored in Lake Monroe, Florida, off the city of Sanford at the head of navigation on the St. Johns River. Lake Monroe is approximately 4-1/2 miles long by 3-1/2 miles wide; the St. Johns River enters at one end and leaves from the other. This morning when I came out on deck soon after sunrise, I noticed toward the east a remarkable phenomenon on the water. A band of prismatic colors, familiarly known as the rainbow seemed to extend on the water, commencing about one third of the way across the lake and reaching to the shore. The atmosphere, while hazy, was in no sense foggy, and the sun was shining brightly. This phenomenon persisted and at the present time, 9:30 a.m., is truly remarkable. The haze has almost entirely disappeared. The lake is perfectly calm. The band of refracted light seems to be much broader than the usual rainbow phenomenon.

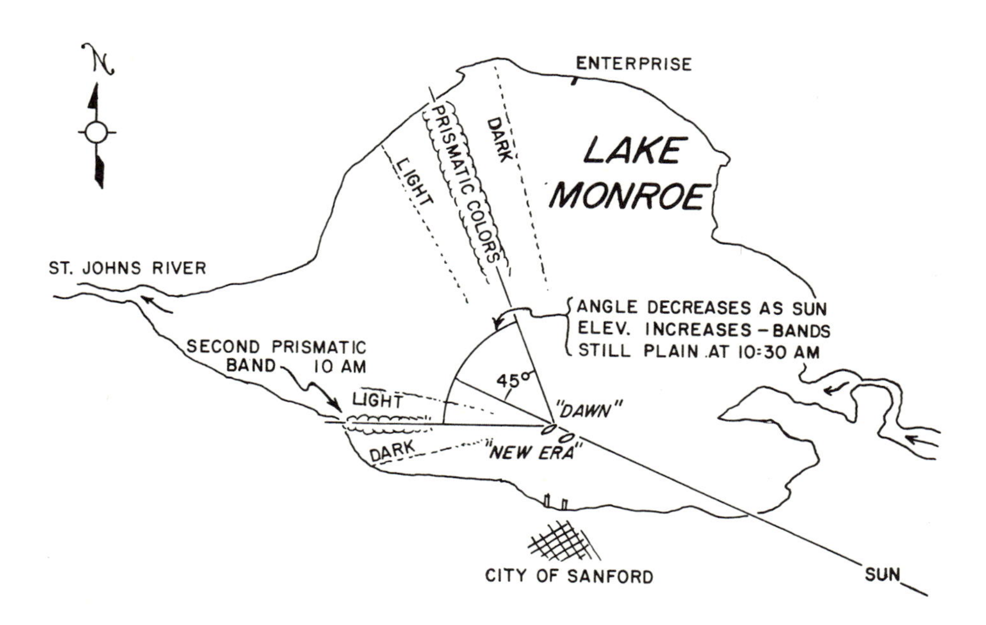


Rainbow-like pattern over Florida lake surface

It occurred to me to measure roughly and record the angular relation of the prismatic band with the sun, and the most ready way of doing so was to take two triangles, sighting along one pointing to the prismatic path on the water and laying the other triangle across it. The result was so geometric as to be almost startling. It will be noted from the sketch that the angle formed by the projection of the sun's rays as compared with the path of the refracted light is an angle of 45 degrees. It was perfectly simple to adjust these angles by sighting down the long side of the 45 degree angle and holding up a finger and adjusting the 60 degree angle until the shadow ran along its edge.

To me the remarkable fact about the phenomenon is its persistency and the brilliance of the refraction all the way across the prismatic scale; an added phenomenon is a very pronouncedly dark band to the right or orange end, of the spectrum, and a bright band on the left, or purple end.

About breakfast time a young native paddled by in his canoe. When I asked him what he knew about the phenomenon, he replied that the rainbow was present every morning when the lake was calm.

Since writing the above, the other end of the rainbow has appeared, relatively as noted on the sketch. It is even more brilliant than the first---in fact, the most brilliant prismatic scale and the most extended that I have ever seen. It is now past ten o'clock; almost every particle of haze has disappeared from the lake, but still the prismatic band persists. At this writing, the tender of the "Dawn," rowed by my daughter, is sailing down this band. Surely it is a land where one can chase the rainbow and, it would almost seem, with some prospect of capturing it, so close and friendly it appears to be. (Natural History, 22:372, 1922)

## UNEXPLAINED PATTERN OF COLORS IN THE SKY
**Mayne, R. J.; *Marine Observer*, 26:147, 1956.**

S. S. Thalamus. Captain R. J. Mayne. Rotterdam to Curacao. Observers, the Master and Mr. D. M. C. Renton, 3rd Officer.

13th September, 1955. At 1115 G. M. T. a light rain shower passed overhead and travelled in a W'ly direction. When about a mile away, bearing 280° (T), a brilliant coloration was observed on the cloud (sketch 1), approximately 300 yd long. In the sketches the letter a denotes a brownish-orange colour, b a greenish tinge and c turquoise blue. After about 3 min the left-hand side of the ellipse seemed to merge into the other side until only a "hump" of one colour remained (sketch 2). After a further 2 min the coloration was observed to be flattening and lengthening into a thin band (sketch 3), which completely disappeared after a further 2 min. The phenomenon lasted for 7 min 1120 to 1127.

Position of ship: 33° 20'N., 32° 18'W.

Note. The observers suggest that this was a rainbow. The sun's altitude at the time of observation was between 41° and 42° and its azimuth approximately 117°. With this altitude the primary rainbow could not have been seen, as its apex would have been on the horizon. The observation does not state the altitude of the apex of the phenomenon, but a secondary rainbow could have been formed with its apex about 13° above the horizon, bearing 297°. The coloration observed does not correspond to that of a secondary rainbow, which has blue and violet on its upper side and red and orange on its lower side. Furthermore, in normal rainbows the arcs of colour are circular and concentric. No explanation can thus be given of this very interesting phenomenon. (Marine Observer, 26, 147, 1956)

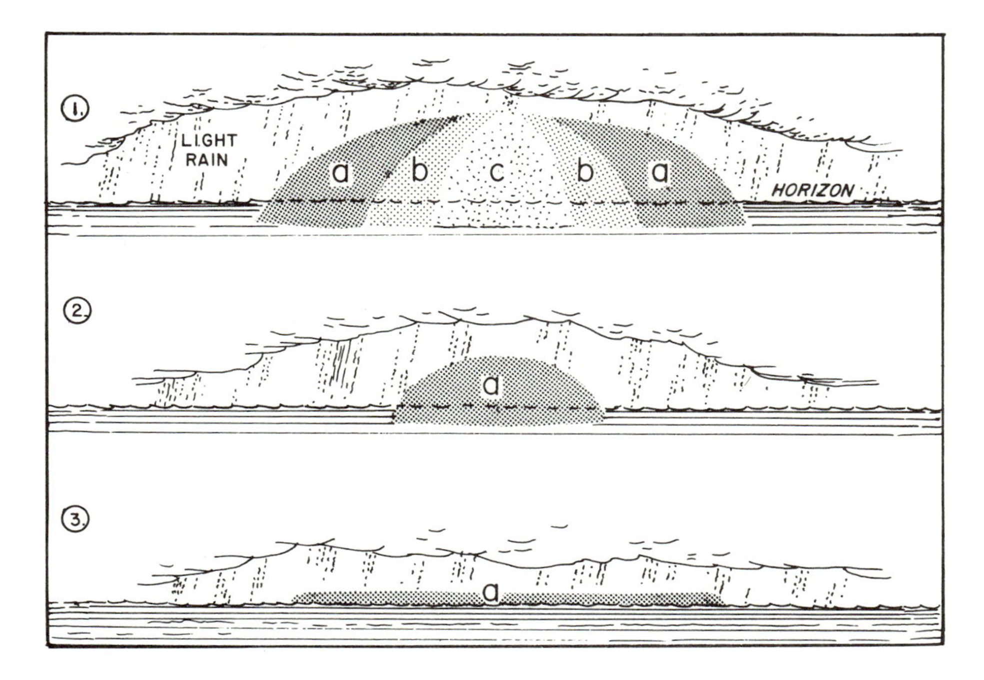


Peculiar pattern of colors in sky in North Atlantic

## A CURIOUS RAINBOW
### Smith, F. B., and Botley, Cicely, M.; *Weather,* 8:30 and 8:160, 1953.

On Wednesday, 2 July 1952 at 8 p.m. BST an interesting rainbow was visible for at least five minutes at Heald Green, Cheshire, two miles from Ringway Airport.

The day had been cool and mainly dry. The sky was 4/8 covered with decaying semi-transparent alto-cumulus, a little cumulus and some cirrus; and the sun, still well above the horizon, was shining brightly through the thin cloud.

The rainbow, superimposed on very thin cloud, was as bright as a normal one, and had its centre just on the opposite side of the zenith to the sun, with radius approximately 10°. It was not complete but consisted of an arc convex towards the sun subtending an approximate angle of 90° at the centre of the arc.

It seems possible that it was due to light reflected by high cirrus being refracted on passing through the thin alto-cumulus. (Weather, 8:30, 1953)

The 'interesting rainbow' reported by Mr. F. B. Smith in the January Weather (p. 30) is almost certainly no rainbow but a fine example of the circumzenithal arc, which Pernter and Exner describe as the 'most splendid' of all halo phenomena, and which appears about 46° above the sun. It appears to be the result of refraction through ice-crystals shaped like plates or umbrellas floating with axes vertical and is proba-

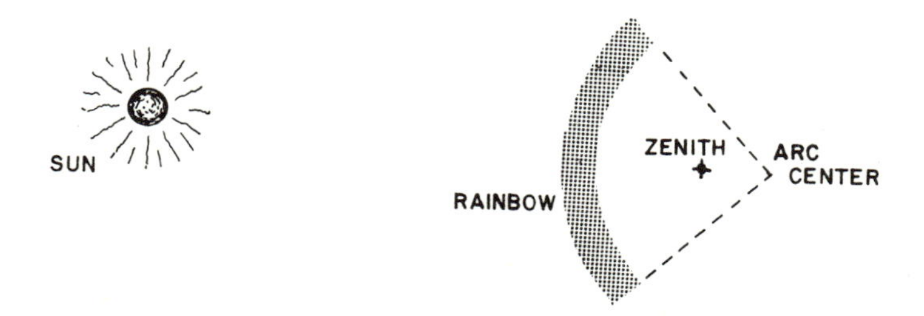


Portion of circumzenithal arc

bly not so rare but rather is transitory.

The alto-cumulus clouds mentioned had nothing to do with the arc, being composed of water-droplets, and not ice crystals. (Weather, 8:160, 1953)

## UNIDENTIFIED PHENOMENON IN THE MOZAMBIQUE CHANNEL
**Poutler, N.; *Marine Observer*, 29:115, 1959.**

M. V. British General. Captain N. Poulter. Aden to Cape Town. Observer, Mr. I. W. Soutar, 2nd Officer.

22nd August, 1958. While moderately heavy showers of rain were falling at 1200 G. M. T., a faint rainbow was seen bearing 110°. Vertically below the apex of the bow, in a seemingly "compressed" state, was a remarkable "block" of light showing the colours of the spectrum, with yellow predominating. As the bow disappeared, the "block" of light rose and expanded until it in turn became a bow, but the intensity of the colouring was by now very much reduced, and there was no predominant tint.

The phenomenon occurred very near to the ship and covered a period of 12 min. Air temp. 75°F, wet bulb 71°, sea 77°. Wind SE., force 3. Visibility very good.

Position of ship: 14° 40'S., 41° 36'E. (Marine Observer, 29:115, 1959)

## UNUSUAL IRIDESCENCE AT SUNSET
**Llewellyn, S. C.; *Marine Observer*, 25:103-104, 1955.**

M. V. Ajax. Captain S. C. Llewellyn. San Francisco to Manila. Observer, Mr. J. K. Marshall, 3rd Officer.

21st April, 1954, 0930 G. M. T. The sun set at 0900, and at 0930 a beautiful rainbow effect was seen in the W, although the arc did not appear to have the sun as centre. It consisted of a greenish patch surrounded by three red quarter arcs enclosing two spectra. These arcs lasted for about 3 min. Apart from a few light Cu on the horizon, there appeared to be no cloud in the vicinity. Later Canopus gave a wonderful exhibition of scintillation, the white flash at times looking like the loom of a light.

Position of ship: 15° 59'N, 135° 38'E.

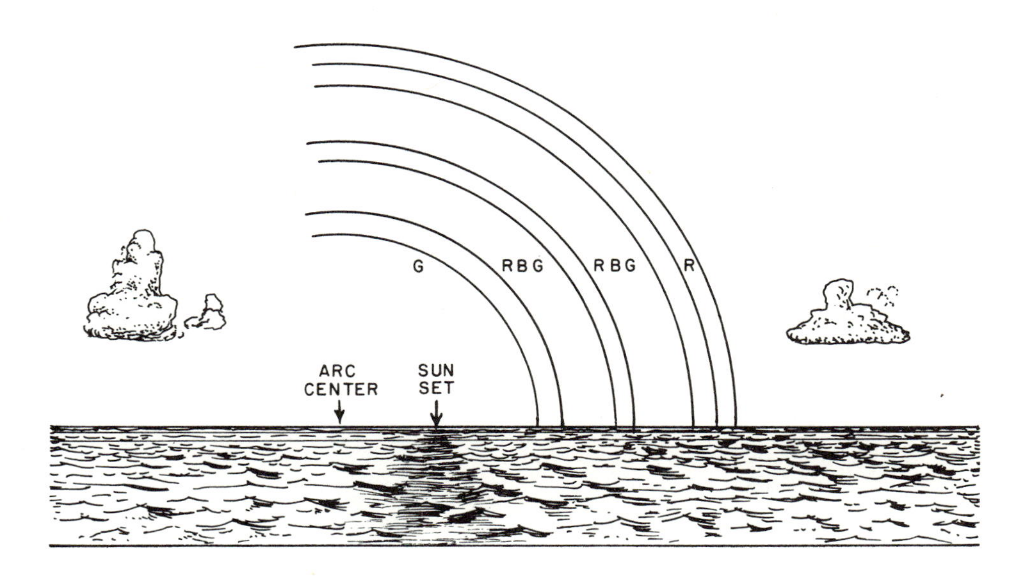


Offset rainbow-like pattern at sunset

Note. This is a very interesting and valuable observation, since it appears to be of something new which cannot yet be explained. There is only one known cause of iridescence in the western sky after sunset. This occurs when the clouds known as mother-of-pearl clouds are present in the stratosphere, at an average height of 15 miles. The iridescence on these may be very brilliant, particularly just before sunset, and may continue visible for about two hours after sunset. In the above observation no clouds were visible. Furthermore, observation of mother-of-pearl clouds has so far been confined to NW Europe, when an extensive deep depression lies over northern Scandinavia in winter. (Marine Observer, 25:103-104, 1955)

# GLORIES, BROCKEN SPECTRES, AND ASSOCIATED PHENOMENA

A wild tale has long circulated about a giant spectre seen by mountain climbers on the Brocken, a German mountain. As the story goes, there was once a climber working his way along a precipice, who suddenly saw an immense human figure rise out of the mists toward him. In his fright he lost his footing and fell to his death. Doubtless just a story, but the Brocken Spectre does exist, not only on the Brocken but anywhere where shadows are cast upon mist and water droplets.

That one's shadow can be seen, even in giant proportions, on a wall of white mist is not surprising. Some features of the phenomenon, however, are more difficult to explain. A common feature of the Brocken Spectre is the "glory" or concentric rings of color seen centered on the observer's line of sight or, equivalently, the shadow of his head. Sometimes several rings are seen, each with the rainbow's succession of colors, and doubtless their origin is the same; that is, the reflection of the solar spectrum by the water droplets. The curious feature concerns the visibility of the observer's shadow and his private set of colored rings to other observers in the vicinity. Some see only their own shadow; others see the shadows of everyone present. Then, too, the circles of color are sometimes elliptical rather than circular. The glories are also apparently distorted by the approach of other observers. Another unusual feature comprises the dark streaks that seem to radiate from the observer's arms. There is much here that is difficult to explain in terms of conventional optics.

The Brocken Spectre type of phenomenon also occurs when a field of dew-laden grass is spread out before a viewer. With the sun at his back, the observer sees his shadow cast upon the field with his head surrounded by a halo. More awe-inspiring is the spectre seen from atop Adam's Peak in Ceylon. When conditions are right, the shadow of the entire peak is thrown upon a wall of vapor. A viewer on the summit sees the peak surrounded by a glory. In addition, there are dark projections that seem analogous to the streaks sometimes seen radiating from the arms of an individual's shadow.

Optical refraction by the atmosphere and the scattering of light by dust also play strange tricks. Double suns and eclipse shadow bands seem to be refraction phenomena, while Bishop's ring likely arises from the scattering of light by volcanic dust. Kaleidoscopic suns, on the other hand, have no easy explanation and their observations may even have strong subjective components. Other optical phenomena in this section are equally baffling but have not yet been recognized with formal names---explanations for them are also missing.

# BROCKEN SPECTRES AND THEIR KIN

## BROCKEN SPECTRE PHENOMENON
**Anonymous; *Nature,* 21:216, 1880.**

I had the opportunity about half-past ten this morning of witnessing from Clifton Down a phenomenon which enjoys the repute of being very rare. The entire gorge of the Avon was filled with mist, so that the river in the bottom and the Leigh Woods opposite were quite obscured. Standing on the western extremity of the Observatory Hill, I observed a dim gigantic figure apparently standing out through the mist upon one of the lower slopes of Clifton Down, where it runs down in undulating ridges from the promenade towards the river. A moment's glance sufficed to show me that it was my own shadow on the mist, and as I waved my arms about the gaunt spectre followed every movement. A gentleman who stood beside me likewise saw his spectre, but not mine, as we ascertained by the movements executed; nor could I see his, unless we stood so close together that the spectres seemed combined into one. The analogy presented by these spectres with the famous Spectre of the Brocken, seen by travellers in the level rays of the morning sun from the summit of that celebrated mountain, and described by Sir David Brewster in his "Letters on Natural Magic," is very striking. (Nature, 21:216, 1880)

## BROCKEN SPECTRE
**Philip, R. W.; *Marine Observer,* 30:131-132, 1960.**

M. V. Apapa. Captain R. W. Philip. Bathurst to Takoradi. Observers, Mr. G. D. Pari-Huws, Chief officer and members of crew.

26th August, 1959. Thick fog giving a visibility of about 150-200 yd, and approximately 150 ft deep lay on the water at 0915 G. M. T. The sun was right astern at 50° altitude and, shining through the fog, it cast upon the water 38 ft below, the shadows of members of the crew working on the forecastle head. Each shadow was surrounded by a miniature coloured halo whose apparent diameter was about 5 ft. Colour was faint, that of the outside of the halo being violet. Air temp. 72-1/2°F, wet bulb 72°, sea 72°.

Position of ship: 4 cables from Takoradi Harbour entrance.

Note. The phenomenon was first noted on the Brocken mountain in Germany, but it is most common in Arctic regions where it is seen on every occasion of simultaneous sunshine and fog. The coloured rings are known as a 'glory', and a typical series of colours seen in a well developed one is as follows: there is a general whitish yellow colour round the shadow, surrounded with rings of colour which, in order outwards, are dull red, bluish-green, reddish-violet, blue, green, red, green, red. A white bow at a considerable distance outside the 'glory' is sometimes also seen.

The shadow of the observer on thick fog may be seen at night, if there is a bright artificial light behind him. (Marine Observer, 30:131-132, 1960)

## THE SUNRISE SHADOW OF ADAM'S PEAK, CEYLON
**Abercromby, Ralph; *Nature,* 33:532-533, 1886.**

Some of the phenomena of the shadow of Adam's Peak in the early morning have been remarked by almost every traveller who has visited this island. The mountain rises to a height of 7352 feet as an isolated cone projecting more than 1000 feet above the main ridge to which it belongs. The appearance which has excited so much comment is that just after sunrise the shadow of the Peak seems to rise up in front of the spectator, and then suddenly either to disappear or fall down to the earth.

Various suggestions have been made as to the source of this curious shadow; among others one, which was published in the Phil. Mag., August 1876, that attributed the rise of the shadow to a kind of mirage effect, on the supposition that the air over the low country was much hotter than on the Peak top.

I determined to attempt the discovery of the true nature of this appearance, and was fortunate to see it under circumstances which left no doubt as to the real origin. Through the courtesy and hospitality of Mr. T. N. Christie, of St. Andrew's Planta-

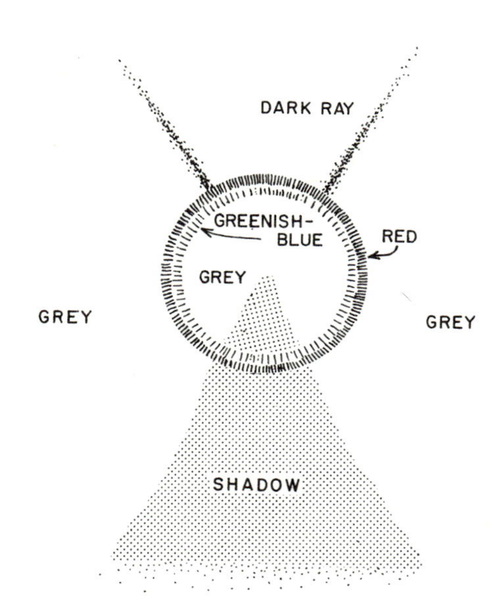


Sunrise shadow of Adam's Peak

tion, I was able to pass the night on the summit, and to carry up a few necessary instruments.

The morning broke in a very unpromising manner. Heavy clouds lay all about, lightning flickered over a dark bank to the right of the rising sun, and at frequent intervals masses of light vapour blew up from the valley and enveloped the summit in their mist. Suddenly, at 6.30 a.m., the sun peeped through a chink in the eastern sky, and we saw a shadow of the Peak projected on the land; then a little mist drove in front of the shadow, and we saw a circular rainbow of perhaps $8^{o}$ or $10^{o}$ diameter surrounding the shadow of the summit, and as we waved our arms we saw the shadow of our limbs moving in the mist. Two dark lines seemed to radiate from the centre of the bow, almost in a prolongation of the slopes of the Peak, as in the figure.

Twice this shadow appeared and vanished as cloud obscured the sun, but the third time we saw what has apparently struck so many observers. The shadow seemed to rise up and stand in front of us in the air, with rainbow and spectral arms, and then to fall down suddenly to the earth as the bow disappeared. The cause of the whole was obvious. As a mass of vapour drove across the shadow, the condensed particles caught the shadow, and in this case were also large enough to form a bow. As the vapour blew past, the shadow fell to its natural level---the surface of the earth.

An hour later, when the sun was well up, we again saw the shadow of the Peak and ourselves, this time encircled by a double bow. Then the shadow was so far down that there was no illusion of standing up in front of us.

I believe that the formation of fog-bow and spectral figures on Adam's Peak is not so common as the simple rising up of the shadow, but one is only a development of the other. In fine weather, when the condensed vapour is thin and the component globules small, there is only enough matter in the air to reflect the Peak shadow in front of the spectator, and no figure is seen unless the arms are waved. In worse weather the globules of mist are large enough to form one or two bows, according to the intensity of the light. We were fortunate to see the lifted shadow accompanied by fog phenomena, which left no doubt as to the cause of the whole appearance.

Any idea of mirage was entirely disproved by my thermometric observations, which cannot be detailed here for want of space. (Nature, 33:532-533)

## HALO IN THE RICEFIELD
**Anonymous; *Nature,* 90:419, 1912.**

The curious phenomenon known in Japan as Inada no goko, or halo in the ricefield, forms the subject of a discussion by Profs. Fuchino and Izu, of the College of Agriculture and Forestry, Kagoshima, in the Journal of the Meteorological Society of Japan (October, 1912). In the early morning, when the dew is on the plants, and the sun is shining, the shadow of the head of a person standing in the fields is surrounded by a luminous halo, elliptic in form, its long axis corresponding with that of the body-shadow. As the sun rises higher in the sky and the dew evaporates the halo vanishes, but reappears on sprinkling the ground with water. The authors describe some experiments which they carried out with blankets, isolated drops of water, and bottles. They conclude from their experiments that the phenomenon of the halo is caused by the reflected light from the sun-images formed on the green blades by the passage of the sun's rays obliquely through the dewdrops. (Nature, 90:419, 1912)

## LUMINOUS HALOS SURROUNDING SHADOWS OF HEADS
**Evershed, J., and Fermor, L. L.; *Nature,* 90:592-593, 1913.**

Exactly a month ago to-day, in the Betul district, Central Provinces, I had set out on field work at dawn, with my colleague, Mr. H. Walker, and two chaprasis (Indian servants). I happened to be watching our shadows as we passed along the edge of a field of young green wheat, when, to my surprise, I noticed a halo of light round the shadow of my own head and neck. Looking at the other shadows, I was still more surprised to see that only my shadow was invested with this halo. I directed the attention of Mr. Walker and the chaprasis to the phenomenon, and found that each could see a halo round his own head only. Whilst we were investigating the matter our camp passed on the march, and inquiries made both from our servants and from local people showed that none of them had previously noticed the phenomenon.

The conditions were obviously special, although frequently obtainable to one who deliberately set out with the purpose of finding them. The sun was at a low altitude on our left, and the wheat was soaking wet with dew on our right. The dew speedily dries up in the morning sun, and although I have kept on the look-out for this phenomenon during the past month I have never happened to pass a wheat-field again with the conditions of time, situation, and wetness repeated.

I had, therefore, intended writing to Nature to inquire whether the occurrence of these halos had been previously recorded, and consequently was greatly interested to read the note on p. 419 of your journal (December 12, 1912) concerning Inada no goko, or halo in the ricefield. I have not seen the Japanese journal referred to, and consequently am not aware if Profs. Fuchino and Izu direct attention to the fact noted above, that each observer sees the halo round his own head only. This fact indicates that the observer perceives those elements of a narrow cylinder of the sun's rays enclosing his head that happen to be reflected back to his eyes by the dewdrops and wheat blades; the major portion of the cylinder of light is reflected back along the cylinder, and consequently a given observer is not in the line of vision for the halo round another observer's head. The explanation advanced by the Japanese observers that the halo "is caused by the reflected light from the sun-images formed on the green blades by the passage of the sun's rays obliquely through the dewdrops" is doubtless correct. I presume that their investigations show that the farther a drop is from the edge of the shadow of the head the smaller is the proportion of the light reflected from the sun-images that can reach the observer's eye; for the boundary of the halo is not sharp, the brightness diminishing somewhat gradually with distance from the shadow. Assigning to the head in the shadow the actual diameter of the head, I estimated the noticeably bright part of the halo as roughly 10 in. wide all round the head, dying out on the shoulders.

A close inspection of the green blades showed that at or near the tip of each blade was one pearl of dew, whilst the whole of the remainder of the blade was coated with a film of minute dewdrops. It is the minute drops that give rise to the major portion of the effect.

The fact that each observer sees only his own halo obviously precludes this phenomenon from having been the origin of the halos recorded in sacred writings round the head of Christ and others. (Nature, 90:592-593, 1913)

## A CURIOUS RAINBOW
**S., W. P. H.; *Nature,* 102:85, 1918.**

In North Wales, on August 20, about two hours before sunset, I saw a rainbow-effect which was quite new to me.

The summit of Tryfaen (some four miles north-east from that of Snowdon) has three sharp, rocky peaks running roughly north and south.

We had climbed up the eastern cliff in a southwesterly gale, which brought up much cloud with some light showers, and were sitting just below the top of the southern peak. The Holyhead road lay north-east and 2000 ft. below us. From it rose the upright portion of a brilliant rainbow. At the centre of its circle was the shadow of our peak with those of the other two peaks to the left of it, all sharply defined. Around the shadow of our peak was a most vivid and persistent bow, the smallest I have ever seen, the radius of the innter edge being about half that of the outer. The central space changed a good deal, being frequently almost filled by a diffused yellow glow, which sometimes appeared to condense towards the centre until it resembled a nebulous sun on a whitish ground, while at intervals little yellow streamers seemed to radiate from it to the inner edge of the bow. Outside this bow (which had the colours in regular rainbow order, red outside) was part of a third bow of perhaps double the diameter, but dim and intermittent.

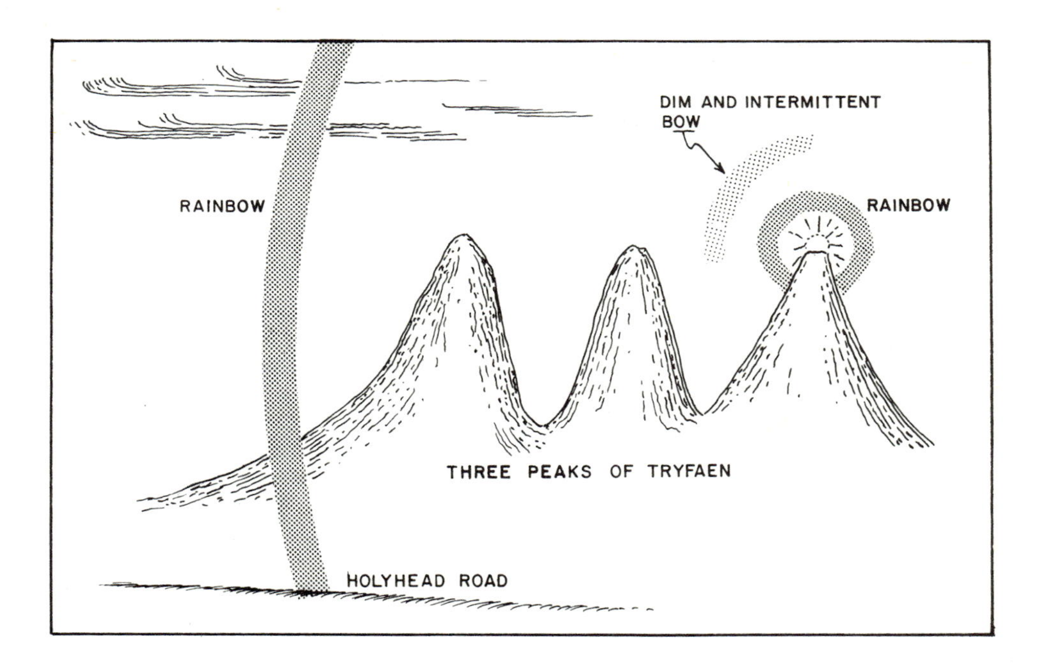


Curious rainbow and glory around mountain-top shadow

We stood up and made gestures, expecting some sort of Brocken effect, but could detect none. However, as we were not on the extreme summit, and the cloud was very distant, our shadows would at best have been extremely minute.

Out of many "Brockens" that I have seen in different parts of the world the most vividly coloured was in Arctic Norway, the most curious and unexpected was on a blazing August day at sea-level in Portugal, and the most realistic on the Mendip Hills in Somerset.

The last was all the more effective for being within an uncoloured and inconspicuous ring. (Nature, 102:85, 1918)

## SELF-CENTERED SHADOW
**Jacobs, Stephen F.; *American Journal of Physics*, 21:234, 1953.**

One of the greatest joys of physics lies in finding explanations for common everyday phenomena which confront all of us, yet which are not quite obvious.

The following example of such a phenomenon excited my curiosity even when I was very young: standing on the end of a diving board over deep water, the sun behind me, I often wondered as I studied my shadow on the water's surface about the rays which seemed to emanate from my head. They radiated from a spot between my eyes, and the rays seemed to glisten and to rotate like the spokes of a wheel, generally rotating continuously in one direction. When a friend would join me, we would both see the rays coming from our own heads. This phenomenon is observed sufficiently seldom to elude investigation, for it is not a very striking effect and requires:

1. deep water; over three feet, depending on the nature of the bottom
2. sunshine
3. suspended particles, such as algae or minerals
4. relatively quiet water.

The explanation is straightforward, though very satisfying: Sunlight produces the same effect in water containing suspended particles as it does in air, where it produces a sunbeam. In the phenomenon described above, the sun, observer's eyes, and shadow comprise the optic axis. Rays of sunlight strike the water parallel to this axis and, if the water be unperturbed, continue through the water parallel to each other. If, however, the surface be disturbed by superposed wave trains, traveling this way and that, the regions where the rays are refracted and continue as through still water (i.e., level patches of water) become spaced according to the surface wave form.

Thus shafts of sunlight within the water are the result of a ruffled surface. An observer looking down into the water sees these shafts but, because of his viewpoint on the optic axis, they appear to radiate from a common centre in much the same manner as the lesser buildings appear to radiate from an observer looking down from atop the Empire State Building.

As for the apparent rotation of the "spokes," this is due to the effect of the moving wave trains. Although the directions of the wave trains may differ, there is generally uniformity in the sense of their travel. As the wave trains progress, they produce the observed shifting in the shafts, causing an appearance of rotation. (American Journal of Physics, 21:234, 1953)

# GREAT VOLCANIC ERUPTIONS AND BISHOP'S RING

## THE REDDISH-BROWN RING AROUND THE SUN
**Davis, W. M.; *Science*, 5:455-456, 1885.**

The ring of reddish-brown colour surrounding the sun, and enclosing a disk of glowing whitish light, to which Prof. G. H. Stone called attention in Science for May 22, p. 415, has been most carefully studied by Kiessling of Hamburg, who has shown clearly that it is due to diffraction on minute particles suspended in the air. Careful observers agree that it was not seen before November, 1883; but since then it has attracted much attention. It is manifestly produced in the upper regions of the atmosphere; for it is best seen from elevated mountains, where it is continually visible in clear weather. It is often hidden at low stations by the plentiful reflected skylight that comes from the coarser dust of the lower atmosphere, even though the sky seem tolerably clear. It has been astonishingly distinct here in Cambridge through the past winter, on the clear anti-cyclonic days, with north-west winds, following the withdrawal of the cyclonic cloud-disk; and it attains its greatest visibility between clouds, because much of the lower dusty air is then in shadow, and does not outshine the delicate colors of the ring. On some recent cloudless but slightly hazy days it has been entirely invisible.

I have not observed the connection between the visibility of the ring and the changes of temperature and formation of clouds noted by Professor Stone, and should be glad to learn more details as to date of observations, and as to closeness of the connection in point of time. A comparison of observations on these questions made at Colorado Springs (where I presume Professor Stone made his records) and on the summit of Pike's Peak would be very instructive in this respect.

The most remarkable point in connection with the ring is its persistence long after the cessation of the brilliant twilights with which it began. How is the volcanic dust or the ice dust that causes it supported so long? It seems incredible that dust could simply float for a year and a half in so thin a medium as the atmosphere at a height of ten or more miles. Electrical repulsion has been suggested as a supporting force, and it may be somewhat effective above the level of storm-circulation; but, besides this, it seems possible that the peculiar properties of water-vapor may give some aid. Wollaston long ago speculated on the limitation of the atmosphere at an altitude where its gases were frozen. The solid particles would there fall till evaporated, when the gases thus formed would rise again till frozen once more by the cold of expansion. Ritter and others have recently reconsidered this process. Whether the theory is applicable or not to oxygen and nitrogen, it certainly is of importance when water-vapor is considered: for, as is well known, the elasticity and condensibility of this constituent of the atmosphere are mutually antagonistic. The vapor tends to diffuse itself to altitudes where the cold caused by its expansion would require the condensation of a part of it; and, although such perfect diffusion is prevented in the lower atmosphere by the friction that the vapor suffers in passing through the air, it does not seem unreasonable to believe it may obtain at great altitudes where a normal distribution of vapor must be more nearly attained, and especially so at times when an extra supply of both vapor and dust is shot high out of volcanic craters. We may therefore believe that at some high level the atmosphere is 'saturated' with vapor: above this there will be continual condensation, supplying a delicate shower of the minutest ice particles; and, if these really need a solic nucleus to freeze upon, the nuclei may be sustained by the continuous upward diffusion of the vapor that rises to take the place of that which has been condensed, only to be condensed itself in its turn. Kiessling's discussion of the diffractive action of particles suspended at considerable altitudes fully accounts for the twilights and the solar ring; and the close agreement in date of occurrence of several great volcanic explosions, and subsequent brilliant twilight displays, naturally leads to the acceptance of the volcano as the source of the diffracting matter. Perhaps the Wollastonian idea may aid in explaining the remaining difficulty; namely, the long-continued suspension of some of the diffracting matter in the upper atmosphere. (Science, 5:455-456, 1885)

## VOLCANIC DUST PHENOMENA
**Backhouse, T. W.; *Nature,* 67:174, 1902.**

The phenomena connected with the volcanic dust are undergoing distinct changes. In common with observers in the south of England, I noted the fresh appearance of the dust phenomena in the end of June, especially on June 26, but they were not very striking until August 1. At first the most decidedly volcanic feature was the great corona round the sun, known in the case of the Krakatoa effects as "Bishop's Ring." Whether this name should be applied to the corona this year is doubtful, as its radius has been fully double that of the Krakatoa corona, having until recently averaged about 70°, measured from the sun to the middle of the reddest part. Yesterday and this morning, however, it averaged only about 40°, and its reddest part was a yellowish-brown rather than a red. The colour of the corona this year has always been much less decidedly pink than was the case with Bishop's ring; indeed, it has sometimes been an absence of blueness in that part of the sky rather than any positive redness.

The pink glows after sunset were very strong in the end of June, but stronger still in November, and on November 1, 17 and 18 there was also a faint second glow, a phenomenon I had not previously seen since the Krakatoa sunsets.

It was not until October 30 that the colouring became very magnificent, and it reached its height about November 1, when the chief feature was an intense fiery orange sky near the west horizon. This was of an unmistakably volcanic character, different from anything that has appeared here since the Krakatoa sunsets, though not equal to those in splendour. Since that maximum, the colouring has been gradually lessening. Yesterday and to-day it was remarkably weak, the chief feature being the dust-wisps, which were more conspicuous than I have previously seen them during this apparition; indeed, I should have at first taken them for clouds had I not previously seen them in feebler form. They were plainest a little after sunrise and before sunset, when they were very bright and of a steely white.

The above descriptions apply to Sunderland; but in visits to Torquay from November 6 to 10 and to Dundee about December 1, the sky effects were not very different, only at Torquay I did not see the fiery orange. (Nature, 67:174, 1902)

## BISHOP'S RING
**Clark, J. Edmund; *Symons's Meteorological Magazine,* 49:33, 1914.**

While at Kandersteg, Canton Berne, recently this interesting appearance was twice clearly seen, namely, on Friday and Monday, January 16th and 19th, in each case from an Alp at 5000 to 5500 ft. elevation, and to less advantage on the 16th from Kandersteg itself, at 4000 ft. It could not be distinguished when the sun was in sight, but directly that was behind a peak the orange-tawny ring round a markedly greenish circle of 15° to 20° radius became sufficiently marked for others of the party to be interested when their attention was called to it. The ring itself stretched over another 15° or so, as roughly measured by the eye.

As yet I have not been able to make, nor have I heard of, an observation of it in England, but its re-appearance, following the striking series of Krakatoa-like sunsets last year, at intervals from early June to mid-December (I have 59 sunrise and sunset records in all), is significant, as pointing to an association of both phenomena with the Aleutian eruption in May, 1912. (Symons's Meteorological Magazine, 49: 33, 1914)

## BISHOP'S RING AND THE ANDEAN ERUPTION
**Anonymous; *Nature,* 130:843, 1932.**

Mr. J. Fraser Paterson, writing from Broken Hill, Australia, says:---"The rare phenomenon known as Bishop's ring was visible in the western sky at 5 p.m. on Saturday, July 23. This date is about ten and a half weeks after the Andean eruptions. The colour of the ring was sepia." (Nature, 130:843, 1932)

# CURIOUS REFLECTION PHENOMENA

## SILVER FOREST
**R., N.; *Knowledge,* 2:439, 1882.**

When in Switzerland last year, in September, in the morning, about 10 o'clock, we saw a beautiful and curious phenomenon, and I should be glad to know if it is common, and its causes. We were en route from Interlaken to Berne, standing on the deck of a steamer at Darligen, waiting for it to start. Above us, to the south or south-east, rose a steep, abrupt hill to a great height, and at the top it was fringed with firs, which stood out like very small Christmas trees, dark green against the sky. In one spot, however, these trees were of a bright, transparent silver, as if turned into crystal, and among and above them there glided, erratically and in all directions, apparently four or five small bright stars (about the size of third or second magnitude stars). These larger stars were plainly visible to the naked eye, but with my glasses I could make out that in this one spot the air was one glinting, glittering mass of tiny bright specks, all moving in every direction. We called the attention of the passengers to it, and many were the conjectures as to what these appearances were, and why they were. Some said birds, others falling snow, but no one gave a satisfactory explanation.

I should add that we were in the shade; that the sun was just, and only just behind and below the fringed sky-line, as we found on starting; that it was a bright day and clear; and that somehow the sun must be the cause of the phenomenon, for directly we moved off he appeared, the little stars disappeared, and the transparent fir trees returned to their natural dark green. N. R.

[Most probably the firs covered with ice-crystals, and the ice-crystals which active air-currents carried round and above the firs, were just in that position (about $22^{o}$ from the sun) at which the arc of a solar halo would be formed.---Ed.] (Knowledge, 2:439, 1882)

## BEAUTIFUL OPTICAL PHENOMENON
**Anonymous; *Nature,* 47:303, 1893.**

A beautiful optical phenomenon, which has not yet been satisfactorily explained, is described by M. F. Folie in the Bulletin of the Belgian Academy. It was observed about a mile from Zermatt on August 13 at 8.30 a.m. "On our right, towards the east, on the steep flanks of the mountains which enclose the valley of the Viege, rose a group of fir trees, the highest of which projected themselves against the azure of the sky, at a height of 500 m. above the road. Whilst I was botanising my son exclaimed: 'Come and look; the firs are as if covered with hoar-frost!' We paid the most scrupulous attention to the phenomenon. To make sure that we were not misled by an illusion we made various observations, both with the naked eye and with an excellent opera-glass." It was observed that not only the distant trees, but those lining the road, glittered in a silvery light, which seemed to belong to the trees themselves, and that the insects and birds playing round the branches were bathed in the same light, forming an aureole round the tops of the trees, somewhat resembling the light effects observed in the Blue Grotto. It is suggested that the light was reflected from the snow. Since it disappeared as soon as the sun rose above the hill, and has never been seen except in the presence of snow, this explanation appears plausible, but it is highly desirable that further and more detailed observations should be made of this spectacle feerique. (Nature, 47:303, 1893)

## AN OPTICAL PHENOMENON
**Murphy, Joseph John; *Nature,* 47:365, 1893.**

In Nature, vol. xlvii, p. 303, you mention that "a beautiful optical phenomenon, which has not yet been satisfactorily explained, is described by M. F. Folie in the Bulletin of the Belgian Academy." From what follows, it is evidently the same as that described in Tyndall's "Glaciers of the Alps" (Murray, 1860), p. 177 et seq. Tyndall gives a description of it in a letter from Prof. Necker to Sir David Brewster, from which I quote the following:---"You must conceive the observer placed at the foot of a hill between him and the place where the sun is rising, and thus entirely in the shade; the upper margin of the mountain is covered with woods, or detached trees and shrubs, which are projected as dark objects on a very bright and clear sky, except at the very place where the sun is just going to rise; for there all the trees and shrubs bordering the margin are of a pure and brilliant white, appearing extremely bright and luminous, although projected on a most brilliant and luminous sky. You would fancy you saw these trees made of the purest silver."

Prof. Necker says that he saw it at the Saleve, which is not so high above the Lake of Geneva as some of our British mountains above the sea, and has no permanent snow near it; so that M. Folie's suggestion, that it is due to light reflected from snow, must be wrong. I have seen it from the Lonig-See, near which I believe there is no permanent snow.

This appearance is always to be seen under the circumstances described, when the sky is clear and bright enough. I had read of it in Tyndall's book, and when in the Alps I sought for and found it. I have often seen a distant approach to it produced by furze bushes, quite near, seen against sunlight, and by leaves against moonlight. (Nature, 47:365, 1893)

## SUN'S IMAGE ON AIRBORNE ICE CRYSTALS?
**Diermendjian, D.; *Applied Optics,* 7:556, 1968.**

On 15 October 1967 at about 1345 h local time, en route from Chicago to Denver by jet airliner, I observed the following interesting optical phenomenon.

A relatively bright spot was seen near the ground, traveling with the aircraft at what appeared the position for specular reflection of the sun's image. That this was indeed precisely the position was corroborated repeatedly by observing that the spot became dazzlingly bright whenever it passed over a small body of water (ponds, rivers, etc.). The sun's zenith distance was estimated at about 55°. There were some broken clouds below, possibly of the altocumulus type; the spot disappeared entirely when passing over these. The glory phenomenon was not seen around the shadow point of the aircraft when passing over these clouds. The spot appeared only between clouds where there might have been some tenuous haze, but disappeared again in larger cloudless areas over bare terrain and vegetation. There was no snow or ice cover apparent on the ground.

I have no ready explanation for the phenomenon nor have I come across references to it by others in print or otherwise. Perhaps the haze (which, if it existed, did not prevent a clear visibility of the terrain below) consisted of a thin layer of otherwise invisible ice crystals in the form of hexagonal platelets such as are thought to produce the well-known light pillars below and above the sun. If so, these crystals must have been quite stable (no wobble) with their flat tops absolutely parallel to the horizontal plane. for the bright spot I observed was circular and of an angular diameter comparable to that of the sun's disk.

One wonders whether the phenomenon is related to a quite similar specular spot appearing in a sequence of time lapse pictures of the sunlit earth, transmitted by the ATS-1 NASA synchronous satellite, a film projection of which I had seen recently. I was also reminded of a reference to a paper by Barabashov in which he reported a quasi-specular brightness maximum observed on the disk of Venus. (Applied Optics, 7:556, 1968)

## SPECULAR REFLECTION FROM A CLOUD
**Houghton, J. T.; *Weather*, 23:292-293, 1968.**

In the issue of Applied Optics for March 1968, Deirmendjian reports the observation of an interesting optical phenomenon from an aircraft. He noticed when flying between clouds over ordinary ground a bright spot in the position for specular reflection of sunlight in a horizontal surface. He points out the surprising nature of this, and suggests that it could have arisen from the reflection of sunlight by ice crystals in the form of hexagonal plates with their flat tops parallel to the horizontal plane.

I am reminded that I observed exactly the same phenomenon when travelling by jet aircraft from Chicago to Boston during the afternoon of 1 December 1965. The aircraft was flying at a height of about 30,000 ft in the presence of a substantial amount of cirrus cloud (one knew this because of the hazy appearance of the ground) and broken low cloud. The bright spot was very little larger than the angular diameter of the sun, and increased intensely in brightness on passing over patches of water, showing that it was accurately in the position for specular reflection from a horizontal surface. Although I have talked to a number of people about this observation, I have no really satisfactory explanation of it. It would not seem very likely that ice crystals should lie with their flat surface horizontal to an accuracy of about 1/2°---which accuracy would be necessary to form the observed bright spot.

I do not know whether it is significant that my observation was near to the same place and about the same time of year as that by Deirmendjian. It would be interesting to know if anyone else has observed it, and whether there is any other explanation of its occurrence. (Weather, 23:292-293, 1968)

## SPECULAR REFLECTION IN A CLOUD
**Fraser, Alistair B.; *Weather,* 24:160-161, 1969.**

In the July 1968 issue of Weather (23 (7), p. 292), J. T. Houghton refers to a specular reflection from clouds. This phenomenon which is variously known as a 'sub-sun' or 'under-sun' seems to be rare only because one is rarely in a position to observe it. However, it is usually to be found on those occasions when one is flying above a cloud of ice crystals (because the crystal has a horizontally stable settling-position), suggesting that it is actually fairly common.

The usual shape of the sub-sun is that of a spot of light elongated towards the observer as can be seen in the two accompanying pictures. Tipping of an ice crystal causes the light to be spread considerably in the direction of an observer, but very little to either side (for an explanation see M. Minneart: Light and Colour in the Open Air, pp. 16-20). This means that Houghton's crystals did not have to 'lie with their flat surface horizontal to an accuracy of about 1/2$^{o}$ to produce the short horizontal axis that he undoubtedly observed.

I do not find it significant that the observations of Houghton and Deirmendajian were near to the same time and place.

One of the accompanying pictures, Fig. 1, was taken (28-mm lens, 35-mm camera) in January 1965 at about 18,000 feet over the Colorado Rockies. We were flying within a thin cloud of crystals as can be seen by the presence of a 22$^{o}$ halo as well as the sub-sun. The other picture, fig. 2, was taken (55-mm lens) in July 1967 at about 30,000 feet over Hudson's Bay. This time the plane was above all the crystals so only a sub-sun was possible. The blotchy quality of the light in the vicinity of the sun is due to the aircraft window.

If you want to observe another fine reflection phenomenon, try looking for Scorer's white horizontal arc through the antisolar point (Quart. J. R. Met. Soc., 89, 1963, p. 151). (Weather, 24:160, 1969)

## ABNORMAL LOOMING AND MIRROR EFFECT
**Etches, J. A.; *Marine Observer,* 34:180-181, 1964.**

s. s. Bravo, Captain J. A. Etches. Lisbon to Gibraltar. Observer, Mr. M. A. Morrill, 2nd Officer.

26th November 1963. At 0315 GMT when the vessel was 4 miles due south of the Cape St. Vincent light a false loom was seen on the reciprocal bearing of the light-house, which was so real in appearance that one would have expected the light itself to have been sighted in that direction at any moment. The loom was flashing every 5 sec. as does the St. Vincent lighthouse. As the nearest land was some 240 miles away in the direction of the loom, it was evident that the St. Vincent light itself was undergoing some effect of abnormal refraction. This was borne out by the fact that as St. Vincent drew astern on our port quarter, the loom drew ahead on our starboard bow. Air temp. 59$^{o}$F, wet bulb 55.5$^{o}$, sea 62$^{o}$. Light W'ly wind. Sky cloudless.

Position of ship: 36$^{o}$ 57'N, 9$^{o}$ 00'W.

Note. We have now had three occasions on which images have been observed in reciprocal direction from the object. On two previous occasions it has been suggested that a sharp horizontal increase in moisture content with a sharp fall in temperature along the line of vision, extending over a large vertical plane or one inclined towards the observer in the atmosphere, can act like a mirror causing an observer to see an image of a scene or object behind him. (See Marine Observer, October 1962, page 183, and January 1964, page 19.) On the occasion described above the phenomena appear to have been associated with anticyclonic atmosphere subsidence behind a cold front. (Marine Observer, 34:180-181, 1964)

# SHADOW BANDS SEEN DURING ECLIPSES AND AT SUNSET

## THE SHADOW BANDS AT THE AUSTRALIAN ECLIPSE
**Chant, Jean L.; *Popular Astronomy,* 32:202, 1924.**

In the Nov.-Dec., 1923, number of the Journal of the Royal Astronomical Society of Canada is given an interesting description of observations of the "shadow bands" by Jean L. Chant at the total eclipse of the sun September 22, 1922, at Wallal, Australia. The following extracts will interest our readers.

"Suddenly, and before I had expected it, the shimmering, elusive, wave-like shadows began to sweep over me and the sheet. I grasped the rod and moved it back and forth until it was parallel to the crests of the waves as they moved forward. At the same time I began counting seconds---'one-and, two-and, three-and, etc.'--- and I continued counting up to 150 before totality was announced. I think perhaps I counted a little too rapidly and, allowing for that, I judge that the bands began approximately 2-1/4 minutes before totality. They continued to move over me for only a short time, perhaps ten seconds, and then they were gone! They were faint, thrilling, ghost-like, but definite enough for me to be sure that I had seen the shadow bands. ....."I took my position at the second sheet to watch for the shadow-bands again. They began at 15 seconds after totality and to my surprise they seemed to come from the opposite direction. They were even fainter than at the beginning and lasted perhaps five seconds. I moved the second rod until I judged it was parallel to the crest of the waves.

"At the beginning of totality the bands came from west to east, moving in the same direction as did the moon's shadow as it swept across the earth; at the end they moved in the opposite direction, or so it seemed to me."

In a note concerning these observations Professor C. A. Chant concludes, with some uncertainty, that the direction of the line of the crests of the shadow bands was at right angles to the direction of advance of the shadow cone and tangent to it. The bands were seen 2-1/4 minutes before totality, and as the cone moved forward with a speed of 25 miles per minute it would appear that the air producing the effect was about 56 miles in front of the cone. Professor Chant thinks it almost certain that the bands are caused by the light from the slender crescent of the sun shining through layers of air of continually changing density. From the fact that the bands lasted only a few seconds after totality he concludes that the atmosphere became quiescent at a shorter distance behind the shadow cone than it was in front of it. (Popular Astronomy, 32:202, 1924)

## THE SHADOW BANDS DURING TOTALITY
**Lewis, Isabel M.; *Popular Astronomy,* 34:279-280, 1926.**

Terrestrial Magnetism for March, 1919, has a very interesting account of meteorological observations made by Prof. E. Waite Elder, head of the physical laboratory of East Side High School, Denver, Colo., at Corona, Colo., elevation 10,000 ft., during the total solar eclipse of June 8, 1918. I quote here the notes opposite the time of mid-totality at Corona, G.M.T. 23h 23m, "Had to use match to read thermometer. Too cloudy to see shadow bands at beginning of totality; shadow bands about 10 cm wide and 20 cm apart." An entry two minutes later reads: Brisk wind from N.W. at end of totality, shadow bands 15 degrees N. of E. traveling 2.5 to 3 m. per sec. toward S.E. by S. Would this not imply the visibility of shadow bands at mid-totality? It is also interesting to note here that although a brisk wind was behind the shadow bands they were hardly moving as fast as the shadow bands that we observed at Beacon, N. Y., last year, although there the air was absolutely calm. If the shadow bands are due to surface air currents why should their motion not be speeded up appreciably by a stiff wind.

Observations of shadow bands during totality have a very important bearing upon the question of the cause of shadow bands in general for if it can be proved beyond any question that shadow bands are seen---if only occasionally---during the total phase of the eclipse then the thin solar crescent immediately preceding and following totality must be eliminated as the source of light in the production of shadow bands and as the cause of their peculiar banded appearance. Eliminating the sun as the source of light in their production there is left only the general sky illumination, the corona or possibly light from conspicuous prominences. Also if the shadow bands are simply the shadows of surface air currents and can be seen occasionally during the darkness of total eclipse, one can think of many other occasions when they should be seen, for example daily around the time of sunrise and sunset when the glare of sunlight is sufficiently reduced. Is it not also true that all of the so-called shadow bands produced by search lights and other artificial means differ in some very important and essential details from the shadow bands that are associated with total solar eclipses? (Popular Astronomy, 34:279-280, 1926)

## PHYSICAL OBSERVATIONS DURING THE TOTAL SOLAR ECLIPSE
**Rotch, A. Lawrence; *Science,* 11:752, 1900.**

While the most important observations during the total eclipse of the sun are, of course, astronomical, some simple physical observations can be made with little or no apparatus and may serve to elucidate two obscure atmospheric phenomena, namely, the so-called 'shadow-bands' and the changes in the direction and velocity of the wind.

Professor R. W. Wood, in Science of April 27, has described the appearance of the shadow-bands and has given instructions for observing them, so that, although I myself had prepared a circular of instructions for co-operating observers, yet, in consequence of the fact that so able a physicist as Professor Wood will study this phenomenon, I shall be glad to send him my own observations and any that I may receive. It may be interesting here to state briefly the results of the observations made and collected by Professor Winslow Upton, Mr. A. E. Douglass and myself during total solar eclipses. In the eclipse of August 19, 1887, observed in Russia, it was cloudy and no shadow-bands were seen, but in the eclipse of January 1, 1889, observed in California with a clear sky, the bands were well defined, though an attempt to photograph them failed. They were more prominent at high altitudes than at low levels, but they seem to have no connection with the position of the stations in or near the shadow-belt. While the reports of the various observers indicated a general agreement for the direction in which the bands lay, yet there was no uniformity in the direction of progression which seemed not to be related to the direction of the wind. In every case the speed of the bands was much less than that of the shadow itself, thus disproving the theory that the bands are diffraction fringes in the shadow of the moon. The observations are discussed by Professor Upton and myself in Vol. XXIX., No. 1, Annals Astron. Observatory of Harvard College. During the eclipse of April 16, 1893, observed in Chile under the most favorable circumstances, the shadow-bands were very generally seen immediately after totality. They lay approximately northwest and southeast, and moved mostly towards the southwest at a speed variously estimated at from three to twenty miles an hour. The width of the bands appeared to vary from one-eighth of an inch to four inches, and their distance apart from one to ten inches. A significant fact was that, contrary to the observations in the previous eclipse, the bands were much less conspicuous on the mountain summit, occupied by the writer, than near sea-level, where they were also coarser, thus indicating the effect of increased thickness of atmosphere. (Science, 11:752, 1900)

## ECLIPSE SHADOW BAND MOTIONS—AN ILLUSION?
**Paulton, Edgar; *Sky and Telescope*, 25:328, 1963.**

At an eclipse of the sun, an observer inside the path of totality who looks at a light-colored surface (such as a sheet pinned to a wall) may see the famous shadow bands. About two minutes before the sun's disk is fully covered by the moon, they become visible as diffuse smoky bands, 1/2 to two inches wide and about two to 10 inches apart. Another set of bands should appear for a similar period after totality.

The motions of shadow bands are very difficult to interpret, because of their seeming inconsistency. For example, observations on the central line of the moon's shadow path show practically the same orientation of their motions before and after totality, while observations made only a few miles away show widely divergent directions. The interpretation of these movements as an illusion seems never to have been given serious consideration.

One school of thought maintains that the changes in direction are caused by variations in atmospheric temperature, humidity, density, and pressure. If so, there is no way of predicting the behavior of the bands. But the consistency of shadow band motions along the central line seems to demand more of an explanation than mere coincidence.

At the eclipse of January 24, 1925, a committee of the American Astronomical Society made a concerted effort to study the shadow band phenomenon. The collected reports were turned over to W. J. Humphreys at the U. S. Weather Bureau. His notes in the November, 1925, Popular Astronomy gave the following generalizations: "(1) that the shadow bands must parallel the solar crescent and (2) move, apparently, at least, normally to their length. It follows that there can be no relation between the direction of their travel and the course of the wind at any level."

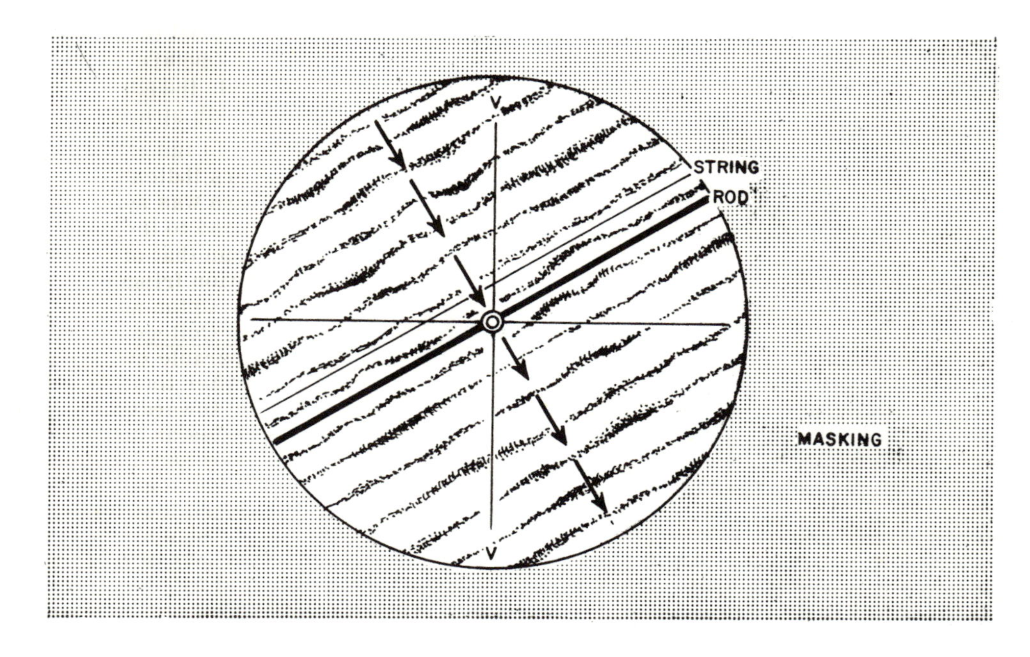


Typical eclipse shadow bands

Humphreys concluded: "Apparently, then, shadow bands are owing to pseudo-total reflections, or mirage effects produced by transition shells between warmer and cooler adjacent masses of air in a state of thermal convection."

This conclusion did not satisfy followers of a diffraction-ring hypothesis that had been presented as early as 1900 in a Popular Astronomy article by H. C. Wilson. This maintains that there is a pattern of concentric rings about the moon's shadow, and the various movements are a result of this pattern traveling across the earth's surface..... (Sky and Telescope, 25:328, 1963)

## SUNSET SHADOW BANDS

**Ives, Ronald L.; *Optical Society of America, Journal,* 35:736, 1945.**

Observed Phenomena. At sunset, when looking eastward over flat barren terrain, from a point having a higher angular elevation than the sun, faint, eastward-moving shadow bands, several miles wide, and extending an unknown, but relatively great, distance to the north and south, have been observed on several occasions. These bands were seen, in each instance, just as the sun dipped below the local horizon. Duration of the phenomena was less than 30 seconds. Estimated rate of eastward motion of the nearest shadow band was 40 miles per hour. Recession of the bands seemed to accelerate with distance. Conditions extant at time of observation are shown in Fig. 1. General similarity of these shadow bands to those occurring during a solar eclipse is notable.

Field Conditions. Sunset shadow bands have been observed six times in fifteen years---once from the Pinacate Peaks, Sonora, Mexico; once from the summit of Navajo Peak (Boulder County), Colorado; twice from an airliner approaching El Paso, Texas, from the west; and twice from high points in the Dugway Mountains (Tooele County), Utah.

Astronomical, topographic, and meteorological conditions were substantially identical in each instance. The angular elevation of the observation point was greater than that of the sun; there were high mountains to the west of the location of the shadow bands, which were seen on level, treeless land, with either sparse or no vegetative cover. The sky was absolutely clear (C. A. V. U.), the relative humidity very low, the temperature high, the wind speed near zero, and the lapse rate steep (nearly adiabatic).

As shadow bands were seen rarely even under these conditions, it is believed that other critical conditions, still unknown, must exist.

Explanatory Hypothesis. Because of their general similarity to eclipse shadow bands, sunset shadow bands are believed to be caused by diffraction of sunlight across mountain crests. Discrepancy between computed diffractive effects and observed phenomena is considerable, and led to a search for other possible causes.

Rather extensive field meteorologic investigations in the Salt Lake Desert failed to disclose any meteorologic cause for the shadow bands. Specifically investigated were air motion patterns, both normal wind and katabatic; regional variations in

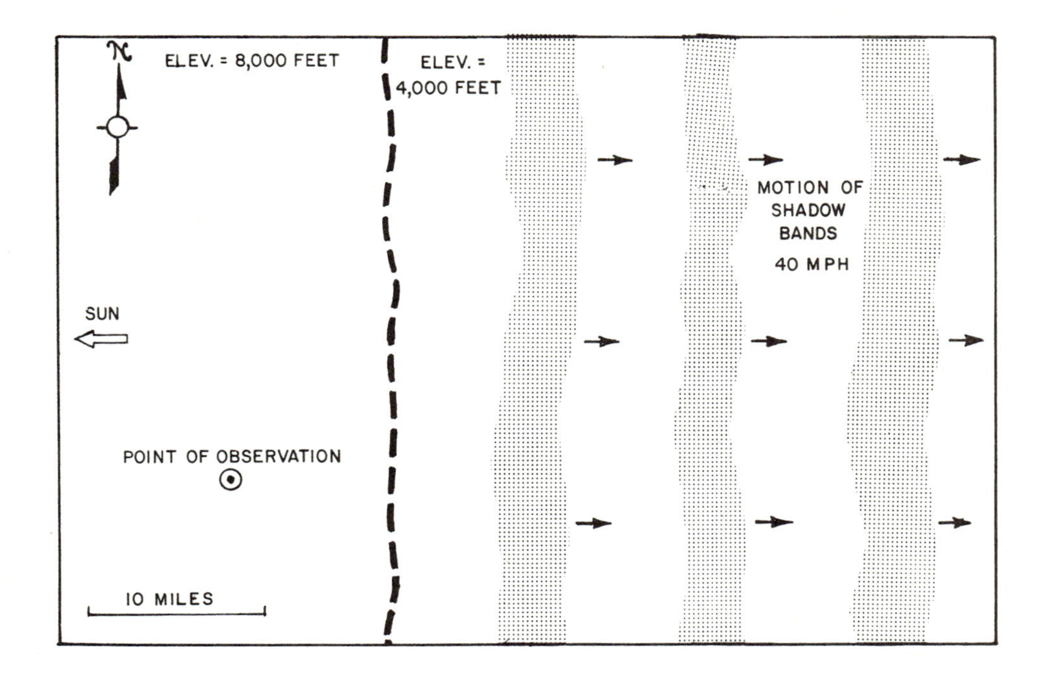


Sunset bands observed on desert terrain

temperature and relative humidity; and the possibility of interference between thermally differing air strata.

Absence of other plausible explanations, and qualitative, but not quantitative, agreement of diffraction theory with field observations, lead to the belief that sunset shadow bands are a diffraction phenomenon.

Conclusions. Field observations suggest that sunset shadow bands are of infrequent occurrence, even under supposedly favorable conditions. Theoretical considerations, and the absence of a better explanation, indicate that they may be diffraction phenomena. Additional attempts to observe them, where terrain and meteorological conditions are supposedly favorable, seem desirable. (Optical Society of America, Journal, 35:736, 1945)

# VISIBLE SOUND WAVES

## VISIBLE SOUND WAVES
**Saunders, Frederick A.; *Science,* 52:442, 1920.**

The following notes, written by Lieutenant Thomas T. Mackie, 123d Field Artillery, A. E. F., describe a phenomenon which must have been observed rarely, if ever before, and it seems to be very much worth while to put the circumstances on record.

On one or two occasions within recent years the occurrence of sound waves visible to the naked eye under peculiar atmospheric conditions has, I believe, been reported; yet the event is so unusual that I have been persuaded to describe a similar one which I witnessed at the front on the opening day of the Meuse-Argonne offensive.

During the days immediately preceding the attack my regiment moved into position in a wooded area opposite Montfaucon, characterized by the roughness of the terrain, a jumble of high hills cut up by narrow and deep valleys. The battery to which I belonged was sent into position at the head of one of these valleys, enclosed by very steep slopes, and having roughly the shape of a V with the open end to the south. Some four or five hundred yards to our rear and approximately on a line with the extremities of the arms of the V was a battery of six-inch rifles.

For several days the weather had been more or less rainy and wet, and the morning of September 26 found us covered by a very heavy bank of fog which entirely excluded the sun. Soon after the attack opened, I had occasion to go to the top of one of the hills which flanked our position, and at a certain definite level above the battery a very considerable disturbance in the fog was noticeable after each discharge of the heavy rifles behind me. The visibility was such that the flash of the discharge could not be seen, but each time before the report reached us a band of greater density was clearly visible in the fog, moving with great rapidity up the valley toward us in the form of an arc. Its arrival was simultaneous with that of the sound of the discharge. This arc of greater fog density was perhaps six feet from its anterior to its posterior edge, and of about the same depth. It followed closely an altitude of some sixty or seventy feet above the floor of the valley and was clearly visible from both above and below that plane, but no similar phenomena were visible in any other plane.

The recent researches of Professor D. C. Miller, and others have shown that the muzzle wave from a large gun carries in its front a narrow region of compression immediately followed by a relatively wide region of expansion. From the above account, it would appear that the air was saturated with water vapor at a particular level, and that the expansion in the wave produced a visible increase in the fog density, the effect disappearing immediately again, owing to the subsequent re-evaporation when the air regained its normal pressure and temperature. The conditions of the terrain were very favorable to the concentration of a great amount of energy into the wave-front, and this was probably assisted by a sound-mirage effect. The upper layers of air being warmer than the lower the sound wave-fronts would be so bent as to tend to keep the energy near the earth's surface. The "experiment" was thus being conducted under such circumstances and on such a scale as can not readily be reproduced in the laboratory, and would rarely occur anywhere. (Science, 52:442, 1920)

## OPTICAL PHENOMENA IN THE ATMOSPHERE
**Anonymous; *Nature*, 155:388, 1945.**

L/Cpl. V. S. Taylor, 6 Field Park Coy., R. E., C. M. F., writes: "The discussion on optical phenomena in the atmosphere in Nature of December 9, 1944, brings to mind an occurrence frequently witnessed at Anzio while it was a beachhead. During, and immediately after, intense A. A. fire under conditions of virtually clear sky, with the sun behind the observer, concentrically disposed wave ripple arcs could be seen passing away from the barrage zone, in the portion of the sky about 45$^{o}$ forward of the observer. The acute compression of the atmosphere peripheral to the bursting shells caused the compression zones to be sufficiently altered in refractive index to produce an optically visible phenomenon when refracting undiffused sunlight." This phenomena would seem to be similar to the concentric waves observed by Dr. A. H. Goldie following a bomb burst (Nature, 154, 738; 1944). (Nature, 155:388, 1945)

## SECONDARY SHOCK WAVES AND AN UNUSUAL PHOTOGRAPH
**Jones, Arthur Taber; *American Journal of Physics*, 15:57, 1947.**

The remarkable photograph reproduced in Fig. 1 [not reproduced] was taken about 15 mi from Bitche, France, in the middle of March 1945, by M/Sgt Howard B. Gray, of the 208th Battalion, U. S. Field Artillery. The air was very clear, and the sky was of a deep blue with cirrus clouds that were estimated to be at an altitude of 20,000 to 30,000 ft. During the preparation firing prior to breaking through the Siegfried Line many men saw dark shadows in the form of huge arcs that passed across the white clouds, and were seen only where there was a background of cloud. Three of these arcs can be seen in Fig. 1. The centers of the arcs were on the German side of the Line, the direction of motion was away from the German positions, and the arcs ceased to be visible after they passed overhead. At first many men thought that the arcs were the visible sign of some new secret weapon, but the general opinion soon came to be that they were sound waves. This latter opinion is doubtless correct. (American Journal of Physics, 15:57, 1947)

# MULTIPLE SUNS AND MOONS: TRICKS OF REFRACTION

## DOUBLE SUN
**Atkinson, W. S.; *Marine Observer*, 25:31, 1955.**

M. V. Lotorium. Captain W. S. Atkinson. Mena-al-Ahmadi to Liverpool. Observer, Mr. J. Behrsing, 3rd Officer.

2nd March, 1954, sunset. Two suns were observed at sunset, the image being above the true sun. The observation was not very good because the true sun was obscured by clouds until a few minutes before sunset, and consequently the period of this phenomenon is not known.

Position of ship: off Cape Finisterre. (Marine Observer, 25:31, 1955)

Double sun off Cape Finisterre

## TWIN SUNS
**Evans, J. E.; *Marine Observer*, 27:84-85, 1957.**

M. V. Salaverry. Captain J. E. Evans. Liverpool to Curacao. Observer, Mr. C. Rowntree, 2nd Officer.

14th June, 1956, 2145 G. M. T. Shortly before sunset, with the sun at an altitude of 2° 35', bearing 298°, twin suns were observed side by side, quite clearly separated. The image was to the right of the true sun, there being no distortion of either. This phenomenon persisted for 2 min, after which the image gradually elongated in the direction of the true sun, eventually merging with it within 1/2 min. The cloud directly above the sun before, during and after the phenomenon was Fn. It was not possible to take a photograph.

Position of ship: 32° 52'N., 39° 22'W.

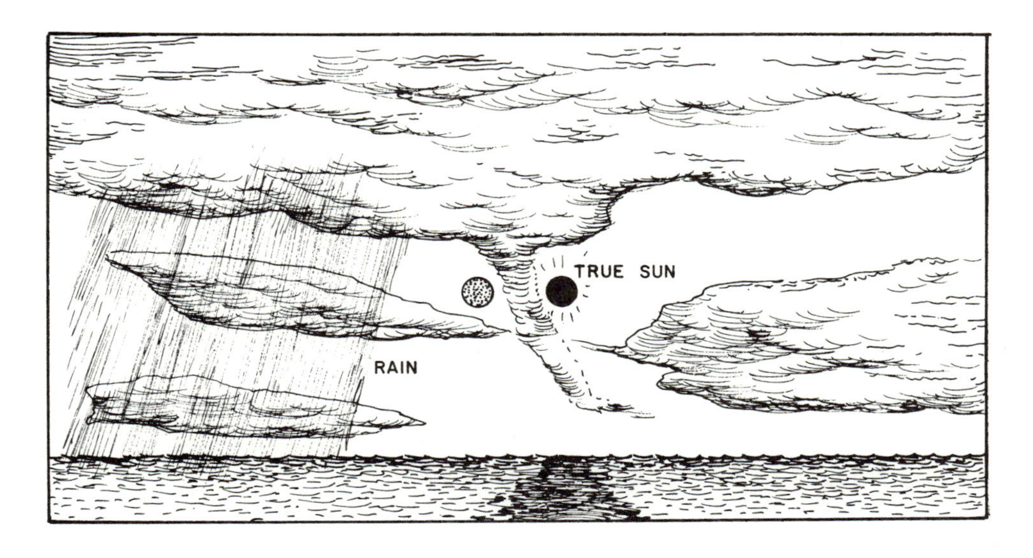


Twin suns created (perhaps) by a lateral mirage

Note. This is an extremely interesting observation of lateral mirage and appears to be a unique one. For lateral mirage it is necessary for the air to be stratified into vertical masses of different densities. Ordinary abnormal refraction, which may produce an image vertically above the object, is produced by a more or less horizontal distribution of layers of air of different densities. The textbooks on meteorological optics give no observations of lateral mirage made in the open sea. Such mirages have been seen on land in the vicinity of walls or cliffs, when the temperature of these differed widely from that of the air close to them. There is also an observation of the duplication of the sails of a vessel on Lake Geneva by lateral mirage. The vessel was just inside the edge of the shadow thrown on the water by adjacent hills and therefore in an air temperature cooler than that above the sunlit lake just beyond the shadow. (Marine Observer, 27:84-85, 1957)

## ABNORMAL REFRACTION
**Fraser, N.; *Marine Observer*, 27:83-84, 1957.**

M. V. Tyrone. Captain N. Fraser. Sydney to Balboa. Observer, Mr. B. P. Telfer, 3rd Officer.

28th June, 1956. At 1415 G. M. T., just prior to moonrise, a bright red glow was observed on the horizon on the bearing on which the moon was expected to rise. A red light suddenly appeared and at first was taken to be a hurricane lamp, lit by a small boat about 2 miles away, on our approach. The light spread to $2^{\circ}$ in width and was obviously in, not on, the water, as it did not lift to the swell; it had the appearance of a flaming log, but the "flames" were motionless. Three minutes later a line of white light appeared over an arc of about 10° of the horizon. The moon then rose and the red, flame-like light disappeared, to be replaced by two half-moon shaped glows in the water. As the moon rose, the irregularities on its upper limb were plainly visible to the naked eye, and when just clear of the horizon the lower limb appeared to have a dent in the edge at about five o'clock. Visibility was exceptionally good---Altair, Enif and Mars were visible on rising. Air temp. 72°F, wet bulb 64°, sea 73°. Barometer 1022.8 mb. Light airs, smooth sea, low SE'ly swell. Clear sky with a trace of Ac lent. ($C_{M4}$).

Position of ship: 24° 24'S., 127° 45'W.

Note. This is a very unusual and interesting observation and we have had only one similar report before, the observation of M. V. Winchester Castle, published on page 78 of the April 1953 number of this journal. In the present observation what was seen was a distorted image of the moon on the sea just before moonrise. In the earlier observation a distorted image of the sun was seen in the sea just after sunset. It is not possible to give any precise explanation of the course of the refracted light rays from the sun or moon below the horizon which would produce such images. (Marine Observer, 27:83-84, 1957)

## UNUSUAL SOLAR PHENOMENON
**Read, Geo. R.; *Meteorological Magazine*, 71:139, 1936.**

At 19h. 45m. G.M.T. on Monday, June 15th, 1936, an unusual solar phenomenon was witnessed at Saughall. At the time the sun was below a band of alto-cumulus lenticularis with the elevation of the centre of the sun about 8°. The sun appeared as a deep red orb while above it were two segments of a circle also deep red in colour. The upper and smaller segment (see diagram) was at a distance of approximately 2° from the upper limit of the sun and the second was about midway between this and the sun.

It is of interest to note that at the same time a particularly well-developed sun pillar was observed from a view point about six miles due east by Mr. W. G. Davies, of Upton Park, Chester. (Meteorological Magazine, 71:139, 1936)

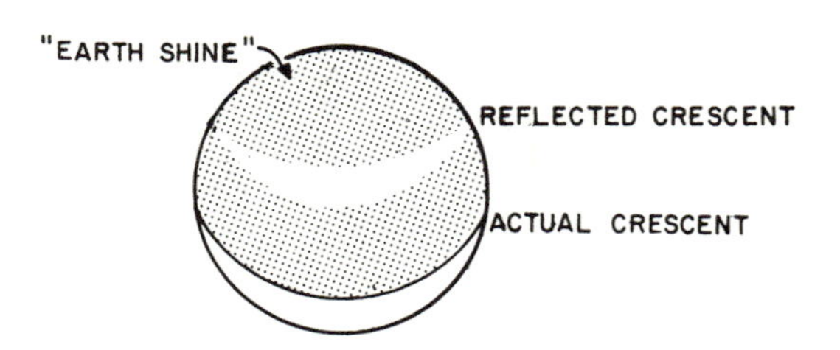


Refraction effects produce detached segments of sun

## UNUSUAL REFRACTION OF A CRESCENT MOON
**Townshend, C. R.; *Marine Observer*, 29:178, 1959.**

M. V. Port Auckland. Captain C. R. Townshend. Cristobal to Suva. Observer, Mr. L. B. Williamson, 3rd Officer.

7th December, 1958. At 1000 G. M. T., when at an altitude of 35°, the whole of the moon's disc was very faintly visible, due to "earth shine", in addition to the crescent which was of normal brilliance (Fig. 1).

At 1007, the crescent appeared to become detached, and occupied a separate position, a small distance away from the moon's disc. At the same time, another crescent, slightly thinner than the original one, took its place; it was just as distinct and had the same colour (Fig. 2). At 1010 the separation between the two crescents increased further and at 1012 a third narrow crescent appeared between the other two. It was just as clearly defined and of the same colour as those which preceded it (Fig. 3). At 1014, low cloud covered the moon and no further observations were possible. Air temp. 71.7°F, wet bulb 67.1°, sea 72.8°. Wind E'ly, force 3.

Position of ship: 4° 06'S., 109° 52'W. (Marine Observer, 29:178, 1959)

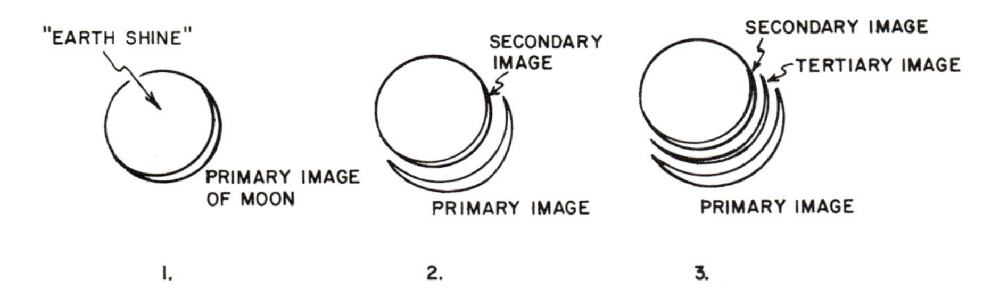


Multiplication of lunar crescents by refraction

# KALEIDOSCOPIC SUNS

## A KALEIDOSCOPIC SUN
**Mintern, Joseph; *Meteorological Magazine*, 58:10-11, 1923.**

A very remarkable optical(?) display was seen here on January the 9th, about 1:30 p.m. local mean time. During the entire forenoon an almost clear sky and sunshine prevailed and up to 1 p.m., when a slight snow-shower fell, clearing again a little later. A friend shouting to me to hurry out, I saw the sun behaving in a most unusual fashion; now surrounded by bright red, flashing rays in all directions, then changing to yellow in which the body of the sun, though more clearly visible, appeared to dance and shift about here and there in a radius of about $5^{o}$; again, changing to green, the rays flashing as in the red---all these changes taking place in less time that it takes to write.

Could this have been caused by a cloud of snow particles crossing the sun in otherwise clear air?

Of all the beauties seen I should think the quickly changing mock suns the most beautiful as they flashed here and there faster than it was possible to count them. The colours were so brilliant and dazzling, that even after I had come indoors anything I looked at appeared a mixture of all the colours seen.

The only thing which I can compare it is Venus "boiling" in the telescope on a hazy windy evening, on a very large scale; but no description is able to convey what was seen in these ten or twelve minutes. (Meteorological Magazine, 58:10-11, 1923)

## REMARKABLE OPTICAL PHENOMENON
**Alexander, S.; *American Meteorological Journal*, 3:486, 1887.**

The following account of a remarkable optical phenomenon was recently related to me by a lady living in this vicinity. She is intelligent and entirely trustworthy. Her statement has been corroborated by others. The occurrence herein related took place from twenty to thirty minutes before sunset in the latter part of June of the year 1885.

The weather was more than usually fine. The sky was clear with the exception of a few clouds of the cumulo-nimbus order a few degrees above and to the northward of the sun. Suddenly there appeared a peculiarly weird and hazy condition of the atmosphere. There was an indescribable commingling and general diffusion of all the hues of the rainbow. During this state of things there appeared in the sky, on the earth, and on the trees, innumerable balls of decomposed light, presenting all imaginable colors and apparently of about the size of a bushel basket. They were uniform in size and appearance.

This phenomenon was confined to that region of the sky about the sun, extending but a few degrees each side of it. It lasted about twenty minutes, when it disappeared as suddenly as it came. (American Meteorological Journal, 3:486, 1887)

# UNIDENTIFIED OPTICAL PHENOMENA IN THE ATMOSPHERE

## PECULIAR ARCS OF COLOR AFTER SUNSET
**McNeil, H. W.; *Marine Observer*, 25:216, 1955.**

M. V. Timaru Star. Captain H. W. McNeil. Melbourne to Aden. Observers, the Master and Mr. N. Johnson, 4th Officer.

31st December, 1954, 1400 G. M. T. Shortly after sunset a very noticeable set of violet rays were seen, which reached an altitude of 47° and extended over a considerable arc.

Immediately above the horizon the sky was bright orange, obscured in parts by Sc and Cu. Above this was an arc of approximately 5° of greenish-white, topped by the violet-rays, which became most noticeable 15 min after sunset; they retained maximum brilliance for a further 10 min and then commenced to fade. Within another 10 min the rays were invisible. Wind calm. Sky throughout very brightly coloured.

Position of ship: 5° 00'S, 66° 10'E.

Note. This is perhaps a phenomenon allied to that variety of the green flash at sunset in which the sky above the point of sunset is coloured green, or green rays are observed to shoot up from this point, several instances of which have been published in this journal. In certain atmospheric conditions the green coloration may be extended to a violet one, just as the green flash proper (the coloration of a setment of the sun) is sometimes seen as a purple or violet flash, when rays of this colour passing through the lower atmosphere are not completely absorbed before they reach the observer. (Marine Observer, 25:216, 1955)

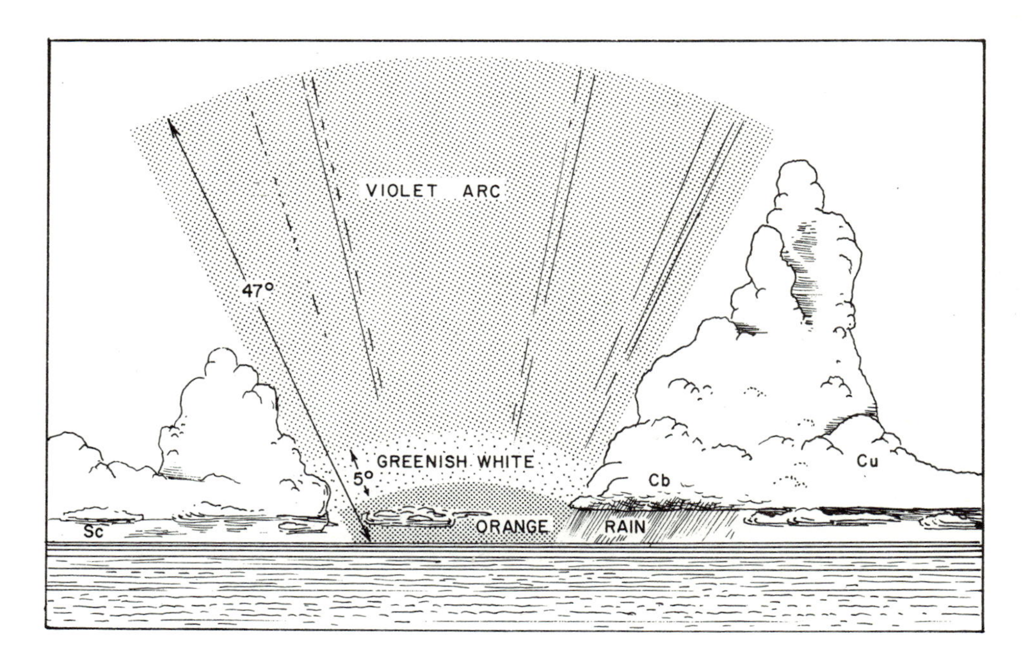


Unusual colored arcs seen over Indian Ocean

## UNIDENTIFIED PHENOMENON
**Vaughn, J.; *Marine Observer*, 25:123-124, 1965.**

m.v. Great City. Captain J. Vaughan. Houston (Texas) to India. Observers, Mr. P. D. Haworth, 3rd Officer and Mr. A. McGrady, Carpenter.

6th July 1964. At 1915 GMT approx., but definitely after sunset, the rings

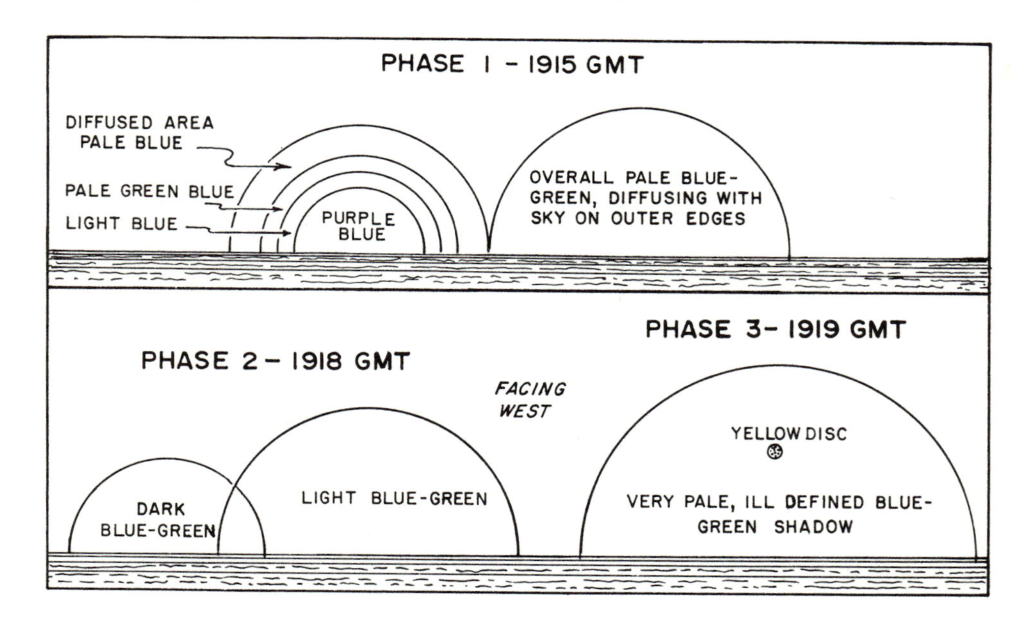


Strange color effects following sunset

shown in the sketches appeared on the western horizon in the direction where the sun had set. They were distinct and the colour differentiation was well defined, but the shadow to the right was less so, though plainly seen. Within a short space of time the whole complex expanded, became lighter in colour and began to merge. With no apparent warning a yellow disc appeared superimposed upon the very hazy shadow. It is stressed that the disc appeared and did not rise from the horizon. It rapidly diffused, increasing in size and becoming pale in colour until it finally merged with the now rapidly dispersing blue green shadow. All traces disappeared by 1920. This disc was estimated to be about 32' in diameter and was seen at an altitude of 10°-15°. Sky cloudless. Air temp. 74.3°F, wet bulb 70.8°, sea 76°. Pressure (corrected) 1014.6 mb. Wind WNW force 2-3.

Position of ship: 34° 28'N, 20° 53'F. (Marine Observer, 25:123-124, 1965)

## MILKY WHITE LIGHT ON HORIZON

**Angus, F. S., and Carling, G.; *Marine Observer,* 40:17, 1970.**

m.v. Otaio. Captain F. S. Angus. Curacao to London. Observers, Mr. W. Marshall, 2nd Officer, Mr. A. J. Davies, 3rd Officer and Mr. G. Adkins, Cadet.

20th March 1969. At 2315 GMT, about an hour after sunset, a semicircle of milky-white light became visible in the western sky and rapidly expanded upward and outward during the next 10 min. When first seen it was quite bright (e.g., as clearly defined as a high cloud across the face of a full moon) but, as it expanded, it became more diffuse. It was possible to follow the expansion to an altitude of 50°

when the circumference of the arc of the light cut the horizon at points bearing 280° and 350° respectively. The sky was cloudless at the time. Air temp. 77.9°F, wet bulb 71.2°, sea 80.9°.

Position of ship: 18° 21'N, 63° 37'W.

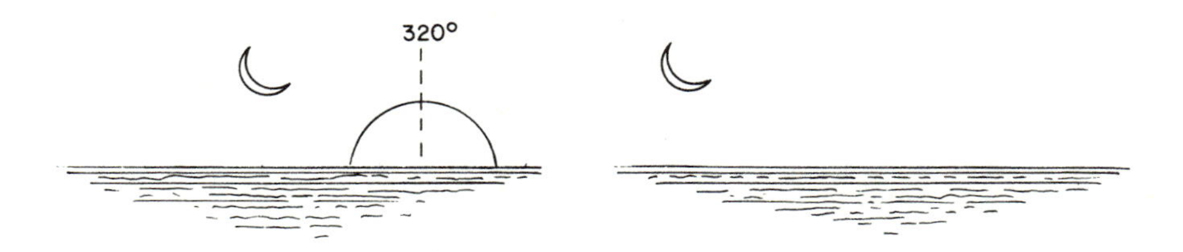


Milky white light appears on horizon

m.v. Port Victor. Captain G. Carling. Curacao to Dunkirk. Observer, Mr. R. A. Cunningham, 3rd Officer.

20th March 1969. At 2315 GMT a sharply-defined globular area of light was seen to rise on the western horizon and enlarge rapidly. By 2319 it had become irregular in shape and spread a general light. The moon, which had a double crescent, bore 282° and was two days old. At first, the moon was above and to the left of the light area but it eventually became enveloped by it. At 0001 on the 21st: Air temp. 78.1°F, wet bulb 74.3°, sea 79°. Cloud 3/8 Se with good visibility.

Position of ship: 20° 36'N, 61° 10'W. (Marine Observer, 40:17, 1970)

# FATA MORGANAS AND OTHER MIRAGES

Strictly speaking, the Fata Morgana is the mirage of a magnificent city seen on the Straits of Messina, but the name is often applied to any spectacular mirage. And why not? The name derives from Margan la Fey, King Arthur's enchantress sister, whose magic could just as easily make a city appear on any shore in the world, luring seafarers to watery fates.

Most mirages are associated with the sea. Ships and landfalls are seen long before they poke above the horizon. Inverted ships and greatly distorted ships are very common. Good mirages appear less frequently over land; they seem to be seen most often on plains and snow fields. The most common mirage of all is that of the sky seen in the air above a hot roadway. But let us talk of less mundane things: armies in the sky, multiple mirages, sideays mirages, and images that are harder to explain.

Mirages sometimes display highly magnified objects. Islands and cities hundreds or possibly thousands of miles away may appear on the horizon. Polar ice may appear as a distant mountain range; a fact which led to the embarrassing "discovery" of nonexistent Crockerland in the Arctic a few decades ago. Stones and hillocks become buildings and great mountains. To magnify in this fashion, the atmosphere must behave like a lens---several lenses in the case of multiple mirages. Just how magnifying air lenses are formed is not well-known. Mirages frequently show distant people and objects with great clarity---something expected only from rigid, precisely ground glass lenses. Like a glass lens, the air lens also focusses scenes in a very restricted area. Moving the head up or down a few feet often destroys the image. Lateral or sideways mirages are also difficult to imagine because the layers of air that create the mirages are at different temperatures and should not be very stable when oriented vertically. Yet sideways mirages are not unknown.

The question that no conventional physicist will admit into a discussion is whether some images seen in the sky might not be mirages at all but rather visual projections through time and/or space completely divorced from real objects. Are all the visions of armies, unidentified cities fair, idyllic countrysides, and ghost ships all real in this space-time continuum? Or do we verge here on the psychic?

## MAGNIFYING MIRAGES

### MIRAGE MAGNIFIES TAHITI
**MacDonald, D.; *Marine Observer*, 15:52, 1938.**

The following is an extract from the Meteorological Log of M.S. Hauraki. Captain D. MacDonald. Suva to Papete. Observer, Mr. J. H. Ibbotson, 2nd Officer.

9th May, 1937. At 5.05 p.m. A.T.S. (0321 G.M.T. 10th) the islands of Tahati, Moorea, Huahine Iti and Raiatea were seen. Only the peaks of Raiatea and

Huahine Iti were visible, but Tahiti and Moorea were plainly recognizable. Very heavy rain had been experienced for four hours previously, ceasing at 4.30 p.m. A.T.S. The horizon cleared from east to north and showed with unusual sharpness and clearness some time before the mirage came into being. Tahiti gave the appearance of being 35 miles distant, when in reality Tahiti Peak was 210 miles away. The sky was overcast except that it had cleared from E. to N. above the horizon. The mirage remained constant throughout, lasting for 15 minutes and disappearing at 5.20 p.m. A.T.S. (0336 G.M.T. 10th).

Wind W.S.W., force 1; barometer 1011.3 mb.; air temperature 76.5°F., sea 81°F.

Position of Ship:---Latitude 17° 39-1/2' S., Longitude 153° 09' W. (Marine Observer, 15:52, 1938)

## A REMARKABLE EXAMPLE OF POLAR MIRAGE
**Hobbs, William H.; *Science,* 90:513-514, 1939.**

In midsummer of the present year those on board the schooner Effie M. Morrissey, while midway between the tip of South Greenland and Iceland, were favored by a remarkable example of superior or polar mirage. Captain Robert A. Bartlett, master of the Morrissey, who has reported the occurrence to me with a view to its publication in Science, has during an experience of more than forty years in the Arctic seen many polar mirages, but, as he says, none so remarkable as this and certainly none so well checked for position and distance.

On July 17, the schooner was from its noon observation in sunshine found to be in latitude 63° 38'N and longitude 33° 42' W. The ship's three chronometers had been checked daily by the Naval Observatory signal, and the air was calm and the sea smooth. At 4 p.m. with sun in the southwest the remarkable mirage appeared in the direction of southwestern Iceland. The Snaefells Jokull (4,715 feet) and other landmarks well known to the captain and the mate were seen as though at a distance of twenty-five or thirty nautical miles, though the position of the schooner showed that these features were actually at a distance of 335 to 350 statute miles. A checking observation of the sun made at 6 p.m. gave the latitude at that time as 63° 42' N and longitude 33° 32' W. It was warm and rainy; the air had throughout been calm and the sea smooth. Captain Bartlett writes: "If I hadn't been sure of my position and had been bound for Rejkjavik, I would have expected to arrive within a few hours. The contours of the land and the snow-covered summit of the Snaefells Jokull showed up almost unbelievably near." (Science, 90:513-514, 1939)

## EXTRAORDINARY MIRAGE
**Anonymous; *Royal Meteorological Society, Quarterly Journal,* 27:158-159, 1901.**

We have received, through Mr. F. Napier Denison, of the Meteorological Office, Victoria, British Columbia, a copy of the Victoria Daily Times, January 26, 1901, which contains an illustration of a remarkable mirage known as "The Silent City of Alaska." It is said to appear every year on the gigantic glacier of Mount Fairweather.

This phenomenon has engaged the attention of scientists, and up to the present time it has baffled all investigation. The scene has been known to the Alaska Indians of the locality for generations, and has been a common subject of speculation among them.

The photograph was shown to some Alaska Indians by Capt. Foot of the Danube, who brought it down to Capt. Walbran, and they instantly recognised it as the famous city in the clouds.

The phenomenon is seen between 7 and 9 o'clock between June 21 and July 10, and the scene never varies excepting for slight changes in the buildings and other prominent landmarks. It is believed that the mirage is a representation of the city of Bristol, England. That it is a seaport is shown by the mast of a vessel, while a tower, an exact duplicate of that of St. Mary Redcliff, appears in the background.

The earthquake of last year broke up the Muir glacier over which the route to the mountain lies. It is a distance of about 15 miles from Muir glacier bay to the scene of the phenomena. A scientific party from San Francisco will investigate it next May.

The distance between Bristol and Mount Fairweather is about 2500 miles. The longest distance a mirage has been seen hitherto is 600 miles. (Royal Meteorological Society, Quarterly Journal, 27:158-159, 1901)

## ATMOSPHERIC MAGNIFICATION ON ENGLISH CHANNEL
**Anonymous; *Nature,* 126:118, 1930.**

July 26, 1798. Atmospheric Refraction. At about 5 p.m. the coast of France became clearly visible from the shore at Hastings, Winchelsea, and neighbouring parts of the south coast of England, and appeared to be only a few miles away, although the distance is actually 40 to 50 miles. The various features of the French coast were easily recognised, and with a telescope even the buildings on shore. This phenomenon continued fully developed until after 8 p.m., when it gradually faded away. (Nature, 126:118, 1930)

# TYPICAL FATA MORGANAS

## THE REAL FATA MORGANA
**Talman, C. Fitzhugh; *Scientific American,* 106:335+, 1912.**

"When the rising sun shines from that point whence its incident ray forms an angle of about 45 degrees on the sea of Reggio, and the bright surface of the water in the bay is not disturbed either by the wind or the current, the spectator being placed on an eminence of the city, with his back to the sun and his face to the sea---on a sudden he sees appear in the water, as in a catoptric theatre, various multiplied objects, such as numberless series of pilasters, arches, castles well delineated, regular columns, lofty towers, superb palaces with balconies and windows, extended alleys of trees, delightful plains with herds and flocks, armies of men on foot and horseback, and many other strange figures, all in their natural colors and proper action, and passing rapidly in succession along the surface of the sea, during the whole short period of time that the above-mentioned causes remain. But if, in addition to the circumstances before described, the atmosphere be highly impregnated with vapor and exhalations not dispersed by the wind nor rarefied by the sun, it then happens that in this vapor, as in a curtain extended along the channel to the height of about thirty palms and nearly down to the sea, the observer will behold the scene of the same objects not only reflected from the surface of the sea, but likewise in the air, though not in so distinct and defined a manner as in the sea. And again, if the air be slightly hazy and opaque, and at the same time dewy and adapted to form the iris, then the objects will appear only at the surface of the sea, but they will be all vividly colored or fringed with red, green, blue, and the other prismatic colors."

So runs the classical description of the Fata Morgana, written in 1773 by the Dominican friar, Antonio Minasi, and since become the common property of encyclopaedists the world over.

Minasi was born not far from Reggio, and saw the Fata Morgana himself three times. His description is probably, in the main, accurate, though, as we shall presently see, he limits the time of occurrence of the phenomenon too narrowly in stating that the sun must be at an altitude of about 45 degrees. He was the first writer to point out that the Fata Morgana occurs in two distinct varieties; viz., the _marine Morgana_, which appears to lie in or beneath the water, and the _aerial Morgana_, which extends upward to a considerably greater apparent altitude. Minasi's third variety, the _iridescent Morgana_, appears not to have been observed since his time. The atmospheric refractions to which the Fata Morgana, in common with other forms of mirage, is due are not usually attended with a sensible dispersion of light (i. e., separation of the prismatic colors), but that such dispersion may sometimes occur is by no means impossible. In fact, we shall have occasion later to mention a form of mirage, seen in another part of Italy, in which iridescent coloring is stated to be a common feature.

Minasi's attempts to explain the Fata Morgana are far less happy than his descriptions. In his day little was really known about abnormal refractions in the atmosphere, though he lived upon the eve of the important discoveries of Gruber, Busch, Monge and Biot. The Straits of Messina are subject to strong tidal currents, which often run in opposite directions at a given time; that is to say, in midchannel the current sets to the north while along shore it is running south, and _vice versa_. Minasi supposed that the surface of the water thus acquired, at times, marked differences of level, so that it behaved like a mirror lying, not horizontal, but tilted at a slight angle, or at several angles in different places, and reflecting objects along the Calabrian shore (i. e., the shore of the Italian mainland, on which Reggio lies). He therefore took the Morgana to be the reflected image of Reggio and the adjoining coast---the same coast from which the phenomenon is seen. In Minasi's picture, which has been the basis of the greater number of the representations of the Fata Morgana shown in textbooks and reference books, the city in the background is Reggio, while in the foreground are shown all three of the forms of the phenomenon that he has described. It was intended to be a generalized diagram of the Fata Morgana, rather than a faithful picture of its appearance at any one time.

Minasi's explanation is of course no longer accepted, and we now know that the terrestrial objects seen in the Morgana are the refracted images of such objects on the Sicilian coast, or in some cases on parts of the Calabrian coast remote from the place of observation. Moreover, if we except the altogether unconvincing narrative of the French traveler Jean Jouel, there are no cases on record in which the Morgana has been seen _from_ the Sicilian side of the straits.

It is most remarkable that a phenomenon so renowned as the Fata Morgana has been the subject of extremely few accurate scientific investigations conducted on the spot. The rarity of the phenomenon, however, serves in a measure to explain this fact. Many people who have spent their whole lives in and about Reggio have never once seen the famous spectacle.

In our own time only three persons, two of whom lived for some years in Reggio and were eye-witnesses of the phenomenon, have attempted to collect and discuss all the existing information concerning it. Pernter, in his great "Meteorologische Optik," not only publishes the principal descriptions of the Morgana that had appeared up to the year 1902, but includes several that he himself gathered by diligent correspondence. In 1902 Dr. Boccara, professor of physics at the Technical Institute at Reggio, published a memoir on the subject, in which he discussed all the earlier observations and three made by himself. Finally, in 1903, Giovanni Costanzo published what is no doubt the most important contribution to the descriptive side of the subject; but postponed, as the theme of the second memoir not yet published, the theoretical discussion of the phenomenon.

The last-named writer has collected, in a convenient table, abstracts of all available trustworthy descriptions of the Morgana, from that of Fazello (1558) down to the present time.

So far as we know, the Fata Morgana is not specifically mentioned by any writer before Fazello, although the phenomenon of mirage, in general, was well known to the ancients; while the name "Fata Morgana" has not been found, in this particular application, in any extant work before that of Marc' Antonio Politi (1617). In other uses the name is, of course, much older; for Morgan le Fay, the fairy sister of King Arthur, was a favorite theme of mediaeval romance. The legend of this fairy, who dwelt in a marvelous palace under the sea, appears to have been carried to southern Italy by the Norman settlers in the 11th century. The phantoms of the Straits of Messina were subsequently alleged to be wrought by her enchantments for the purpose of luring mariners to destruction. According to this tale, the seaman would mistake the aerial city for a safe harbor, and would be led hopelessly astray in endeavoring to reach it. Hence this enchantress was the mediaeval successor of Scylla and Charybdis, who also had their home in the Straits of Messina. (Scientific American, 106:335+, 1912)

## EXTRAORDINARY MIRAGE IN THE FIRTH OF FORTH

Anonymous; *Symons's Monthly Meteorological Magazine,* 6:98-99, 1871.

For some time past the atmospheric phenomena at the mouth of the Firth of Forth have been of a remarkably vivid and interesting character, and have attracted a great deal of attention. During the past week especially, scarcely a day has passed without exhibiting extraordinary optical illusions in connection with the surrounding scenery, both at sea and on shore. As an instance of the unusual nature of these phenomena, the whole of the Broxmouth policies, mansion-house, and plantations, were one day apparently removed out to sea. One of the finest displays of mirage, however, occurred on Saturday afternoon. The early part of the day had been warm, and there was the usual dull, deceptive haze extending about half-way across the Firth, rendering the Fife coast invisible. The only object on the Fife coast, indeed, which was brought within the range of the refraction was Balconic Castle on the "East Neuk," which appeared half-way up the horizon, and in a line with the Isle of May. The most extraordinary illusions, however, were those presented by the May island, which, from a mere speck on the water, suddenly shot up in the form of a huge perpendicular wall, apparently 800 or 900 feet high, with a smooth and unbroken front to the sea. On the east side lay a long low range of rocks, apparently detached from the island at various points, and it was on these that the most fantastic exhibitions took place. Besides assuming the most diversified and fantastic shapes, the rocks were constantly changing their positions, now moving off, and again approaching each other. At one time, a beautiful columnar circle, the column seemingly from 20 to 30 feet high, appeared on the outermost rock. Presently the figure was changed to a clump of trees, whose green umbrageous foliage had a very vivid appearance. By and bye the clump of trees increased to a large plantation, which gradually approached the main portion of the island, until within 300 or 400 feet, when the intervening space was spanned by a beautiful arch. Another and another arch was afterwards formed in the same way, the spans being nearly of the same width, while the whole length of the island, from east to west, seemed as flat and smooth as the top of a table. At a later period the phenomena, which were constantly changing, showed huge jagged rifts and ravines in the face of the high wall, through which the light came and went as they opened and shut, while trees and towers, columns and arches, sprang up and disappeared as if by magic. It is a singular fact, that during the four hours the mirage lasted, the lighthouse, usually the most prominent object from the south side of the Firth, was wholly invisible. The last appearance which the island assumed was that of a thin blue line halfway up the horizon, with the lighthouse as a small pivot in the centre; and the extraordinary phantasmagoria were brought to a close about seven o'clock by a drenching rain, which fell for two hours. (Symons's Monthly Meteorological Magazine, 6:98-99, 1871)

## FATA MORGANA
**Macray, J.; *Notes and Queries,* 1:9:267, 1854.**

Not having met with the following account in any English newspaper, of a phenomenon said to have been witnessed quite recently in Germany, I beg to send you a translation from the Allgemeine Zeitung (generally quoted in England by the name of the Augsburgh Gazette) of February 13, detailing, in a communication from Westphalia, the particulars of the phenomenon, new, perhaps, to your pages, but by no means new to the world.

"Westphalia.---If the east has its Fata Morgana, we, in Westphalia, have also quite peculiar natural phenomena, which, hitherto, it has been as impossible to explain satisfactorily, as to deny. A rare and striking appearance of this description forms now the subject of universal talk and comment in our province. On the 22nd of last month a surprising prodigy of nature was seen by many persons at Buderich, a village between Unna and Werl. Shortly before sunset, an army, of boundless extent, and consisting of infantry-cavalry, and an enormous number of waggons, was observed to proceed across the country in marching order. So distinctly seen were all these appearances, that even the flashing of the firelocks, and the colour of the cavalry uniform, which was white, could be distinguished. This whole array advanced in the direction of the wood of Schafhauser, and as the infantry entered the thicket, and the cavalry drew near, they were hid all at once, with the trees, in a thick smoke. Two houses, also, in flames, were seen with the same distinctness. At sunset the whole phenomenon vanished. As respects the fact, government has taken the evidence of fifty eye-witnesses, who have deposed to a universal agreement respecting this most remarkable appearance. Individuals are not wanting who affirm that similar phenomena were observed in former times in this region. As the fact is so well attested as to place the phenomenon beyond the possibility of successful disproof, people have not been slow in giving a meaning to it, and in referring it to the great battle of the nations at Birkenbaum, to which the old legend, particularly since 1848, again points." (Notes and Queries, 1:9:267, 1854)

## MIRAGE ON THE CORNISH COAST
**Horton, Percy H.; *Royal Meteorological Society, Quarterly Journal,* 41:71, 1915.**

A most curious phenomenon was to be seen from the Cornish coast at Mawnan, a headland at the mouth of Helford River, five miles from Falmouth, from 3.15 to 4.15 in the afternoon of Saturday October 24, 1914.

In a shallow cloudbank stretching across the horizon from southeast to south gradually appeared what seemed to be a line of coast, with woods, trees, fields, hedges, and houses in their natural colours.

At first I thought it was merely a peculiar arrangement of clouds, but soon I recognised that it was a complete reflected panorama of this coast, and I gradually identified St. Keverne with its church spire, the Helford River, Mawnan Church, Rosemullion, Falmouth Harbour, and St. Anthony.

Shortly before the whole scene faded away, a reflection of Pendennis Head with the castle and the military huts, emerging like some huge ship from a fog bank, appeared most distinctly and was visible for several minutes. Every detail of the whole scene was reversed as in a looking-glass and slightly magnified, for although the panorama seemed to be some twenty miles away, everything appeared as it would to an observer from the distance of only a mile or so. The colours were a little less bright than those of the real coast at the time.

The scene as a whole was visible all the time, but the details of a small portion only were quite distinct at any one time, commencing in the south and gradually moving towards the east, the effect being that of a powerful searchlight being very slowly turned from right to left on to the opposite shore of a large lake. Nothing of it could be seen through a telescope; at least I found it impossible to get it into focus.

I presume that the reason that the reflected objects were seen reversed and not inverted, as is usual in mirages, was the fact that my point of observation was about 300 feet above sea-level, and I was therefore, as it were, looking down upon the mirage, the reflecting medium of which was between the horizon and myself, whereas in most instances mirages are the reflection in the sky of objects beyond the horizon.

A curious point I noticed was that the reflection of some of the objects was not reversed precisely to correspond with my view of the real objects. For instance, I could see St. Keverne Church itself, but its reflection seemed to be from an entirely different point of view. This would, I think, mean either that the cloudbank or medium of reflection was irregular in its reflecting surface and reflected different points at different angles; or that, perhaps more probably, the cloudbank acted as a large mirror from far beyond the object, and showed me the reversed view of the coast as it would appear if observed from the point of reflection; and, since most of the scenery, viz. Helford River, Rosemullion, Pendennis Castle, and the fields, etc., inland were quite out of my range of vision, and yet were all reflected, I am inclined to think that the latter theory is the correct one.

Another curious point was that although the sun was behind me and therefore the coast was in the shade, the reflection was not a silhouette, as one would have expected, but the details showed up distinctly as though a brilliant light were shining on them. (Royal Meteorological Society, Quarterly Journal, 41:71, 1915)

# DOUBLE MIRAGES

## A DOUBLE VERTICAL-REFLECTION MIRAGE AT CAPE WRATH
**Brunt, D.; *Nature*, 111:222-223, 1923.**

On the morning of December 5, 1922, about 10.30 a.m. G.M.T., Mr. John Anderson, lightkeeper at the Cape Wrath Lighthouse, Durness, observed a mirage of an unusual character. Mr. Anderson focussed his telescope on a sheep which was grazing on top of a conical hill (height about 200 feet) about a quarter of a mile away, and immediately noticed an unusual appearance in the atmosphere around. On swinging the telescope slightly upward, he observed that a belt of the atmosphere appeared to be land and sea, giving a perfect representation of the whole of the coast line from Cape Wrath to Dunnet Head.

The appearance in the mirage was an exact replica of what would have been seen from a distance of about 10 miles out to sea. In a direction south of the lighthouse there were three repetitions of the mirage one above the other with sea separating each pair. The entrance to Loch Eriboll and the other bays could be seen and easily recognised in the main mirage, though Cape Wrath itself was rather indistinct.

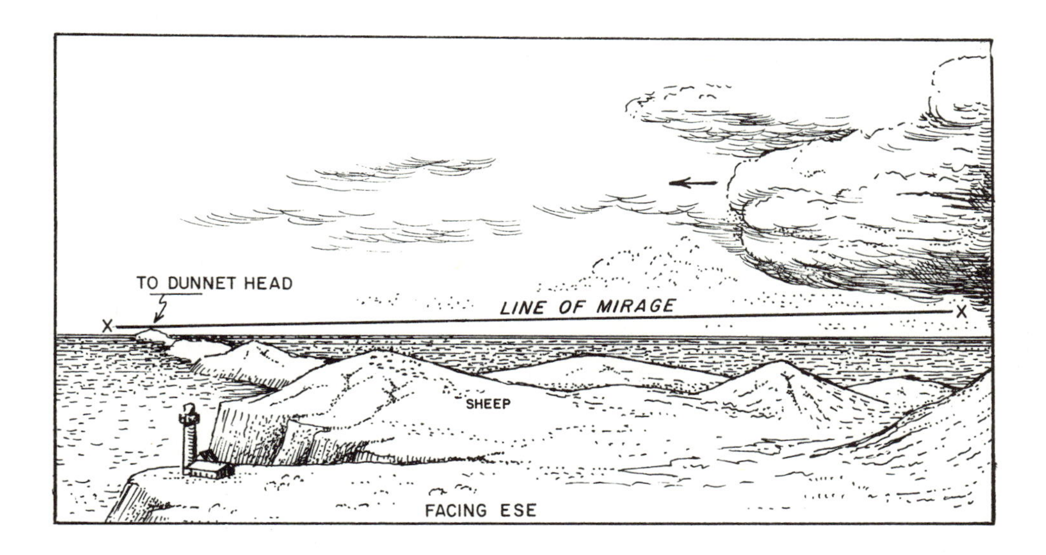


Double mirage at Cape Wrath

The accompanying map shows the apparent position of the mirage and the outline of the coast, while the sketch gives a rough idea of how the country appeared to the observer. The mirage was hidden at one point by a hill.

The mirage was practically invisible to the naked eye, and was only visible from a very restricted area. Mr. Anderson states that it was not visible at a distance of 20 yards either way from his original position, but was still visible 4 or 5 yards from that point. Mr. Anderson estimates the apparent height of the image above the ground as about 1000 feet, in a southerly direction, while the distance from Cape Wrath of the triple image shown on the map is about 12 miles.

The phenomenon was observed for about thirty minutes, when it was blotted out by heavy, dark clouds from the south-west. Within a short time the sky was darkly overcast and rain began to fall, lightly at first, accompanied by slight fog; later rain fell very heavily, the raingauge giving a total of 1.97 inches for the afternoon.

The mirage was seen by practically all the residents at the station. (Nature, 111:222-223, 1923)

# SIDEWAYS MIRAGES

## A SIDEWAYS MIRAGE
**Emmett, W. G., and Corless, R.; *Meteorological Magazine*, 63:15-16, 1928.**

Perhaps the following phenomena of the sky which I observed during a walk from Cape Curig to Pen-y-Gwryd may be of interest to you, while if you can suggest an explanation it will satisfy a natural academic curiosity of the writer.

At 17h. 5m. G.M.T. on October 11th, the sun had fallen behind the Snowdon range, and the range was backed by a large golden, fleecy cloud, on which was outlined a dark, well-defined shadow of Snowdon, persisting for some five or ten minutes. I find it difficult to understand how the shadow was cast in the direction of the sun, as the cloud was manifestly on the far side of the hills, the atmosphere was clear, and the eastern sky was almost cloudless.

At 17h. 20m. the cloud had disappeared, and the western horizon was clear except for some distant striations, when suddenly there appeared over Crib Goch a streaky golden outline of its summit, which faded out after perhaps 15 seconds or half a minute. (W. G. Emmett)

Cases are on record in which sideways mirages have been observed. A mirage is the appearance assumed by a single object which can be seen in two different directions simultaneously; in one direction the rays of light may proceed in straight lines from the object to the eye; in the other they may be bent slightly, owing to the refraction effects in air of varying density. The second image of Snowdon would in that case be reversed (i.e., right to left and left to right).

Sideways mirages are much rarer than vertically displaced mirages, and the appearance of Crib Goch seen later might perhaps be explained as an ordinary mirage as often seen in air arranged in layers of varying density.

It would be difficult to say how the varying densities were produced, but it may perhaps be tentatively suggested that as the air was apparently clear, the temperature of the air in the shadow of the mountains might fall rapidly as the shadow increased in extent, so that appreciable temperature differences would form near the boundary of the shadow. (R. Corless) (Meteorological Magazine, 63:15-16, 1928)

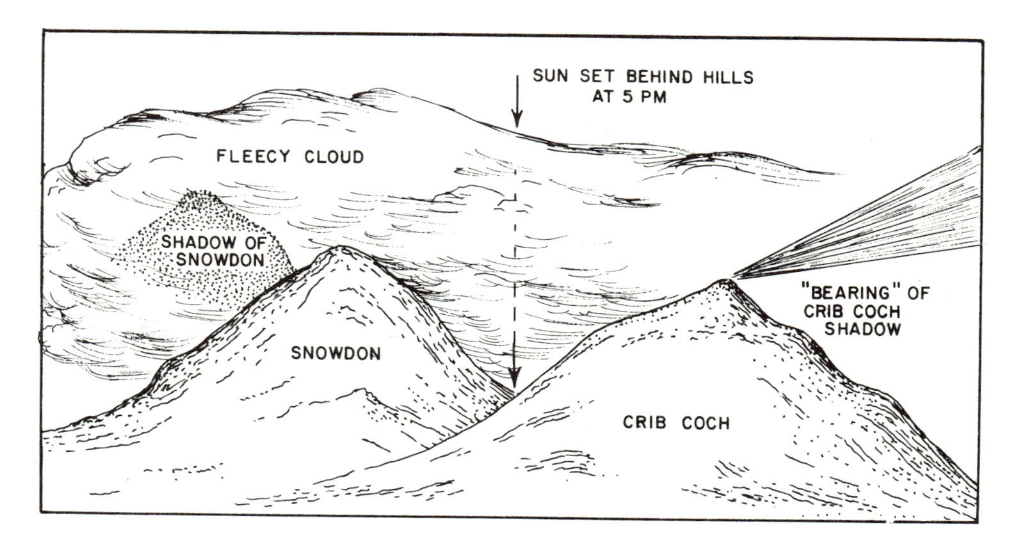


Sideways mirage

# MIRAGES AND THE SENSITIVITY OF THE OBSERVER'S POSITION

## MIRAGES AND THE POSITION OF THE OBSERVER
**MacMahon, P. A.; *Nature*, 59:259, 1899.**

One very curious thing about mirage is that it depends very much upon the position of the eye; a few inches in the height of the eye may make all the difference. I remember myself, on the plains of India, observing a mirage which was only evident when I was at a particular height; there was only a vertical space of two or three inches in which the effect could be seen, so that these phenomena may easily escape notice. A singular effect may sometimes be observed at a particular spot on the south coast, and very likely at other places; when the waves come in on to a very hot beach, if you place the eye within about a foot from the ground and look parallel to the wave-fronts, you can see an image of the wave two or three feet above the real wave. This may conceivably arise in this way: the wave may bring in some cold air, and if the wind were blowing a little off the heated beach there might be some heated air brought in as a layer above the cold air; that would give that rapidly diminishing density upwards which gives a ray with considerable curvature and with concavity presented downwards, and would certainly result in an image of the wave above the real wave. (Nature, 59: 259, 1899)

# RADIO AND RADAR ANOMALIES

Our eyes see only a narrow portion of the electromagnetic spectrum and the major share of atmospheric anomalies thus go undetected by the unaided observer. When radio and radar came along earlier in this century, the operators of this equipment discovered a whole new universe of anomalies. These idiosyncracies are not well known and understood even less, although many do seem explainable in terms of irregularities in the atmosphere and the effects of reflection, refraction, and frequency dispersion.

The long-delayed radio echoes first noticed in the 1920s have been the most sensationalized of these irregularities. Even in the overcrowded electronic environment of today, such echoes are still detected. The delays are so long that any radio reflector (if that is what causes them) must be located far outside the moon's orbit! Alien space probes carrying radio repeaters have even been proposed to explain these echoes. Other solutions proposed below seem at least as likely.

The apparent effects of lunar and planetary position upon radio transmission are also controversial. For the planets in particular, there seems no direct mechanism by which they can affect terrestrial radio propagation. More likely, solar activity, which can also be correlated with planetary position, is the culprit.

In the case of radar, human blindness at long wavelengths is particularly frustrating because radars are always detecting "angels" and spurious targets unaccounted for by aircraft, ships, or even topographic and meteorological features. Birds and insects, singly and en masse, are occasionally the causes of angels. In addition, the bubble-like constitution of the atmosphere---quite invisible at visible wavelengths ---is the creator of other radar anomalies. Still, many solid-appearing radar targets remain unaccounted for.

## LONG-DELAYED RADIO ECHOES

### SHORT WAVE ECHOES AND THE AURORA BOREALIS
**Van der Pol, Balth.; *Nature,* 122:878-879, 1928.**

In the issue of Nature for Nov. 3 a short note by Prof. Carl Stormer appeared under this title. Prof. Stormer there described the observations made by him and Engineer Jorgen Hals, Oslo, of some remarkable echoes heard several seconds after the original signals---which were emitted from the short wave transmitter PCJJ ($\lambda$ = 31.4 metre) Hilversum specially for the experiment---reached the receiver at Oslo. These special signals were first sent in March 1928. Since then the experiment has been repeated over and over again, sometimes twice and often four times a week. A continuous watch for these echoes was also kept at Eindhoven, Holland, in two different places, either by myself or an assistant, or by both of us. We did not hear any of these long period echoes for several months.

Then suddenly, on Oct. 11, I got a telegram from Prof. Stormer stating that very fine echoes had been heard that afternoon. Thereupon I immediately arranged the same night a series of test signals to be sent consisting of three short dots in rapid succession given every 30 seconds between 20 and 21 o'clock local time. I listened with my assistant to the 120 signals. Thirteen echoes were observed by both of us, the times between the signals and the echoes being: 8, 11, 15, 8, 13, 3, 8, 8, 8, 12, 15, 13, 8, 8 sec. The (radio) frequency of an echo was always exactly equal to the frequency of the signal, which fact could be easily verified as the signals were 'unmodulated,' and therefore the receiver was kept oscillating. The combination tone thus formed had exactly the same pitch, whether the original signal was received or the echo. The frequency of the local oscillator was slightly changed a few times after a signal was received, and then the echo came in causing a slightly varied pitch of the combination tone. When thereupon the receiver was left unaltered, the next real signal caused exactly the same pitch in the receiver as the last echo.

The echoes I heard were rather weak, and though their oscillation frequency could be easily identified to be the same as the frequency of the direct signals, the three dots of the original signal could not be recognised in the echo, the latter being of a blurred nature, except in the one case where the echo came in 3 seconds after the signal, when the three dots of the original signal were very plainly audible in the echo as well.

Thereupon I suggested to Prof. Stormer to count the signals in the further experiments, so that echoes heard in Oslo and in Eindhoven could perhaps be identified. Up to Oct. 24 neither in Oslo nor in Eindhoven were echoes heard. However, on that date, between 16 and 17 G.M.T., echoes were again observed both in Oslo and at two different places (3 km. apart) at Eindhoven. The frequencies of the two oscillating receivers at Eindhoven were adjusted at different sides of the carrier frequency of the signal in order to eliminate so far as possible the risk of regarding stray signals as echoes. Prof. Stormer kindly sent me the observations made that day at Oslo where 48 echoes were noted. Receiver No. 1 at Eindhoven (with two observers) noted 4 very weak echoes, and receiver No. 2 at Eindhoven registered 5 echoes. A part of the simultaneous observations are plotted in the accompanying graph (Fig. 1). The timing of the ◉ observations was done with a stop watch, while for the observations ◎ the second hand of an ordinary watch was used. As the echoes often lasted more than 1.5 sec., there is no doubt that some echoes were heard practically

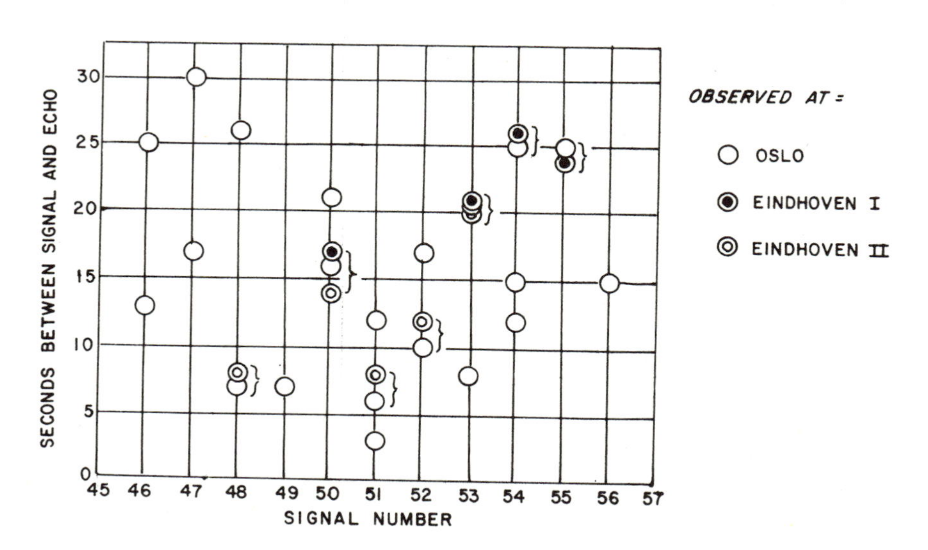


Pattern of delayed short-wave echoes

simultaneously in the three places referred to above. Therefore, though they are often difficult to observe, there is no doubt that the echoes really exist, as they have been heard by several observers at different places and a few times even simultaneously.

As an explanation of these echoes, Prof. Stormer in his letter suggests that the waves are reflected from the streams and surfaces of electrons which he has postulated as the result of his researches on the aurora borealis. According to this view, the waves would have to penetrate the Kennelly-Heaviside layer and travel distances outside the earth's atmosphere comparable with the distance to the moon. Now fortunately wireless waves, even short ones, usually do not penetrate the Kennelly-Heaviside layer, otherwise long-distance communication would be impossible, and an alternative explanation for the occurrence of these long interval echoes may be found in the fact that the waves may penetrate well into but not through the layer. Usually, as Prof. Appleton has shown, the layer has a relatively well-marked lower boundary against which waves travelling nearly vertically are sharply reflected. Now the apparent dielectric constant $e = 1 - \frac{4\pi^2 Ne^2}{mw^2}$ (where N is the density of electrons) diminishes with N, and even becomes zero for waves of 31.4 metre length and a density of circa $10^6$ electrons per c.c. Moreover, with the dispersion law expressed by e we easily obtain for the phase and group velocity: v(phase) x v(group) $= c^2$, so that at the places where the electron density is near the critical one, the phase velocity becomes infinite, but at the same time the group velocity approaches zero. When it now happens that the relative variation of the electron density with height over a distance of a wave-length is small, then the waves may penetrate and soak well into the Kennelly-Heaviside layer and travel in regions where the group velocity is small: they will thereupon be reflected at the region where e approaches zero.

It is obvious that in these circumstances a considerable time may elapse before the echo is received, though the waves have never travelled outside the earth's atmosphere. This point of view would also explain the curious echoes observed by A. Hoyt Taylor and L. C. Young (Proc. Inst. Radio Eng., 16, 561; 1928) which were distinct from the well-known round-the-world echoes (as was also remarked by Prof. Appleton at the last U. R. S. I. meeting). In fact, according to this explanation, any time-interval between signal and echo can be expected to occur, the phenomenon being wholly governed by the gradient of the electron density. This explanation fits in well with the fact that the time interval between signal and echo is extremely variable.

Our view is, therefore, that the group is compressed and 'bottled' for some time in those regions where the group velocity approaches zero. (Nature, 122: 878-879, 1928)

## DELAYED WIRELESS ECHOES
**Anonymous; *Nature,* 126:381, 1930.**

The address given by Prof. Carl Stormer to the Royal Society of Edinburgh on Feb. 17 has now been published in the Society's Proceedings (vol. 50, p. 187). He discusses the problem of whether the 'wireless echoes of long delay' come from space outside the moon's orbit or not. In a communication to Nature of Jan. 5, 1929, he said: "The mathematical theory of the motion of electric corpuscles around a magnetised sphere shows that the chances of obtaining a well-defined toroidal space round the earth are good when the direction to the sun lies near the magnetic equatorial plane (perpendicular to the magnetic axis)." He predicted that it was very improbable that echoes would recur before the middle of February. This prediction was duly verified by several physicists. In particular, two observers in Indo-China observed two thousand echoes from a relatively small emitter station. The echoes came about 30 sec. after the signal and their amplitude was sometimes as great as one-third of the signal. Some of the experiments recorded prove conclusively that they were echoes. It seems as if the space outside the earth's orbit was traversed intermittently by very unstable streams of electrons. This may explain the great variety of echo times observed. It is also possible that multiple echoes may be caused by reflection between the inner walls of the toroidal space. The great variety of echoes is similar to the great variations in aurora phenomena and magnetic perturbations. If this explanation is correct, these wireless echoes give a striking proof of the corpuscular theory of aurora and a valuable method for exploring electron currents in cosmic space. (Nature, 126:381, 1930)

## POSSIBLE OBSERVATIONS AND MECHANISM OF VERY LONG DELAYED RADIO ECHOES

**Crawford, F. W., et al; *Journal of Geophysical Research*, 75:7326, 1970.**

Summary. Observations have been made under controlled conditions of what are believed to be very long delayed radio echoes returning from the ionosphere about 10 sec after transmission. Ionosonde records have suggested a new mechanism for the phenomenon. It is proposed that the signals travel through the ionosphere at very low group velocity under conditions in which collisional attenuation is offset by weak beam-plasma instability. (Journal of Geophysical Research, 75:7326, 1970)

## GHOST ECHOES ON THE EARTH-MOON PATH

**Rasmussen, Hans Lohmann; *Nature*, 257:36, 1975.**

On July 7, 1974 while using a Moon Bounce technique on 1,296 MHz I observed the appearance of strange, delayed echoes. My equipment consists of a parabolic antenna 26 feet in diameter with a circularly polarised feed horn driven with 500-W continuous wave from a transmitter. The receiver has a noise figure of 2 dB and a bandpass of 500 cycles and the equipment had a very distinct note because of a spurious frequency near the fundamental; on the Moon-Earth circuit it is very easy to identify this signal because of this unique characteristic. On the day in question a series of dots or a single dash were being reflected back from the Moon after 2.6 s. Suddenly there appeared a second signal delayed by approximately 2 s. This signal had the same characteristics of the Moon Bounce signal except that it was weaker.

At the time of the observations it was afternoon, the Sun was almost due west and the Moon was to the south-west with an altitude of about 30°. Throughout a series of transmissions the returning Moon signal was followed about 2 s later by the delayed ghost signal with the same characteristic note of the transmitter. Unfortunately, I could not record the signals though they continued for 20 min. When I failed to track the Moon with my antenna, the Moon signal would fade but the echo remained at about the same strength.

The following day a severe radio blackout occurred, and lasted for several days, coincidentally with the appearance of a large sunspot. Relating my reception of the ghost echo with this violent solar eruption I suggest that the large streamer of gas from the corona of the Sun produced a highly ionised cloud which reflected the radio signals I had directed towards the Moon. If this streamer approached with a speed of 1,000 km $s^{-1}$, with a front like a shock wave, it could have acted as a good reflector. The cloud would have been about 800,000 km out in space, as indicated from the delay times of 4-5 s. (Nature, 257:36, 1975)

## LONG-DELAYED ECHOES—RADIO'S "FLYING SAUCER" EFFECT
**Villard, O. G., et al; *QST*, 53:38-41, 1969.**

Have you ever had the experience of hearing your own voice repeat the last couple of words of your transmission, after you have switched over to receive? Or have you been aware, after another station stands by, that a weaker signal on the same frequency is repeating the last few words of the transmission, with exactly the same "fist"?

Well, believe it or not, some amateurs have. If you, dear reader, think us out of our minds to even bring this matter up, rest assured that there are many others who share your view and would cheerfully consign us to the booby hatch. If you haven't tuned out by now, you are undoubtedly asking: just who are the folk who have had this experience? Are they emotionally unstable types, prone to LSD-style hallucination? But hear this: one is a professor of mathematics at a well-known West Coast university; another is a physicist at a midwest research foundation; still another has managerial responsibility for important communication satellite programs at a prominent West Coast aerospace corporation, and most of the rest have a professional connection with electronics in some way. . .

Hard to discount their reports, it appears. Were these men hoaxed, you ask? That's always a possibility, and it apparently has happened in the past. But what about the instances where the echo was heard both on the ham's own signal and on the signal of the station being worked. It would take a pretty clever spoof to simulate both the sound of long-distance transmission and the transmit-receive timing. Still, it could be done, just as a photograph of a flying saucer can be handily simulated with the aid of ordinary crockery.

That's what makes the study of long-delay echoes (LDEs) exciting. At the moment, there is no really indisputable proof that they exist. Scientists remain unconvinced about UFOs, and LDEs are in the same category. However, an increasing body of experimental evidence argues for the reality of LDEs, and it is interesting that a number of new ideas for possible theoretical explanations have come to light only within the last couple of years.

Scientific research is placed under great handicaps when the effect being studied is highly infrequent in occurrence. The handicap is even worse when there is no satisfactory theory to guide experimentation. In these circumstances it hardly pays to set up a special test if a useful result is achieved only once a year on the average. The problem is well known to astronomers, who depend almost entirely on amateur reports to locate comets which pop into view in unannounced places and at unannounced times. Busy professionals simply cannot devote that many hours per year to scanning the skies. LDEs provide an analogous opportunity for hams to be of service to the professional community. Reports on LDEs, with time logged accurately, should be invaluable in helping to solve this particular puzzle.

Background. Echoes of very long delay were first reported in 1928 (References 1 and 2), not long after international short-wave broadcasting got under way. Transmitter powers were around ten kilowatts; antennas were tilted wires; the radio frequency used was around ten megacycles, and receivers were for the most part generators; oscilloscopes and tape recorders were unheard of. On the other hand, interference levels were far below those of today. The experiment consisted of transmitting one or more dots or dashes, and timing the received signals with the aid of a stop watch. Delays ranged from 2 to 30 seconds. Echoes were heard at locations both close to and distant from the transmitter, sometimes apparently at the same time.

A number of theories in explanation of the observations were tried and discarded. The basic difficulty is that radio waves in most circumstances travel at the velocity of light (186,000 miles per second), so that a complete transit of the earth takes only one-seventh of a second. A trip to the moon and back takes roughly two seconds. One theory held that the waves might be slowed down sufficiently if they happened to be close to the ionospheric "critical frequency;" however, it soon became obvious that the accompanying losses would inevitably swallow them up. Loss also makes the possibility of multiple passes around the earth unlikely (210 are required for a 10-second delay)---for the ionospheric gas is by its very nature a lossy dielectric. The hypothesis that echoes might be returned from uncharted clouds of electrons far distant from the earth was seriously considered at the time; today, of course, we know that deep space holds no surprises of that particular sort.

By the middle 1930s few echoes were being received, and the matter remained dormant until the Cavendish Laboratory of Cambridge University undertook a study in 1948 (Reference 3). In a careful year-long test involving transmission of about 27,000 test signals at 13.4 and 20.6 MHz., not one LDE was recorded. No further published scientific activity seems to have taken place since that time. In the intervening years there appears to have been at least one amateur report which was discovered to be a hoax, and in another instance a mechanical fault in a recording was responsible for reports of "delayed echoes" audible on a standard-frequency-station time announcement.

In scientific work when none of the postulated explanations satisfactorily explains a reported effect, and when a reputable scientific organization attempts to find it experimentally and doesn't succeed, there is an understandable and almost overpowering impulse on the part of other members of the scientific fraternity not to become further involved. This is how LDEs came to have roughly the same dubious status as UFOs.

More Recent Experiments. In 1958, W5LFM drew W6QYT's attention to field-strength recordings in which there was an apparent decay of received-signal energy during the 30-second interval of carrier interruption for identification purposes. This behavior, which could have been ascribed to weak (perhaps incoherent) long-delayed echo energy, turned out in the end to be due to the effect of mechanical "stiction" on operation of the pens of the then-standard Esterline-Angus paper-chart recorders. The observation did, however, suggest an expensive means for collecting data on possible LDEs: use a more suitable recorder and see what is left behind on the frequency when WWV's carriers leave the air once an hour. Studies of this sort were made by W6QYT with the help of various part-time graduate-student assistants at Stanford University in the period 1958-1960 (Reference 4). The following suspicious circumstances were---very occasionally---noted:

1) extra noise, decaying exponentially for tens of seconds,

2) extra noise of roughly constant intensity, enduring for about the same period of time, and

3) instances where the same noise actually contained a weak signal similar to the WWV carrier. Some 18 of the type 3 events were observed in a period of about a year. These findings were reported to the Office of Naval Research under whose contact the work was performed, but they were never published because it could not be proved beyond reasonable doubt that the observed signals were in reality caused by the WWV transmissions. They could, for example, have been the result of an obscure fault in the transmitter, although this is considered highly unlikely. WWV frequencies are shared by other standard-frequency stations throughout the world: this introduces troublesome uncertainty. (So does harmonic radiation from 100-KHz. crystal oscillators on the Hewlett-Packard Palo Alto production line, as WB6FDV found out in a classic bit of detective work.) A more sophisticated experiment was clearly needed to decide the matter one way or another, and the effort was sidetracked owing to the pressure of other activities.

Possible Theoretical Explanations. If h. f. signals are to endure for tens of seconds, a way must be found for ionospheric loss to be overcome. In the 1930s the possibility of signal amplification in the ionosphere had not occurred to anyone, but today we can visualize a number of means by which this might take place. Parametric amplification has been suggested: the ionosphere is not a perfectly linear dielectric, and if we could exploit this property, one signal---in principle---could "pump" the other. Another new development is maser amplification; the ionospheric plasma is acted upon to a whole spectrum of radiation from the sun; is it possible that amplification-producing population inversion somehow takes place? Still another explanation has to do with signal storage in the ordered motion of electrons spinning around magnetic field lines; for example, there might be an ionospheric analogue of the phenomenon of spin echoes in nuclear magnetic resonance.

Professor F. W. Crawford of Stanford University has been studying---on paper and in the laboratory---plasmas that "talk back," almost like Edison's original phonograph. A complex signal is fed in, which then disappears insofar as the external circuit is concerned. To call it out, the plasma is pulsed; a replica reversed in time then appears. These "plasmas with memory"---and the above is only one scheme of many---are most readily studied when comparatively high pressures and gigahertz radio frequencies are used. The tantalizing feature of these experiments is that if they could be extended to ionospheric pressures and h. i. frequencies, the indicated time delays fall right in the 3-30 second ball park.

Another remarkable and comparatively recent finding is the so-called "stimulated natural emission" observable at v. l. f. At very low frequencies (on the order of 15 kHz.), radio signals both travel underneath the ionosphere and penetrate it. Those which penetrate are guided by the magnetic field lines and travel from northern to southern hemispheres at phenomenally high altitudes over the equator (one or two earth radii). During their travel, these waves actually rearrange the ambient electrons and store energy in them. This energy is available to amplify any signals of the same frequency after the causative wave is shut off. As a result, an unstable but recognizable replica of the signal is heard after the original transmission stops. This mechanism most emphatically will not work at h. f., since the circumstances are then wholly different. But the fact that radio signal amplification in the ionosphere can happen at all, makes the possibility that something analogous might happen at h. f. seem more likely.

These new developments in the understanding of plasmas stimulated W6QYT to ask for reports of LDEs at a recent get-together of the Northern and Southern California DX Clubs; to his surprise five excellent ones were received; they are included in the summary on the next page.

W5LFM, who has also been interested in this subject since 1958, has collected reports from W5VY and W5LUU, and has himself observed a difficult-to-explain half-second time delay on the time ticks of a Russian standard-frequency station.

Summary of LDE Reports

| Date | Call | Band, MHz. | Approx. duration, seconds | Time, GMT | Phone/ c.w. | Audible on Own/Other |
|---|---|---|---|---|---|---|
| Oct. 16, 1932 | W6ADP | 28 | 18 | 1800 | c.w. | Own |
| Winter, 1950-51 | W5LUU | 7 | 5 | 0300 | c.w. | Own |
| Winter, 1965 | K6EV | 14 | 3-4 | 0600-0700 | s.s.b. | Own |
| Dec. 2, 1967 | W5VY | 28 | 3 | 1328 | s.s.b. | Own |
| Jan. 27, 1968 | W5LFM | 10 | 1/2 | 1400-1430 | Time Ticks | Station RID |
| Dec. 18, 1968 | W6KPC | 28 | 1 | 2000 | s.s.b. | Other |
| Jan. 21, 1969 | W6OL | 14 | 6-10 | 1536 | c.w. | Other |
| Feb. 17, 1969 | K6CAZ | 2 | 2 | 1430-1500 | s.s.b. | Own and Other |

Summary of Characteristics. The Stanford recordings suggested---but did not prove---that incoherent noise "echoes" may exist, as well as coherent ones containing a replica of the signal. The amateur and the early reports, of course, deal only with the coherent variety, which seem to be appreciably less frequent in occurrence. Following is a summary of the conclusions which can be derived from the ham reports taken as a group:

1) multiple-second "coherent" signal echoes, either phone or c.w., appear to be real, and are observable for short periods of time at highly infrequent intervals.

2) they are audible both on a station's own signals, and on signals of other stations,

3) they have been observed at 7, 14, 21, and 28 MHz., but apparently not at higher frequencies,

4) They either occur most frequently (or perhaps are most easily heard) when a given band is just "opening up"---i.e., when skywave propagation to some point on earth is just becoming possible.

5) They seem to be audible when long-distance propagation is good, and when geomagnetic activity is low. (The presence of long-path as well as short-path propagation, or signals from stations at antipodal locations, is apparently a good omen.)

6) Stations reporting LDEs typically have been ones having antennas well up in the air, at locations reasonably good for DX, but other than that no exceptional facilities seem to be required.

7) An active ham who DXes one or two hours a day, may expect to hear an LDE once a year, on the average.

8) The LDEs appear to be one single echo, rather than several successive ones.

9) No Doppler shift is perceptible.

10) The sound of the echo resembles that of a DX signal (i.e., it apparently involves long-distance multipath propagation.)

11) The strength is usually weak, although some reports have put it at S3 or more.

12) Echo strength always decays with time, rather than the other way around.

13) The total time interval during which the echo effect can be heard is remarkably

short---usually no more than a few minutes.

14) There is some indication that LDEs may be heard more frequently on signals which have travelled through the northern and southern auroral zones.
(QST, 53:38-43, 1969)

## STUDIES OF HIGH-FREQUENCY RADIO WAVE PROPAGATION

**Taylor, A. Hoyt, and Young, L. C.; *Institute of Radio Engineers, Proceedings,* 16:561, 1928.**

Summary. Studies of multiple signals of high frequency have been made upon a quantitative basis with reference not only to the round-the-world signals (sometimes called echo signals) but to nearby echoes which have a very much shorter time of arrival. A method of predicting in advance the likelihood of round-the-world echoes occurring between any two different stations has been worked out.

Further studies of nearby echoes show a remarkable retardation on very high-frequency signals coming from the Rocky Point stations to Washington, these signals coming from the Rocky Point stations to Washington, these signals traveling an actual distance varying from 2900 kms. to over 10, 000 kms., although the great circle distance is only 420 kms. The apparent violation of the skip distance law by these stations as observed in Washington has been explained. The nearby echo signals have been tentatively assumed to be due either to reflections from a heavily ionized region in the neighborhood of the magnetic poles or more likely to be due to scattered reflections thrown backwards from the first and second zones of reception, which follow the skip distance region. This throwing back of the signal thus permits under certain conditions the reception of the signal on very high frequency within what is really the skipped zone, the signal having entered this zone by a very indirect route and with a considerable time retardation as compared with the direct route.

Influence of both nearby and round-the-world echo signals upon various types of radio communication have been briefly discussed. (Institute of Radio Engineers, Proceedings, 16:561, 1928)

# EFFECTS OF THE SUN, MOON, AND PLANETS UPON RADIO TRANSMISSIONS

## PLANETARY POSITION EFFECT ON SHORT-WAVE SIGNAL QUALITY

**Nelson, J. H.; *Electrical Engineering,* 71:421-424, 1952.**

At the Central Radio Office of RCA Communications, Inc., in lower Manhattan, an observatory housing a 6-inch refracting telescope is maintained for the observation of sunspots. The purpose of erecting this observatory in 1946 was to develop a method of forecasting radio storms from the study of sunspots. After about one year of experimenting, a forecasting system of short-wave conditions was inaugurated based upon the age, position, classification, and activity of sunspots. Satisfactory results were obtained, but failure of this system from time to time, indicated that phenomena other than sunspots needed to be studied. The first article by the author on this subject appeared in March 1951; the current article is in part a review of that article, and in part will submit additional evidence supporting deductions made at that time.

Study of Planet Positions. Cyclic variations in sunspot activity have been studied by many solar investigators in the past and attempts were made by some, notably Huntington, Clayton, and Sanford, to connect these variations to planetary influences. The books of these three investigators were studied and their results found sufficiently encouraging to warrant correlating similar planetary interrelationships with radio signal behavior. However, it was decided to investigate the effects of all the planets from Mercury to Saturn instead of only the major planets as they had done. The same heliocentric angular relationships of 0, 90, 180, and 270 degrees were used and dates when any two or more planets were separated by one of these angles were recorded.

Investigation quickly showed there was positive correlation between these planetary angles and transatlantic short-wave signal variations. Radio signals showed a tendency to become degraded within a day or two of planetary configurations of the type being studied. However, all configurations did not correspond to signal degradation. Certain configurations showed better correlation than others.

Considerable study was devoted to the most severe degradations and led to the discovery that when three planets held a "multiple of 90 degrees" arrangement among themselves, the correlation was more pronounced. These arrangements were called "multiple configurations" and exist when two planets are at 0 degree with each other and a third planet is 90 degrees or 180 degrees away from them. Also, a multiple exists when two planets are separated by 180 degrees with a third planet 90 degrees from each. These multiples are quite common. A more uncommon type of multiple is the case where all three planets are at 0 degree with each other. From the few cases recorded, this type of multiple shows the least correlation.

Many of the multiples are completed in the space of a few hours, being accompanied by sharp severe signal degradation. At other times, the multiple may take several days to pass, being accompanied by generally erratic conditions during the period. The time needed to complete the multiple depends on the relative speeds between the three or more planets involved. These multiples show correlation for plus and minus about 5 degrees from the exact arrangements previously mentioned.

Configurations of this type actually can be considered as cycles and when several cycles peak at the same time there should be maximum effects. The records for 1948, 1949, and 1950 indicate that such has been the result. Specific instances are demonstrated in Cases 1 to 9 in Table 1. Since consistency of data is of paramount importance in an article of this type, the same cycles between the same three planets have been selected. We may refer to these arrangements as multicycles.

All the close multicycles made between Mercury-Venus-Jupiter were extracted from the records of 1948, 1949, 1950 and correlated with existing radio conditions. Multicycles for these three planets were found in nine cases which are listed in Table 1. Case 10 is a triple multicycle that coincided with an extremely severe signal degradation in 1951. The heliocentric arrangements of the planets involved are shown in Figures 1 and 2 for Cases 1 and 10. **[Figures not reproduced]**

Single configurations (one cycle) between only two isolated planets show the least correlation, but often when several single cycles between several isolated planets coincide in time the correlation is quite pronounced. Most of the single cycles, however, do correlate with at least slight signal degradation. Some correlate with severe degradation.

At times, two, and sometimes three, complete multicycles occur in the space of a few days. At other times a multicycle will occur mingled with one or more single cycles between other planets.

Theoretically, if these planetary arrangements do have the effect that correlation indicates, the cycles between the slow planets should have gradual long-term effect establishing an over-all standard. The most degraded periods should come as the faster planets come into cycle with them or among themselves.

Table 1. Multicycles Among Planets Affecting Radio Signal

| Cases | Dates | Time Consumed | Results in Signal Degradation |
|---|---|---|---|
| 1 | Feb 23/48 | 1 day | Severe 23d and 24th |
| 2 | Apr 18-22/48 | 5 days | Severe 19th to 22d |
| 3 | June 19-23/48 | 5 days | Slight 19th to 22d |
| 4 | Aug 18-21/48 | 4 days | Slight 19th to 21st |
| 5 | Oct 15/48 | 1 day | Very severe 14th and 15th |
| 6 | Apr 12/49 | 1 day | Very severe 11th to 13th |
| 7 | Oct 6-8/49 | 3 days | Very severe 7th and 8th |
| 8 | Apr 2-5/50 | 4 days | Very severe 1st to 6th |
| 9 | Sept 28-Oct 1/50 | 4 days | Very severe 30th to 4th |
| 10 | Sept 21-23/51 | 3 days | Extremely severe 20th to 26th |

Jupiter and Saturn, the largest planets in the solar system, are the most important. Due to their great size and slow motion, they can exercise the predominant influence on the sun for prolonged periods of time and therefore establish an overall standard of disturbed or quiet conditions. However, the arrangements of the other slow planets can add to or take away from their effectiveness to some extent. Therefore, when Jupiter and Saturn are spaced near any multiple of 90 degrees, we should find the most degraded years with a high percentage of the radio disturbances severe.

The year 1951, which was a very degraded year, is an example of this. A slow planet multiple existed between Jupiter, Saturn, and Uranus and Jupiter and Saturn at nearly 180 degrees and Uranus almost 90 degrees from each. This arrangement set a low standard for 1951 and even normally weak cycles between isolated planets showed an effect. The radio disturbances were prolonged and severe. This multiple between these three slow planets permitted a great increase in fast planet multiples and semimultiples. A semimultiple took place each time a fast planet cycled with Jupiter, Saturn, and Uranus successively within a few days. This happened frequently in 1951.

The records indicate that when Jupiter and Saturn were spaced by a multiple of 60 degrees, radio signals were of better quality than when spaced by multiples of 90 multicycles. During such years a high percentage of the single cycles show no important correlation except on the normally weaker circuits. Only the stronger groups of cycles are then accompanied by significant degradation.

It is worthy of note that in 1948 when Jupiter and Saturn were spaced by 120 degrees, and solar activity was at a maximum, radio signals averaged of far higher quality for the year than in 1951 with Jupiter and Saturn at 180 degrees and a considerable decline in solar activity. In other words, the average quality curve of radio signals followed the cycle curve between Jupiter and Saturn rather than the sunspot curve.

When all nine planets of the solar system are considered, we find a great many multicycles and individual cycles which would make correlation very difficult if these cycles were evenly spaced in time. However, the cycles are not evenly spaced, there being a general tendency for the cycles to occur in groups. There are, however, exceptions to this at time.

Conclusion. The author chooses to look upon this hypothesis of a planetary-positions effect upon the quality of short-wave signals as a new approach to the problem and it should be considered as one more tool with which a researcher might work. A tremendous amount of work needs yet to be done.

The correlation found between signal degradations and these planetary arrangements in the past has been sufficiently consistent to indicate that under these arrangements, particularly in the case of multicycles, the planets possibly influence the sun in such a manner as to cause a temporary change in its radiation characteristics. The ionosphere of the earth is apparently particularly sensitive to these changes and reacts accordingly.

By combining planetary indications with solar observations and a day-to-day signal analysis, a 24-hour forecasting system has been developed which averaged close to 85 per cent accuracy throughout 1950 and 1951 as reported by RCA Communications at Riverhead, L.I., N. Y. (Electrical Engineering, 71:421-424, 1952)

## EFFECT OF THE MOON ON RADIO WAVE PROPAGATION
**Anonymous; *Nature*, 159:396, 1947.**

From time to time, some observational evidence has been forthcoming suggesting that the strength of signals received from radio transmitting stations over fixed paths is under certain conditions dependent upon the position of the moon at the time of observation. In particular, H. T. Stetson has investigated this matter, and in 1944 (Terr. Mag. and Atmos. Elec., 49, 9) described the results of the analysis of some eight years measurements taken at Boston on the strength of signals received from the Chicago broadcasting station. Contrary to his expectations, the evidence strongly indicated that the strength of the signals was dependent upon the age of the moon, although the effect was of a complex nature. It will also be recalled that in a paper published in 1939 (Proc. Roy. Soc., A, 171, 171), E. V. Appleton and K. Weekes showed the existence of a lunar tide effect in the height of Region E of the ionosphere. The tide was found to be semi-diurnal, with an amplitude variation of about 1 km. If these results are interpreted according to the simple theory of the formation of the ionized regions, they indicate a relative air-pressure oscillation several thousand times the measured relative pressure oscillation at ground level.

At a recent communication to the Editors, Mr. P. A. de G. Howell, 77 Glandovey Road, Fendalton Christchurch, N. W., New Zealand, claims to have observed during 1938-39 and 1944-45, a correlation between the variation in long-distance transmission conditions at short wave-lengths and the phases of the moon. It was observed that there was a minimum of background noise and high signal strength with little tendency to fade for about two or three days on either side of full phase, these conditions changing to a maximum of noise with poor signals and fading around the time of new moon. The cycle of occurrences is easily confused with those associated with the solar period of similar duration, with which, however, it moves in and out of phase as the months progress. While both Dr. Stetson and Mr. Howell suggest that the effects are the result of changes in ionization in the earth's upper atmosphere, it is clear that the whole subject requires further investigation before definite conclusions can be reached as to the cause of the effect. (Nature, 159:396, 1947)

# THUNDERSTORMS AND RADIO RECEPTION

## EFFECT OF THUNDERSTORMS UP ON THE IONOSPHERE
**Colwell, R. C.; *Nature*, 133:948, 1934.**

Morgantown, West Virginia, U. S. A., is situated on the western slope of the Appalachian Mountains, which run in a south-westerly direction. These mountains cause a great deal of variation in the signal strength of the broadcasting stations along the Atlantic coast as received in Morgantown, and also affect the short-wave band.

One of my students, Mr. A. W. Friend, has been operating a short-wave station here for many years. He informs me that on account of the high hills near his home, his station cannot be heard in the southeastern sector of the United States except after a thunderstorm. He can hear the amateur short-wave stations in the southern States, but they can never hear him if the weather is fine; but after a thunderstorm he can remain in contact with them for several hours.

At my suggestion, Mr. Friend made out the following table from the log of his station. It shows the times at which the stations were able to hear him.

| Station | Date | Time | Freq. Kc. | Output, watts |
|---|---|---|---|---|
| Greensboro, N. C. | April 29, 1931 | 11.16 p. m. | 7,000 | 12 |
| Wilmington, N. C. | April 30, 1931 | 6.15 p. m. | 7,000 | 12 |
| Marion, Ala. | June 2, 1931 | 12.55 a. m. | 7,000 | 12 |
| Atlanta, Ga. | June 5, 1931 | 10.38 p. m. | 14,000 | 8 |
| Salisbury, N. C. | Aug. 24, 1931 | 1.05 a. m. | 7,000 | 15 |

These stations are all located several hundred miles south or south-east of Morgantown, and two-way communication was never possible under normal atmospheric conditions. Mr. Friend's observations strongly support Prof. C. T. R. Wilson's theory that some of the ionisation in the ionosphere is due to thunderstorms. Not all of the abnormal ionisation arises from local thunderstorms, for I have often observed increased ionisation in the *E*-layer after sunset during the winter months when there were no thunderstorms within a thousand miles. (*Nature*, 133:948, 1934)

# RADIO PROPAGATION PATTERNS

## AN ELECTROMAGNETIC RADIATION PATTERN OVER THE OCEAN
**Curtis, George D.; *Undersea Technology*, 5:29-30, 1964.**

During a flight test program performed by LTV for another purpose, it was recognized that there exists a unique low and medium frequency electromagnetic radiation pattern over the surface of the ocean. This radiation pattern appears to be quite stable and is found as variations in the signal from radio stations. Statistical analysis showed correlation between patterns taken on consecutive runs over the same area to be as high as 0.9* over a one-hour period. Figures 1 and 2 clearly show the repeatability of the pattern during successive runs over the same areas.

Figure 1 shows two patterns (runs 2 and 5) from two broadcast stations, recorded approximately twenty minutes apart, while Figure 2 shows a set of successive runs, arranged one above the other, from one station. The day-to-day repeatability is not known.

The amplitude of the variations shown in the figures represents only the top few percent of the signal strength in the area; zero has been greatly suppressed. The equipment and experimental techniques will be explained before the "wiggles" are discussed further.

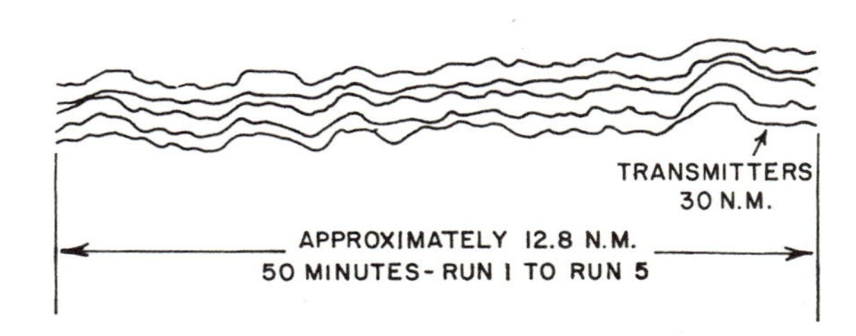


Comparison of five separate runs showing the stability of electromagnetic pattern over ocean

These patterns were obtained by accurately flying back and forth over the same track over the open ocean. A pair of Collins URR-23A multiband AM receivers, modified to output their delayed AGC voltage to a seven-channel Ampex FM tape recorder (with response to DC), and a Texas Instruments dual pen recorder, were used as a sensitive field strength recording system. A dead reckoning computer, radar, careful drift corrections, and smokelights were used as navigational aids to obtain repeat accuracies of a few hundred yards, at altitudes of around 1000 feet. The Navy supplied the aircraft and crew.

The data were played back from the magnetic tape through a 0.1 cps low pass filter, onto rectilinear Sanborn charts. Large photo positives were then made from these, in which the downwind (fast) runs were enlarged slightly to match the upwind (slow) runs, thus providing a series of signal strength patterns scaled to a constant ground speed. The transparencies permit the traces to be overlaid or compared in any fashion (Figures 1 and 2) or digitized for computer analyses.

What produces this pattern? It cannot be considered to be random noise, due to the close correlation of the patterns on the different runs. This pattern is considered to be a natural phenomenon which has never been recognized before to our knowledge, but which can be duplicated by anyone using adequate equipment and techniques.

The exact cause of this radiation pattern is presently unknown. It was noticed during the program that the nature of the pattern varies with the transmitted frequencies which are used; see Figures 1 and 3. The effects are not particularly noticeable above 5,000 feet, but the pattern is generally consistent (on the same flight path) in the range from 500 to 2000 feet as shown in Figure 3. The pattern has been recorded primarily from two, clear channel, 50 KW Los Angeles broadcast stations, but is equally noticeable from other stations nearer our usual flight area, southwest of San Diego. It has also been observed, but not mathematically correlated, on the Atlantic and Gulf Coasts.

Spectral analyses were made on a number of sets of data from the original program by LCdr. L. R. Roberts of the U. S. Navy Post Graduate School. Using an IBM 704 computer, LCdr. Roberts found that a major portion of the repeatable "pattern energy" lies in the 0.005 to 0.2 cps band, which corresponds to a spacing of 3800 to 15,500 yards (approximately 10 to 50 wavelengths). Spectral bands appear above 0.02 cps, but are much more variable. It should be mentioned that the maximum length of the flight runs---20 miles---set a low frequency limit on this analysis and that field variations of even longer periods may exist.

Several explanations have been proposed for this radiation pattern. These suggestions have included:

1. Undersea or Bottom Anomalies---This is a possible explanation of the characteristic of the radiation pattern in shallow water. However, good patterns are also observed in deep water (300 fathoms and deeper).

2. Shore Line Effect---The pattern has been observed consistently at ranges over 15 miles from islands and the continental shore line. Since 15 miles ≈80 wave lengths at 1 mc, the pattern should not be considered shore line effect

3. Clouds---It has been suggested that the variations of the pattern are due to cloud cover. However, as stated previously, the pattern has been consistent for periods as long as one hour. This would suggest that cloud cover is not a major contributor to the radiation pattern.

4. Ionospheric Reflections---Typical patterns (Figures 1 and 2) were made midday in relatively high signal strength areas (order of $\mu$v/m), 130 and 30 miles from the transmitters. Theoretically, under these conditions, ionospheric reflection should not produce even the small variations which constitute the observed pattern. However, recent work has revealed weak MF reflecting ionization between 50 and 100 KM. It is possible that these layers or minor backscatter, could weakly reflect into the test area. Reflection from any continuous ionized layer would, however, produce a pattern symmetrical about the transmitter. This limitation has not been observed.

5. Anomalies near the transmitters which are "reflected". This seems unlikely, due to the fine structure and distance involved, but can be investigated with a three-dimension plot of a large area. The pattern occurs on station radials as well as normal thereto, and is not considered due to antenna directivity.

6. Ocean currents, temperature, salinity, etc.---The effects on radio propagation due to these variables are relatively unknown at the present time and considerable effort is required in the oceanographic field to investigate the effects of these parameters. Figure 4 shows a run over the edge of the Gulf Stream, representing a temperature difference of about 4°F at the surface, plus possible changes in waves and swells. This strongly suggests an oceanographic effect, but the repeatability of this particular pattern is not known. Zero is not suppressed so strongly in this run, made off Cape Hatteras. It is felt that (6) is the most likely cause, but no possible origin has been ruled out for further investigation.

By what mechanism could anomalies in the ocean surface produce an effect such as Figure 4 or the other patterns? When traversing an area of altered conductivity, the tilt of a vertically polarized electromagnetic wave is altered. However, this angular change would be extremely small for the change in the conductivity of sea water caused by a change in its temperature of 4°F. Now, an area of altered conductivity can also exhibit re-radiation, resembling an antenna in a lossy medium. This could give rise to a minor interference pattern. Quite possibly, the recorded signal is due to a combination of such phenomena. (Undersea Technology, 5:29-30, 1964)

*About an 80 percent chance that the wiggles are the identical---an excellent correlation.

# UNIDENTIFIED INTERFERING SIGNALS

## SUDDEN CHANGES IN AMPLITUDE AND PHASE OF THE VERY-LOW-FREQUENCY SIGNAL . . . . .
**Rohan, P., et al; *Nature,* 197:783, 1963.**

A peculiar phenomenon, to our knowledge not reported before, has been observed on the signals of Station GBR on 16 kc/s.

The signals of this Station monitored for frequency standardization purposes, are mixed with a suitable frequency derived from a 100 kc/s standard oscillator and a record, as the one shown in Fig. 1, is produced. The number of beat cycles over 24-h intervals is counted and the drift of the local source determined.

At about 12.30 h local time (9-1/2 h ahead of U. T.) on May 31, 1962, the number of beat cycles suddenly increased and this faster beat persisted for about half an hour, as shown in the figure.

Similar occurrences were recorded on June 11 at approximately 12.30 h local time, on June 12 at 09.45 h and on June 25 at 09.40 h. These lasted approximately 15, 30 and 8 min respectively.

All possibility of instability of the local source must be ruled out as the same derived signal is mixed also with the signals of the United States transmitter NBA on 18 kc/s and no trace of a similar effect has been noticed on the record of this station.

The beat frequency, while the described phenomenon lasted, was of a high stability not likely to be possible to derive from a local source switched on for this period. (Nature, 197:783, 1963)

## INTERFERING RADIO SIGNALS ON 18 KC/S RECEIVED IN NEW ZEALAND
**Allan, A. H.; *Nature,* 201:1016-1017, 1964.**

On three occasions in April 1963, an interfering signal on 18 kc/s with a frequency deviation from nominal of $6 \pm 1$ parts in $10^7$ received at approximately the same signal strength as NBA (Panama Canal Zone) was observed at the Dominion Physical Laboratory, Lower Hutt, New Zealand. The times of occurrence to the nearest 5 min were:

| | | G. M. T. |
|---|---|---|
| April 12-13, 1963. | Intermittently from | 23.35-01.00 |
| April 14, 1963 | " " | 14.20-14.25 |
| April 16, 1963 | " " | 12.10-12.20 |

The beat frequency record for April 13, 1963, is reproduced in Fig. 1. [Not reproduced].

It is of interest to compare these results with those reported in Nature for somewhat similar signals received on 16 kc/s.

The U.S. National Bureau of Standards, Boulder Laboratories, did not observe anything unusual on 18 kc/s on the dates given. The U.S. Naval Observatory, Washington have stated their stations were not scheduled to transmit simultaneously on 18 kc/s in April 1963. Their records show no indication of any interfering signal on the dates indicated but we are confident that it was not a local effect.

If very-low-frequency transmissions are to be used for navigation and standard frequency purposes, it is disturbing to think that the signals may suffer from interference so close to the nominal frequencies. (Nature, 201:1016-1017, 1964)

## INTERFERING VLF RADIO SIGNALS OBSERVED . . . . Reder, F., et al; *Nature,* 213:584, 1967.

In May and June 1962, Rohan et al, observed unexplained sudden amplitude and phase changes on GBR signals at Salisbury, South Australia. Allan reported similar observations made simultaneously at Lower Hutt, New Zealand. He suspected interfering signals from a powerful transmitter in the vicinity of Australia and operating near 16.0 Kc/s. In April 1963, Allan recorded anomalies on the 18.0 Kc/s very low frequency transmissions emitted by NBA and in 1964 Isted proposed an explanation of these "peculiar phenomena" in terms of antipodal signal interference and changes in propagation conditions over one or both great circle routes.

Our NBA records of April 1963, taken at Deal (New Jersey), also show phase anomalies with times of observation which agree with the times given by Allen. The anomalies of April 12, 13 and 14 were almost unnoticeable, but the anomaly of April 16 was quite strong. If it was caused by an interfering signal, its frequency had an offset of approximately $7 \times 10^{-7}$ and its amplitude at Deal could have been as large as 6 dB below that of the signal from NBA which is located only 3,500 km south of Deal.

This simultaneous observation of an anomaly at Lower Hutt and Deal strongly reduces the probability of an antipodal signal interference because an antipodal NBA signal to Deal crosses 4,500 km of Antarctic and Greenland ice and has a path length ten times greater than that of the short-path signal.

Malfunction of the transmitter control circuit as a cause of the anomalies was dismissed by Allan because the British Post Office had allegedly found no evidence of an anomaly on its May/June 1962 GBR records; also the U.S. National Bureau of Standards at Boulder found no NBA phase anomalies in April 1963. The discrepancy between the Boulder and Deal records can be readily resolved by the fact that the Deal tracking station had a time constant which was probably five times shorter than that of the Boulder equipment---an important consideration with fast phase variations.

Simultaneous observation of numerous phase and amplitude anomalies on the 16.0 Kc/s GBR signal with practically identical equipment (servo time constant 50 sec) at eight widely separated locations in November and December 1965 confirms beyond doubt that the anomalies were caused by an interfering transmitter. Fig. 1 shows two typical interference records. In all cases the anomalies were strongest at Tokyo, followed in general by those observed at Brisbane; Oslo II (a temporary aurora research site in the Arctic); Kiruna, Sweden; Tananarive, Malagasy; Beirut, Lebanon; Deal; and C. Rivadavia, Argentina. The interference always saturated the amplitude channels at Tokyo and (except in December 1965) at Brisbane, indicating an interference signal level of between 20 dB and more than 35 dB above the local GBR signal strength at Tokyo. The interferences, however, were often barely detectable at Deal and C. Rivadavia. All stations received GBR with tuned loop antennae.

The frequency of the interfering signal was on average either approximately $10^{-7}$ parts above or $10^{-7}$ parts below the highly stable GBR carrier frequency, and often changed by several parts in $10^8$ during the few minutes of interference. (Nature, 213:584, 1967)

## RADIO ANOMALIES ASSOCIATED WITH AN EPHEMERAL SATELLITE STILL IN ORBIT
**Bagby, John F.; *Nature*, 215:1050-1052, 1967.**

Attention was recently directed in Nature to some radio reception anomalies observed in 1962, 1963 and 1965 at several different locations receiving the VLF stations GBR (Rugby) on 16.0 Kc/s and NBA (Panama Canal Zone) on 18.0 Kc/s. These anomalies consisted of an interfering signal just off the nominal transmitted frequency by some $10^{-7}$ parts, sometimes above the nominal value and sometimes below it. The interfering signal was often as strong as or stronger than the GBR or NBA transmissions, depending on the location of the receiving station. Two of the possible explanations for these anomalies have been discussed---nearby unknown transmitters operating on the same frequencies, and an alternate antipodal transmission path. I have studied the Nature report and others and offer some independent evidence of high frequency radio reception anomalies and a third possible explanation.

Early in January 1961 I listened to the high frequency time signals of CHU (Ottawa, Canada) on 7.33 Mc/s at Pacific Palisades, California. Occasionally, for periods of 30 min and more the signal, usually weak at my location, was strong enough not to require an external antenna. To examine this phenomenon further, I monitored CHU on 7.33 Mc/s almost continuously each evening during February and March 1962. There were many brief periods when the signal-to-noise ratio became very large, sometimes reaching 18 dB (Table 1, line 1, except last column).

Table 1

| Duration of Observation period | Observed between (U. T.) | Average period between HF enhancements (h) | Average daily advance (h) | Average duration of HF enhancements (h) | Implied sidereal period (h) |
|---|---|---|---|---|---|
| Feb. 9, 1962 | 2200 | 4.25 | -0.40 | 1.0 | 4.72 |
| Mar. 12, 1962 | 0700 | | | | |
| Jan. 20, 1963 | 0100 | 4.38 | -0.63 | 0.9 | 4.67 |
| Mar. 3, 1963 | 0700 | | | | |
| Nov. 13, 1965 | 2200 | 3.78 | 0.00 | 1.3 | 4.00 |
| Dec. 12, 1965 | 0700 | | | | |
| Dec. 5, 1966 | 0100 | 2.53 | -1.20 | 1.0 | 3.26 |
| Feb. 8, 1967 | 0700 | | | | |
| April 24, 1967 | 0100 | 2.76 | -0.50 | 0.8 | 2.94 |
| May 5, 1967 | 1330 | | | | |

All observations were made at 118.5 W. longitude, 34.0 N. latitude. A positive daily advance signifies a later time on each succeeding day.

At first, I assumed these enhancements would correspond with the onset of evening twilight at a skip zone, but my early 1962 records clearly demonstrated that this was not the case for more than 80 per cent of the enhancements. CHU was monitored again in early 1963 between January 20 and March 3. Once more I found a definite pattern of reception enhancements, but it differed from that of early 1962 (Table 1, line 2). I tried unsuccessfully to correlate the times of enhancements with overhead passages of ECHO I and other artificial satellites.

During November and December 1965 I made reception measurements of CHU on 7.33 Mc/s and again found periodic enhancements. More than 90 per cent of them had a very definite pattern which differed from the previous patterns (Table 1, line 3). Additional data for CHU were taken during December 1966 and January 1967 by B. A. Bagby and myself at Pacific Palisades. Again there was a very definite pattern to most of the enhancements, which differed from all preceding ones (Table 1, line 4). It was obvious that the time between enhancement periods was different at each succeeding epoch, and that the daily advance was also changing. If these enhancements were due to reflexions from the passage of a particular Earth satellite through the upper atmosphere, then that satellite had a changing orbital period. I tried to correlate the various reception enhancements with the osculating orbit of the natural Earth satellite previously discussed in Nature, but because I was unable to account for all the various radio enhancement periods as being due to the natural satellite, I put the task aside temporarily.

I have re-evaluated the CHU radio data, and have found that the source of the earlier difficulty was the very elliptical orbit of the natural object. I had assumed that the apparent period deduced from the radio anomalies would always be the simple reciprocal addition of the object's sidereal period and that of the Earth's rotation, which is true only for a circular orbit. Allowing for the orbital situation at each radio epoch, I related the CHU radio enhancement records on 7.33 Mc/s in Table 1 to the osculating orbit of the natural satellite on each occasion (Table 1, last column).

I recently attempted to find a correlation between the GBR and NBA radio reception anomalies and those found for CHU, and tentatively concluded that the VLF radio anomalies could also be caused by the natural satellite. This would be confined to radio reflexions taking place near the object's perigee. Here the charged cloud associated with the passage of the body through the upper atmosphere would be large enough to reflect 16 and 18 Kc/s signals. This would temporarily provide an alternate transmission path similar to, but shorter than, that suggested by Isted as the cause of the VLF interferences. I computed the geocentric latitude and longitude of the perigee of this satellite on the various occasions of VLF anomalies and found that exceptionally favourable conditions were present during both the May-June 1962 and November-December 1965 periods. Fair conditions were present during the April 1963 period. For June 11, 1962, the perigee dip occurred near 14°S., 167°W. at 0300 U.T. For November 13, 1965, the perigee dip was near 30°N., 149°E. at 0800 U.T. These positions are approximate to the estimated locations of the postulated second transmitter in each case. The frequency shifts observed ($\cong 10^{-7}$ parts) on 16 and 18 Kc/s for the secondary VLF signals could have been due to the satellite's approaching or recessional velocity relative to both the receiving and transmitting stations. The rarely occurring maximum possible displacement, due to this Doppler effect, would be $\Delta f/f \;\; 5.4 \times 10^{-5}$. More often this value would be much lower, and would have varied slightly during the brief enhancement period, as was observed.

Allowing for differences in the orbital situation, the GBR records of November 1965 are almost replicas of the CHU records for that same period, with reference to time of day and periodicity. The 1962 and 1963 VLF data and the 1961 HF data are too meagre for such a positive comparison.

In order to compare the radio anomalies with the orbit of the natural satellite beyond the period of time covered in my earlier paper, I made more optical observations. With the aid of the indirect CHU radio evidence, this was accomplished photographically near the object's apogee on both January 18 and 20, 1967. Further optical observations are given in Table 2. Because the CHU radio experiment was also being run during January 1967 the two sets of data could be directly compared. The successful pairs of plates (out of sixty pairs taken) were each exposed shortly after a radio enhancement period had begun at Pacific Palisades.

With the recent optical observations, orbital elements have been derived to cover the time period since October 25, 1965 (Table 3). The solution was quite unique, converging very rapidly during the trial and error computational process. The slowly changing inclination is not without precedent. This effect, due to the upper atmosphere near perigee, has been discussed at length by Nigam and others. After the orbit solution was completed a radio enhancement check of the orbit was made between April 24, 1967, and May 5, 1967 (Table 1, line 5). This confirmed the orbital parameters as extrapolated from Table 3 for that epoch. (Nature, 215: 1050-1052, 1967)

Table 2

| Date | Time (U. T.) | Right ascension | Declination | Magnitude |
|---|---|---|---|---|
| Jan. 18, 1967 | 05.18.35 | 5h 00m | $+46.0^{\circ}$ | 8 |
| | 05.23.00 | 4h 30m | $+42.5^{\circ}$ | 8 |
| Jan. 20, 1967 | 05.24.45 | 4h 25m | $+41.0^{\circ}$ | 8 |
| | 05.28.45 | 3h 57m | $+38.2^{\circ}$ | 8 |
| | | Azimuth | Elevation | |
| Mar. 22, 1967 | 13.17.20 | $205^{\circ}$ | $6.5^{\circ}$ | 1 |
| | 13.18.25 | $231^{\circ}$ | $6.2^{\circ}$ | 3 |

All observations were made from 118.5 W. longitude, 34.0 N. latitude. The epoch of the celestial co-ordinates is 1920.

Table 3

$$a = 12,860 - 0.784\,(t)^{1,27}$$
$$e = 1.000 - (6,530/a)$$
$$i = 137.0 - 6.64 \times 10^{-4}\, t$$
$$\Omega = 340 + 1.096t + 1.49 \times 10^{-5}\,(t)^{2,53}$$
$$\pi = 316 - 0.159t - 3.64 \times 10^{-7}\,(t)^{2,75}$$
$$\text{Period} = 10.72\,(7.993 - 4.87 \times 10^{-4} t^{1,27})^{3/2}$$
$$\text{Epoch} = \text{October 25.0, 1965}$$

In the equations: the semi-major axis, *a*, is in kilometres; the eccentricity, *e*, would be 0.00 for a circular orbit and 1.00 for a parabolic orbit; the inclination, *i*, exceeds $90^{\circ}$ due to the retrograde sense of the orbit; the longitude of perigee, $\pi$, is the algebraic sum of the right ascension of the node, $\Omega$, and the argument of perigee; the anomalistic period is in mean solar minutes; and the symbol *t* is the number of mean solar days elapsed since the epoch.

# RADAR "ANGELS" AND ANOMALOUS RADAR TARGETS

## ANGELS IN FOCUS
**Atlas, D.; *Journal of Research NBS,* 69D:871, 1965.**

Abstract. Recent independent observations of "dot" angel echoes are drawn together to provide a coherent picture of their physical structure. The angels act as point targets with durations proportional to beam width (for vertically pointing radars). The echoes are strongly coherent, indicating a smooth specular-like surface. Dual frequency measurements in the band 1 to 3 cm indicate that their radar cross sections are proportional to between the first and second power of wavelength. A range square dependence is also indicated. When tracked, the cross sections show strong enhancement at the zenith, decreasing roughly symmetrically on either side. Doppler measurements indicate that they almost always ascend with speeds of about 1 m sec.

The observations are shown to be consistent with specular reflections from the hemispherical, concave-downward cap of a rising thermal or convective bubble. The specular point of the cap appears only briefly in a vertically pointing beam and so the echo appears to be from a point source. The concave-downward surface provides at least partial focusing, thus accounting for both the magnitude of the cross sections and their range square dependence as the bubble expands with altitude. Furthermore, only rising bubbles have such favorably disposed upper surfaces. The dual frequency data indicate that the transition zone across the bubble cap is extremely sharp, of the order of 0.5 cm. The sharp transition, the smoothness of the cap, and the slow ascent rates suggest that the flow is laminar. (Journal of Research NBS, 69D:871, 1965)

## UNUSUAL REPORT ON RADAR PERFORMANCE FROM THE T.S.S. CLAN DAVIDSON
**Anonymous; *Marine Observer,* 20:212-216, 1950.**

8th May, 1949, 0200 to 0400 hours ship's time. Speed 15 knots. Course 146$^{o}$ until 0310 hours, when course was altered to 151$^{o}$.

At approximately 0200 hours the radar was switched on. With the "Heading Marker" checked echoes were recorded of all ships in the vicinity, but additional echoes were also observed. The false echoes were clear, firm and constant. One such echo was picked up 10$^{o}$ on the starboard bow at a distance of 10 miles. This echo remained on the same bearing as the distance decreased, though the size of the echo became smaller until at a distance of 3 miles it finally disappeared altogether. Although a strict visual watch was maintained no object could be seen on this bearing. Other false echoes were also apparent at the time and all proper objects, such as ships and eventually the lighthouse, were seen in their correct relative positions. As implied from the paragraph below, the visibility was good and even a darkened native sailing craft would have easily been seen.

Later, on nearing the beam bearing of Daedalus Reef Lighthouse, an echo was again picked up 10 miles distant and 10$^{o}$ on the starboard bow. The bearing of this echo also remained constant but its size gradually diminished until at approximately

3 miles distant it too disappeared. The accompanying sketch gives an indication of the relative positions at this moment. All echoes, false and true, changed their bearing and distance on the screen as might be anticipated. Bearings and distances of Daedalus Reef Lighthouse agreed with those obtained by normal navigational methods.

Temperatures: dry bulb 75.5°, wet bulb 75.5°, dew point 75.5°, sea 77°. Corrected barometer reading, 1008.1 mb. Weather: cloudless, no wind, calm sea. Very heavy dew which formed a slight mist above the surface of the water to a height of approximately 150 ft. Visibility good, Daedalus Reef Light, which should have been sighted at a distance of 19-1/2 miles, was observed 14 miles distant.

During the previous twenty-four hours on the passage from Suez the radar had given excellent results, average-sized vessels being picked up at a distance of 15 to 18 miles on the 30-mile range, but the previous two watches had also reported observing false, but nevertheless distinct, echoes. All the false echoes observed had

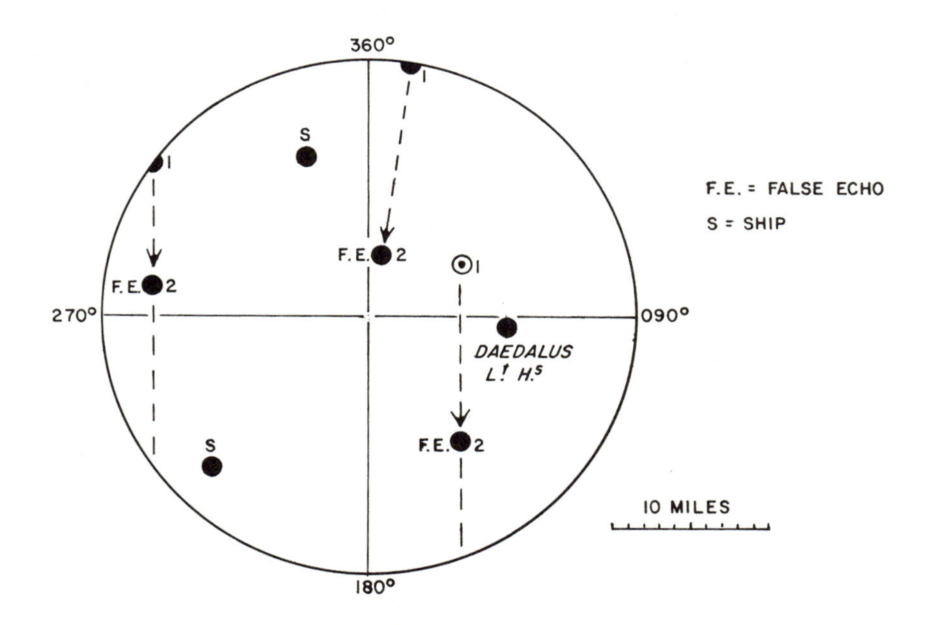


Targets seen on radar screen at 0310

these points in common, namely: (a) at a distance of 10 miles they appeared on the screen larger than the normal echoes obtained from a vessel or similar target; (b) they diminished in size with a decrease of distance; (c) no such echoes were observed at a distance of less than 3 miles (even when the PPI range scale was altered to 3 miles), and any such echoes which "approached" and reached that range then disappeared.

With regard to (a) it may be noted that the 10-mile range is used as the normal "working" range, and that this distance marks approximately the usual visual range of most surface objects from the bridge. It is of interest, too, that though false echoes had been observed during all three watches they only took place between the hours of sunset on the 7th and sunrise on the 8th May. These false echoes were observed by the master and radio officer, as well as the navigating officers. No similar occurrence has taken place to the knowledge of those on board and one cannot help but speculate whether the atmospheric conditions were responsible. The radio officer, who maintains the equipment, is unable to give an explanation.

During the remainder of the passage to Aden the radar gave satisfactory results whenever used.

The following comments upon this report were received from Dr. Hopkins, Radio Research Station, Slough. I do not think with the facts provided that there is any obvious explanation, but this hardly justifies the conclusion that the causes are meteorological. There are two points which would have to be cleared up before such an hypothesis could be safely advanced---and there may be others.

(1) What about birds? Flocks of birds might give echoes at the ranges quoted, but it is curious that the birds showed up, at any rate in two cases, at the same bearing and distance. This hypothesis might explain the "large" nature of the echo and the disappearance might perhaps be connected with the birds flying over the beam. But marine radar sets have broad vertical beams, and for this to happen the birds would have to be at a few thousand feet for disappearance at 3 miles. It is also curious that, in two cases, the echo stayed on the same relative bearing as distance decreased, i.e. the echo "steered" a collision course at some unspecified speed. Would birds come along to have a look at a ship? Talking of birds, what about aircraft? Some idea of apparent speed of echo would either rule out or lend support to this possibility.

(2) What about indirect echoes, due to reflection from largest structures on the ship? These are often troublesome on board ship and are caused, for example, by reflections from the large metal crosstrees of cargo ships. On this type of effect a spurious echo is produced whenever an actual target exists at some appropriate azimuth relative to the line joining radar scanner to the structure. The spurious echo has the same range as that of the wanted target, but its apparent bearing is roughly along the scanner-reflecting structure line. It is impossible to say with certainty from the data whether any or all of the echoes might be due to indirect effects from one or more reflecting structures. But it is perhaps significant that in two cases the echo always stayed at $10^{\circ}$ starboard. On the ship concerned is there a reflecting structure on this bearing? The disappearance at 3 miles might be connected with vertical directivity of the structure. The fact that these peculiar echoes had not been noted before might have been due to (a) fortuitous arrangement of ships, (b) rearrangement of derricks or cargo at, e.g. Suez.

Mr. LePage, of the Marine (Navigational Aids) Division, Ministry of Transport, comments upon this report as follows: This case, of which a good and detailed account is given, is of unusual interest. Before advancing any explanation for those echoes observed but unaccounted for, it should be noted that the set in use, an early Marconi Radiolocator, possesses two design features which have been taken into account in assisting explanations. These features are a pulse repetition frequency rather higher than usual (1,500 p.p.s.) and a pre-set swept-gain or anti-clutter circuit. It should also be noted that in the Red Sea in summer conditions favourable to super-refraction are common, and in this case almost certainly existed, so that the occurrence of anomalous propagation could be expected.

In attempting to account for the echoes, three possibilities should be considered. The echoes might be:

(1) true echoes reflected in the ship's superstructure;

(2) "second-trace" echoes occurring under conditions of anomalous propagation;

(3) echoes from targets unseen by the observers, though present at the ranges and bearings quoted.

Consider first the possibility of reflections from the ship's superstructure. A false echo formed in this manner is invariably at the same range as a true echo, but on a different bearing. This bearing, which will be that of a fixed obstruction such as a mast, samson post or funnel, will evidently remain nearly constant, though range will alter. In the instances quoted, however, there is no reference to any ship echoes maintaining the same ranges from the ship as the "false" echoes. Nor, apart from the Daedalus Reef, is any charted land within 50 miles which could give rise to such echoes. It is practically certain, therefore, that this cause can be discounted.

"Second-trace" echoes can occur if the time taken for the reflection to be returned to the radar receiver is greater than that between one pulse and the next. On this set the time interval between pulses corresponds to a distance of only 54 miles, because of the high pulse repetition rate. Echoes from this or greater ranges might be received in conditions of anomalous propagation, such as were believed to have existed at the time. It should also be noted that because of their distance the rate of change of bearing of echoes at these ranges, and due to their own movement, will be low. Assuming the existence of anomalous propagation, it is possible that the echo shown at $10^{o}$ to starboard in the diagram in the report, was that of a ship on a reciprocal course well over the radar horizon. When first detected, the echo appearing at 10 miles, the ship would in fact be at a range of 64 miles. Were the ship on an approximately reciprocal course, then when last seen, the echo being at 3 miles, the bearing of the ship, and consequently of the echo, would have changed by about $1^{o}$ only, i. e. sufficiently close to the reading accuracy to have been regarded as constant as was reported.

To account for the disappearance of the echo at 3 miles, some other explanation must be sought than the collapse of super-refraction conditions, since the phenomenon repeated itself. Because of the pre-set swept-gain previously referred to as being fitted to this type of set, the sensitivity will fall off toward the center of the screen, so as to keep the screen clear of sea clutter. Normally a ship echo will increase in strength very rapidly as the ship closes in, more than enough to overcome this fall in sensitivity, and so maintain its echo on the screen. But for the ship closing in 7 miles from 64 miles the magnitude of the returned signal would not be increased much in intensity, since the range altered by only 11 per cent approximately. Hence, with gain automatically controlled as described, the echo size would decrease as the echo approached the centre of the P. P. I., probably disappearing altogether at short range.

The above explanation seems to fit the observed facts in most respects. The repetition of the phenomenon can only be explained by the occurrence of more ships following approximately the same course, which would presumably be quite possible in this area.

The other two echoes, one to port and one to starboard, cannot be explained by the above theory, since they apparently followed straight tracks parallel to that of the ship. A "second-trace" echo on the beam, originating from an object whose movement relative to that of the ship is in a straight line, may be shown by simple geometry to trace out a curved path on the display, approaching the centre while forward of the beam, and receding from the centre when aft of the beam (see diagram). To make the second-trace echo describe a straight path the target would have to move in a particular geometrical curve, which seems unlikely. It would seem therefore, that if these echoes did, in fact, follow a straight track, they must have been due to targets at the measured ranges; if not on the sea surface such targets might have been airborne, e.g. aircraft or flocks of birds. Were they aircraft, one would expect them to have been observed. A knowledge of the time taken for the echoes to move from one position to another on the P. P. I. would have materially helped in the search for an explanation.

On the information available in the report it seems, therefore, that although echoes on the bow may be accounted for by assuming anomalous propagation, the only reasonable explanation for those on the beam is that they were from targets unseen by the observers though present at the ranges and bearings reported. (Marine Observer, 20:212-216, 1950)

## UNIDENTIFIED RADAR CONTACT
**North, J.; *Marine Observer*, 33:66, 1963.**

m. v. Cornwall. Captain J. North. Suez to Aden. Observers, Mr. J. Bayliss, Cb. Officer and Mr. P. King, 3rd Officer.

18th June 1962. At 1720 GMT shortly after moonrise an echo was observed on the PPI (Radiolocator Mk. IV), 12 miles off the starboard quarter of the ship, similar to that returned by a rain squall. It increased in size and closed the ship, so that by 1750 it stretched from either beam in a semi-circle round our stern at a range of about 10 miles. At 1820 the echo suddenly split astern of the ship, the strongly defined left arm sweeping away in a clockwise direction, while the nearer right hand arm continued to close our starboard quarter, coming to within 9 miles of the ship. By 1850 the two arms had joined together again but the strength of the echo as a whole had diminished, and it gradually receded astern, being completely lost on the PPI by 1915. In view of the dusty condition of the atmosphere at the time as indicated by the rising moon, the echo was assumed at first to be due to sand, but this idea was discounted when the m. v. Border Castle, known to be some 15 miles away on our starboard quarter, was contacted by W/T. This vessel reported that it had encountered no standstorms and that she had experienced a similar echo astern of her, at the same time as our contact. The possibility of the image being a second trace echo from land seems unlikely, for the sides of the Red Sea are between 64 and 80 miles away from the ship at this position. The radar set in use has a PRF on the 40 mile range of 700, so that equivalent target ranges would be approx. 0-40 miles, 130-170 miles, 250-290 miles, etc. Thus the sides of the Red Sea would fall into the silent period from 40 miles to 130 miles. Further to that, the rapid changes of shape and the generally concave form of the image would seem to preclude second-trace echoes. (N. B. When the contact came within 10 miles it was clearly visible on the 10 miles display.) One theory put forward to account for the contact was a swarm of locusts. Air temp. 81°F, wet bulb 79.3°, sea 81.9°. Cloud 3/8 Ci. Visibility over 10 miles. Wind N'W, force 3. Course 149°. Speed 15-1/2 kt.

Position of ship: 23° 30'N, 36° 46'E-23° 10'N, 36° 58'E.

Note. Captain F. J. Wylie, Director of the Radio Advisory Service, comments:

"This is an excellent report with much valuable detail which enables a number of probabilities to be eliminated, as has been done by the observer.

"One vital piece of information, not available or obtainable, is the vertical lapse rate or a rate of change of temperature or humidity a few hundred feet above sea level.

"Sharp changes in the lapse rate may give rise to echoes of an indefinite character, rather like second-trace echoes of land, but extending over several miles. These echoes have been observed when there has been no possibility of multiple-trace echoes of land nor echoes from rain or cloud.

"These echoes are usually seen at ranges of around 10 miles and may be 20 miles or more in length. They are echoes arising from what are called atmospheric discontinuities and they frequently form an arc with the ship as centre.

"Reports of echoes from swarms of locusts suggest the appearance of a large cloud. One such report of a swarm in the Persian Gulf area said it was 30 miles in diameter.

"It seems, therefore, that the echoes observed by m. v. Cornwall were almost certainly due to atmospheric discontinuities." (Marine Observer, 33:66, 1963)

## RADAR INTERFERENCE
**Clarke, H. S.; *Marine Observer*, 34:65-66, 1964.**

m.v. Glenorchy. Captain H. S. Clarke. Aden to Suez. Observer, Mr. P. A. Leece, 3rd Officer.

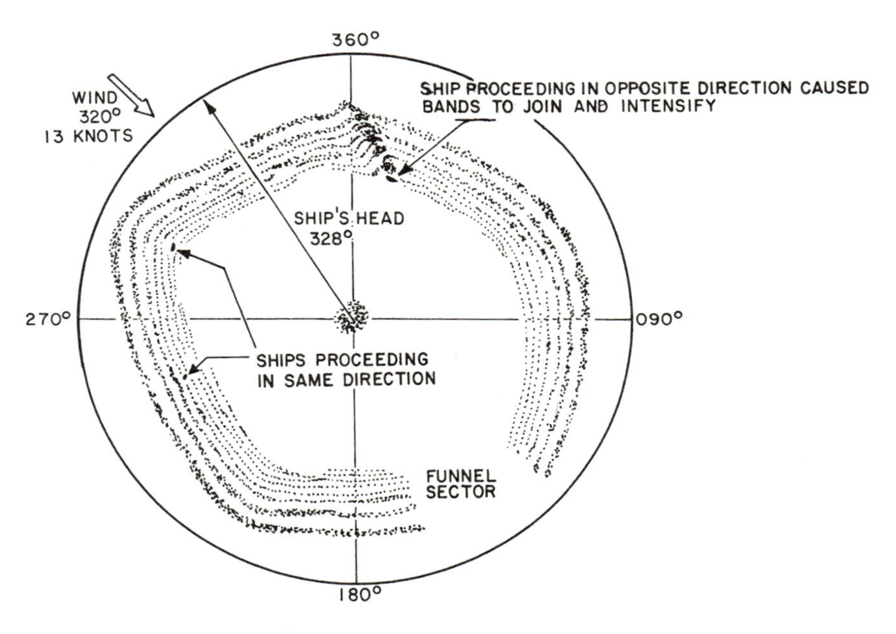


Curious pattern of radar interference in Red Sea

29th June 1963. Between 1430 and 1630 GMT while on passage through the Red Sea, the radar interference shown in the diagram was experienced. It took the form of concentric bands about the ship, generally at a distance of about 6-8 miles away. For the most part there were 8 bands but in some places 12 could be counted. The furthest away was the broadest, 3-4 cables, while the nearest was the narrowest, being about 1/4 cable wide. The spaces in between the bands also increased with increasing distance away. The exact shape of the bands was never constant but was at all times changing slowly. Sometimes a part moved towards the ship, sometimes it moved away; occasionally a part would fade, or bands would virtually join, at times becoming quite solid. This form of interference persisted for about two hours. Both before and after this time, the interference showed no definite form but quite often it resembled second trace echoes from the mountains on either side, though we doubted that this was really the case as the shape was continually changing. At one stage, the passing of a ship in the opposite direction caused the bands to join and become solid, as shown in the diagram. Ships going in the same direction as ours made no difference to the shape of the bands.

Visibility was excellent, mountain tops 3,000 ft. high being seen at 50 miles distance; there was a slight yellow haze opposite to the sun. Smoke from ships spread out into a layer, indicating the presence of a temperature inversion which we thought was probably the cuase of the phenomena. Air temp. 87°F, wet bulb 76°, sea 82°. Wind NW, force 4. Sea slight to moderate.

Position of ship from 1430-1630: 26°N, 35°E.

Note. Captain Wylie, Director of the Radio Advisory Service, comments:

"The explanation is in terms of back-scattering from the sea at ranges down to the 'skip' distance associated with one hop ducting.

Glenorchy's observation is most interesting in that two rays must have been present to produce 'bands', one reflected from a layer which works out to be about 200 ft. high (assuming X-band radar) and one travelling along the surface. The interference between these two rays produces zones of strong field at the sea alternating with zones of weak field as the distance from the ship increases. These zones will be concentric with the ship if the duct is uniform over the sea, and back-scattering from the sea is seen in the strong-field positions. This mechanism explains the increasing separation between bands as the radius is increased.

The modification of the range of these bands by other ships is interesting and is probably due to the alteration of the level of the sea in the ship's wake; a slight lowering of level would produce a much greater change in range owing to the small elevation-angle of the ducted ray. Presumably a bigger and faster ship would produce much bigger effects of this kind than smaller, slower ships. In addition, the apparent effect would be greater for ships crossing the bands than those running more nearly parallel.

The skip zone will of course move exactly with the ship if the duct spreads uniformly over the ocean, but if the height of the duct varies then the skip-radius will alter as the ship moves, and the back-scatter zone will appear to move at a different speed from the ship." (Marine Observer, 34:65-66, 1964)

## UNIDENTIFIED RADAR ECHOES
**Goodman, F.; *Marine Observer*, 24:81, 1954.**

S. S. Rialto. Captain F. Goodman. Middlesbrough to New York.

11th May, 1953. Vessel steaming in dense fog. At approximately 2100 S. M. T. Radar gave echoes on apparent objects that we were unable to identify. Echoes were similar to those given by sea clutter but were in two distinct groups. Firstly, one group to port, distant 1-1/2 miles, and about half an hour later another group, this time to starboard, distant 1/2 mile. Speed was reduced each time accordingly. Nothing was sighted. After the objects (apparent) had been astern for about half an hour, the sea temperature started to rise until reaching 65° at 2300.

Air temperature 53°F throughout. Sea temperature 53° at 2000, 47° at 2130, 45° at 2140, 55° at 2230 and 65° at 2300.

Position of ship: 41° 20'N, 51° 03'W. Course 268°, 14 kt.

Note. The above observations were forwarded to the Director, Naval Weather Service who comments as follows:

The sea temperature observations show a temporary drop of 8°F followed by a sharp rise of 20°F in approximately 19 miles on steaming further westward. This suggests a southward penetration of a tongue of cold water with a local eddy forming in the broad Gulf Stream current. It would seem possible that a current rip might be formed and the echoes (reported as similar to sea clutter) may be due to reflections from areas of very disturbed water. Reports of echoes occurring with current rips would be welcomed, especially when visual observations can also be made. (Marine Observer, 24:81, 1954)

## UNIDENTIFIED RADAR ECHO
**Marrow, J. K.; *Marine Observer,* 31:61, 1961.**

s.s. Marengo. Captain J. K. Marrow. Hull to Halifax.

25th May 1960. A radar echo was plotted at 0340 GMT and it was found to be that of a stationary object; no lights were seen, nor were flashing signals answered. The echo was picked up at 9 miles and showed fairly well: it was intermittent at 5.2 miles, the nearest approach, and when abaft the beam it was not detectable beyond 7 miles. The weather was cloudless and there was an auroral glow, but no moon. Visibility was very good, with ships' lights being seen up to a distance of 20 miles. The sea was rippled with a low swell.

Position of ship: 45$^{o}$ 42'N, 52$^{o}$ 57'W.

Note. This observation was probably of icebergs. However, the extreme uncertainty of the nature of the target is typical of the difficulties of interpreting radar records when navigating in areas where ice is likely to be a serious hazard. (Marine Observer, 31:61, 1961)

## RADAR PHENOMENON ON THE SULU SEA
**Jones, Digby; *Marine Observer,* 25:219, 1955.**

M.V. Calchas. Captain Digby Jones. Hong Kong to Manila. Observers, the Master, Mr. E. Farmer, 2nd Officer, and Mr. M. J. Jones, 3rd Officer.

30th November, 1954, 0130 G.M.T. An image was seen on the radar right ahead about 5 miles. This consisted at first of an ordinary line with indentations similar to a picture of a low coast line. The ship was in radar contact with Baguan Island, and it was realised that this could not possibly be the coastline to the E of Sandakan, which should have been 33 miles ahead of the ship. The image ran in a slight curve from approximately E-W. The ship reduced speed to about 8 kt and approached to within about 2-1/2 miles of the object on the radar. The image had by now become clearer with fewer indentations but with a mass of specks all around it. Its appearance was now very similar to a mass of iron filings with a line of iron filings running through it. It was not light enough to see anything so we waited until daylight. In the meantime the ship steered N and S, turning round every half hour. Each time the ship approached on a S'ly course the same image was seen on the screen. The image varied from about 3 to 5 miles in width.

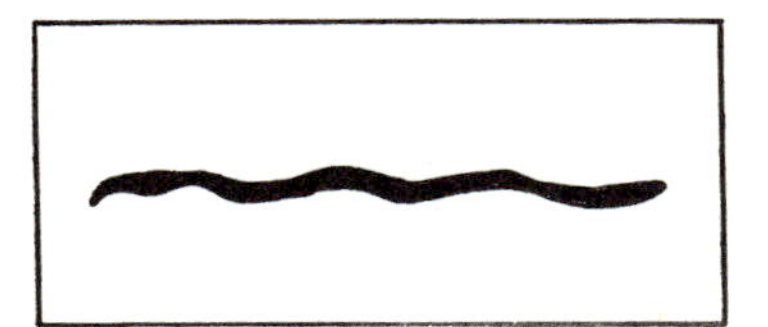

Strange radar patterns observed in Sulu Sea

When daylight came nothing was seen so we steamed towards the image on the screen. When we came within a mile it was much clearer, and we could see just ahead of the ship a strong tide rip. This consisted of a tidal line around which, in our vicinity, was a circular patch of tide rips. We compared the image on the radar with the tide rip and it was without doubt the tide rip which had caused the image on the screen. During this time the sea was smooth, with only light airs. Sky about 5/8 covered, but night dark.

Position of ship at 0130: 6° 12'N, 118° 38'E. (Marine Observer, 25:219, 1955)

## RADAR PHENOMENON
**Cowen, E.; *Marine Observer*, 41:94-95, 1971.**

s.s. Arcadia. Captain E. Cowen. Los Angeles to Vancouver. Observers, the Master, Mr. P. F. Johnson, Jnr. 2nd Officer and Mr. C. Mendoza, Jnr. 4th Officer.

20th July 1970. At 1705 GMT a long line was seen on the radar running 005–185°, extending for over 40 miles in length at a distance of 10 miles from the ship on the port side. At 1714 it was 8 miles away. Visibility at the time was very good, no special cloud formation or change in the sea was observed and there was no precipi-

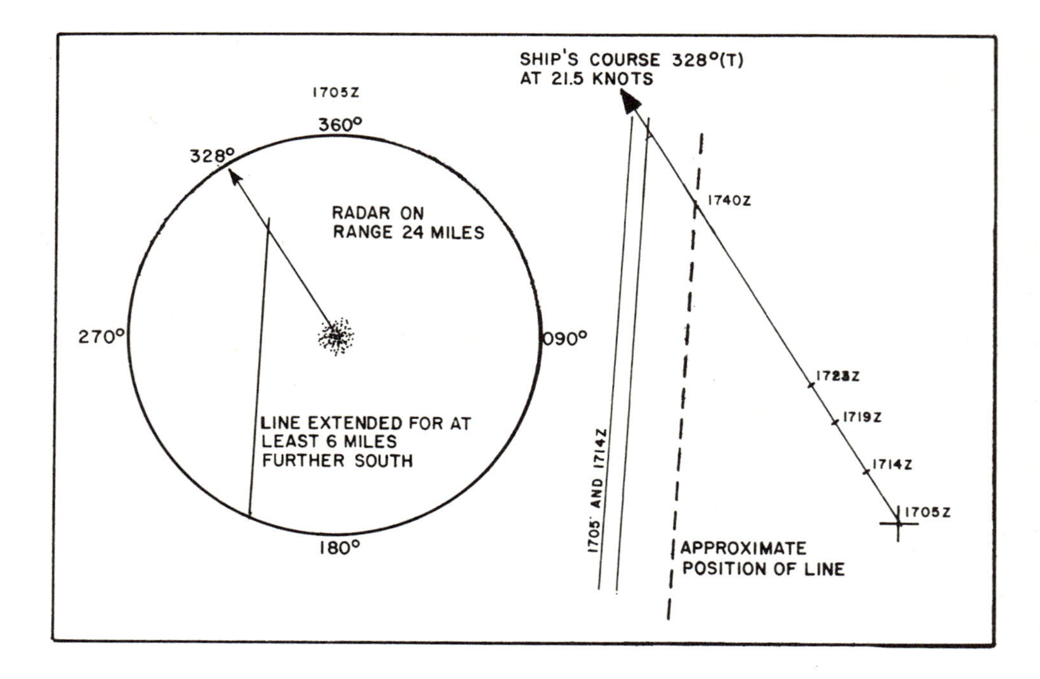


Diagram of radar phenomenon

tation. At 1719 the line was 6.4 miles from the ship and at 1723 it was 5.5 miles. On getting a little closer the width of the line was approx. 1 cable, fixed by radar range. At 1740 the vessel entered the line; there was no change in wind direction or speed, in sea or air temperature, or in pressure. The only visible change was a distinct break in the cloud which had been a continuous layer of St at 300 ft with ragged shreds. Through the break could be seen Cc cloud. The radar continued to show this line until 1810 when it slowly faded out.

The several points of interest are the two breaks in the line, nothing unusual being actually seen except for the break in the cloud, no change in wind or temperatures on passing through the line and, of course, its very great length.

Afterwards a true plot was done of the line's movements and it can be seen that the speed was at some stages nil and at others very slow. The conclusion is that it was not atmospheric but how can one explain the break in the clouds? Air temp. 15.3°C, wet bulb 14.1°, sea 14.4°. Wind NNW, force 5. Pressure 1013.4 mb. Course 328° at 21-1/2 kt.

Position of ship: 36° 24'N, 123° 04'W.

Note 1. When forwarding the above report, the Administrator of the Canadian Meteorological Service enclosed the following comments from his Research Division:

"Of at least three possible explanations for the line echo observed the most likely would seem to be a sharp discontinuity in the atmosphere in the horizontal sufficient to cause a so-called clear-air return (or angel echo). The configuration is somewhat unusual and this seems to be confirmed by the Officers' notes and their interest in reporting the event. The phenomenon was in all probability relatively short-lived and there is insufficient observational data on this scale to determine the exact cause.

"The observation of a 'distinct break in cloud' at the radar echo line supports the idea that there was some form of discontinuity present. Although a thorough study has not been made of the synoptic situation, the current weather maps indicate that at least the remnants of a cold frontal system were in the vicinity of the ship at that time on 20th July 1970. The orientation of the front is not unlike that of the line echo described in the report. It would therefore seem plausible that a sharp discontinuity could be present as either a direct or indirect result of the cold front. Although there was no visible sign of weather or heavy cloud at the surface sufficient to produce an echo, there could be strong gradients of temperature and humidity in the horizontal at elevations above the surface (an explanation supported by the cloud observation) which could act as an electromagnetic wave reflector.

"Radar echoes from relatively clear air have been recognized for many years and theoretical explanations for their existence have been made. Detailed observations in the areas of clear air echoes (on a scale of a few cm to a few metres) are difficult to obtain and rather scarce and not too much is known about the possible frequency and distribution of such atmospheric discontinuities. Thus, while the event described may be rather rare, there would appear to be a reasonable explanation. From the standpoint of the ship's crew, I suppose the important aspect is to recognize the echo and differentiate it from actual weather. Keen observation and experience are the best teachers. Text books such as Captain F. J. Wylie's The Use of Radar at Sea (Hollis & Carter, London, 1952) will provide help to the neophyte." (Marine Observer, 41:94-95, 1971)

## FALSE IMAGES ON RADAR
**Mullock, E. A.; *Marine Observer*, 23:207, 1953.**

S.S. Rincon Hills. Captain E. A. Mullock. Puerto La Cruz to Portland (Me). Observer, the Master.

1st November, 1952, 2345 G. M. T. When in position 41° 42'N, 69° 27'W, ships were observed at a distance of 50 miles (maximum radar range). Nantucket Island, Cape Cod and other land in the vicinity showed a clear outline having their correct bearing and distance from the vessel. This condition was observed for about 14 hours. Odd reflections appeared, having the character of normal vessels on the 15 mile range. About four or five images of vessels would appear at a distance of 8 to 10 miles in a straight line bearing NW, each image having smaller images on each side. These images would approach the PPI centre until they were lost in the sea return and then similar images would appear at a distance of 10 miles.

These ghost images continued for about 8 hours, a distance of 125 miles from Nantucket Shoals Buoy No. 10, on a course of 343° towards Portland. Radar interference was also present in the vicinity of these images, extending from the outer edge of the screen to the centre, having an arc of 10° to 15° width. The interference appeared as small white dots arranged in even rows all the same size.

Wind WSW force 2, shifting to NW force 2. Barometer 30. 10 in, air temp. 53°F, sea 52°, wet bulb 51-1/2°. Clear sky, low smoke haze with poor visibility. Moderate sea and swell.

The following day, 2nd November, when departing from Portland, similar images were observed, only this time they were bearing in a SW'ly direction. (Marine Observer, 23:207, 1953)

## PHOSPHORESCENCE AND RADAR
**Hilder, Brett; *Marine Observer*, 25:119-120, 1955.**

M. V. Malaita. Captain Brett Hilder. Rabaul to Sydney. Observers, the Captain and Mr. T. A. Colquhoun, 2nd Officer.

5th September, 1954, 0100 S. M. T. Off the eastern point of New Guinea approaching China Strait, where the waters are dangerous with coral islets and submerged reefs, with the radar on continuously, the Second Officer reported a loom of light to the SW, extending over about 15°. There were reefs in that direction, distant about 5 miles, and the loom remained apparently in coincidence with the reefs as they came abeam and dropped astern. This loom appeared to be more brilliant than I have observed before.

At 0200 we were on a course of 210°T, heading for the light on Bright Island, when a loom appeared in that direction and extended rapidly from ahead to some points on the port bow. This was likewise in coincidence with some submerged reefs, but as the lighted area came over the horizon towards us it was closer to us than the reefs should have been. We were fixing the ship's position by cross bearings as well as by radar, and I noticed that the radar screen showed what appeared to be a rain squall in the area of phosphorescence. To the eye the light did have rather the appearance of a white squall of fierce wind and rain, except for the illumination. As we got closer the light showed up in horizontal streaks, like sand-banks or submerged reefs, and as we were going to meet the area before we could alter course it was quite alarming. The masses of phosphorescence and the echo on the radar both covered an area of about 2 miles square. The set was Marconi Radiolocator Mark IV, 3-mile range, sweep speed 21 r. p. m.

At 0220 the ship met the area, and the nearest patch passed under the ship. Our breaking bow wave showed up dark against the light, which seemed to be at least 2 fathoms down. The long patches appeared to be round in section and had rounded ends, and while most of them were parallel to the horizon, some curved sharply away, as shown in the sketch made by the master. The islands in sight are Deedes, Good and Bright Islands with Engineer Group in the background.

Light rain was falling from an overcast sky, with a lot of fractonimbus about. The ship reached the radar echo at the same time as the phosphorescence, but there was no noticeable increase in precipitation or wind force. The illumination was very brilliant but did not vary or pulsate, nor did it have any proper motion except relagive motion caused by the ship's speed of 10 kt. The wind was light, about SE, force 2, with slight sea and swell. Air temp. 78$^{o}$F, wet bulb 76$^{o}$, sea temp. 80$^{o}$.

In the circumstances I came to the conclusion that the phosphorescence was actually shown on the radar screen. This is scarcely credible, but if radar can be shown to stimulate this effect in the sea perhaps it can be stimulated in return, possibly through the agency of a low cloud in both directions. (Marine Observer, 25:119-120, 1955)

## RADAR 'RING ANGELS' AND THE ROOSTING MOVEMENTS OF STARLINGS

**Eastwood, E., et al; *Nature*, 186:112-114, 1960.**

Radar observations upon the ring angels reported by Eastwood, Bell and Phelp have been continued throughout 1959. Analysis of the film records obtained has yielded a total of approximately three hundred such incidents and the geographical distribution of the active centres is shown on Fig. 1. The dots upon this map mark the centres of ring angels which have been observed on one occasion only. The small circles show the positions of ring sources which have been observed to be active on several occasions: the number of observed events is shown within the circle.

As the number of observations accumulated, it became possible to pin-point the centres of the persistent ring angel sources more accurately; further investigation established that starling roosts were situated at these centres. A ring source of special interest is that located within the London region; it is roughly centred upon Trafalgar Square and must arise from dispersal of the starlings which roost upon the buildings grouped about the square. This observation was rendered possible by the Doppler device, which discriminates between a moving target and the fixed ground echoes. [Figures not reproduced]

The fact that starlings leave a roost about sunrise, in discrete flocks separated at intervals of a few minutes, has long been known to ornithologists; but the manner of the subsequent dispersal of the birds could not be precisely determined by ordinary bird-watching methods. By the use of radar Ligda has shown that masses of red-winged blackbirds, flying out from a common roosting ground at Texarkana, show up on the plan position indicator as an expanding single diffuse ring. Harper has reported radar echoes in the form of expanding arcs of circles and has attributed them to the movement of starlings from a roost located six miles from his radar station at East Hill, Hertfordshire. Echoes in the form of curved lines moving seawards have been described by Lack and associated by him with starling flocks migrating direct from roosts.

In the present series of radar experiments observers, equipped with radio-telephones, were located at a number of ring angel sources selected from Fig. 1. These observers signalled over the radio link to the radar station at the precise moments of emergence from the associated roost of the successive flocks of starlings. These signals were impressed upon the film which was recording the radar plan position indicator picture. Subsequent projection of the film establishes perfect correlation between the emergence of the successive rings from a point source on the plan position indicator and the eruption of the birds from the roost. In some incidents the echoes are of almost uniform amplitude around the ring, indicating uniform distribution of the birds; in other cases incomplete circles show preferred directions of flight, but this pattern is not constant for a given roost. Fig. 2 shows a spectacular display of ring angels.

This radar experiment clearly establishes that flocks of starlings emerging from a roost may afterwards take the form of a series of expanding, concentric rings of birds flying radially outwards from the roost. It follows that the ring angel source should remain fixed and independent of the wind; this we have confirmed. It also follows that the expanding ring of birds will suffer displacement by the wind relative to the point of emergence. This is shown on Fig. 2 where the Canterbury ring (33 miles to south-east) shows a displacement to the south-west by an amount corresponding to the wind speed.

The manner of assembly of the birds into the roost is scarcely less remarkable than the ring dispersal although markedly different in form. Fig. 3 shows the radar record of the evening assembly in progress at Matching Green, Essex; the birds are extended over a funnel of five miles and are approaching from the south-west. As noted by Witherby the progress of birds towards the roost normally terminates at a pre-assembly point near to the roost and this process may take as long as an hour. Our observations upon the Essex roosts show that the final entry into the roost normally takes place about sunset and is completed within a few minutes. The entry may take the form of a sequence of departures from the pre-assembly region by discrete flocks of birds at intervals of about half a minute; these flocks fly straight into the roost. This disciplined behaviour of final entry is obviously related to the ring dispersal phenomenon. Even more spectacular is the continuous entry from a vast flock of birds circling the roost.

The pre-assembly process can only be studied by radar and shows wide variation in form, with some dependence upon the direction and speed of the wind. The assembly pattern shown on Fig. 3 does not happen every evening and sometimes the assembly process is quite formless with a general dribbling in of flocks of birds from all directions---which is what would be expected from the nature of the morning dispersal. The Fig. 3 type of approach, on the other hand, demands that the birds must perform a circular movement during the day in order to bring them into the approach funnel. This operation, and the circumstances that provoke it, remain obscure.

The essential feature of the ring dispersal phenomenon is the pulsed departure of the birds in successive groups, the rings being the consequence of the constant headings flown by the birds. This periodic eruption of the birds from the roost possesses some of the features of a relaxation oscillation and suggests an excited, emotional state of the birds. We have recorded the noise-level of the starling song from the roost, using a 4-ft. paraboloidal mirror fitted with a microphone at its focus. The total noise emitted by the roost decreases with the departure of the successive flocks; but the most striking feature of the noise record is the sudden quenching which occurs at the moment of departure of each of the waves; an abrupt diminution of noise-level by as much as 10 decibels may then occur which is rapidly restored after the departure of the group.

The roost organization and flight behaviour appear all the more remarkable when it is remembered that such communities may contain hundreds of thousands of birds. We have estimated the minimum population of the Matching Green roost in Essex by assessing the density and area of field covered by the birds when they are following the field mode of pre-assembly. This population figure has agreed in order of magnitude with that derived from radar measurement of the echo amplitude of the rings. In this calculation we have taken the radar cross-section of a starling as $2.5 \times 10^{-3}$ sq. m., as suggested by Edwards and Houghton.

Analysis of the time variation of ring angel incidents observed from the Essex radar site during 1959 shows a maximum during July-August; no ring dispersals were observed during the four weeks mid-April-mid-May; but some activity was recorded for all other weeks of the year. These results may be related to the breeding habits of the starling but it must be remembered that the frequency of occurrence of ring dispersals, while related to the number of active roosts, is also dependent upon the roost organization, flight behaviour and feeding habits of the birds, and these factors may also be subject to seasonal variation. Having regard to the number of permanent roosts, and also the presence of temporary roosts of immigrants during the autumn and winter, it is in fact rather strange that the ring dispersals do not occur much more frequently. (Nature, 186:112-114, 1960)

# Chapter 3
# UNUSUAL WEATHER PHENOMENA

## INTRODUCTION

"Weather" is a composite experience of sight, sound, temperature, and sensation of the elements. The phenomena of this chapter are not primarily luminous or acoustic or confined to the sensory channels that form the bases for other chapters. Here, we deal with rain, fog, wind, clouds, precipitation, and sunshine---but only when they are anomalous.

Our weather experiences are so variable that a phenomenon has to be very strange indeed to gain access to this chapter. The criteria are not so much related to intensity or rarity as they are to unexpectedness and deviations from predictions of meteorological theory. Thus, very heavy general rains, even with catastrophic flooding, are excluded, while rain without clouds or "point" rainfall are welcome additions to the chapter.

Several themes that thread their way through most of this book recur once again in this chapter:

1. The importance of electricity in geophysical phenomena
2. The possibility that some electrical influences may originate in outer space
3. The influence of the sun, moon, and inbound meteoric material on terrestrial weather
4. The occasional prankish, frivolous behavior of some meteorological phenomena. This judgment is, of course, highly subjective but it may nevertheless reveal something of importance about man's interface with weather.

# EXTRAORDINARY PRECIPITATION

What makes precipitation anomalous? The major variables involved are color, shape, size, and intensity. But, as the first group of quotations indicate, any form of precipitation without clouds must also be considered anomalous. True, very fine ice crystals can be precipitated out of clear air when the cold is intense, but where does cloudless rain originate? Throughout this section on precipitation, the very remote possibility that some ice and water may be meteoric should be considered. In other words, some precipitation may originate much higher than normal cloud levels, perhaps even from outside the atmosphere.

Hail is notable if it is unusually large or contains inclusions that "shouldn't be there." (Note that isolated ice falls or "hydrometeors" are dealt with in the chapter 7 on falling material.) The great variety of hailstone shapes: crystals, disks, plates, spiked spheres, etc.; indicate the work of unrecognized forces in hail formation. Some of these forces may be electrical in nature. On a larger scale, the peculiar mesh patterns of some hail falls call for explanation in terms of hailstorm dynamics. Finally, the phenomenon of slowly falling hail seems well verified as well as particularly mystifying.

Colored rains and snows are rather common. The European "rains of blood" described since history began to be written down are generally blamed on dust blown north from Africa. Black rain is even more common and is ascribed to forest fires or, more lately, urban pollution. The forest fires involved are seldom identified, the same situation prevailing with the explanation of dark days. The conflagrations should be traced via a trail of black rain, but they rarely are.

Uncolored rain can, on occasion, be no less intriguing. Fantastic examples of "point rainfall" or cloudbursts are often noted in the meteorological magazines. The naming of the phenomenon does not explain the unbelievable quantities of water sometimes dumped in very small areas. After reading some of the examples from the literature, the reader may want to classify point rainfall along with falling material, for the traditional meteorological mechanisms seem wanting.

## RAIN AND SNOW FROM CLOUDLESS SKIES

### RAIN FROM A BLUE SKY
**Applegate, T. H.; *Meteorological Magazine,* 60:193-194, 1925.**

I was surprised on July 18th at 15h. 16m. G. M. T. to find a shower falling from what to all appearances was a perfectly blue sky. I was at Boveysand (2-1/2 miles from Cattewater) when large drops of rain caused me to look up and observe that the sky was clear with the exception of cumulo-nimbus on the horizon and a small amount of alto-cumulus some distance ahead and receding towards the east. The latter was at a height of, approximately, 12,000 ft. Only about twenty yards of the ground, which was cemented concrete, showed the track of the shower. It was, then, a slight very short and very local fall of rain. (Meteorological Magazine, 60:194, 1925)

## RAIN FROM A CLOUDLESS SKY ("SEREIN") AT STEVENAGE, JULY 22nd, 1888

**Jones, Samuel; *Royal Meteorological Society, Quarterly Journal,* 15:123, 1889.**

I beg to submit to your notice certain phenomena observed by me at Stevenage, on Sunday, July 22nd, 1888, about an hour and a half before the approach of a violent thunderstorm. I was standing in the court-yard of my dwelling and talking with a friend who was a farmer; the weather being then a rather ticklish question with him. I observed what appeared to be the commencement of a shower, and I asked him if it did not rain, to which he said 'No.' We continued the conversation for about the space of five minutes, when he suddenly broke it off by ejaculating 'It does rain.' I said 'How can that be when there are no clouds?' I knew the appearance was of an unusual character, and my question and answer were directed with the view of leaving him an unconscious witness of the cause of it, so that he might give his testimony unbiassed by anything I might have said.

The appearance was first seen about 7 p.m., and it looked like a shower of luminous particles of a somewhat skim-milky appearance. As soon as I realised the fact of its electrical nature I endeavoured to ascertain if the movement of the particles was upwards or downwards, but without effect. But what I did more successfully observe was that the particles were of the size of tare-seed, and that they moved in remarkably straight lines, so much so that they resembled the warp in a weaver's frame. My view extended over about two-fifths of the circle of the heavens, and the appearance within that area was precisely the same. My observation lasted a quarter of an hour. The sky was cloudless but hazy looking to the North-west. The subsequent storm came in that direction, being very severe at Sandy. (Royal Meteorological Society, Quarterly Journal, 15:123, 1889)

## SNOW FROM A CLOUDLESS SKY

**Maclear, J. P.; *Symons's Monthly Meteorological Magazine,* 30:27, 1895.**

On the 6th February, at 9 a.m., light snow began to fall, sparkling in sunshine from a cloudless sky. Thinking that the snow might be blown off a roof, I went out on the common, clear of houses, and made no doubt that the fall was from the sky. Gradually it clouded over, and at 10 a.m. was quite overcast, the snow continuing as a natural shower for a short time longer. (Symons's Monthly Meteorological Magazine, 30:27, 1895)

## SNOW WITHOUT CLOUDS

**Anonymous; *Monthly Weather Review,* 45:13, 1917.**

A phenomenon much more frequent and much better known than rain from a clear sky is the formation and slow fall of snow in the lowest layers of the air under a cloudless sky. This phenomenon occurs only during severe cold and in calm air.

The snow crystals and ice particles, sparkling in the sunlight, are particularly small and sparsely distributed, so that they do not darken the air. There are large numbers of ice needles among the forms, and therefore the phenomenon has been called simply ice needles (Eisnadeln) and even has been given an independent symbol; but there also occur beautifully formed stellar and tabular snow crystals, together with structureless ice granules.

The phenomenon is most frequent in the polar regions, where it early attracted the attention of explorers and more particularly because it is often associated with halo phenomena. It has received the name "diamond dust" (Diamantstaub), a name known to the whaling master Martens (1671) mentioned on page 388 and footnote 20. In Germany many a winter passes without developing the phenomenon, but Bodman observed it 28 times within 18 months on the Swedish Antarctic Expedition, and Heim saw it even 26 times in 9 months while on the second German Antarctic Expedition. It appears from the photomicrographs of the "diamond dust," made on the latter expedition and also by Dobrowolski, of the Belgian Antarctic Expedition, that besides needles and tablets prisms are abundantly present, but the latter only at very low temperatures. (Monthly Weather Review, 45:13, 1917)

## DRIZZLE FALLING FROM A CLEAR SKY
**Dines, J. S.; *Meteorological Magazine,* 70:16, 1935.**

A remarkable case of drizzle falling without cloud occurred at Benson, Oxfordshire, on Sunday morning, January 20th, 1935. At 9.20 a.m. drizzle was observed, and as the sun was shining brightly the state of the sky was carefully studied. No cloud was visible although the blue of the sky was somewhat pale and watery, an effect which may well have been produced by the drizzle itself. The drizzle continued for ten minutes and during this period some cloud formed though it was not overhead and did not appear to be the source of the drizzle. Some idea of the intensity of the precipitation can be formed by the fact that after its cessation a paving stone in the ground appeared moist and the sloping glass roof of a verandah was thickly covered with water particles. These, however, did not coalesce and run down the glass. The sun shining on the drizzle formed a rainbow though not of any particular brilliance. The cloud which began to form was probably of fracto-stratus type, though it was extremely difficult to form any idea as to its height above the ground. It continued to develop rapidly and three-quarters of an hour later the whole sky was heavily overcast and further drizzle occurred. As the first drizzle was unconnected with cloud it seems by no means improbable that this later drizzle may also have formed in clear air in the region between the cloud and the ground and the question naturally arises whether precipitation (rain or drizzle) may not be formed in clear air without passing through the initial form of cloud particles more frequently than is commonly supposed. It may be added that the conditions prevailing at Benson on the morning of January 20th were markedly anticyclonic. A ground level there was a calm though the clouds showed a light north-easterly wind above, the air temperature was 40°F. at the time and pressure approximately 1,040 millibars. (Meteorological Magazine, 70:16, 1935)

## DRIZZLE FALLING FROM A CLEAR SKY
**Ashmore, S. E.; *Meteorological Magazine,* 70:116, 1935.**

I read with interest the article by J. S. Dines on this occurrence in the February Meteorological Magazine. I have one or two records of the same kind of thing occurring at Grayshott. On December 29th, 1929, following on a very stormy day, with gale, hail sleet, heavy rain and lightning, the evening was fine, and the stars brilliant. At 21h. 40m. with the stars as bright as ever and not a trace of cloud visible, palpable drizzle fell for a few minutes. Again on January 3rd, 1933, after a similar kind of day, the evening provided a beautiful display of middle-level cloud, which disappeared before 18h. At 19h. 19m. light drizzle as in the previous case, occurred, and it was some time before cloud reappeared.

The best case was one of which I have unfortunately lost the date, but I am fairly certain that it was during the first ten days of January, 1931. The evening was practically cloudless since sunset, and the temperature near freezing point. At 19h. 35m. copious drizzle fell, enough to wet the face and clothing. A gentleman on coming out of church put up his umbrella, and persisted in keeping it up even when it was pointed out to him that the stars were shining as on the most brilliant of winter evenings.

I think these three observations tend to support Dines's assumption that it is not a very uncommon occurrence for rain to form in clear air without passing through the intermediate stage of cloud particles. (Meteorological Magazine, 70:116, 1935)

# GIANT SNOWFLAKES

## GIANT SNOWFLAKES
**Anonymous; *Monthly Weather Review,* 43:73, 1915.**

About 4:30 p.m. on January 10, 1915, Berlin experienced a brief fall of snow during which snowflakes of very considerable dimensions occurred simultaneously with those of more usual size. On this occasion a large number of snowflakes had diameters of 8 to 10 centimeters, and these giant flakes fell with both a greater speed and more definite paths than did the smaller flakes. They did not have the complicated, fluttering flight of the latter. In form the great flakes resembled a round or oval dish with its edges bent upward. During flight they rocked to this side or that, but none were observed to turn quite over so that the concave side became directed downward. The temperature was but little above freezing.

Dr. Baschin refers in his report to analogous occurrences observed in 1887. On January 7 of that year, at Chepston, England, very large flakes were observed by E. J. Lowe from 12:12 p.m. to 12:20 p.m. At that time the temperature was 0.3°C. and the relative humidity 100 per cent. At first the flakes were 6-1/2 centimeters, then 7 centimeters, and finally 9 centimeters in length; and even larger ones were observed to fall outside the cooled dish in which the observer was attempting to catch them. The weight of 10 flakes varied between 1.1 grams and 1.4 grams; in water content one flake yielded 14 drops, a second yielded 15 drops, and another 16 drops. A falling flake 9 centimeters long, 6.5 centimeters wide, and about 4 centimeters thick, was compressed to 0.6 centimeter thickness by its own impact upon the little glass dish that caught it. The snowflakes under consideration were not composed of fragments, but of hundreds of undamaged crystals set together at all angles in a manner which Lowe illustrated. He believed that he could observe the greater flakes exerting an attraction on the small ones; but in part he explained the increasing size of the big flakes by their rapid fall whereby they could overtake a large number of the small flakes. Mr. Lowe also reported that he had seen such large flakes but once before, viz, in January, 1838, when the flakes had attained a length of 5 centimeters (2-1/2 inches).

In the same winter of 1887 very large snowflakes were reported as having fallen on January 28, near Matt. Coleman's ranch at Fort Keogh, Mont. In this case, which is not recorded in the Monthly Weather Review for that year, the great flakes were described as being "larger than milk pans" and measuring 38 centimeters (15 inches) across by 20 centimeters (8 inches) thick. A mail carrier who was caught in the storm verified the occurrence. These tremendous flakes made patches all over the fields within an area of several square miles.

Snowflakes "larger than milk pans" fall in Montana

On January 24, 1891, during one of the heaviest snowstorms known up to that time in Nashville, Tenn., very large snowflakes were observed there. "Many flakes were as large as a dollar (3.8 centimeters) and some nearly as large as a saucer (14 centimeters)."

On March 25, 1900, Richmond, Va., also enjoyed the spectacle of falling snowflakes of very large size. On the morning of that date Richmond had cloudy weather with a fresh, chilling wind from the northeast. The temperature rose slowly during the forenoon, and at 1:17 p.m. a light rain began falling. Soon sleet accompanied the rain, and later the rain ceased so that sleet alone fell. Some of the icy particles were nearly cubiform, measuring about one-fourth inch (0.64 centimeter) either way and mixed with them was the usual sleet---small spheres of frozen rain. At 5:25 p.m. moist snow fell with sleet. There was nothing unusual about the first falls of flakes, but the sleet immediately diminished in volume and as this occurred the flakes increased in size until they attained unusually large dimensions. They were of irregular shape, usually oblong: several were observed whose greatest diameters could hardly be covered by a teacup (perhaps about 7.6 centimeters). Some of these flakes were caught upon a piece of dry wood and examined: in every instance they showed the center to consist of a soft mass of snow about one-half inch (1.3 centimeters) in diameter, while the outer edges were thin, as though they were separate flakes that had attached themselves to the central mass while it was falling. The greater weight of the center caused the larger flakes to assume the form of an inverted cone as they fell, the outer, thinner edges being bent upward by the resistance of the air. Three of the large flakes were caught in a bowl and when melted yielded nearly a tablespoonful (14-1/2 cubic centimeters?) of water. The flakes were widely separated from one another and did not obscure the vision when looking upward toward the sky. (Monthly Weather Review, 43:73, 1915)

### OUTSIZE SNOWFLAKES
**Hawke, E. L.; *Weather,* 6:254, 1951.**

Snowflakes of exceptional size, variously reported by postmen and other early risers to have been 'enormous', 'as big as saucers', and 'about five inches across', fell in Berkhamsted town and neighbouring districts of the Chiltern Hills 400-600 ft above sealevel between 0430 and 0445 GMT on 13 April 1951. A well-marked cold front associated with a deep depression near the Shetlands had passed south-eastward across the area earlier in the night. In the Chilterns steady rain appears to have given place to sleet around 0400 GMT, the snow following just before 0430 when the surface air temperature stood at 33$^{o}$F.

It seems to be uncommon for agglomerations of snow crystals to attain average diameters exceeding 3 in., though on rare occasions much larger ones have been observed. In his book The Elements Rage, F. W. Lane cites from the Monthly Weather Review of February 1915 a record of the fall of aggregated masses of snow measuring 15 in. in width and 8 in. thick. Presumably this happened in the U. S. A. where things tend to occur on the grand scale. What, one wonders, would be the liquid equivalent of such a downfall per minute?

In British Rainfall 1888 it is stated that during a snowstorm at Chepstow on 24 March of that year there were flakes 3-3/4 in. in diameter and 1/4 in. thick, 'falling like plates'. The storm lasted for only two minutes, but covered the ground to 2 in. deep. Its yield in terms of water was 0.33 in., giving a rate of 9.90 in. per hour, which is seldom matched in Britain's heaviest thunderstorm rains. On this occasion the snow must have been of unusual density (Weather, 6:254, 1951)

## CONICAL SNOWFLAKES

### CONICAL SNOWFLAKES
**Moore, A. D.; *Science,* 73:642, 1931.**

Within the writer's observation, at least, a very unusual snowfall occurred in Ann Arbor on Sunday afternoon, April 26, 1931; during the afternoon three main falls occurred, each lasting for only a short while, and each yielding but little actual precipitation.

The crystal formation was that of a solid cone with a round base. The side of the cone made an angle of about 30 degrees with the axis. The round base was part of a spherical surface. The shape was exactly like that of a conical section of a sphere.

As nearly as the eye could see into the formation of a snowdrop, the whole structure seemed to be made up of conical needles, packed together, with the upper ends forming the pointed tip of the cone, and the larger lower ends forming the rounded base.

The density was very high; a handful of snow, slightly compressed, immediately formed soft ice. The ratio of snow volume to volume when compressed to ice without air included would hardly have been higher than four to one.

Hundreds if not thousands of individual snowdrops came under observation on a window sill; irrespective of size, from tiniest to largest, the crystals were of the same form. In spite of landing on the stone sill at high speed in the high wind, nearly all crystals were tough enough to retain their shape; such few observed crystals as were shapeless were almost certainly shapeless only due to damage from impact.

The shape of the snowdrops was the same in all three falls, although the three falls occurred from a half hour to an hour apart. Many large crystals fell; one of the largest, by actual measurement, was three eights of an inch across the base.

The day was very gusty, with rapid changes in wind velocity. Wind velocity at times during the snowfall became very high. There were rapid alternations between brilliant sunshine and considerable cloudiness. The temperature was mild. Most of the snow would melt in a few minutes, although an inch or two that became piled on the running board of a car was there the next morning, partly melted, but with somewhat globular outlines still visible. (Science, 73:642, 1931)

# STRANGELY SHAPED HAILSTONES

## REMARKABLE FORMS OF HAILSTONES RECENTLY OBSERVED IN GEORGIA

**Abich, Staatsrath; *Smithsonian Institution Annual Report,* 1869, 420-421, 1871.**

I take this opportunity of giving you a preliminary notice of two hailstorms, of both of which I was fortunate enough to be a witness. The phenomena were of so unusual a character that they are well worthy of a full and precise account.

They took place within fourteen days of each other; the first on the 27th May last, at 3 p. m., the second on the 9th June, at 6 p. m. The localities were not far asunder, being both in the neighborhood of Tiflis, near Beloi Kliutsch. The morphological characters of the hailstones, which were very large, as much as sixty or seventy millimetres (2-1/2 inches) in diameter, were as remarkable as they were dissimilar. On the first occasion they were oblate spheroids, resembling Mandarin oranges, while their structure seemed almost organic. On the second there was a fall of actual ice crystals, an occurrence which has never before been noticed, at least as far as I could discover from the literature within my reach.

The stones were not mere lumps, exhibiting indistinct crystalline forms, but spheroidal bodies of definite crystalline structure, overgrown along the plane of the major axis by a series of clear crystals exhibiting various combinations belonging to the hexagonal system. The commonest forms were those which occur in calcite and specular iron. Of the former type, by far the most abundant were combinations of the scalenohedron, with rhombohedral faces; crystals of fifteen to twenty millimetres (3/4 inch) in height, and corresponding thickness, prettily grouped with combinations of the prism and obtuse rhombohedra. The terminal plane was also occasionally noticeable. Some which fell at the beginning of the storm were flat, tabular, crystalline masses, thirty to forty millimetres (1-1/2 inch) in diameter, resembling the so-called "cisen-rose," which occurs at St. Gotthardt.

The stones, when picked up quite fresh, showed sharp edges, with faces which were for the most part slightly curved like those of diamond; however, those which I took to belong to the scalenohedron were perfectly plane.

Crystalline hailstones observed at Tiflis

I was in the open air when each of the storms began, and was able to gain shelter before I received any injury. This was fortunate, for the damage done, even to large trees, was very serious.

I reached home in a quarter of an hour, and found a pail full of the largest stones, which had been collected as soon as the first fright had passed over. My house was not much damaged. I sat down at once and drew ten of these remarkable forms, which had scarcely undergone any alteration.

I have often thought over our conversations about hail, and I see that if I now applied all the theories which have ever been broached to the facts which have come under my own notice, not a single one of them will give me any light toward their explanation. I would ask how such a regular growth of crystalline masses, reminding us in their character of the drusy crystals of calcite from Andreasberg, can be reconciled with the violent atmospheric commotion which we suppose to accompany the formation of hail. We say in natura nihil fit per saltus, and I believe it. The growing crystalline mass must have been suspended for a long time in a very cold stratum of aqueous vapor before it reached the earth. (Smithsonian Institution Annual Report, 420-421, 1869)

## HAILSTORM IN LINCOLNSHIRE, AUGUST 22nd, 1893.
**Anonymous; *Royal Meteorological Society, Quarterly Journal,* 20:72-73, 1894.**

On this date a remarkable storm of rain and ice occurred between 4 and 4.30 p.m. over a small district on the Lincolnshire Wolds bounded on the south by Fotherby, South Elkington, Welton, Biscathorpe, and Hainton, and on the north by Ludford, Kelstern, North Ormsby, and Utterby. In this space of about 10 miles east to west, and 2 to 3 miles north to south, much damage was done to corn crops, gardens, and glass. A party driving barely escaped an upset, as the horse was paralysed by the force of the storm. Their dog was killed on the road, and many

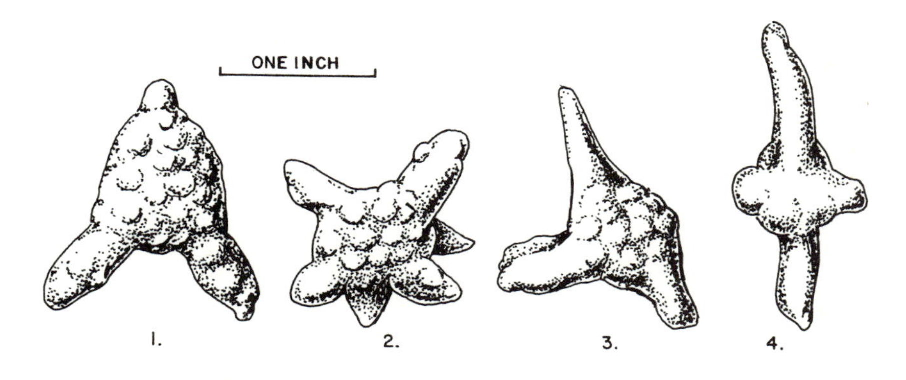


Irregularly shaped hailstones

chickens were beaten to death in the farmyards. At Hainton Hall gardens the ice was gathered up by the barrow load, pieces in some parts remaining unmelted until next morning.

The Rev. J. M. Coates made the accompanying sketches of some of the stones which fell at Welton-le-Wold at 4.30 p.m.

Fig. 1 weighed 90 grains 15 minutes after it fell.

Fig. 2 was drawn 42 minutes after it fell.

Fig. 4 measured 2-3/16 ins. in length 50 minutes after it fell.

In Figs. 1, 3, and 4 the projections were all roughly in one plane. In fig. 2 the projections extended from all sides.

The largest stones remaining on the grass lawn were not quite melted away after two hours. The temperature of the air was probably about 70° at 4.30 p.m. There was no appearance of crystalline structure in any of the hailstones examined, but rather of aggregation of globular stones of small size in the projections as well as in the globular nuclei. (Royal Meteorological Society, Quarterly Journal, 20: 72-73, 1894)

## ON SOME CURIOUS AND SINGULAR APPEARANCES OF SNOW AND HAIL
**Clark, Daniel A.; *American Journal of Science,* 1:2:134, 1820.**

There was also, two years before, a fall of hail in the same county [Morris County, N. J.], which was to me in some respects new.---The hail stones were generally about one fourth or three eights of an inch thick and of sufficient dimension in length and breadth to hide a shilling, and in many cases a cent, and almost every one perforated in the middle as if they had been held between the fingers, till the fingers by their warmth had melted away the middle and had met. When the perforation was not complete, there was in every case an inclination to perforation. The storm was tremendous, but of short duration and took place in the heat of summer. (American Journal of Science, 1:2:134, 1820)

## REMARKABLE HAILSTONES
**Anonymous; *Nature,* 36:44-45, 1887.**

On April 24, about 12.30," Mr. Durrant writes, "while walking between Melrose and Kelso, a friend and myself were overtaken by a sudden and very violent hail-storm, accompanied by thunder. The violent burst lasted about two minutes, in which time the ground was completely covered with large hailstones rather more than half an inch long. I say 'long' advisedly, for all the specimens I examined were conical, and were all of them formed in the same way. The points had all the appearance of snow, being softer than the main bulk of the 'stones.' These snow portions occupied about one-third of the whole length, being white and non-transparent. The main portions of the hailstones were hard and ice-like, stranded lengthwise with from forty to fifty fibres of ice---each fibre curved separately at the top---and together forming a curved surface, as of a sphere having the snow point for its centre. Thus---

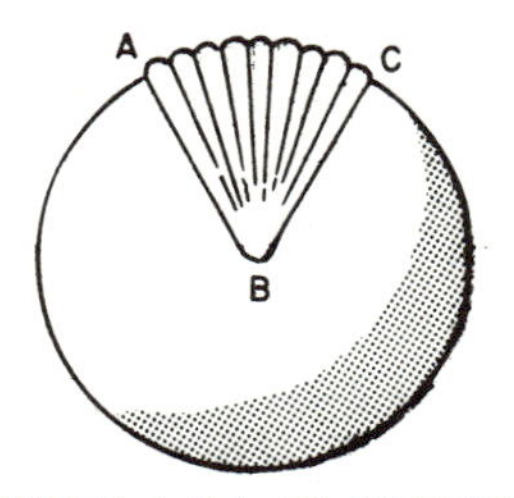


ANGLE A B C OF SECTION
BETWEEN 50° AND 60°

"On melting, the pointed part became translucent, while the other part became more opaque than at first, strands often remaining for a time, partially separated and curving outwards, as though they had been freed from compression in their lower extremities. Thus---

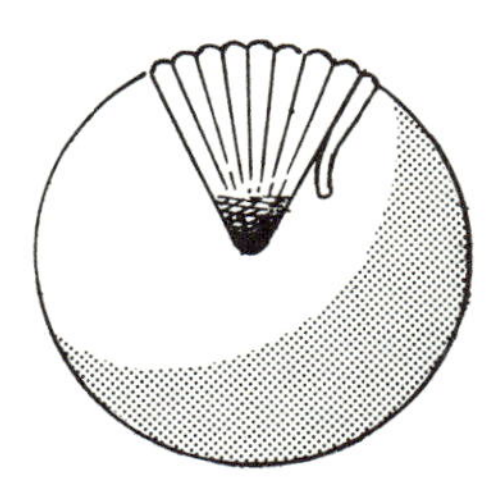

"The above appearances might admit of the hypothesis that these hailstones were fragments of radiated crystalline spheres, but one would expect in that case to find pyramidal rather than conical shapes, or at least to find some shaped so as to complement the cones. I failed to notice any indications of such shapes in the specimens (about thirty) which I examined. I should be inclined to believe that the soft, snow-like portions had been formed during the passage of the harder stranded stones through a moist and possibly clouded stratum of air.

"I was unable to see how they reached the ground, whether point or blunt end downwards. If in the latter way, one could account for the soft part, as being formed from previously unfrozen particles, cooled by contact with the nucleus, and, so to speak, sliding back to a position sheltered from the air, as it swept by the sides of the cone. (Nature, 36:44-45, 1887)

## HAILSTORM AT LEAMINGTON
**Black, W. T.; *Symons's Monthly Meteorological Magazine,* 11:55-56, 1876.**

I beg to send you the accompanying memoranda respecting the heavy fall of hail at Leamington, on the afternoon of March 31st last, which I had the opportunity of witnessing, when present there for a few days. The hail fell from a thick heavy nimbus, coming from the S. E., while the sun in the W. was not obscured. It fell thickly for about 20 minutes, with much noise, both in the air, and from the roofs and trees, and splashed the river considerably on the surface. Everybody ran into shelter, as well they might, considering the size of the stones, which were as big as pebbles, angular, and pointed. They appeared to be shaped like pyramids, with points, and convex bases, studded with tubercles, and had plain sides, either square, pentagonal, or hexagonal. They would have probably belonged to complete spheres, when their sides were adjusted together, and which would have been about 1-1/4 in. in diameter, and these segments were, therefore, about 1/2 in. long. They were of crystalline ice, with concentric bands, and probably the perfect pyramids weighed from 10 to 20 grains, and, as 24 might be estimated to make up a sphere, this, at the least, might have been half-an-ounce in weight, and at the

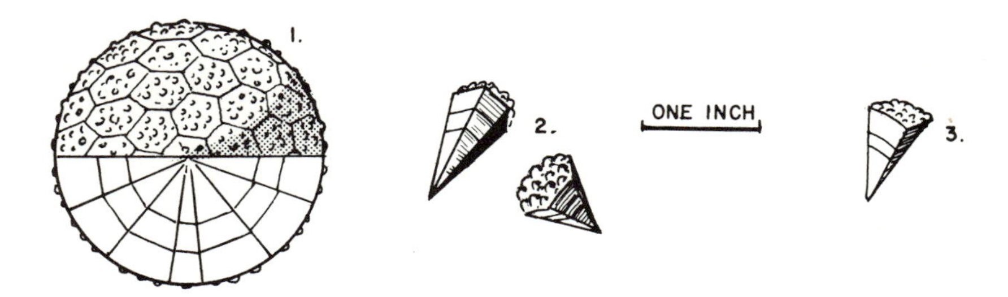


most about one ounce. As they were tuberculated on the convex bases, the whole sphere would have looked very much like a raspberry, or an oxalate of lime calculus. These spheres of ice must have burst, or become split up into segments somewhere in the fall through the air, as none were actually found perfect on the grass in the gardens. The primitive spheres must have been formed at great elevations, to have allowed of such a large accretion of ice from the original snow in the clouds. (Symons's Monthly Meteorological Magazine, 11:55-56, 1876)

## STRANGELY SHAPED HAILSTONES
**Anonymous; *Nature,* 23:233, 1881.**

To the October number of Symons's Monthly Meteorological Magazine Col. Foster Ward writes describing some remarkable hailstones that fell during a slight thunderstorm at Partenkirchen, Bavaria, at 6 p.m. on August 21. He was on a mountain about 3000 feet above the village, and saw the cloud (a small one) pass over the valley below. There were several peals of thunder, but there was no visible lightning, owing, he concludes, to the sun's brightness. "On arriving near home, I met a friend who told me it had been hailing 'tadpoles' and 'acidulated drops.' There had been little or no rain and no visible lightning, and the hailstones fell at intervals and about six feet apart. There were very few of them, my family only picking up twenty in a space occupied by a full-sized lawn tennis court. My son made a sketch of their shape and size, which I inclose. The greater

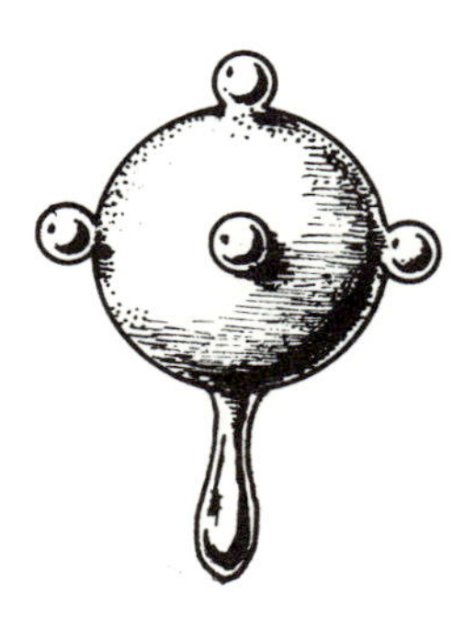

part were of the 'tadpole' shape and were clear as glass, perfectly round, the five knobs being at equal distance from one another. The flat stones had more or less a slight nucleus of snow in the convex portion of the stone. My wife and three daughters, and two ladies staying with us, say that the stones looked just like a lady's hand looking-glass, with a knob at the top and on either side for ornament. More than twenty, perhaps thirty, were picked up of this shape. Of these about two-thirds were studded, the rest plain, with only the tail or handle, the thinnest part of it being near the body of the stone, as in sketch. The studs were all symmetrically placed. There were from three to five in each stone besides the handle. When there were less than five they occupied the same positions as if the five had been complete. In some cases the handle appeared to have been knocked off. The drops were more numerous, were all of same shape, convex at the top, the bottom being concave (like a small china painting palette)." (Nature, 23:233, 1881)

## FALL OF PECULIAR HAILSTONES IN KINGSTON, JAMAICA
**Bowrey, James John; *Nature,* 36:153-154, 1887.**

Shortly after midday on the 2nd inst. a thunderstorm visited this city; the rain began with the wind from the east, as is usual with our May seasons, but it speedily changed to the west, accompanied with much lightning and thunder. Immediately hailstones became mingled with the rain, attention being drawn to their advent by the sharpness with which they struck on the shingled roofs. The west door of the laboratory being open to the air, the hail came in freely, nearly covering the floor

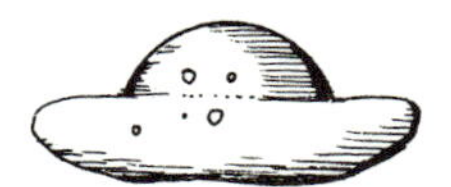

for more than 12 feet. The hailstones were of clear ice, inclosing a few bubbles of air, varying from mere points to bubbles of the size of a split pea. The shape of the stones was singular. Suppose a shallow and very thick saucer to have a shallow cup, without a handle, inserted in it, and you will have a good idea of the form of the hailstones when unbroken. Many had more or less lost the "saucer" by violence, while some were entirely without it, presenting the appearance of a double convex lens with faces of different curvature.

By actual measurement the hailstones were found to vary from one-quarter to three-quarters of an inch in diameter, and from one-eighth to one-quarter of an inch in thickness at the thickest part. I observed that in very many of the larger stones the air-bubbles could move about, showing the interior to be still liquid; as melting proceeded the bottom of the "saucer" would suddenly give way and become concave. The storm lasted about 15 or 20 minutes, hail falling for the greater part of the time. The hail which fell on grass remained unmelted for ten or fifteen minutes after the rain ceased. The fall of hail was very local, none falling at my house a mile away. I am informed that hail last fell in Kingston in 1839. (Nature, 36:153-154, 1887)

## "FLATTISH" HAILSTONES OBSERVED IN THUNDERSTORM OVER MT. OLYMPUS, CYPRUS
**Moorhead, H.; *Meteorological Magazine*, 65:289-290, 1931.**

At Cyprus on September 12th, 1930, flat and other peculiar-shaped hailstones were observed by numerous spectators up Mt. Olympus, about 6,000 ft. above sea level.

During the early part of the day the weather showed no unusual symptoms, being fine during the forenoon with a moderate southerly breeze at sea level. During the forenoon slight cumulus cloud commenced to form over the land, principally on the westward side. One observer stated that at about 14h. (East European time) he was sitting inside a tent when he noticed a roaring sound which had been going on for about 10 minutes; he imagined the noise to be rather like what Niagara Falls would sound like from a distance of about 3 miles. The occupants of the tent went outside and noticed a large thunder cloud in the vicinity from the edge of which some hailstones fell amongst them. No thunder or lightning was observed and the cloud gradually moved away. He states that the hailstones were flattish or rather like the underside of a button, and in size about the circumference of a penny. The four local residents with him at the time had never previously seen hailstones of a similar type or heard a noise to correspond to the roaring sound.

Another observer about two miles distant and approximately 400 feet higher up, states that at 14h. a violent hailstorm occurred, accompanied by thunder and lightning which lasted about a quarter of an hour. There were a few round hailstones of normal size and some others of peculiar shape, like teeth, with needle-sharp points, but for the most part they were "flattish," rather like an acid drop and about the circumference of a shilling. He cut open a round one and found the usual formation which he described as like a section of an onion. The "flattish" ones when cut open at the sides or centre appeared as one piece of solid ice. The hailstones did not melt rapidly and when they did commence to do so, the flat ones dissolved first from the centre and eventually left a ring of ice on the outside which was the last to go. (Meteorological Magazine, 65:289-290, 1931)

## HAILSTORM ON THE ST. LAWRENCE
**Anonymous; *Monthly Weather Review*, 29:506-507, 1901.**

Next came a heavy fall of rain, which was followed by hail. The hailstones to fall were formed as though icicles the size and shape of lead pencils had been cut into sections about three-eighths of an inch in length. These were soon followed by others as large as walnuts, and later by still others slightly disk shaped, measuring fully 3 inches in diameter by 2 inches in thickness. The ground was covered with them, and several branches of trees were broken off. They were exceedingly hard and would rebound, when falling on the rocks, without breaking. They melted very slowly even when placed in the sun. When half melted many had the appearance of the human eye---a pupil in the center and a ring surrounding it, with fine lines radiating in all directions. Others were composed of hard crystals of ice, several stones often being frozen together; and still others were of frozen snow. The next morning at 8 o'clock remnants of hailstones as large as peas were lying on the ground.

During the storm the river presented a beautiful appearance, there being thousands of miniature fountains from a foot to 6 feet in height spurting up where the hail plunged in. (Monthly Weather Review, 29:506-507, 1901)

## REMARKABLE HAILSTONES

**Winston, Charles H.; *Symons's Monthly Meteorological Magazine*, 32:171-172, 1898.**

About 5 o'clock in the afternoon of August 10th I was at Manassas depot, in Prince William County, Va., near the famous battlefield, waiting for a train. There was some pretty severe thunder and lightning for a half-hour or so, and then came a heavy shower of rain, during which there was the most remarkable fall of hail I have ever witnessed. I hurried out in the rain to examine the stones and picked up several. These were nearly square flattish blocks, say from 3/4 to 1 inch in length and breadth, and from 1/4 to 1/2 an inch in thickness. They suggested, by both shape and size, the ordinary "chocolate caramels" of the confectioner. There were some 8 or 10 persons, I think, in the station house with me, and several of these, observing my interest and enthusiasm, began to pick up the larger stones and bring them in to me and to my friend, Professor Hargrove, of Luray, Va. Soon larger and larger ones were thus collected, and I sought for means of measuring or weighing them. No rule or scales could be found, and so we set ourselves, several of us, conjointly and carefully to estimate the dimensions. I recorded at the time one as being, honestly estimated, "2 inches long, 1-1/2 inches wide and 3/4 of an inch thick," these being rather the average than the extreme dimensions.

..........

I think that very few of these stones or blocks fell. Perhaps they would have been a yard or two apart as they lay on the ground. I think it likely also that the storm of hail was of brief duration, say 10 or 15 minutes, and that it embraced a very limited area.

It was, perhaps, about over when I took the train, as I infer from the fact that I have seen no account of it in the papers; and I found at the next station, only 5 miles off, that the road was dry and dusty. (Symons's Monthly Meteorological Magazine, 32:171-172, 1898)

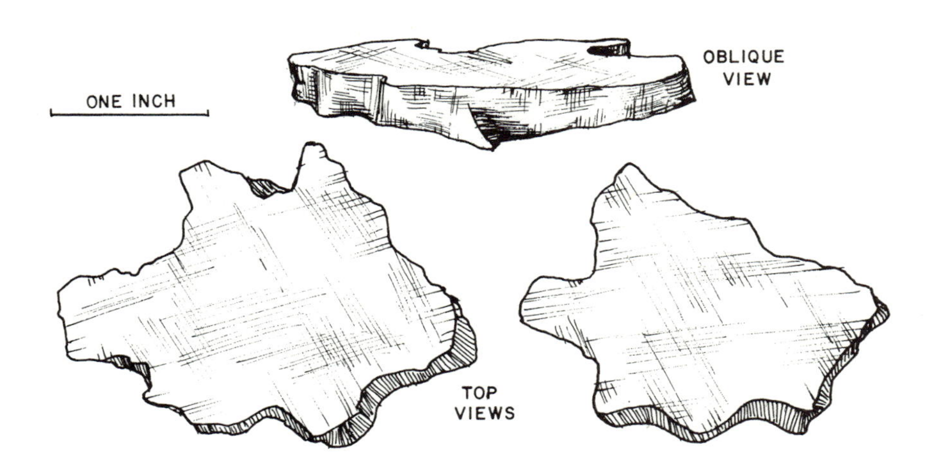


Flattish ice fragments fall in hailstorm

## HUGE HAILSTONES
**Anonymous; *Scientific American*, 71:371, 1894.**

Prof. Cleveland Abbe includes the following among his notes in the Monthly Weather Review for July: On June 3 a tornado passed northeastward through the counties of Harney, Grant, and Union, in eastern Oregon. The most novel feature attending the disturbance was the hail. It is stated that the formation was more in the nature of sheets of ice than simple hailstones. The sheets of ice averaged three to four inches square, and from three-fourths of an inch to one and a half inches in thickness. They had a smooth surface, and in falling gave the impression of a vast field or sheet of ice suspended in the atmosphere, and suddenly broken into fragments about the size of the palm of the hand. During the progress of the tornado at Long Creek a piano was taken up and carried about a hundred yards. (Scientific American, 71:371, 1894)

## UNUSUAL HAILSTONES
**Grinsted, W. A.; *Meteorological Magazine*, 78:55, 1949.**

The following account of unusual hailstones, sent to us by Mrs. P. O. Bickford of Kapsaret Estate, Eldoret (0° 35'N. 35° 20'E. 7,000 ft.), may be of interest.

"On Friday August 6 on this Estate, 8 miles from Eldoret on the Kapsaret road, at 5 p.m., there occurred a rather singular fall of hail which I thought might be of interest to you, though it may be quite familiar to you.

The sky became very dark and large hailstones fell, but in the shape of stars similar to those of snow crystals in shape (when seen under the microscope). They nearly all measured 1-1/2 in. in width. They consisted of large ice balls the size of a very large pea with fine radiating arms of ice. Some of these landed on their sides and stuck sideways on in the grass. These soon gave way to rain. Some were of a squarish shape but these were not perfect."

There was a zone of convergence over this area between south-easterly and south-westerly winds on August 6 with the freezing level at 15,000 ft. (Meteorological Magazine, 78:55, 1949)

## LUMPS OF ICE AS HAILSTONES
**Anonymous; *Monthly Weather Review*, 22:293, 1894.**

From Mr. S. M. Blandford, temporarily in charge of the Weather Bureau office in Portland, Oreg., there was received too late for the June Review a report of the tornado that occurred June 3, 1894, passing northeastward through the counties of Harney, Grant, and Union, in eastern Oregon. He states that the most novel feature was the hail. One correspondent states that the formation was more in the nature of sheets of ice than simple hailstones. The sheets of ice averaged 3 to 4 inches square and from three-fourths of an inch to 1-1/2 inches in thickness. They had a smooth surface and in falling gave the impression of a vast field or sheet of ice suspended in the atmosphere and suddenly broken into fragments about the size of the palm of the hand. During the progress of the tornado at Long Creek a piano was taken up and carried about a hundred yards. (Monthly Weather Review, 22:293, 1894)

## HAILSTONES OF IRREGULAR SHAPE
**Schleusener, Richard A.; *American Meteorological Society, Bulletin*, 40:29, 1959.**

In a recent hailstorm on the campus of Colorado State University at Fort Collins, the writer observed a number of hailstones having the shape of a triangular pyramid. The stones were formed from layers of ice of varying opacity in concentric layers about the tip of the pyramid.

The measured dimensions of the largest stones are given on the accompanying sketch. The shape of the stone indicated that about 1/4 inch had melted from the tip of the pyramid before measurements were made. The smooth sides of the pyramid indicated that the stone probably was fractured from a sphere that initially was about 3-1/2 inches in diameter.

This is the same type of stone that has been observed by Beckwith in the Denver area. (American Meteorological Society, Bulletin, 40:29, 1959)

# EXTRAORDINARILY LARGE HAILSTONES

## HAIL FIVE INCHES IN DIAMETER
**Calvert, D.; *Marine Observer*, 32:112-13, 1962.**

s.s. Afghanistan, Captain D. Calvert. At anchor off Umm Said. Observers, the Master and all deck officers.

7th April 1961. At 1445 GMT the wind veered from ENE, force 3, to ESE, force 4, and precipitation in the form of heavy hail commenced; within a few minutes the hail was very dense and of a size larger than tennis balls, some hailstones being estimated to be at least 5 in. in diameter. The visibility by this time had decreased to about 350-400 yd. By 1500 GMT the hail had given place to torrential rain and visibility had fallen still further to about 100 yd; the wind had increased to about force 8 with gusts up to force 12. The air temp. was 75°F and the sky was covered with Cb.

At 1510 GMT the wind backed to NE, force 9-10, but by 1512 it had decreased to force 6. Patches of blue sky appeared and the rain became intermittent, although still heavy. Much lightning had been visible for the previous 2 hours and it was seen to the eastward for the following 7 hours, during which time the wind was backing and veering between NW and E through N, finally becoming E, force 5.

The size of the hail may be judged from the fact that the sea was a mass of white foam caused by the splashes from the hail, the splashes being from 2 to 3 ft. high; the damage done to the brass binnacle covers for both gyro and magnetic compasses, which were dented to a depth of 3/4 in., is another indication. Over 80 holes were found in the boat covers the following morning.

Position of ship: 24° 56'N, 51° 36'E. (Marine Observer, 32:112-113, 1962)

## DESTRUCTIVE HAILSTORM IN THE TRANSVAAL
**Anonymous; *Nature,* 137:219-220,1936.**

Six years ago, in a weekly column devoted to remarkable "Historic Natural Events", many records were given in Nature of great hailstorms and damage done by them. There is an authentic record, for example, of a hailstone 17 inches in circumference and weighing 1-1/2 lb. having fallen in Nebraska in July 1928 during a storm when the hailstones were "as large as grapefruit". A hailstorm of this character is reported by The Times correspondent at Johannesburg to have occurred on February 1 in a native area of the Northern Transvaal when nineteen natives of the Barolong tribe were killed. The report states: "About 3 in. of rain fell in a few minutes, and then came the hail, which consisted of jagged lumps of ice. In 30 minutes the hail was lying everywhere to a depth of 3 ft., and in some cases the dead natives had to be dug out of it. There were many cattle killed, which the natives afterwards dragged away on sleighs. Whole crops were obliterated, and there are said to be over 1,000 native families afflicted in the area". (Nature, 137:219-220, 1936)

## AN EIGHTY-POUND HAILSTONE
**Anonymous; *Scientific American,* 47:119, 1882.**

Considerable excitement was caused in our city last Tuesday evening by the announcement that a hailstone weighing eighty pounds had fallen six miles west of Salina, near the railroad track. An inquiry into the matter revealed the following facts: A party of railroad section men were at work Tuesday afternoon, several miles west of town, when the hailstorm came upon them. Mr. Martin Ellwood, the foreman of the party, relates that near when they were at work hailstones of the weight of four or five pounds were falling, and that returning toward Salina the stones increased in size, until his party discovered a huge mass of ice weighing, as near as he could judge, in the neighborhood of eighty pounds. At this place the party found the ground covered with hail as if a wintry storm had passed over the land. Besides securing the mammoth chunk of ice, Mr. Ellwood secured a hailstone something over a foot long, three or four inches in diameter, and shaped like a cigar. These "specimens" were placed upon a hand car and brought to Salina. Mr. W. J. Hagler, the North Santa Fe merchant, became the possessor of the larger piece, and saved it from dissolving by placing it in sawdust at his store. Crowds of people went down to see it Tuesday afternoon, and many were the theories concerning the mysterious visitor. At evening its dimensions were 29 x 16 x 2 inches. (Scientific American, 47:119, 1882)

# ODD HAIL STREW PATTERNS

## REMARKABLE HAILSTORM
**Brunt, D.; *Meteorological Magazine,* 63:14-15, 1928.**

Mr. C. S. Durst has drawn my attention to an interesting paper by Becquerel communicated to the Academie des Sciences on November 13th, 1865, entitled "Memoire sur les zones d'Orages a Grele." In this paper is given a description of a remarkable hailstorm observed at Clermont-Ferrand on July 3rd, 1863. The day in question was exceedingly hot, and by 3 p.m. the sky was covered by an enormous nimbus cloud, with flashes of lightning in quick succession. About 6 p.m. there rapidly approached from west a cloud whose height was estimated to be about 1,500 metres, whose form resembled a huge net. The portions of the network showed violent agitation, and soon after the arrival of the cloud there was a violent hailstorm lasting for about five minutes, the hailstones having the size of nuts. During the fall of the hail there was no wind. The most remarkable feature of this particular storm was the distribution of hail over the ground. The fall of hail was so violent that it caused considerable damage wherever it fell. M. Lecoq, who observed the storm, and described it in the Comptes Rendus, Vol. XXXVII, p. 75, stated that the damage produced by the hail was limited to small patches, which were surrounded by undamaged zones, forming a network whose meshes were irregular, but roughly 60 to 100 metres apart.

The distribution of hail corresponds to the form of the lower cloud which form recalls Benard's cellular divisions in unstable liquids, a note on which appeared in the Meteorological Magazine for February 1925, p. 1. I should be very much interested to know whether any further observations of similar distribution of hail or rain have ever been noted, since they have a very definite bearing on the physics and dynamics of the atmosphere. (Meteorological Magazine, 63:14-15, 1928)

## HAILSTORM IN GUILFORD COUNTY, N.C.
**Anonymous; *American Journal of Science,* 2:22:298, 1856.**

On the 9th of June, 1856, a hailstorm of unusual violence passed over a portion of Guilford County, N. C. An observer at Hillsdale in that county, gives the following description. "The cloud came up from the SW about 12. The storm began with rain, thunder and lightning. In a few minutes hailstones of great size began to fall, dashing in exposed windows, and splitting the shingles on the roof of the building in which I was. The rain continued an hour after the hail ceased. As soon as it was safe to go out, some of the hailstones were brought in. One measured eight inches in circumference, and I concluded it must have been nine or ten when it fell, as there had been so much rain and that a very warm one. The weather was very hot, and there was no change of temperature during the week following. This hailstone was a perfect globe. Others measured as large in one direction, but they were flat."

"The grounds around us were so completely covered with leaves and boughs of trees from the oak grove in which we were, that we had little chance to know what actually fell about us. A mile westward the storm was still more severe. The trees have a strong appearance of winter, and fields of wheat have been turned over to the use of cattle. There was destruction of windows and of small animals, and a few wayfarers were severely beaten. The storm extended about fifteen miles in one direction and five or six in the other. The hail fell in lines, a field here and a garden there being destroyed, while intermediate ones were left uninjured. The hail had a strong flavor of turpentine. This is the testimony of persons testing it at different and distant localities." (American Journal of Science, 2:22-298, 1856)

# SLOW-FALLING HAIL

## REMARKABLE HAILSTORM IN IRAQ
**Anonymous; *Meteorological Magazine,* 65:143-144, 1930.**

The report of the Meteorological Section in Iraq for April, 1930, contains an account of a remarkable hailstorm which visited Hinaidi and Bagdad on April 24th. On the morning of that day the weather conditions were unsettled, the morning chart showing a primary depression extending eastwards along the eastern Mediterranean into Syria and a secondary depression south of Rutbah. The primary depression moved slowly northeastwards, but the secondary travelled rapidly eastwards across western Persia. At Hinaidi the morning began with light northeast to east winds, which became calm at 3h. G. M. T. (6 a. m. local mean time). At 3h. 35m. a line-squall occurred, the wind reaching a velocity of 36 m. p. h. and veering from south-east to west. A second line-squall followed at 4h. 34m., in which the wind rose to 30 m. p. h. and veered from north-west to north-east. Subsequently the wind gradually became more easterly and maintained an average velocity of 32 m. p. h. At 14h. 15m. G. M. T. slight rain commenced, and at 14h. 30m. the wind began to drop to calm and a heavy fall of hail occurred, continuing for seven minutes. The most remarkable features of the hailstorm were the large size and slow terminal velocity of the hailstones and the absence of dust in their composition. Some idea of their size is given by the illustration forming part of the frontispiece of this number of the magazine.

An average specimen was cut open and was found to consist of five layers of alternating hard transparent ice and soft white ice, arranged as follows:---

Core of hard transparent ice . . . . . . . . . . . . . 1/8 inch diameter
Layer of soft white ice . . . . . . . . . . . . . . . . . 1/4 inch thick
Layer of hard transparent ice . . . . . . . . . . . . 1/8 inch
Layer of soft white ice . . . . . . . . . . . . . . . . . 1/4 inch
Outer crust of hard transparent ice . . . . . . . . . 1/8 inch

The total diameter was one and five-eighths inch, and in calm air the terminal velocity of hailstones of this size would be nearly 30 miles per hour. The descent of several specimens was actually timed against the wall of a building, and it was found that they fell 40 feet in about three seconds giving a velocity of only 9 miles per hour. The observer is to be congratulated on his enterprise in obtaining this measurement, which appears to indicate an upward current of air of at least 30 m.p.h. at a comparatively low elevation. (Meteorological Magazine, 65:143-144, 1930)

## THE STORMS OF JUNE 27th, 1866
**Anonymous; *Symons's Monthly Meteorological Magazine,* 1:44, 1866.**

Some of our great authorities on wind have urged the existence of vertical rotary motion---(i.e. if a cyclone under ordinary circumstances is represented by a wheel revolving while lying on its side, a vertical cyclone would be represented by the same wheel revolving in its proper position on the axle of a carriage.) I have mentioned this because it seems to me the readiest solution of a puzzling fact, namely, that the hargest hailstones did not do most mischief. [During Cricklewood storm of June 27, 1866] Doubtless this was partly due to the relative hardness of the stones, but it was not wholly this, as many of the larger stones at Hampstead were of the hard species, yet the damage there was trifling, as it was also beyond Cricklewood, although the hail was large and hard, they fell "quite gentle like," as a native told me at Willesden Green. Now supposing (for it is only a bit of theory) there was such a vertical cyclone, its rotation being such that the motion of its lower half was in the same direction as the general course of the storm, it would solve at once several difficulties; thus---let its lowest portion be assumed to have swept near the surface while crossing over Cricklewood, it would then give wind force sufficient to break the poplar, and by increasing the velocity of the hailstones would produce equal or greater destructive power than where they fell without the wind and yet were larger; moreover, if this theory be accepted, it will also explain how large and heavy hailstones could fall "quite gentle like" a mile or two further on, for there the upward motion of the lower advancing quadrant of the cyclone would, by its tendency to lift the stones, partly counteract the force of gravity and so let them fall "quite gentle like" to the earth. (Symons's Monthly Meteorological Magazine, 1:44, 1866)

# PECULIAR INCLUSIONS IN HAIL

## SALT HAIL
**Anonymous; *Nature,* 5:211, 1872.**

Prof. Kengott, of Zurich, states that a hail-storm lasting five minutes occurred at eleven o'clock in the morning of August 20, 1871, the stones from which were found to possess a salty taste. Some of them weighed twelve grains. They were found to consist essentially of true salt, such as occurs in Northern Africa on the surface of the plains, mainly in hexahedric crystals or their fragments, of a white colour, with partly sharp and partly rounded grains and edges. None of the crystals were entirely perfect, but appeared as if they had been roughly developed on some surface. They had probably been taken up and brought over the Mediterranean from some part of Africa, just as sand is occasionally transported thence to the European continent and the Canaries by means of hurricanes. A still more remarkable phenomenon has been recently recorded by Prof. Eversmann, of Kasan---namely, the occurrence of hailstones, each containing a small crystal of sulphuret of iron. These crystals were probably weathered from some rocks in large quantity, and were then taken up from the surface of the ground by a storm, and when carried into the hail-forming clouds served as a nucleus for the formation of hailstones. (Nature, 5:211, 1872)

## ON THE ORIGIN OF HAIL

**Schwedoff, Theodore; *Symons's Monthly Meteorological Magazine,* 17:151, 1882.**

But the case which, above all, merits our attention, is that of the shower at Padua, on the 26th of August 1834, of hailstones with nuclei of a dark grey colour. These nuclei, examined by Cozari, consisted of grains of different sizes, the largest of which could be attracted by the magnet, and were found to be composed of iron and nickel. The identity of this matter with that of aerolites can scarcely be open to doubt." An analogous observation has been made at Stockholm by Nordenskiold, who proved the presence of little dark grains of metallic iron in some hailstones. I think it appropriate to note here a fact relating to the colour of hailstones. According to M. Lagounowitche, many of the hailstones that fell on the 14th (2nd O. S.) of June 1880, in the province of Minsk, were visibly coloured, some rose, others clear blue, recalling the colours of solutions of salts of nickel and cobalt, bodies very frequent amongst meteoric masses.

It has been sought to explain the presence of stony masses in hailstones, by the supposition that hurricanes raised the stony matters from the surface of the soil, and carried them up into the clouds; and it is this way of regarding them that has caused so general an indifference with respect to the stones that accompany hail. In 1815 the Academy of Sciences of St. Petersburg received a case containing specimens of stones that fell during a hailstorm at Wilna, of which some hundreds weighed as much as a pound. It is not known what subsequently became of these stones, at least no traces of them are to be found in the museums of the Academy. Nor is better information to be obtained about the stony masses which accompanied showers of hail at Permj in 1809, at Fatesch in 1844, and at Nachraschinsk in 1833 (Russia). Later, when the cosmic origin of such masses had become evident, it has been sought to explain their presence in hailstones by supposing that atmospheric vapours congeal around stony masses of cosmic origin floating in the air.

All these suppositions become unnecessary from the moment that one adopts a cosmic origin for hailstones themselves. That which appears astonishing, improbable and doubtful in hailstones of terrestrial origin, becomes natural, logical, and necessary in hail of cosmic origin. Hailstones sometimes have enormous dimensions, for there are no limits to the size of celestial bodies. The quantity of them is often extraordinary, for celestial space has no bounds. The form of them is most often spheroidal, for that form is the typical one for celestial bodies. Certain hailstones offer a development of ice-crystals, unknown on the surface of our globe, for the crystallisation of these masses is effected during thousands of years, and under conditions of repose unknown on the surface of the earth. The temperature of hail is very low, because the temperature of celestial space is so also. Lastly, hailstones are sometimes accompanied by meteorites, because these two kinds of bodies belong to the same family, and travel together through the depths of the sky. (Symons's Monthly Meteorological Magazine, 17:151, 1882)

## REMARKABLE HAIL
**Anonymous; *Monthly Weather Review*, 22:215, 1894.**

During a severe hailstorm at Vicksburg on the afternoon of Friday, May 11, a remarkably large hailstone was found to have a solid nucleus, consisting of a piece of alabaster from one-half to three-quarters of an inch. During the same storm at Bovina, 8 miles east of Vicksburg, a gopher turtle, 6 by 8 inches, and entirely encased in ice, fell with the hail.

An examination of the weather map shows that these hail storms occurred on the south side of a region of cold northerly winds, and were but a small portion of a series of similar storms; apparently some special local whirls or gusts carried heavy objects from the earth's surface up to the cloud region, where they were encased by successive layers of snow and ice, until they fell as hailstones. The fact that hailstones, as well as drops of water and flakes of snow, often contain nuclei that must have been carried up from the earth's surface, is entirely in accord with the general principle that ascending currents precede the formation of cloud and rain, and that solid nuclei are needed to initiate the ordinary precipitation of moisture. (Monthly Weather Review, 22:215, 1894)

Hailstone enclosing gopher turtle

# EXPLOSIVE HAIL

## EXPLOSIVE HAIL
**Brown, W. G.; *Nature,* 88:350, 1912.**

On the afternoon of November 11, 1911, there was a brief storm of explosive hail at this place.

The morning had been unseasonably warm; about noon there were the usual signs of a coming thunderstorm---heavy cumulo-nimbus clouds with a gusty wind---which began about 2.30 p.m. with a slight slower of heavy raindrops; shortly afterwards there were two or three flashes of lightning and thunder, followed by a fall of large hailstones, which on coming in contact with the windows or walls or pavement in many instances exploded with a sharp report, so loud as to be mistaken for breaking window panes or a pistol shot. As the hail fell, the fragments sprang up from the ground and flew in all directions, looking like a mass of "popping corn" on a large scale.

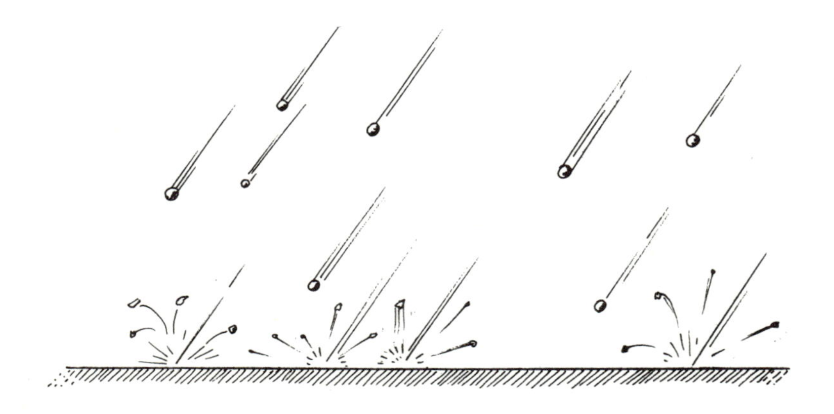

The fall lasted two or three minutes, about half the hailstones being shattered, the ground in some places being nearly covered white with the stones and fragments.

Of the unbroken stones, seventy were gathered. They weighed, roughly, 225 grams. A few were ellipsoidal, the longest axis about 25 mm. in length; most of them, however, were nearly spherical, and somewhat smaller, from 15 to 20 mm. in diameter.

Practically all of them contained a nucleus. In a few of the stones the nucleus was porcelain-like, raspberry-shaped, surrounded by almost colourless spherical layers of ice, for about five-sevenths of the diameter, and then a shell of porcelain-like, snowy ice.

A fair proportion of the stones showed, in addition to the spherical, a radiate structure, which was very apparent as the stones melted in a flat dish, showing the crosssection with great distinctness.

The writer noticed a similar fall of explosive hail about eighteen years ago at Lexington, Virginia. The stones in this fall were much smaller, and attention was directed to the stones by the peculiar way in which they seemed to rebound on striking the ground, which was also due on that occasion to their breaking into fragments, without, however, any noticeable explosion. (Nature, 88:350, 1912)

# COLORED PRECIPITATION

## RED RAIN FALLS OVER 20,000 SQUARE MILES
**Anonymous; *Nature,* 125:256, 1930.**

Feb. 21-23, 1903. Red Rain.---Dust or 'red rain' fell over an area of 20,000 square miles in the southern half of England and Wales as well as in many countries on the Continent. It is estimated that in England and Wales alone the total quantity of dust was not less than 10 million tons. It was traced back to the Sahara, south of Morocco, where it was raised by a strong north-east wind; it travelled on the western side of an anticyclone over south-west Europe for a distance of at least 2000 miles in a wide sweep around Spain and Portugal, probably across the Azores. In Europe the fall was associated with oppressive heat, and visibility was limited to short distances. (Nature, 125:256, 1930)

## PHENOMENAL WEATHER IN VICTORIA
**Anonymous; *Symons's Monthly Meteorological Magazine,* 32:27, 1897.**

Dr. Argyle, writing from Melbourne, on the 17th of January, to his uncle in Tamworth, says, "Our climate, at which we growl quite as much as you do in England, has gone quite mad, I think. The temperature on Thursday week was 101 deg. in the shade and terribly oppressive. On Saturday and Sunday there was a terrific storm, the thermometer dropped nearly 60 deg., and we had 4-1/2 inches of rain in 48 hours. (This in January, usually our driest and hottest month.) To-day it has been hailing and is bitterly cold, but the sun has now come out, and I should not be surprised if it turned out hot again to-morrow. We also had an interesting phenomenon about a fortnight ago; there was a heavy fall of rain, which was full of a red dust, and the next morning the whole landscape was red. The "rain of blood," it was called, and really it looked like it. The funny thing was that it occurred on a public holiday and on a hot day, so that all the holiday-makers who were caught in it had their clothes stained a deep red, and as many people were dressed in white, you may guess what they looked like." (Symons's Monthly Meteorological Magazine, 32:27, 1897)

## A FALL OF YELLOW RAIN
**Anonymous; *Nature,* 2:166, 1870.**

On the 14th of February a remarkable yellow rain fell at Genes. The following details respecting it are given in a letter addressed to M. Ad. Quetelet by M. G. Boccardo, director of the Technical Institute of Genes, who examined it in concert with Dr. Castellani, professor of chemistry. The quantitative analysis gave the following results:---

| | |
|---|---|
| Water . . . . . . . . . . . . . . . . . . . . . . . . . . . . . 6.490 | per cent |
| Nitrogenous organic substances . . . . . . . . . . . . . 6.611 | " |
| Sand and clay . . . . . . . . . . . . . . . . . . . . . . . . .65.618 | " |
| Oxide of iron . . . . . . . . . . . . . . . . . . . . . . . . .14.692 | " |
| Carbonate of lime . . . . . . . . . . . . . . . . . . . . . . 8.589 | " |

Observed narrowly under the microscope, the presence was revealed of a number of spherical or irregular ovoid substances of a cobalt blue colour; corpuscles similar to the spores of Peziza or Permospora; spores of Demaziacece or Spheriacece; a fragment of a Torulacea (?); corpuscles of a pearly colour, concentrically zoned, probably small grains of fecula; gonidia of lichens; very scarce fragments of Diatomacece; spores of an olive brown colour; a few fragments of filaments of Oscillaria, Ulothrix, and Melosira varians; a fragment of Synedra; a peltate hair from an olive leaf. If, instead of collecting the earth on the morning of the 11th, when it had already been subjected to the action of rain falling for several hours, I had been able (writes M. Boccardo) to observe the phenomenon during the night, at the moment when it was produced, it is very probable that the microscope would have shown the existence of several kinds of Infusoria, as has been the case in several similar instances.

The author notes that the direction of the wind at Genes during the night of the 13th and 14th was from the south-east, and without being exactly a hurricane as on the preceding few days, was still very strong. (Nature, 2:166, 1870)

## BLACK SNOW AT ESKDALEMUIR IN 1897
**James, G. B. Kingston; *Meteorological Magazine*, 68:20-21, 1933.**

Mr. Richard Bell, of Castle O'er, in his "My strange Pets and Other Memories of Country Life" (Blackwood, 1905), describes a fall of black snow in the parish of Eskdalemuir, and an account of it may prove of interest.

Mr. Bell tells us that a fall of black snow took place on the evening of January 30th, 1897. On the following morning, whilst taking a walk in his avenue, he noticed that the marks made by his footsteps "were pure white below the coloured surface of the snow, which was of quite a dark colour in comparison." On investigating the peculiarity, he found that the surface of the unbroken snow to the depth of about a quarter of an inch was deeply coloured as if mixed with soot. The author goes on to say---and I cannot do better than to quote the actual words of this careful observer---"Having gone further afield during the day I found the whole of the snow on the road and hillsides was of the same black colour; and on looking across the valley the hills appeared as if they were bathed in the lights and shadows, as seen during bright sunshine. As the day was dull, sunshine could not account for this peculiar appearance; and it was evident that it was caused by the wind during the time the snow fell. The wind swept the more salient parts of the hill and deposited the fall of coloured snow in the hollows, thus giving the country the appearance of light and shadow. The area over which the snow fell was, to my own knowledge, four miles long by about a mile and a half wide."

It appears that soot from the chimneys of some manufacturing town must have been responsible for the discoloration of the snow, but as the prevailing winds were of a light order, it is impossible to say whether an English or Scottish town caused the phenomenon. (Meteorological Magazine, 68:20-21, 1933)

## BLACK RAIN

**Bonney, T. G.; *Symons's Meteorological Magazine*, 53:42, 1918.**

A black rain is recorded in the minutes of the Philosophical Club of the Royal Society which I have recently been editing. It was described by Mr. James Rust, Minister of Slains, in Aberdeenshire, and Colonel Sykes, who made the communication, had satisfied himself of his trustworthiness. The rain fell not only at that place but also along the whole eastern coast of the county on January 14th, 1862. The morning, about 8.30 a.m., was clear, then the sky darkened, threatening rain. About an hour later, a "large, dense, black, smoky-looking cloud came driving over the sea from the S.S.E. and discharged a shower of rain, with drops like ink, which blackened all the water collected in cisterns from the roofs of houses and dirtied clothes put out to bleach, so effectively that warm water was needed to wash out the spots." Mr. Rust suggested that a recent eruption of Vesuvius might be the cause; Sir R. Murchison thought this origin impossible, because of the distance, and that it must be smoke, while Professor Tyndall stated that his own experiments made him doubtful whether a sufficient quantity of soot could have been distributed through the atmosphere to produce the blackness described. (Symons's Meteorological Magazine, 53:42, 1918)

## ANCIENT METEOROLOGICAL NOTICES

**H., E. C.; *American Journal of Science*, 1:43:399, 1842.**

Moreover it is credibly affirmed that in the Winter of the Year 1688, there fell a Red Snow, which lay like Blood on a spot of Ground, not many miles from Boston; but the Dissolution of it by a Thaw, which within a few hours melted it, made it not capable of lying under the contemplation of so many Witnesses as it might have been worthy of. The Bloody Shower that went before the suffering of the ancient Britains from the Picts, (a sort of People that painted themselves like our Indians,) this Prodigy seemed a second Edition of. (American Journal of Science, 1:43:398, 1842)

# POINT RAINFALL

## A WATERSPOUT

**Wethered, E.; *Nature*, 18:194-195, 1878.**

Among the meteoric phenomena of which we have heard recently, not the least interesting occurred on Thursday the 14th near the Kelston Round Hill, about three miles to the west of Bath. Shortly after five o'clock in the evening the inhabitants of the village of Weston, which lies between Kelston Hill and Bath, were startled by a volume of water advancing like a tidal wave along the Kelston Road; in a minute the water was upon them, flooding the houses and laying the main street four feet deep under water; with such force did it come that a stone weighing five hundred-weight was carried several yards, while smaller ones were taken a much greater distance.

It was not known in the village from where the water had come, but it so happened that about five o'clock I was proceeding to Weston Station by the Midland Railway from Bristol to Bath, and when in sight of the Round Hill I was struck by the blackness and lowness of the clouds in its vicinity. Suddenly there was a flash of lightning, and immediately after the Hill was enveloped in what appeared to be a storm of rain of unusual density.

On arriving home I was not altogether surprised to find the commotion in the village, and I at once attributed the source of the water to the cloud which I had seen; I therefore made my way in the direction of Kelston Hill.

On arriving under the brow of the Hill it was very clear that something more than an ordinary storm had occurred. Near the end of a lane (Northbrook) leading to some fields, the hedge on the right for some yards was lying in the road, but the field beyond at this point presented only the appearance of an ordinary storm, while the lane itself was like the bed of a river. To the left was a field of standing grass; for about twelve feet from the hedge the grass remained intact, then for about the same distance it was as though it had been mown down. This torrent, for such it might have been compared to, came to almost a sudden termination a little above the end of the lane, but it extended down the Hill till it was joined by two other, one of which had carried a hedge away bodily.

The increased volume of water then poured down over some gardens, uprooting trees and vegetables; in less than ten minutes the hedges were lost sight of, and the water rose to a height of eight feet. This was occasioned by a block caused by an arch, which carried off the water from a small stream, not being large enough to take the increased volume. Finally it burst over, scooping the ground out in front of some cottages several feet deep and flowed on as a river some yards wide, again destroying gardens in which were valuable stocks of vegetables.

Near the point the volume of water was again increased; in all five distinct water courses could be made out, all of which had done considerable damage to grass, cornfields, and gardens. Finally, all united in one body and poured into the village of Weston, levelling three walls as it came, and thence passed into the river Avon.

I gather from spectators at Kelston Hill that it began to be cloudy at half-past four in the afternoon; at five there was a rattling clap of thunder, followed by a downpour of rain---in "bucket-fulls," as one expressed it; but all seemed to agree that the greater portion of the water fell under the brow of the hill, where it came down in several columns. There were no houses close to the spot; had there been they must have been washed away.

The atmosphere had been perfectly still all day, but very sultry. Heavy rain fell in the neighbourhood, and the storm to which I have referred specially was accompanied with hail, which in a few minutes covered the ground some inches deep.

What I have described is no doubt what is popularly termed a waterspout. (Nature, 18:194-195, 1878)

## REMARKABLE POINT RAINFALL AT GREENFIELD, N.H., EVENING OF AUGUST 2, 1966
**Lautzenheiser, R. E., et al; *Monthly Weather Review,* 98:164-165, 1970.**

An excessive rain of amazingly small areal extent fell late on Aug. 2, 1966, at Greenfield, N. H. This note describes the storm, presents related storm statistics, reviews briefly the synoptic situation, and mentions the danger of interpreting point rainfall data as being representative of an area.

Mr. Robert H. Stanley, of Pine Ridge Road, Greenfield, in southern New Hampshire, reported a remarkable rainstorm occurring late on Aug. 2, 1966. A total of 5.75 in. was measured in a V-type plastic gage of 6-in. capacity. This type of gage is of reasonable accuracy in comparison with standard ESSA-Weather Bureau standard rain gages (Huff 1956). Mr. Stanley has observed weather for many years and is conscientious about the accuracy of his records. While 5.75 in. may not be an exact figure, it is believed to be substantially correct.

Mr. Stanley's locale is 1.5 mi northeast of Greenfield, of about 900 ft. at an elevation above sea level. It is situated on the southern slope of a gentle ridge running generally east-west and lies about 2.4 mi south-southeast of Crotched Mountain, which has peaks with elevations just over 2,000 ft.

Rain began at about 1900 EST, or about an hour before the outbreak of more generalized showers in the region. It soon became a downpour, continuing until about 2300 EST, at which time Mr. Stanley went to bed. It was then still raining, but had slackened noticeably. The rain may have stopped by midnight. A remarkable nonvariability of the intense rain was noted by Mr. Stanley. There was little slackening, even for brief intervals, during the period of heaviest fall, which was from about 1945 to 2215 EST. There was practically no wind. Neither thunder nor lightning was observed. The noise on the roof was terrific, like that of a continuous waterfall. A plastic bird feeder mounted on the side of the house was broken by the impact of sheets of water from the eaves. Looking out the window, Mr. Stanley could see stones and gravel from the roadway, south of the house, being washed away by torrents of water.

Upon rising in the morning, Mr. Stanley noted that the weather had cleared, with a brisk westerly wind. After finding the 5.75 in. of rain in the gage, he inquired from a neighbor 0.3 mi to the east. He found that the neighbor had but 0.50 in. in his gage. He thereupon examined the countryside for visible effects. The road washout extended for only a few hundred feet. Upon going one-half mile in either direction, no evidence of rain erosion of sand or gravel could be found. South of the house, beginning at the gage which was mounted on a pole, well distant from structures or trees, there stretches a 10-acre field. The knee-high grass therein was beaten down flat. By afternoon it began to revive. By the following noon it was erect. To the west of the house, a dry-wash brook running bankful at dawn was empty by 0800 EST.

Drawing a line around the traces of erosion, one obtains an oval area about a mile north-south and about three-fourths of a mile east-west. Within this area, rain varied from the order of 1 in. on the limits to almost 6 in. in the center. Outside this limit, rain is believed to have fallen off sharply to less than one-fourth of an inch, generally within a few thousand feet. (Monthly Weather Review, 98:164-168, 1970)

## WELL-DEFINED RAIN AREA
**Hinkson, G. A.; *Meteorological Magazine,* 67:159-160, 1932.**

I witnessed a rather remarkable phenomenon during the thunderstorm this afternoon (August 1st).

I was standing under the porch of the Metropolitan Railway Station in Queen's Road, Bayswater, at 2.30 p.m. The sky was suddenly overcast with dark clouds. The prevailing wind was from SW. Towards the north-east I saw small stringy clouds approaching (from the north-east) and seeming to rise in the sky---this is perhaps a normal prelude to a thunderstorm.

It began to rain---heavily at first and then softly. Suddenly torrential rain fell only one hundred yards from where I was standing. The nearest trees in Kensington Gardens were almost hidden behind a milky mist of heavy rain. The rain-drops rebounding off the street created a layer of spray as high as the tops of the wheels of the taxis standing in the street. Where I was sheltering, hardly a drop of rain was falling. The contrast was so striking that I called the attention of an unknown bystander (as though I could not believe my eyes). Then the spray on the ground came nearer like a wave, and receded. Suddenly it vanished completely, and the trees behind in Kensington Gardens stood out black against the sky. It was over. Afterwards the storm continued in a normal manner.

I have witnessed phenomena something like this in the Tropics; but never before have I seen in England really torrential rain a hundred yards away, with only a few drops falling where I was standing. It was extremely local, and, I suppose, was something in the nature of a waterspout. (Meteorological Magazine, 67:159-160, 1932)

# SPARKLING RAIN

## SPARKLING RAIN
**Anonymous; *Symons's Monthly Meteorological Magazine,* 27:171, 1892.**

Rain which on touching the ground crackles and emits electric sparks is a very uncommon but not unknown phenomenon. An instance of the kind was recently reported from Cordova, in Spain, by an electrical engineer who witnessed the occurrence. The weather had been warm and undisturbed by wind, and soon after dark the sky became overcast by clouds. At about 8 o'clock there came a flash of lightning, followed by great drops of electrical rain, each one of which, on touching the ground, walls, or trees, gave a faint crack, and emitted a spark of light. The phenomenon continued for several seconds, and apparently ceased as soon as the atmosphere was saturated with moisture. (Symons's Monthly Meteorological Magazine, 27:171, 1892)

# PRECIPITATION AND ASTRONOMY

Folklore is emphatic that the moon affects rainfall, and scientists have long suspected a solar influence on terrestrial weather. Neither effect is overwhelming, whatever the mechanisms may be, assuming they really do exist. In fact, the effects are so small that most scientists have pooh-poohed both ideas for many years. Nevertheless, an immense literature favors definite lunar and solar effects on terrestrial precipitation.

Recently, with spacecraft and rocket measurements, it has become apparent that solar activity does indeed affect the earth's atmosphere through sun-emitted streams of plasma. These atmospheric changes manifest themselves as weather modifications. This new creditability of sun-weather relationships removes them from the "strange' category and thus outside the purview of this book.

The moon's influence on weather, however, is still highly controversial. Favorable and unfavorable data exist in quantity. Two kinds of effects are noted, each with a proposed physical mechanism: (1) lunar effects on rainfall through the gravitational modification of the influx of meteoric dust that serves as a source of condensation nuclei; and (2) lunar effects on thunderstorms through the moon's interaction with the earth's magnetic tail that, in turn, modify the number of electrified particles entering the atmosphere and the planet's electrical energy balance.

## THE MOON AND RAINFALL

### COSMIC METEOROLOGY
**Broun, John Allan; *Nature,* 18:126-128, 1878.**

The popular beliefs in the moon's influence on the weather are first disposed of; they are conclusions from unrecorded observations where the coincidences are remembered and the oppositions are forgotten; and they are opposed to strict deductions when all the facts are employed.

Agreeing, as all men of science do, with this decision, the question remains, Whether the moon may not have some slight effect in producing meteorological variations? She reflects, absorbs, and radiates the solar heat; may this heat, in accordance with the thesis, not produce some effect on our atmosphere?

Sir John Herschel had observed the tendency to disappearance of clouds under the full moon; this he considered a fact which might be explained by the absorption of the radiated lunar heat in the upper strata of our atmosphere. He cited Humboldt's statement as to the fact being well known to pilots and seamen of Spanish America. I may add the testimony of Barnardin de St. Pierre, who, in his "Voyage a l' Ile de Reunion," says: "I remarked constantly that the rising of the moon dissipated the clouds in a marked way. Two hours after rising, the sky is perfectly clear" ("Avril,

1768"). Herschel also cited in favour of his "meteorological fact," a result supported by the authority of Arago, that rather more rain falls near new than near full moon.

Arago's conclusion that the phenomenon was "incontestable of a connection existing between the number of rainy days and the phases of the moon" was founded on the observations of Schubler, of Bouvard and of Eisenlohr, three series which, on the whole, confirmed each other. Schubler also, as Arago showed, had found that the quantity of rain which fell was greater near new than near full moon. These results, accepted by Arago, have not been noticed by M. Faye when he cites Herschel only, as one of those "men of science who interest themselves in popular prejudices, take bravely their defence in hand and exert themselves to furnish not facts but arguments in their favour." It seems, indeed, to have been forgotten that Herschel's argument was given to explain what he considered a meteorological fact.

M. Faye founds his argument wholly on the conclusions of M. Schiaparelli from a weather register kept at Vigevano by Dr. Serafini during thirty-eight years (1827-1864). The Italian physician entered the weather as clear, cloudy or mixed (misti), or rainy from morning to evening. M. Schiaparelli finds from this register that the sky was clearest in the first quarter of the moon. It has not been remarked that if the moon's heat has any effect in dissipating clouds, as Herschel and others believed, this must be seen best when the moon is near full, that is to say, during the night hours, for which Dr. Serafini's register has nothing to say. In confirmation of the conclusion that the moon does not dissipate the clouds, another result from the Vigevano weather register is cited, namely, that the greatest number of rainy days happens near full moon. This result is opposed to that derived from the observations of Schubler, Bouvard, and Eisenlohr.

The value to be given to observations of the number of rainy days must evidently depend on whether the observations include the rainy nights; and an investigation on this question, to have any considerable weight, should depend rather on the measured rainfall than on the term "rainy day," for which no distinct definition is given.

The great objection to M. Faye's conclusions, as far as the facts go, is to be found in their entire dependence on the Vigevano weather register (da mane a sera). No notice is taken of other observations and results showing a lunar action on our atmosphere, such as those already mentioned, which Arago considered incontestable, those of Madler and Kreil, and the more recent investigations of Mr. Park Harrison and Prof. Balfour Stewart. All of these, and many others, must be carefully considered before we can accept the conclusion that the moon has no influence on our atmosphere. The subject is, however, too large to be entered into here at present, and it will be possible to study it better after other conclusions of the learned French Academician have been examined.

There is, however, a part of the argument, whatever the results obtained may say, which merits particular consideration; and that is, that the moon's heat cannot produce the phenomena in question. M. Faye shows that if the moon's reflected heat is in the same proportion as the reflected light, such heat cannot produce a change of temperature of a thousandth of a degree Fahrenheit. I would remark that Lord Rosse's carefully-made experiments with the most delicate apparatus have shown for the total heat radiated and reflected, nearly ten times the proportion given by the reflected light; but, as M. Faye observes, if the proportion were increased a hundredfold the effect would still be insensible. "How then," it is added, "can we expect such an action to dissipate the clouds when that of the sun does not always succeed?"

If, however, we can establish that real lunar actions exist which cannot be explained by the moon's heat reflected or radiated, the only philosophical conclusion will be that the moon must act in some other way, which it will be in the interests of science to seek out. (Nature, 18:126-127, 1878)

## LUNAR COMPONENT IN PRECIPITATION DATA

**Adderley, E. E., and Bowen, E. G.; *Science,* 137:749-750, 1962. (Copyright 1962 by the American Association for the Advancement of Science)**

The influence of the moon in producing tides in the upper atmosphere and the appearance of a lunar component in daily temperature in certain parts of the world are comparatively well known, but the effects are extremely small and difficult to detect. The possibility of a large effect on rainfall first came to our notice in 1960 with the chance reading of a paper by Rodes. Rodes showed what appeared to be a connection between rainfall in the Spanish peninsula and both the declination and the position of apogee and perigee of lunar motion. An investigation was therefore made of the rainfall data in our possession, and it was apparent that it contained a strong lunar component. However, it was also clear that this was connected with the phase of the moon rather than with the parameters used by Rodes.

At this point a decision was taken not to publish the data immediately, but to reserve it for a later date. The reason for doing so was that our work on singularities in rainfall was still being treated with disbelief in meteorological circles, and to suggest a lunar effect on rainfall would simply not have met with the right response.

We were not surprised therefore when we received a communication from Bradley, Woodbury, and Brier indicating a pronounced lunar effect in U.S. rainfall. The effect is matched by similar effects in the Southern Hemisphere. A curve showing the heaviest falls of the month for 50 stations in New Zealand, plotted in the same way as the U.S. data are plotted, is given in Fig. 1. It shows variations of a magnitude comparable with those in the U.S. and closely related in phase.

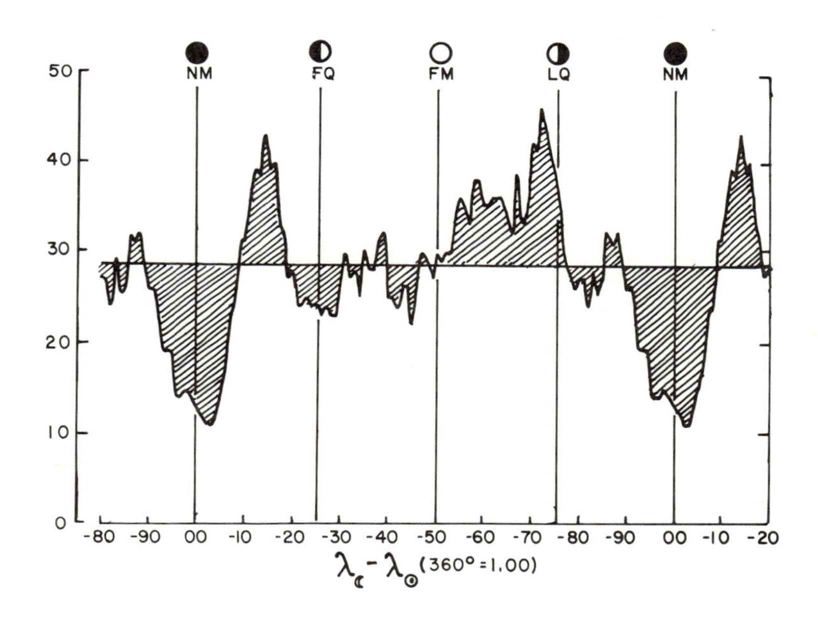


Effect of the moon's position on rainfall

It should not be assumed from this that a lunar component will be found in all rainfall records. It is already apparent that there are distinct variations in harmonic content of the curves with geography and, in addition, some shifts of phase. But the effect is sufficiently widespread in the land masses of the Southern Hemisphere to warrant further investigation.

It is not yet possible to advance a physical explanation of the phenomenon, but it is clearly not incompatible with the meteor hypothesis. Meteoritic dust reaching the earth is known to be distributed in orbits, the majority of which are in the plane of the ecliptic. The moon's orbit is also close to the ecliptic and, as the moon revolves around the earth, it could impose a lunar modulation on the amount of dust reaching the earth. However, a calculation of the magnitude of such an effect shows that it is unlikely that gravitational forces alone could produce a variation in rainfall as large as that shown by the accompanying curves. It may therefore be necessary to look for some other explanation for the phenomenon. (Science, 137:749-750, 1962.

## FORMATION OF DEPRESSIONS IN THE INDIAN SEAS AND LUNAR PHASE

**Visvanathan, T. R.; *Nature,* 210:406-407, 1966.**

It was recently shown that the occurrence of heavy rainfall (10 in. and above in 24h) is related to lunar phase. It will be evident that such heavy falls are mostly associated with depressions, some of which intensify into cyclones and severe cyclones. A study was therefore undertaken to examine the dates of formation of depressions in the Indian Seas vis-a-vis the lunar phase. This communication was prompted by Bradley's article "Tidal Components in Hurricane Development" and is intended to provide some additional information to support his conclusions.

The data for this investigation were taken from a publication issued by the Indian Meteorological Department which catalogues depressions formed in the Indian Seas, and gives the date on which the depression stage was reached and its corresponding position.

The phase of the Moon for each day is given in terms of synodic decimals (hundredths of a synodic month) in the ephemeris by Carpenter. The synodic decimal corresponding to each date was taken from that source and the frequency of formation of depressions corresponding to each synodic decimal was found. Successive ten-unit moving totals of the frequencies were then made, and of the hundred such totals thirty were chosen to correspond to the thirty days of the lunar month. For purposes of comparison the frequencies were expressed as percentages of the mean.

Fig. 1 shows the curve, prepared in this manner, of depressions which formed in the latitude band $19^{o}$N-$22^{o}$N in the Bay of Bengal and Arabian Sea during the period 1877-1960. Fig. 2 shows the frequencies of occurrence of heavy rainfall (10 in. and more in 24 h) along the latitude band $20^{o}$ N-$22^{o}$ N during the period 1891-1950, and is reproduced from my recent article. A similarity is noticed in the trends of the curves in Figs. 1 and 2; their range is also comparable. The correlation between the two curves is 0.50. The correlation between the curves with 1 day lag (heavy rainfall following depressions) is 0.57 with $\phi = 0.07$. There is a strong indication that the heavy falls were associated with depressions. The tidal influence on the formation of depressions is also indicated, as in both the curves the 14.765 day wave accounts for 48.1 of the total variance and is the largest component.

Any extra-terrestrial effect on meteorological phenomena should be established on a global scale. This calls for an analysis of data pertaining to various parts of the globe. This note, it is hoped, will stimulate interest in further studies of this kind. (Nature, 210:406-407, 1966)

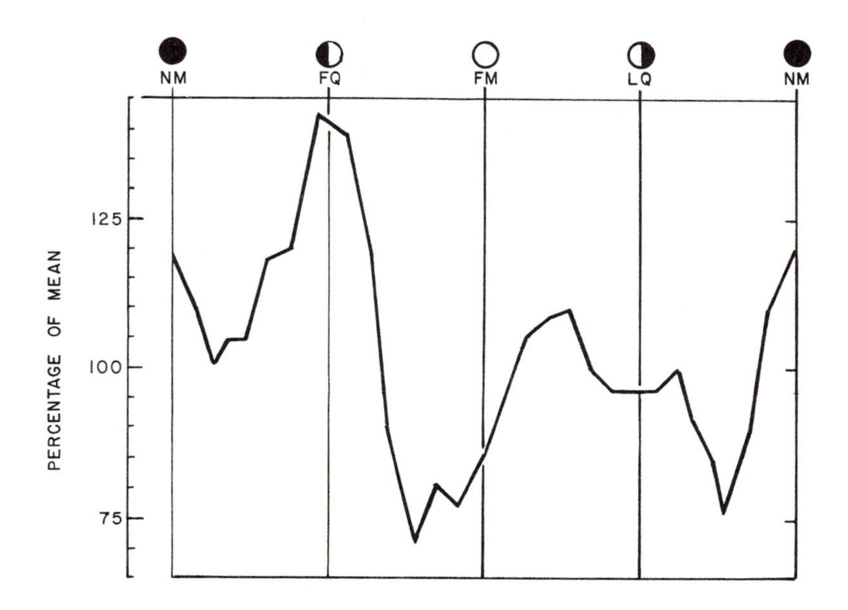


Atmospheric depressions versus lunar phase

# LUNAR EFFECT ON THUNDERSTORMS

## THE MOON AND THUNDER STORMS
**Seagrave, F. E.; *Popular Astronomy*, 10:332, 1902.**

I read two or three weeks ago the article in the April 1902 Popular Astronomy on "The Moon and Thunder Storms," page 252. Was much interested in same and also the one referred to in the February 20 copy of Nature. Since that time I have been looking over our thunderstorm records here and comparing them with the phases of the Moon. I have examined the records for the past six years taking in the years 1896, 1897, 1898, 1899, 1900 and 1901. In the records here all days were considered thunderstorm days when lightning was seen or thunder heard during the 24 hours. During these past six years there were 159 thunder storm days here, or 26-1/2 days per year on the average. They were divided up as follows in relation to the phases of the Moon:

49 days when the Moon was between New and First Quarter
29 " " " " " " First Quarter and Full
43 " " " " " " Full and Last Quarter
38 " " " " " " Last Quarter and New
159 Total number of thunderstorm days in 6 years at Providence, R.I.

If the Moon has any influence at all in producing thunderstorms that influence is certainly of a different nature here than at Sidmouth or Greenwich. The records here seem to show that thunder was heard most frequently when the Moon was between New and First Quarter, and least often from First Quarter to Full Moon. I really do not believe that the phases of the Moon or its position in its orbit has anything to do with it. (Popular Astronomy, 10:332, 1902)

## THE MOON'S PHASES AND THUNDERSTORMS
**Anonymous; *Nature,* 68:232, 1903.**

Evidence of a connection between the occurrence of thunderstorms and the moon's age has been referred to in Nature on several occasions. Prof. W. H. Pickering gives a table in Popular Astronomy to show the results of investigations of this relationship by various observers. From this table, which is abridged below, it will be seen that, with one exception, the number of thunderstorms occurring near the first two phases of the moon is greater than the number occurring near the last two.

The Moon's Phases and Thunderstorms.

| Station | Authority | Years | New and First Quarter | Full and Last Quarter |
|---|---|---|---|---|
| Kremsmunster | Wagner | 86 | 54 | 46 |
| Aix la Chapelle | Pohs | 60 | 54 | 46 |
| Batavia, Java | Van d. Stok | 9 | 52 | 48 |
| Gotha | Lendicke | 9 | 73 | 27 |
| Germany | Koppen | 5 | 56 | 44 |
| Glatz County | Richter | 8 | 62 | 38 |
| N. America | Hazen | 1 | 57 | 43 |
| Prague | Gruss | 20 | 51 | 49 |
| " | " | 20 | 53 | 47 |
| Gottingen | Meyer | 24 | 54 | 46 |
| Greenwich | MacDowall | 13 | 54 | 46 |
| Madrid | Ventatasta | 20 | 52 | 48 |
| Providence, R.I. | Seagrave | 6 | 49 | 51 |

Prof. Pickering adds:---"The number of observations here collected seems to be large enough to enable us to draw definite conclusions, without fear that further records will revise or neutralise them. From these observations we conclude that there really is a greater number of thunderstorms during the first half of the lunar month than during the last half, also that the liability to storms is greatest between new moon and the first quarter, and least between full moon and last quarter. Also we may add that while theoretically very interesting, the difference is not large enough to be of any practical consequence. Thus it would seem that, besides the tides and certain magnetic disturbances, there is a third influence that we must in future attribute to the moon." (Nature, 68:232. 1903)

## RELATIONSHIP BETWEEN THUNDERSTORM FREQUENCY AND LUNAR PHASE AND DECLINATION

**Lethbridge, Mae DeVoe; *Journal of Geophysical Research,* 75:5149, 1970.**

Abstract. Using the superposed-epoch method, we statistically analyzed thunderstorm frequencies for 108 stations in eastern and central United States in relation to lunar positions for the years 1930-1033 and 1942-1965. The results are as follows. With full moon as key day, a peak in thunderstorm frequency occurs for 1953-1963 two days after full moon, the mean being 2.7$\sigma$ above average level. This mean frequency is 4.8$\sigma$ above the mean when computed only for the full moons as key days that have a declination of 17$^{\circ}$ north or more. We suggest that these increases near full moon may be related to the earth's magnetic tail and the neutral sheet. (Journal of Geophysical Research, 75:5149, 1970)

# FAMOUS DARK DAYS

A dark day is not merely the veiling of the sun by ordinary clouds---they at least let through enough light to conduct the business of the world. On a dark day the pall of darkest night descends suddenly, chickens go to roost, and men and women pray and grope for candles. New England's famous Dark Day of 1780 is typical of the genre.

The standard explanation for a dark day is smoke from forest fires in the hinterlands. This is reasonable because dark days are frequently accompanied by black rains that are thick with ashes and soot. Wind drives these heavy clouds across the affected region obscuring the sun more completely than ordinary rain clouds. Sometimes, though, the darkness seems more irrational, descending suddenly without the approach of ominous black clouds. And where did the supposed forest fires rage? In America, they were out west somewhere, set by Indians. In pioneer days, of course, "out west" was devoid of newspaper stringers and the forests of an entire state could be consumed without easterners knowing of it. While admitting forest fires to be the cause of most dark days, we should also ask whether some may not owe their origins to stratospheric or even extraterrestrial clouds of obscuring matter.

## DARK DAYS AND FOREST FIRES
**Talman, C. F.; *Scientific American,* 112:229, 1915.**

Instances of daytime darkness are recorded in the old chronicles along with such other "prodigies" as multiple suns, showers of blood, and warring armies in the sky---all of which can easily be identified to-day with well-known meteorological phenomena (parhelia, rain reddened with desert dust, and the aurora). The two famous cases mentioned in the Bible---the plague of darkness in Egypt and the darkness attending the crucifixion---illustrate the fact that such occurrences were once universally assumed to be miraculous.

Some of the early cases of daytime darkness mentioned in history are doubtless attributable to solar eclipses, and must, accordingly, have been restricted to a small part of the earth's surface, and have been of but a few minutes' duration. The majority of the famous "dark days" were, however, the result of an abnormal accumulation of smoke or dust in the air, sometimes arising from burning forests, moors, or prairies, sometimes from volcanic eruptions, and in many instances covering vast areas of the globe.

In a recent publication on "Forest Fires" (Forest Service Bulletin 117), Mr. F. G. Plummer gives the following list of dark days in the United States and Canada:

1706---May 12th, 10 a.m., New England.
1716---October 21st, 11 a.m. to 11:30 a.m., New England.
1732---August 9th, New England.
1762---October 19th, Detroit.
1780---May 19th, New England. (Black Friday. The Dark Day.)
1785---October 16th, Canada.

1814---July 3rd, New England to Newfoundland.
1819---November 6th to 10th, New England and Canada.
1836---July 8th, New England.
1863---October 16th, Canada. ("Brief duration.")
1868---September 15th to October 20th, Western Oregon and Washington.
1881---September 6th, New England. (The Yellow Day.)
1887---November 19th, Ohio River Valley. ("Smoky Day.")
1894---September 2nd, New England.
1902---September 12th, Western Washington.
1903---June 5th, Saratoga, N. Y.
1904---December 2nd, 10 a.m., for 15 minutes, Memphis, Tenn.
1910---August 20th to 25th, Northern United States, from Idaho and Northern Utah eastward to St. Lawrence River.

Forest fires are the common cause of dark days in this country. The fact that such days are most frequent in the Northeastern United States and Eastern Canada is evidently related to the fact that practically all barometric depressions ("lows"), with their attendant whirl and indraft of the surface air, pass down the St. Lawrence Valley on their way to the ocean, and usually become intensified and sharply defined in this region. The smoke from a conflagration anywhere on the periphery of a "low" is drawn into the vortex along more or less converging lines, and at the same time rises to a considerable altitude. Eddies in the circulation of the "low" will result in a dense accumulation of the smoke in places, and this may occur above the level of the lower clouds, which thus mask the cause of the phenomenon. Hence the startling effect of darkness in the daytime, often with little or no turbidity of the air near the earth's surface. Mere smokiness of the air near the ground or a fog heavily charged with smoke (as in the case of the London fogs), however great the obscurity produced, would hardly be placed in the same class with the awe-inspiring dark days of the chroniclers. If, however, showers occur during one of these occurrences, a large amount of soot is likely to be brought down, and thus we have another "prodigy"; viz., "black rain." A very recent case of this sort is reported in the Quarterly Journal of the Royal Meteorological Society for October, 1912; during a thunderstorm in Eastern Hampshire darkness almost like that of night occurred in the early afternoon, and inky rain fell. The phenomenon was due to soot carried from London, fifty miles away.

When the pall of smoke is rather thin a certain amount of sunlight struggles through, and owing to the same process that gives us the golden glow of sunset a yellow or coppery tinge is cast over the landscape. This effect has been noted in connection with several dark days, including the most famous of all, that of May 19th, 1780. It was the principal feature of the dark day of September 6th, 1881, in New England, which is accordingly known as "the yellow day."

The great Idaho fire of August, 1910, was responsible for dark days over an area larger than in any other case on record in this country. The accompanying chart, from the Forest Service Bulletin above mentioned, shows the area in which artificial light was used in the daytime, but smoke was observed far beyond these limits. The British ship "Dunfermline" reported that on the Pacific Ocean, 500 miles west of San Francisco, the smell of smoke was noticed, and haze prevented observations for about ten days. (Scientific American, 112:229, 1915)

## ATMOSPHERIC PHENOMENON
**Anonymous; *Monthly Weather Review,* 14:79, 1886.**

The following is taken from the LaCrosse, Wisconsin, "Daily Republican," of March 20, 1886:

Oshkosh, Wisconsin, March 19th.---A most remarkable atmospheric phenomenon occurred here at 3 p. m. The day was light, though cloudy, when suddenly darkness commenced settling down, and in five minutes it was as dark as midnight. General consternation prevailed; people on the streets rushed to and fro; teams dashed along, and women and children ran into cellars; all business operations ceased until lights could be lighted. Not a breath of air was stirring on the surface of the earth. The darkness lasted from eight to ten minutes, when it passed off, seemingly from west to east, and brightness followed. News from cities to the west say the same phenomenon was observed there in advance of its appearance here, showing that the wave of darkness passed from west to east. Nothing could be seen to indicate any air currents overhead. It seemed to be a wave of total darkness passing along without wind. (Monthly Weather Review, 14:79, 1886)

## THE DARK DAY IN CANADA
**Anonymous; *Scientific American*, 44:329, 1881.**

What was the strangest occurrence of that time, or rather the strangest thing that ever happened in the history of this country, was what has been always known as the "Phenomenon of 1819. On the morning of Sunday, November 8, 1819, the sun rose upon a cloudy sky, which assumed, as the light grew upon it, a strange greenish tint, varying in places to an inky blackness. After a short time the whole sky became terribly dark, dense black clouds filling the atmosphere, and there followed a heavy shower of rain, which appeared to be something of the nature of soapsuds, and was found to have deposited after settling a substance in all its qualities resembling soot. Late in the afternoon the sky cleared to its natural aspect, and the next day was fine and frosty. On the morning of Tuesday, the 10th, heavy clouds again covered the sky, and changed rapidly from a deep green to a pitchy black, and the sun, when occasionally seen through them, was sometimes of a dark brown or an unearthly yellow color, and again bright orange, and even blood red. The clouds constantly deepened in color and density, and later on a heavy vapor seemed to descend to the earth, and the day became almost as dark as night, the gloom increasing and diminishing most fitfully. At noon lights had to be burned in the courthouse, the banks, and public offices of the city. Everybody was more or less alarmed, and many were the conjectures as to the cause of the remarkable occurrence. The more sensible thought that immense woods or prairies were on fire somewhere to the west; others said that a great volcano must have broken out in the Province; still others asserted that our mountain was an extinct crater about to resume operations and to make of the city a second Pompeii; the superstitious quoted an old Indian prophecy that one day the Island of Montreal was to be destroyed by an earthquake, and some even cried that the world was about to come to an end.

About the middle of the afternoon a great body of clouds seemed to rush suddenly over the city, and the darkness became that of night. A pause and hush for a moment or two succeeded, and then one of the most glaring flashes of lightning ever beheld flamed over the country, accompanied by a clap of thunder which seemed to shake the city to its foundations. Another pause followed, and then came a light shower of rain of the same soapy and sooty nature as that of two days before. After that it appeared to grow brighter, but an hour later it was as dark as ever. Another rush of clouds came, and another vivid flash of lightning, which was seen to strike the spire of the old French parish church and to play curiously about the large iron cross at its summit before descending to the ground. A moment later came the climax of the day. Every bell in the city suddenly rang out the alarm of fire, and the affrighted citizens rushed out from their houses into the streets and made their way in the gloom toward the church, until Place d'Armes was crowded with people, their nerves all unstrung by the awful events of the day, gazing at, but scarcely daring to approach the strange sight before them. The sky above and around was as black as ink, but right in one spot in mid-air above them was the summit of the spire, with the lightning playing about it shining like a sun. Directly the great iron cross, together with the ball at its foot, fell to the ground with a crash, and was shivered to pieces. But the darkest hour comes just before the dawn. The glow above gradually subsided and died out, the people grew less fearful and returned to their homes, the real night came on, and when next morning dawned everything was bright and clear, and the world was as natural as before. The phenomenon was noticed in a greater or less degree from Quebec to Kingston, and far into the States, but Montreal seemed its center. It has never yet been explained. (Scientific American, 44:329, 1881)

## DARK DAY IN NEW ENGLAND
**Anonymous; *Nature,* 24:540, 1881.**

A remarkable phenomenon occurred in New England on September 6, almost exactly similar to one that occurred in the same region on May 19, 1780. The Springfield Daily Republican describes it as follows:---In this city the day began with a slow gathering of fog from all the watercourses in the early hours, the thin clouds that covered the sky at midnight seemed to crowd together and descend upon the earth, and by sunrise the atmosphere was dense with rapour, which limited vision to very short distances, and made those distances illusory; and as the sun rose invisibly behind, the vapours became a thick, brassy canopy, through which a strange yellow light pervaded the air and produced the most peculiar effects on the surface of the earth. This colour and darkness lasted until about three o'clock in the afternoon, once in a while lightening, and then again deepening, so that during a large part of the time nothing could be done conveniently indoors without artificial light. The unusual complexion of the air wearied and pained the eyes. The grass assumed a singular bluish brightness, as if every blade were tipped with light. Yellow blossoms turned pale and gray; a row of sunflowers looked ghastly; orange nasturtiums lightened; pink roses flamed; lilac-hued phlox grew pink; and blue flowers were transformed into red. Luxuriant morning-glories that have been blossoming in deep blue during the season now were dressed in splendid magenta; rich blue clematis donned an equally rich maroon; fringed gentians were crimson in the fields. There was a singular luminousness on every fence and roof-ridge, and the trees seemed to be ready to fly into fire. The light was mysteriously devoid of refraction. One sitting with his back to a window could not read the newspaper if his shadow fell upon it---he was obliged to turn the paper aside to the light. Gas was lighted all over the city, and it burned with a sparkling pallor, like the electric light. The electric lights themselves burned blue, and were perfectly useless, giving a more unearthly look to everything around. The darkness was not at all like that of night, nor were animals affected by it to any remarkable extent. The birds kept still, it is true, the pigeons roosting on ridge-poles instead of flying about, but generally the chickens were abroad. A singular uncertainty of distance prevailed, and commonly the distances seemed shorter than in reality. When in the afternoon the sun began to be visible through the strange mists, it was like a pink ball amidst yellow cushions---just the colour of one of those mysterious balls of rouge which we see at the drug-stores, and which no woman ever buys. It was not till between five and six o'clock that the sun had sufficiently dissipated the mists to resume its usual clear gold, and the earth returned to its everyday aspect; the grass resigning its unnatural brilliancy and the purple daisies no longer fainting into pink. The temperature throughout the day was very close and oppressive, and the physical effect was one of heaviness and depression. What was observed here was the experience of all New England, so far as heard from, of Albany and New York city, and also in Central and Northern New York. In reference to this phenomenon the New York Nation suggests that it may be worth the while of weather-observers to note the approximate coincidence between the interval separating the two dark days in New England (May 19, 1780, and September 6, 1881) and nine times the sun-spot cycle of eleven years. (Nature, 24:540, 1881)

## SUN OBSCURATION IN BRAZIL
**Anonymous; *Scientific American,* 3:122, 1860.**

On the 18th of April last, the sun became obscured about noon, in Brazil, although no clouds were visible in the sky. The darkness continued several minutes, and Venus became quite visible to the naked eye. Historians relate that, in 1547 and 1706, like phenomena were witnessed. The cause has been attributed to the passage of clusters of asteroids across the sun's disk. (Scientific American, 3:122, 1860)

# STRANGE FOGS

The folklore of many countries contains accounts of fogs that shriveled up vegetation and left a trail of death---"blasting fogs" they were often called. The scientific literature has little to say about these lethal fogs, but a few curious items have been collected.

## THE FATAL BELGIAN FOG
**Alexander, Jerome; *Science,* 73:96, 1931.**

About the week-end of December 7, an extremely heavy fog prevailed in Belgium and England, and the daily press reported that in the neighborhood of Liege more than forty persons and a considerable number of cattle died, exhibiting symptoms of asphyxiation. Autopsies performed on twelve cows indicated that they had died from pulmonary edema. Although final judgment on this phenomenon must await the results of the investigation which the Belgian government has undertaken, there are certain aspects of the situation which might be here referred to.

According to the New York Times, of November 30, 1930, a terrific sandstorm and hurricane blew over French Morocco on November 27. The following night, yellow sand was heavily deposited on the streets and on the foliage at Barcelona, Spain. On the morning of November 28, a "mud rain" fell in Paris and all over northern France as far west as Granville, on the southern Brittany coast and along the English Channel.

It seems probable that the more finely dispersed material carried by this storm reached considerable heights and settled down slowly over Belgium and England. These colloidal or semicolloidal dust particles served to reinforce the normally high atmospheric nucleation of the winter season; for, as Carl Barus has shown (Smithsonian publication 1309, published 1905) the products of combustion (burning coal, etc.) furnished highly efficient nuclei, and this atmospheric nucleation is especially marked in industrial neighborhoods and in the winter season. The tiny nuclei serve as centers about which, under suitable atmospheric conditions, moisture will deposit to form fog droplets. The extremely fine dust will continue to settle down for days and, given a sufficiently still and moisture-depositing atmosphere, prolonged and dense fogs would be expected. (Science, 73:96, 1931)

## THE POGONIP
**Anonymous; *American Meteorological Journal*, 4:105, 1887.**

A curious phenomenon is often witnessed in the mountainous districts of Nevada. Mountaineers call it "pogonip," and describe it as being a sort of frozen fog that appears sometimes in winter, even on the clearest and brightest of days. In an instant the air is filled with floating needles of ice. To breathe the pogonip is death to the lungs. When it comes, people rush to cover. The Indians dread it as much as the whites. It appears to be caused by the sudden freezing in the air of the moisture which collects about the summits of the high peaks. (American Meteorological Journal, 4:105, 1887)

## THE POGONIP FOG
**Anonymous; *Scientific American,* 66:240, 1892.**

The city of Carson, Nev., experienced the other evening the thickest and coldest pogonip fog "in the memory of the oldest inhabitant," says a writer in a recent issue of the Evening Post. The pogonip fog is peculiar to elevated altitudes in the Nevada Sierras. It ascends from the valleys, and its chill embrace is so much feared by the Indians, who are predisposed to affections of the lungs, that they change their camp if apprised by the atmospheric conditions that the dreaded fog is approaching. Mr. Ogden, a chemist of the Nevada Mining Bureau, furnishes this pleasing description of the pogonip:

"In the White Pine Mountains, the Toyabi, the Hyko, and the Pahranagat ranges it is quite common to see the trees, houses, and everything out in the open gradually become white without any apparent cause. There is no perceptible fog, but the hot air from the valleys gradually ascends up the mountain side, and, becoming crystallized, the minute crystals attach themselves to anything in sight. This phenomenon affects human beings in just the same manner, and when the fog passes by, the frozen particles will adhere to the hair and clothing, producing a very grotesque effect. Hot Creek Valley is situated right in the center of the mining district, and is so called because of the warm springs that are always to be found there. These springs cause a pogonip in that district every night, and for this reason: The wind in the valley always blows from one direction in the daytime, and after sunset it invariably blows from the opposite point. The effect of the cooler air passing over the hot valley is to force the heated air to rise. When it reaches a temperature of about $25^{o}$, the result is a pogonip." (Scientific American, 66:240, 1892)

## STRANGE AND EXTRAORDINARY FOG
**Anonymous; *Annual Register,* 1:90-91, 1758.**

Extract of a letter from Kensington, in Connecticut.

"On the third instant, about sun-rise, at this place was a fog of so strange and extraordinary appearance, that it filled us all with amazement. It came in great bodies, like thick clouds, down to the earth, and in its way, striking against the houses, would break and fall down the sides in great bodies, rolling over and over. It resembled the thick steam rising from boiling wort, and was attended with such heat that we could hardly breathe. When first I saw it I really thought my house had been on fire, and ran out to see if it was so; but many people thought the world was on fire, and the last day come. One of our neighbours was then at Sutton, 100 miles to the eastward, and reports it was much the same there. (Annual Register, 1:90-91, 1758)

## REMARKABLE "DUST FOG"
**Anonymous; *Science,* 23:193, 1906.**

A remarkable 'dust fog,' observed in the Malay Archipelago in October, 1902, of such density as to interfere with navigation, has been investigated by the Batavia Observatory staff, large numbers of circular letters of inquiry having been sent to ship captains whose business took them to those seas at the time in question. The causes of this remarkable dust fog have been sought in the deficient rainfall of the year 1902; in extended forest fires, especially in Borneo and southern Sumatra, and in the transportation of dust by the southeast trade from Australia. As the progress of the dust from Australia could be followed, by successive stages, northward, the latter cause was doubtless the most important one. (Science, 23:193, 1906)

# UNUSUAL CLOUDS

Clouds can be classified as unusual if they have peculiar shapes, motion, orientation, or emit unexpected light or noise. Meteorologists recognize a wide variety of clouds as "usual" even though they are not fully understood. Lenticular and mammatus clouds look very strange but are considered recognized species and thus omitted here. Rotating ring-shaped clouds, on the other hand, are included for, although they probably originate through the action of atmospheric vortices, they are neither common nor simple to explain. Cloud arches connecting seemingly independent clouds also pose theoretical problems. Waterspouts have occasionally been reported as joining two clouds without touching the ground. Are cloud arches of a similar umbilical nature? We also have clouds, particularly polar-oriented cirrus bands, that seem to interact with the earth's magnetic field. The mechanism here is obscure unless the water droplet nuclei can be affected magnetically. All in all, there are many imponderables in cloud formation and dynamics.

## CURIOUSLY SHAPED CLOUDS

### AN UNUSUAL SKY EFFECT
**Anonymous; *Meteorological Magazine*, 58:13-14, 1923.**

On the evening of December 5th, 1922, Mr. T. H. Applegate observed a curious phenomenon from the meteorological station at Biggin Hill Aerodrome. At 20h. 30m. a squally west wind was blowing at a speed varying between 20 and 30 miles per hour. Light cloud of the strato-cumulus type, with frequent rifts, was moving from the west so that the sky was mainly covered with detached cloud but varied from clear to overcast. The cloud height was estimated to range from 2,500 ft. to 5,000 ft.

Near the west, at a point about 15° above the horizon, a white elliptical ring of light was clearly visible. The ring was complete except at its lowest extremity and appeared to move in the same direction as the cloud but with a somewhat less speed. The interior of the ring seemed to be clear in spite of the passage of clouds which were seen to enter and leave it. Only a few stars, however, were actually visible within the ring and these only at occasional intervals.

As the ring approached it increased in size and became a true circle with a diameter of about 60° just before it was centered overhead. The lowest segment remained broken throughout. After passing the zenith the ring broke up and disappeared. For the whole period of its visibility the plane of the ring seemed to be inclined at an angle of 25° to the vertical, the upper segment being the nearer.

The moon was at an altitude of 27° and situated nearly due east. It was covered with thin cloud throughout the duration of the ring.

The phenomenon resembled a great smoke-ring. It has been suggested that it was the result of "sky-writing," but the time and other circumstances do not seem to render this probable. Information would be welcome from anyone who saw anything unusual on December 5th in the region of south-east England, or from those who have seen similar phenomena, unconnected with sky-writing, on previous occasions. (Meteorological Magazine, 58:13-14, 1923)

## EXTRACT FROM THE LOG OF BARQUE "LADY OF THE LAKE'

**Banner, Frederick W.; *Royal Meteorological Society, Quarterly Journal,* 1:157, 1873.**

22nd March, 1870. Lat. N. 5° 47', long. W. 27° 52'. At from 6.30 to 7 p.m. a curious-shaped cloud appeared in the S.S.E. quarter, first appearing distinct at about 25° from the horizon, from where it moved steadily forward, or rather upward, to about 80°, when it settled down bodily to the N.E. Its form was circular, with a semicircle to the northern face near its centre, and with four rays or arms extending from centre to edge of circle. From the centre to about 6° beyond the circle was a fifth ray broader and more distinct than the others, with a curved end: ---diameter of circle 11°, and of semicircle 2-1/2°. The weather was fine, and the atmosphere remarkably clear, with the usual Trade sky. It was of a light grey

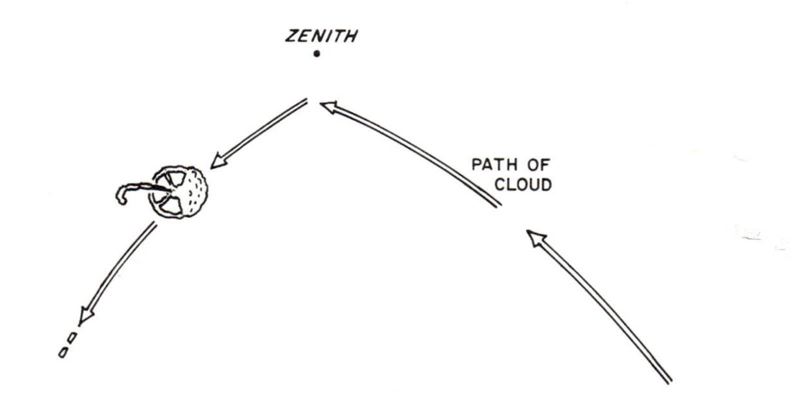


Curious cloud with curved tail

colour, and though distinctly defined in shape, the patches of cirro-cumulus at the back could be clearly seen through. It was very much lower than the other clouds; the shape was plainest seen when about 55° to 60° high. The wind at the time was N.N.E., so that it came up obliquely against the wind, and finally settled down right in the wind's eye; finally lost sight of it through darkness, about 30° from the horizon at about 7.20 p.m. Its tail was very similar to that of a comet. The men forward saw it nearly 10 minutes before I did, and came aft to tell me of it. This may give a rough idea of its shape and track; its general appearance was similar to that of a halo round the sun or moon. (Royal Meteorological Society, Quarterly Journal, 1:157, 1873)

## UNIDENTIFIED PHENOMENON
**Dennis, T. S.; *Marine Observer,* 33:190, 1963.**

s.s. City of Liverpool. Captain T. S. Dennis. Aden to Port Said. Observers, Mr. F. Keith, Extra 3rd Officer and many of the ship's company.

1st November 1962. At 2005 SMT a ball of what seemed to be dense white cloud was seen on a bearing of 260° at an altitude of about 7°. As it approached, and passed ahead of the vessel, moving in a NE direction, it assumed the form of a smoke ring, the apparent diameter of which, when bearing 330°, was about 5 or 6 times that of the full moon. The ring, which became elliptical in shape, was thought to be rotating in an anticlockwise direction. By 2015 it had completely disappeared, having become increasingly indistinct as it receded from the ship. The sky was cloudless and visibility was very good. The moon's age 4 days, was setting on a bearing of about 252°. Air temp. 86.2°F, wet bulb 77.3°F, sea 88.1°. Wind, light NW'ly airs. Position of ship: 19° 37'N, 39° 10'E. (Marine Observer, 33:190, 1963.

## BALLOON-LIKE CLOUD
**Anonymous; *Nature,* 58:353, 1898.**

In connection with the reports which have appeared from time to time that Andree's and other balloons have been sighted in the distance, it is worth while to direct attention to an observation recorded by Mr. F. F. Payne in the Canadian Monthly Weather Review. Looking at the sky one afternoon, Mr. Payne saw a large, grey, pear-shaped object sailing rapidly across, immediately behind a thin stratum of cirro-stratus cloud. At first the object was taken for a balloon, its outline being sharply defined, and its shape and size exactly corresponding to one; but as no cage was seen, it was concluded that it must be a mass of cloud, and after watching it for about six minutes, its mass became less dense and finally it disappeared. Whilst no whirling motion could be noticed, this balloon-like mass was undoubtedly of cyclonic formation, appearing less elongated when viewed at a distance probably of a mile and only about 30° from the zenith. The observation suggests an origin for strange war balloons and other aerial machines occasionally reported as having been sighted. (Nature, 58:353, 1898)

## HOLE-IN-CLOUD
**Kinney, John R.; *American Meteorological Society, Bulletin,* 49:990-991, 1968.**

On 23 February 1968, a very unusual cloud phenomenon occurred over Vandenberg AFB, California. It consisted of a near circular hole in a rather homogeneous cloud layer (apparently cirro-cumulus) with a virga-type fibrous veil cloud formation under the hole. Calculated diameter of the hole was six miles. Rawinsonde data taken about two hours prior to appearance of the hole indicate the main cloud deck to be at an altitude of about 26,000 ft.

The hole was first sighted at 1110 PST to the west and last seen at about 1210 PST to the east. It moved toward a heading of 070$^{o}$, consistent with upper winds at these levels, passing to the north of the base directly over the city of Santa Maria. There was no visible change in its size or structure during traverse of the area. The official observation for Vandenberg AFB, taken at 1156 PST, was high broken. The upper-air analyses for that day reveal that the main cloud deck moved in behind a rather marked ridge line passage from west to east.

The appearance of this phenomenon suggests that the cloud layer was composed of supercooled water and was seeded either intentionally or unintentionally. Photographs of artificially seeded clouds characteristically show the cirroform veil in the middle of the hole.

Similar appearing phenomena were observed recently near Corpus Christi, Tex., by observers from Air Force Cambridge Research Laboratories. In that situation about five irregularly shaped holes with accompanying cirroform tails formed in a high alto-cumulus layer. Waves could be seen forming and disappearing in the layer and occasionally a new cirrus streamer would form in the thickest part of the wave. Based on these observations, AFCRL meteorologists suggest a local thickening of the cloud triggered the California hole.

Other seeding mechanisms that have been suggested are aircraft or rocket exhausts and even debris from a meteoroid. The symmetry of the hole and its appearance in the Air Force Western Test Range make the rocket hypothesis attractive, but checks with range officials revealed no launches that day.

The AFCRL suggestion of a natural cause seems to be the most logical explanation put forth so far, but the fact that there was only one hole and that it was nearly perfectly round is disturbing. It would be interesting to hear readers' accounts of similar phenomena and possible explanations of the California hole. (American Meteorological Society, Bulletin, 49:990-991, 1968)

## SHIP TRAILS OR ANOMALOUS CLOUD LINES
**Parmenter, Frances C.; *Monthly Weather Review*, 100:646-647, 1972.**

Numerous cases of anomalous lines have appeared in satellite pictures. These lines have been observed in both the Atlantic and the Pacific Oceans, but most frequently appear off the California coast in late spring and early summer.

Conover (1966, 1969) presented an interesting discussion on the cause or origin of anomalous lines in satellite photographs. In 1967, Weather Bureau Western Region offices, with the cooperation of commercial airline pilots, began to investigate these lines.

One documented case is shown here. On June 27, 1967, the ESSA 2, automatic picture transmission (APT) photograph, taken at 1720 GMT, showed anomalous cloud lines. B. J. Haley, pilot of United Air Lines flight 185-27, flew over this area at 1945 GMT. He reported no clouds or vapor trails at his flight level of 31,000 ft. Below him, he observed a layer of low, thin stratus through which sunlight, reflected from the ocean, could be seen. He estimated the height of the stratus layer to be 1,000 ft above the surface. Haley also reported that a ship located near 35$^{o}$N, 131$^{o}$W and traveling due east was clearly visible through the stratus layer. The smoke from the ship's stack was forming a "definite line of thicker clouds" in the layer of scattered and broken stratus. This condensation trail was estimated to be 2-4 mi wide at a point 10 mi behind the ship, and to extend 125-150 mi before fading out. It was drifting southward with the cloud layer.

This observed condensation trail had the same orientation as a line (a-b) in figure 1. The 1800 GMT surface reports shown in figure 2 indicate that two ships, WHEX and WAZB, were in this area. Ship WHEX was located to the east of the cloud line and was traveling northeastward at 20 kt. Ship WAZB was traveling westward at 15 kt away from this area.

The surface analysis (fig. 2) shows these ships to be located in the eastern sector of a large high-pressure area. At this time, 5-10-kt northwesterly winds, fog, and stratus, typical of the ship trail-producing regime, were reported.

Recent observations of these anomalous lines in the Applications Technology Satellite 1 data show that they are distorted and propagated by low-level winds. Once the synoptic regime becomes established, these lines will appear in various configurations for 2 or 3 successive days. (Monthly Weather Review, 100:646-647, 1972)

# CLOUD ARCHES

## UNUSUAL CLOUD FORMATION
**Townshend, C. R.; *Marine Observer,* 23:76, 1953.**

M. V. Port Adelaide. Captain C. R. Townshend. Sydney to Aden. Observer, Mr. G. B. Bonds, 3rd Officer.

31st May, 1952, 1900 G. M. T. Two very dark towering Cu clouds bearing 034° appeared to be joined to a single dark towering Cu cloud, bearing 214°, by a narrow band of Cc which stretched across the sky over the ship. This formation lasted about 2 hours before disintegrating. The moon was not visible at the time and the sky was cloudless except for the above formation.

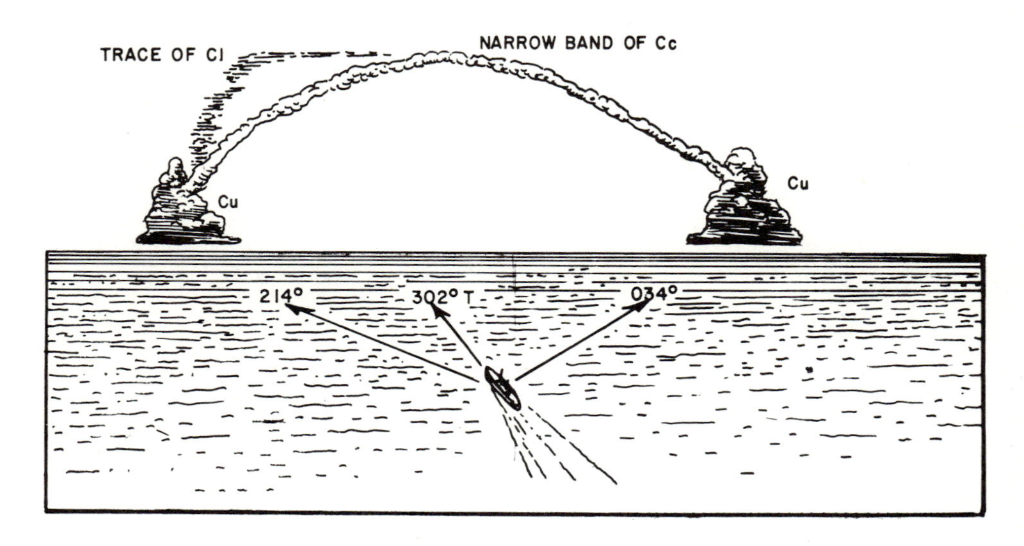


Cloud arch connecting two separated clouds

Position of ship: 14° 35'S, 78° 43'E.

A somewhat similar cloud formation was observed from the same ship by Mr. E. E. Chapman, 4th Officer, on 3rd June, 1952, at 1600. On this occasion the moon was practically overhead, and slight drizzle typical of "monsoon weather" was experienced as one of the Cu clouds passed over the ship. The phenomenon lasted for 1 hour 15 minutes.

Position of ship at commencement: 03° 08'S, 62° 40'E. (Marine Observer, 23:76, 1953)

## CLOUD FORMATION
**Nelson, A. L.; *Marine Observer,* 11:138-139, 1934.**

The following is an extract from the Meteorological Log of R. R. S. Discovery II. Captain A. L. Nelson. Tristan da Cunha to South Georgia. Observer, Mr. L. C. Hill.

"November 24th, 1933. Throughout this day conditions had been steadily improving after a fresh N. W. gale. The wind had slowly backed to west and remained steady in direction while the sky remained veiled with a thick film of Cirro-Stratus cloud. About one hour before sunset an arch of hard clear sky commenced to form slowly and to spread from the S. E. to the N. W. quadrants. This arch had assumed a regular outline by 2115 G. M. T. when the following observations were recorded:---

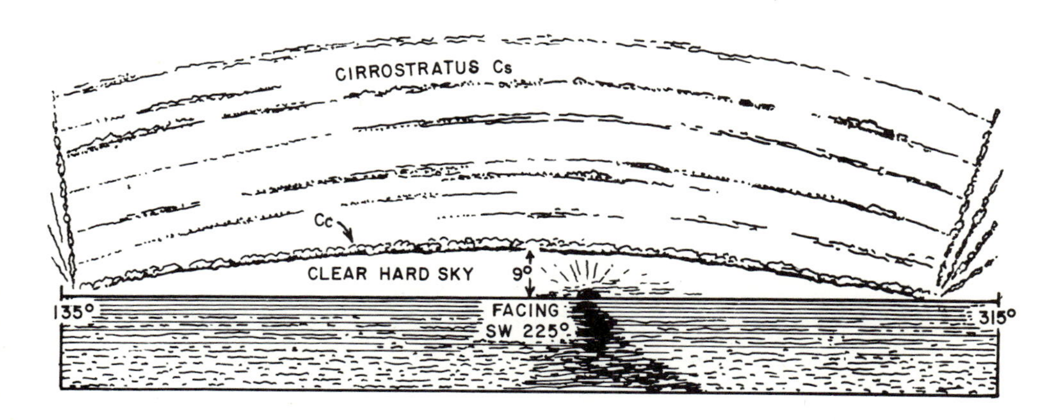


Strange cloud arch in South Atlantic

"From two points on the horizon bearing 138° and 304° respectively an area of hard clear sky arched itself to an altitude of 9°. This maximum angular height was approximately above the Sun which set bearing 237-1/2°. Maintaining an almost equal distance with the rim of this arch was another smaller arch with an angular height of 2°. The space between these two arches was packed with Cirro-Cumulus globules which appeared to be closely stowed between well-defined layers of Cirro-

Stratus clouds. Another patch of Cirro-Stratus clouds, low in the sky, temporarily dimmed the setting sun. From the two points on the horizon, bands of cloud radiated and passed obliquely into the sky, the Northerly band in both cases being composed of Cirro-Cumulus clouds. These bands became fainter with altitude and finally diffused into the upper sky. With approaching sunset the whole formation became suffused with mixed colorations which faded softly with waning twilight.

"The accompanying photographs taken by Mr. A. Saunders will better describe the cloud formation above the sun, while the following small sketch might serve to show the nature of the radiating bands and the completeness of the arch.

"At midnight the arch still persisted with the easterly point still bearing 138°. The Westerly point had closed and bore 250°. With approaching sunrise the lower part of the clear sky became banked up with Stratus clouds, to about 4° altitude. The Eastern convergence still remained very definite and at sunrise radiating bands of Cirro-Cumulus again passed into the upper sky. By 0145 (daylight) the formation commenced to break and by 0215 the whole Southern sky from that 138° bearing had cleared to an overhead point. (Marine Observer, 11:138-139, 1934)

## LUMINOUS NIGHT CLOUD
**Horrex, E. J.; *Meteorological Magazine*, 68:186, 1933.**

During the evening of July 24th, 1933, an unusual cloud formation of cirrus was seen at Westcliff-on-Sea, Essex. At 21h. 20m. G. M. T. a belt of cirrus cloud was observed stretching across the western sky at an elevation of 35°, from north to

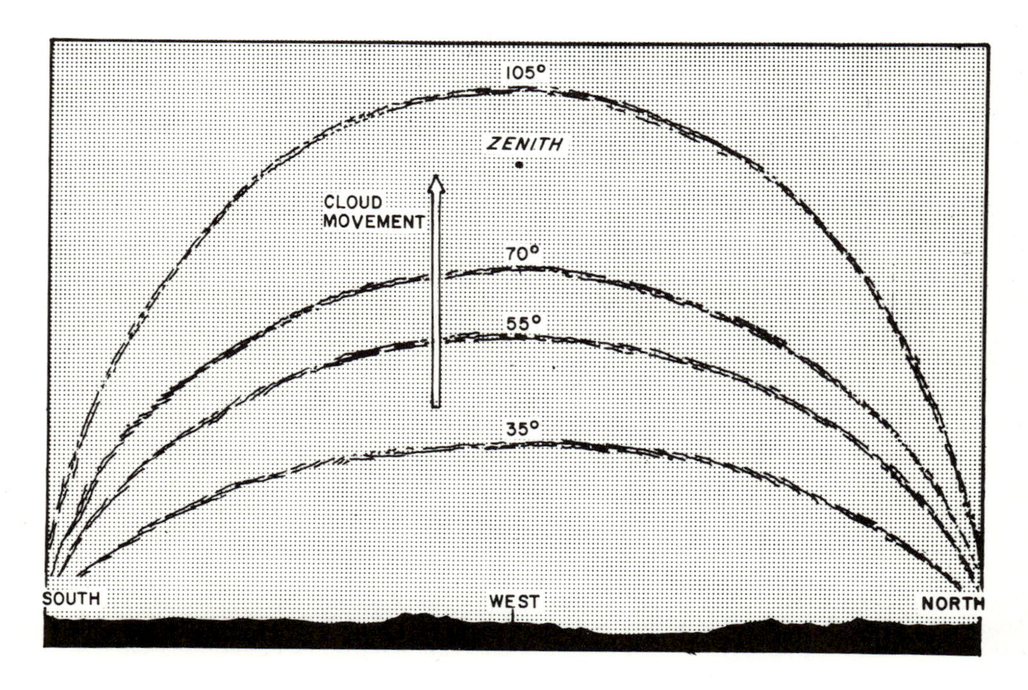


Luminous cloud belts cross sky

south, forming a bow, and moving in an easterly direction. This was rapidly followed by a second belt, and between 21h. 20m. to 21h. 50m. eight belts of cloud travelled across the sky from due west to east, the eighth belt disappearing at 21h. 50m. at an elevation of 105°. None of the belts appeared from the west until at an elevation of 35°, probably because the western sky was too bright at the time to allow of any contrast at lower elevations, nor were any visible after reaching the elevation at 105°. At 21h. 30m. four belts were visible simultaneously at elevations of 35°, 55°, 70° and 105° from west. All the belts were approximately 1-1/2° in width, the width being even throughout their length.

The belts of cloud were whitish in appearance, the three lower being very clear, whilst that which had passed overhead was considerably less clear. The brilliancy of the belts was similar to that of the Milky Way on a clear moonless night. As the belts of cloud travelled across the sky they presented a striking appearance. As each attained the 70° elevation and beyond, stars in the path of travel were plainly seen through the cloud, which clearly denotes that the cloud belts were very tenuous.

The sky during the time of observation was very translucent, no other clouds were visible; in fact, a perfect summer night, and a calm.

Weather conditions during the preceding two or three days had been very fine. (Meteorological Magazine, 68:186, 1933)

# POLAR-ALIGNED CIRRUS BANDS

## ALIGNMENT OF CIRRUS CLOUDS ALONG THE MAGNETIC MERIDIAN
**Kellner, L.; *Nature,* 199:900, 1963.**

Cirrus clouds appear very often in a regular array of parallel bands. Alexander von Humboldt discovered in the course of his meteorological observations in South America and Siberia that these arrays follow frequently the line of the geomagnetic meridian of the locality; he invented the term "bandes polaires" for this phenomenon. Even very casual observations confirm the truth of his statement. He suspected some connexion between the appearance of these 'polar bands' and that of the aurora borealis. In the light of present-day knowledge of the composition of cirrus clouds, it is possible to put forward an explanation of this alignment. Since the constituent ice particles are known to form around dust particles of partly meteoric origin, they will align along the lines of force of the geomagnetic field in those cases in which the dust core consists of a magnetic material. It follows incidentally that the concentration of magnetic meteoric dust in the upper atmosphere should increase from the equator towards the poles, a hypothesis which is confirmed by Sobermann's recent work on the composition of noctilucent clouds above northern Sweden.

Humboldt observed, furthermore, that the alignment of cirrus clouds is more often seen and is of longer duration near the equator than in northern latitudes where their direction gradually changes from that of the magnetic meridian to an east-west line. This movement of the cirrus bands may be due to the action of the jet streams which occur at approximately the same height as cirrus clouds. It seems therefore that observations of the occurrence and movement of aligned cirrus bands may be of use in the investigation of the force and direction of jet streams. (Nature, 199:900, 1963)

## RADIATING CIRRUS CLOUD
**Bedwell, L. A.; *Marine Observer,* 5:203, 1928.**

The following is an extract from the Meteorological Report of S. S. Nagoya, Captain L. A. Bedwell, London to Calcutta via Suez. Observer, Mr. T. A. Sergeant:---

"On 22nd October, 1927, off Cape Garde at about 5.30 a.m., observed remarkable cloud formation consisting of radiating streaks of Cirrus cloud forming complete evenly spaced arcs terminating in approximately the true north and south points of the horizon. It gave the sky a wonderful domed appearance and lasted until after sunrise (about 6.45 a.m.), the formation gradually broke up merging into Alto Stratus.

"I have endeavoured to give a rough idea of the above in the accompanying sketch." (Marine Observer, 5:203, 1928)

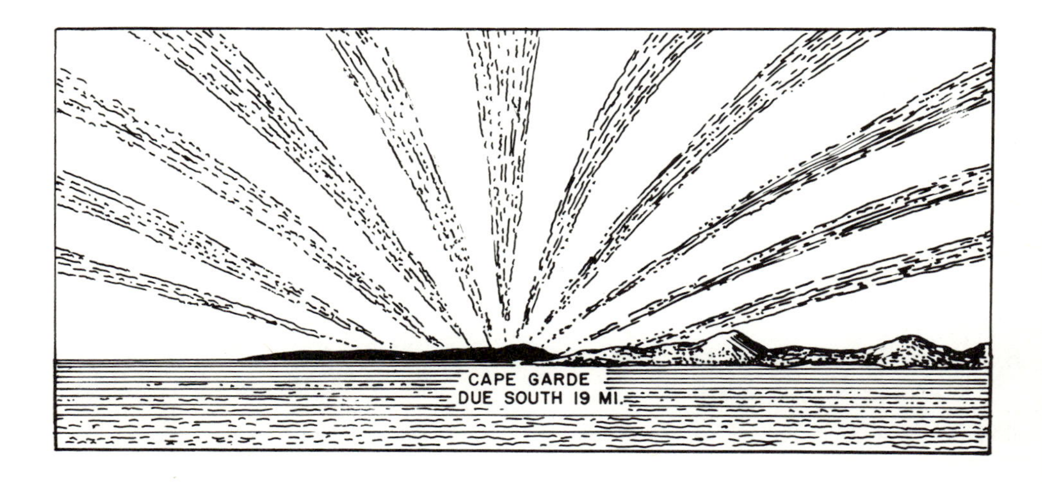


Radiating cirrus clouds in Mediterranean

## A REMARKABLE CIRRUS BAND
**Jupp, C. A.; *Meteorological Magazine,* 67:93-94, 1932.**

At 5h. 40m. G.M.T. on Sunday, March 20th, 1932, approximately half an hour before sunrise, I observed what appeared to me to be a definite and single band of cirrus stretching from due north to south in a complete semicircle. The band tapered slightly at each extremity and its width was estimated to be about $5^{o}$. It was of a brilliant pink colour and the glow gave rise to considerable abnormal illumination. This glow was especially bright near the northern end, but both extremities were brighter than the portion in the zenith. The cirrus band appeared quite thick and even dense in places and was not of the thin fibrous type usually associated with such bands.

Apart from the band the sky was clear but for 2/10ths of alto-cumulus lenticularis situated due east with its base about 2° above the horizon. Other conditions were as follows: temperature 35°F; wind NW 1 m.p.h.; average visibility three miles, but ground mist in valleys. Previous to the appearance of the band the sky had been cloudless for some hours.

When the sun rose at 6h. 10m. G.M.T. the band rapidly faded and was followed by the gradual appearance from north-north-west of a sheet of cirro-stratus.

The sketch shows the southern extremity of the "arc." (Meteorological Magazine, 67:93-94, 1932)

## CURIOUS CLOUD FORMATION

**Anonymous; *Meteorological Magazine*, 73:45-46, 1938.**

We are indebted to Sergeant Winton, of the Royal Air Force Station, Boscombe Down, Wiltshire, for the accompanying sketch of a curious cloud formation observed from the air off the South Devon coast on November 1st, 1937. Mr. C. V. Oekenden, Meteorological Officer at Boscombe Down, supplies the following particulars:--- "A shallow depression was centred in the western English Channel. Conditions were mainly fair in the west, but much low cloud and rain were encountered on the return journey eastward of Yeovil. It is not improbable that the large cumulus up to 9,000 ft. was quite close to the centre of the depression. The pilot was flying at 10,000 ft., and he estimates his distance from the towering cloud to be 12 to 15 miles."

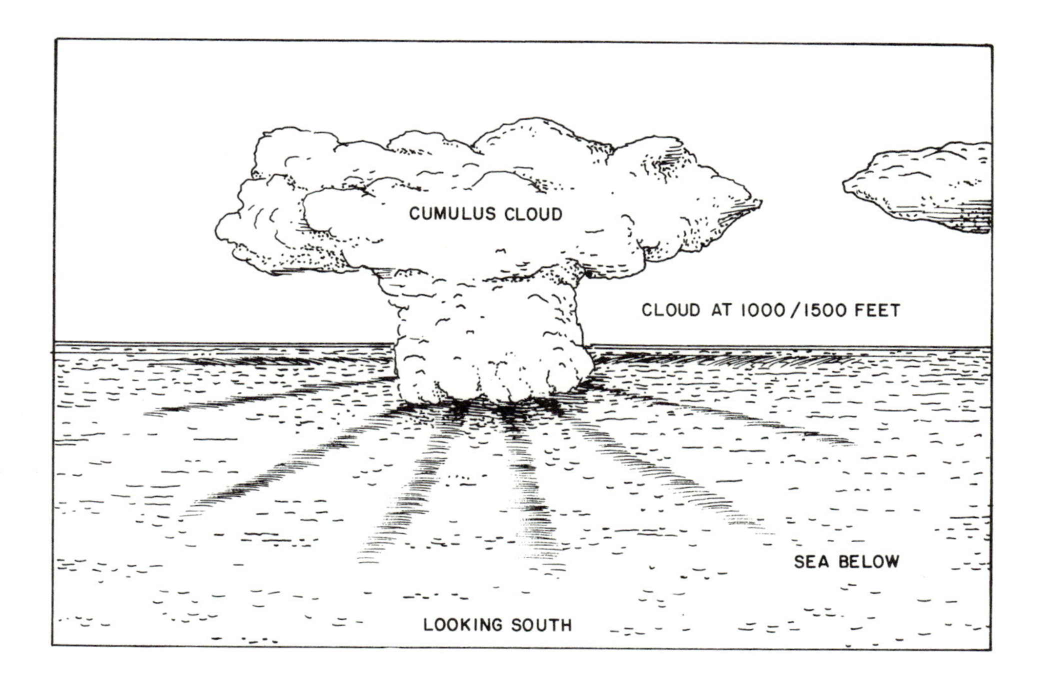


Alignment of clouds over England

The main body of the cloud with its anvil-like development is typical of a large cumulus cloud, but the radiating bands extending horizontally from the cloud base are remarkable. A continuous layer of low cloud was observed farther east and it is reasonable to suppose that the air at this level was almost saturated, and so required little upward motion to produce condensation. One is tempted to visualize a wave motion circulating round the cloud base although it is doubtful whether such an explanation would be dynamically acceptable. Judging from the diagram the bands might, however, be regarded as radiating from a point on the horizon rather than from the cloud base, in which case they would be parallel. The formation might therefore have to be explained as merely an illusion of perspective. (Meteorological Magazine, 73:45-46, 1938)

# MINIATURE CLOUDS

## A MINIATURE THUNDERSTORM
**Bailey, C. S.; *Weather,* 4:267, 1949.**

With regard to the frequent discussions on strange thunderstorms occurring from time to time, I wonder if anyone can quote experiences similar to a strange occurrence witnessed by myself a year or two after the first World War.

I was staying at Stockton Heath, Cheshire, in July, about 100 yards from the Manchester Ship Canal. The evening was somewhat oppressive, and the air had become strangely still. Gazing down the road, I saw a small black thundercloud gathering along the length of the Canal, and about 30 or 40 ft. above it. It was approximately 100 yards long and perhaps 6 ft. thick. As I gazed at this strange formation, a dazzling lightning flash raced through the entire cloud, i. e. parallel to the water, and a bang like the discharge of field artillery followed immediately. About 40 seconds later, another flash and report occurred: then the cloud thinned and dispersed in about four minutes.

I might add, that at least in those days an air-current of varying intensity moved up that Canal almost incessantly, i. e. inland towards Manchester, at the Stockton Heath section---one felt it on the neighbouring bridge. (Incidentally, a year or two later, a "thunderbolt" fell in an adjacent road.) I have often wondered since, to what extent, and in what way, water evaporation at that point contributed to the occurrence. The evening (7 p. m.) was dry. (Weather, 4:267, 1949)

# NOISY CLOUDS

## RUMBLING CLOUDS AND LUMINOUS CLOUDS
**Zeleny, John; *Science,* 75:80-81, 1932.**

A brief description of two rather unusual cloud phenomena which have come to my notice may be of some interest. One of these was observed from the east shore of a narrow bay of Cache Lake in Algonquin Park, Ontario, on an early morning during the latter part of July of this year. It was a chilly morning and the sky was completely overcast with clouds. My attention was attracted by a rumbling sound coming from the west, such as heralds the approach of a heavy thunderstorm. As I watched, a very long, low, narrow, tenuous cloud, resembling a squall cloud, appeared above the trees on the opposite shore, moving at right angles to its length. The continuous, rumbling noise, now grown remarkably loud, seemed to come unmistakably from this cloud, whose cross-sectional diameter was only about 200 feet. The cloud passed overhead eastward and was not followed by the expected rain storm. The cloud apparently marked the meeting place of two oppositely directed currents of air that differed in temperature. It seems almost incredible, however, that so much sound could have arisen from the agitated air alone, and yet this seems to be the only plausible explanation of its origin. I steadfastly looked for small lightning flashes in the cloud and saw none, although they would have had to come in rapid succession to produce the persistent sound which was heard. The noise could not have come from the rattle of hail because the cross-section of the cloud was too small to give time for hail formation; and in any case no hail fell.

The other cloud I wish to describe was a solitary, brightly luminous, cumulus cloud which I saw on a clear summer night at Hutchinson, Minnesota, some thirty-five years ago. The cloud had a horizontal diameter of about a third of a mile and a thickness of about one fourth of that distance. It rose majestically from the eastern horizon, shone with a uniform, steady, vivid, whitish light and passed directly over the town. When the cloud was overhead a great shower of insects descended to earth covering the ground all around to the number of about 50 to 100 per square foot. These insects proved to be a species of hemiptera and were non-luminous. They had apparently been induced to take wing by the bright object in the sky. I have been at some loss to account for the luminosity of the cloud. It could not have been due to reflected light coming from a city. It might be postulated that the cloud consisted of a mass of organic vapor that was slowly oxidizing, being in fact a case of an extended will-o'-the-wisp, but for several reasons this seems to be an unlikely hypothesis. At the time the cloud was observed, it was thought to be far too late in the evening for its light to be reflected sunlight. There is a possibility that a bright moon below the horizon might have been the source of the light, although I have no recollection of having seen the moon rise later. (Science, 75:80-81, 1932)

## RATTLING CLOUDS
**Anonymous; *Pursuit,* 2:32, 1969.**

Jacksonville Beach, Florida (AP)---Hundreds of persons---including Police Chief James Alford---reported strange sounds coming from two clouds. One man described the sound as like "someone rattling cellophane." A woman said it was more like "someone walking on pebbles." Alford ordered Capt. Harold Bryan to follow the first cloud. Bryan did so---to the edge of the Atlantic where the cloud dissipated. The listeners started to go back inside their homes when, they said, another cloud repeated the performance. Bryan also followed it to dissipation over the Atlantic. Officials at the Mayport Naval Air Station said they could offer no explanation; neither could other officials. (Pursuit, 2:32, 1969)

# IDIOSYNCRACIES OF TORNADOES AND WATERSPOUTS

Ostensibly, tornadoes and waterspouts are the same species of rotary storm. The passage of a tornado from land to water converts it automatically to a waterspout, or so the prevailing theory goes.

A controversial hypothesis that does unite the two phenomena is that of their purported electric origin. Both tornadoes and waterspouts are usually accompanied by severe lightning, but the hypothesis of electrical origin requires powerful earth-cloud currents to keep the storm's motor running. The effects of this large-scale flow of electricity should be observable. The "tornado lights" introduced in Chapter 1 may be considered evidence of strong electrical action. There are other effects mentioned below that are also confirmatory: large terrestrial current flows in the neighborhoods of tornadoes, the occasional smell of ozone and/or sulphur, and the burning and dehydrating effects sometimes noticed along the paths of tornadoes. The latter effects, however, are attributed more conventionally to the aerodynamic heat generated by tornadoes. Wherever the truth lies, the auxiliary phenomena accompany tornadoes and waterspouts are most curious.

The physical mechanism of the waterspout is still a puzzle to some meteorologists who find the peculiar tubular vortex structure different from what theory requires. Along this vein, conventional wisdom also has trouble dealing with forked waterspouts, waterspouts stretching between clouds, and waterspouts with two concentric tubes.

## ELECTRICAL PROPERTIES OF TORNADOES AND WATERSPOUTS

### UNUSUAL DAMAGE BY A TORNADO
**Davy, E. S.; *Weather,* 4:156-157, 1949.**

On May 24, 1948, at Curepipe, Mauritius (approximately 1,850 feet above sea level), a phenomenon of the tornado type exhibited a very interesting feature which the writer has not seen mentioned in reports of other tornadoes. During most of the day the weather was mainly fair at Curepipe and convection appeared to be no more intensive than is normal for this region. In the middle of the afternoon a whirlwind touched down to earth over a hard tennis court, causing considerable damage to its hard compact surface. A trench running in a north-south direction, 60 feet long and 1 to 2-1/2 feet wide, was cut in the bare surface of the court to a depth varying from 1 to 4 inches. The material lifted from the trench was all thrown to the west to a distance of 50 feet; pieces weighing about one pound were thrown as far as 30 feet. The surface material was slightly blackened as if by heating, and a crackling like that of a sugar-cane fire was heard for two or three minutes. The court was made of a ferruginous clay, which packs down to a surface more smooth than that of the hard tennis courts usually made in Great Britain.

Unfortunately, there were no very reliable witnesses of the phenomenon. The impressions gained by two servants of the tennis club who saw the incident differed considerably on important details. One claims to have seen a ball of fire about two feet in diameter which crossed from a football pitch to the tennis court through a wire-netting fence without leaving any evidence of its passage until it bounced along the court, making the trench in the surface before disappearing completely.

The high winds and ascending currents in this small tornado were unusually irregular in their action; on the tennis court an umpire's chair weighing about 50 pounds was picked up from 50 yards to the west of the trench, carried upwards to an estimated height of 60 feet, said to have been broken whilst in the air, and dropped in pieces about 20 feet to the east side of the trench, that is, on the side opposite to the pieces of court surface. A hundred yards to the south of the tennis court the roof of a small building was lifted off and dropped nearby. About two miles to the north a rather insecure building was blown over as the whirlwind moved in a northerly direction. No other evidence of damage to buildings or vegetation was reported nor could any be seen from a hill-top which was in the track of the phenomenon, between the damaged sites of which a good view was obtained.

At the Aerological Station, Vacoas (1,393 feet above sea-level), 1-1/2 miles to the west of the destroyed building, there was no wind at 1 p.m.; between 2 and 3 p.m. the wind rose gradually to an average of 7 m.p.h. (highest gust 15 m.p.h.) and fell again to very light; during the same period the wind direction changed from west-south-west through south to east-south-east. There was no wind after 4 p.m. The Aerological Station was therefore under the influence of the whirlwind and the circulation was clockwise, as one would expect in a low-pressure system in the southern hemisphere. At the tennis court there was no rain at the time the whirlwind passed, but heavy rain started five minutes later and continued for twenty minutes. At the Aerological Station no rain fell: there was no thunder reported.

An explanation of the mechanism by which the bare surface of the tennis court could be lifted out is difficult to find. It seems most probable that it was an electrical discharge which disrupted the court surface; this would also account for the light, the blackening of the surface and the crackling noise. It seems improbable that high winds and a sudden atmospheric pressure change could have much effect on a closely packed earthy material with a smooth surface. (Weather, 4:156-157, 1949)

## ELECTRIC CURRENTS ACCOMPANYING TORNADO ACTIVITY
**Brook, Marx; *Science*, 157:1434, 1967.**

Abstract. Measurements of the magnetic field and earth current in the vicinity of a tornado show large step-like deflections coincident with the touching down of the funnel. Calculations with a simple current model indicate that a minimum current of several hundred amperes must be postulated to account for the observed deflection in magnetic field. The existence of a steady current of 225 amperes for a period of about 10 minutes provides joule heat at the rate of approximately $10^{10}$ joules per second, and involves a total charge transfer of 135,000 coulombs. The calculations imply that a tornado is electrically equivalent to several hundred isolated thunderstorm cells active simultaneously. (Science, 157:1434, 1967)

## ELECTRIC THEORY OF TORNADOES

**Vonnegut, Bernard; *Journal of Geophysical Research,* 65:205-206, 1960. (Copyright 1960 by the American Geophysical Union)**

Presence of Energetic Electrical Phenomena in Tornadoes. If the high velocity of the tornado is to be explained on the basis of thunderstorm electrical energy, one must first prove that intense electrical activity is closely associated with tornadoes. That this is the case is clearly evident from many different accounts that have been written about tornadoes. Flora quotes Harrison's conclusion 'that lightning of a peculiar and intense character is an almost invariable accompaniment of tornadoes, and that airplane pilots frequently so identify tornado clouds.

The visual evidence of the close association between lightning and tornadoes is confirmed by the radio 'static' measurements made by Jones. He concluded from sferics measurements that in tornado-producing storms lightning discharges occur at the rate of 10 or 20 per second, which is about ten times the rate in ordinary storms.

Further evidence of the close association between electricity and tornadoes is contained in reports of those who have looked up into the interior of a funnel and lived to tell about it. These observers report a variety of electrical phenomena, such as incessant lightning, a brilliant luminous cloud, a ball of fire, or a display "like a Fourth of July pinwheel" in the tornado tube.

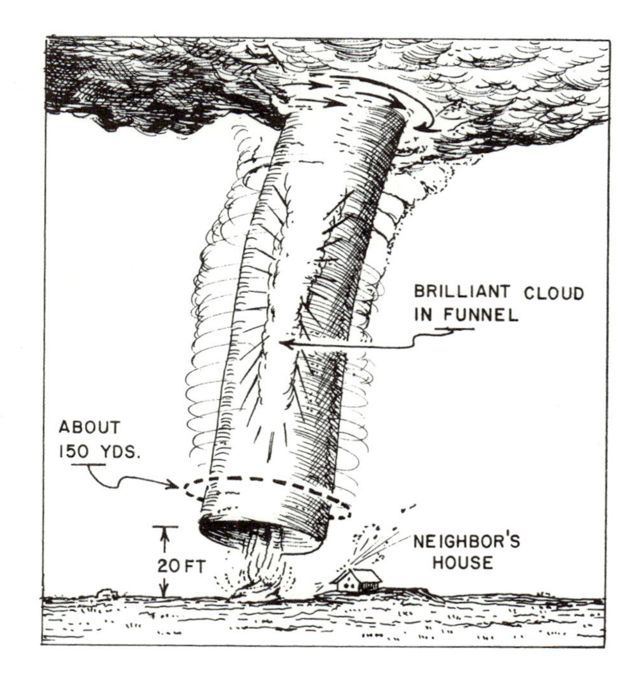


Bright cloud inside tornado funnel

Although no one has as yet photographed these striking displays, two different observers have made drawings of what they saw. Hall observed a tornado in Texas at close range and published the sketch shown as Figure 2. Montgomery watched the devastating Blackwell, Oklahoma, tornado of 1955 from a distance of about half a mile and prepared the drawing shown as Figure 3. The approximate dimensions indicated were computed from angles subtended by the tornado relative to structures near the point of observation. There are other phenomena suggestive of electrical effects. Intense St. Elmo's fire is frequently observed near the funnel, and odors, probably of ozone and nitrogen oxides, have been described. Buzzing and hissing noises suggestive of electrical discharges have been reported near the funnel. After its passage, dehydration of vegetation and the surface soil has been noted along the path.

In addition to the fairly well understood primary and secondary electrical effects discussed above, accounts of tornadoes rather frequently include mention of 'beaded lightning' and glowing or exploding fireballs. These phenomena are apparently the same as or are closely related to the controversial 'ball lightning' whose existence and nature are still debated. In view of our present almost complete ignorance, we shall make no attempt to discuss this class of observations. It is worth remarking, however, that an understanding of ball lightning may very well be necessary if the tornado puzzle is to be solved. (Journal of Geophysical Research, 65:205-206, 1960. Copyright 1960 American Geophysical Union)

## THE TORNADO PULSE GENERATOR
**Jones, Herbert L.; *Weatherwise;* 18:78-79, 1965.**

The tornado pulse generator was observed for the first time about 2025 CST on the evening of 25 May 1955, when the thunderstorm that developed into the Blackwell, Oklahoma, tornado passed some 12 or 15 miles west of the Tornado Laboratory at Stillwater. For some time the 150 KC direction finder had been showing abnormal directional pip activity, first to the southwest, and then as time passed, indicated directions of arrival were from points close to, and finally at, 270° azimuth, or due west of the laboratory. The 150 KC directional activity was much greater than usually experienced, it being obvious that some type of unusual thunderstorm was active in the vicinity: first to the southwest of the laboratory and then west of the laboratory.

At the same time an echo on the radar had gradually been moving northward from a point some 15 miles northeast of Oklahoma City along a line through a point approximately 15 miles west of the Tornado Laboratory and from there north to the Kansas border. The sketch showing the details of this movement is included in a paper by Jones in 1958. From the indications on the 150 KC direction finder, it was evident that considerable electrical activity was taking place at the 270° azimuth. In order to verify that observation, the area to the west of the laboratory was observed visually. Since the radar showed the precipitation echoes some 15 miles to the west, it was thought that there would be a considerable number of cloud-to-ground and cloud-to-cloud lightning discharges. Much to the disappointment of the author at that time, however, there was no apparent electrical activity whatsoever except for an odd flashing of a patch of illumination in the top section of the thundercloud, as shown in figure 1, a flashing that appeared as a circular patch of light on the side of the cloud structure between the observer and the center of action where the elec-

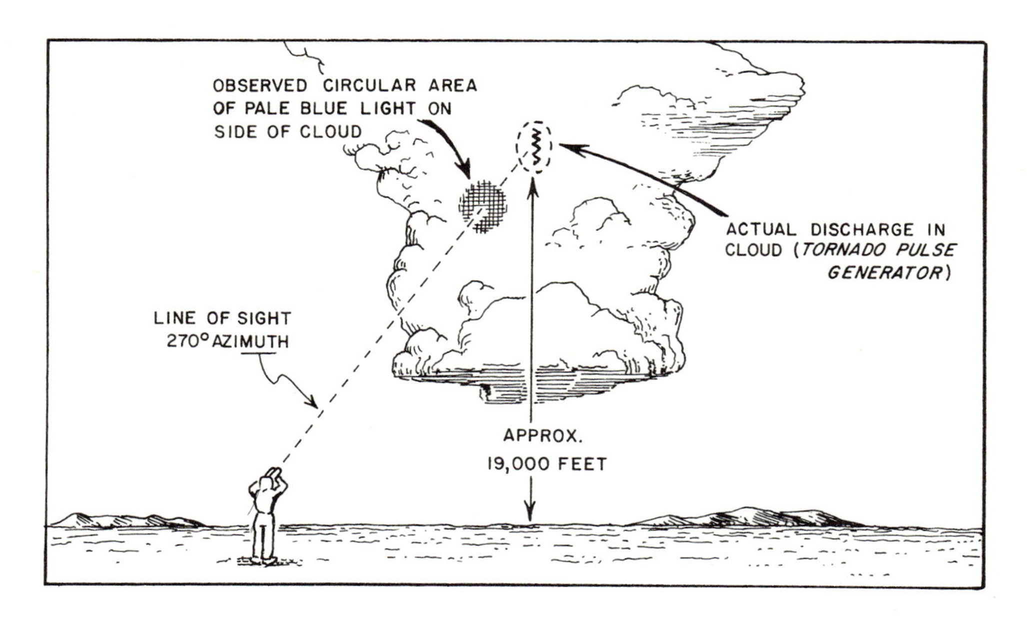


Tornado pulse generator observed at Blackwell

trical activity appeared to be. At that time these patches of light appeared as circular patches of pale blue illumination for an estimated time of two seconds; this circular area would then become dark for another estimated period of two seconds, and again would reappear. The on-and-off flashing continued for the entire time of visual observation. It was definitely obvious that there was absolutely no cloud-to-ground or cloud-to-cloud lightning visible or even any lighting of the sky in that particular region. But the staff at the Tornado Laboratory was forced to the conclusion that the electrical activity appearing on the Cathode Ray Oscilloscope of the 150 KC direction finder was due either to the flashing patch of light high in the cloud or to some center of electrical activity many more miles to the west.

This latter conclusion was abandoned almost immediately because as the radar echo of the thunderstorm continued to move northward, the angles of the 150 KC directional pips continued to increase from 270° to larger and larger values. This increase in angle corresponded to the increase in the azimuth of the center of the precipitation echo of the storm that was then moving to the northwest of the laboratory; it was necessarily concluded that the 150 KC directional pips were coming from the actual precipitation echo that was showing on the radar as moving almost due north towards Blackwell where the roaring funnel of destruction would strike in the darkness, and then move onward to devastate the sleeping town, Udall, across the border in Kansas.

While the tornado pulse generator has been observed a number of times from the laboratory, one of the outstanding examples was a double generator system observed at 0150 CST on 31 May 1956, as shown in figure 2. When observed these two centers were moving east, and approximately four miles apart, one to the northeast and one to the southeast. As observed on the 150 KC directional finder, the action was intermittant, and this observation was confirmed by the typical "blinking" nature of the two visible circular patches of luminescence. (Weatherwise, 18:78-79, 1965)

## WATERSPOUT AT CLOSE RANGE
**Anonymous; *Royal Meteorological Society, Quarterly Journal* 40:78, 1914.**

On the U.S. Hydrographic Office Pilot Chart of the North Atlantic Ocean for December 1913 is given the following account of the formation and characteristics of a waterspout observed on October 2, 1913, by Captain H. C. Hansen, of the Norwegian ship Majorka:---

"Latitude 45° 37' north, longitude 14° 0' west from Greenwich, wind North-north-east, sky overcast, with a dark cloud formation in the north-west, threatening rain.

"At 7.30 a.m. a whirlwind was seen about 1/8 mile to the leeward of the ship, which was headed north-west, and when it became apparent that it was an incipient waterspout, the wheel was put hard to starboard and the ship was gotten off toward the west-south-west, and immediately the whirlwind passed the ship's stern in a north-south direction at a distance of about 10 feet. It had taken the form of a thin spout, which increased in diameter and became denser. It was accompanied by a strong puff of wind, a heavy downpour of rain, and roaring sound.

"Within a circle of about 25 feet in circumference the water was in a violent uproar similar to a large waterfall. The spout grew until it was about 5 feet in diameter and had a lighter appearance in the centre about 3 feet in diameter, in which the water steamed up, and on each side of this a darker band about 1 foot in breadth, in which the water was descending. The motion in both the upward and downward currents was very rapid. The spout lifted and bent forward about half-way between the cloud at the top and the surface of the water. At that time it was about 1500 feet from the ship. The duration of the spout from the beginning of the formation until it broke loose was about ten minutes.

"The barometer was 30.10 in. immediately before the whirlwind was seen, and stood at 30.16 in. immediately after the passage of the spout. There was no marked change in temperature, but one could notice an odour of sulphur in the air which was dissipated when the wind shifted to North-north-east. The barometer rose from that time on until 9 o'clock, when it stood at 30.20 in." (Royal Meteorological Society, Quarterly Journal, 40:78, 1914)

## RELATION BETWEEN TORNADOES AND VARIATIONS OF THE COMPASS
**Llewellyn; J. F.; *American Meteorological Journal,* 4:135, 1887.**

The following may be of interest as to the possible relation between tornadoes and whatever causes variation of compass. The variation in Missouri was determined by a magnetic survey made by Prof. F. E. Nipher. The number of tornadoes was obtained from Signal Service publications, Missouri Weather Reports, newspapers and individuals. None but genuine twisters were listed. The greatest variation is in the extreme N. W. part of the state, the least in S. W.

On the line of

| | | | | | | |
|---|---|---|---|---|---|---|
| 11° | variation | 3 | counties | had | 7 | tornadoes. |
| 10° | " | 5 | " | " | 15 | " |
| 9° | " | 20 | " | " | 22 | " |
| 8.30° | " | 18 | " | " | 13 | " |
| 8° | " | 14 | " | " | 7 | " |
| 7.30° | " | 23 | " | " | 8 | " |
| 7° | " | 11 | " | " | 5 | " |
| 6.30° | " | 8 | " | " | 4 | " |
| 6° | " | 4 | " | " | 1 | " |

(American Meteorological Journal, 4:135, 1887)

# BURNING AND DEHYDRATING EFFECTS OF TORNADOES

## A FIERY WIND
**Anonymous; *Symons's Monthly Meteorological Magazine*, 4:123-124, 1869.**

Our in Cheatham county [Tennessee] about noon on Wednesday---a remarkably hot day---on the farm of Ed. Sharp, five miles from Ashland, a sort of whirlwind came along over the neighbouring woods, taking up small branches and leaves of trees and burning them in a sort of flaming cylinder that travelled at the rate of about five miles an hour, developing size as it travelled. It passed directly over the spot where a team of horses were feeding and singed their manes and tails up to the roots; it then swept towards the house, taking a stack of hay in its course. It seemed to increase in heat as it went, and by the time it reached the house it immediately fired the shingles from end to end of the building, so that in ten minutes the whole dwelling was wrapped in flames. The tall column of travelling caloric then continued its course over a wheat field that had been recently cradled, setting fire to all the stacks that happened to be in its course. Passing from the field, its path lay over a stretch of woods which reached the river. The green leaves on the trees were crisped to a cinder for a breadth of 20 yards, in a straight line to the Cumberland. When the "pillar of fire" reached the water, it suddenly changed its route down the river, raising a column of steam which went up to the clouds for about half-a-mile, when it finally died out. Not less than 200 people witnessed this strangest of strange phenomena, and all of them tell substantially the same story about it. The farmer, Sharp, was left houseless by the devouring element, and his two horses were so affected that no good is expected to be got out of them in future. Several withered trees in the woods through which it passed were set on fire, and continue burning still. (Symons's Monthly Meteorological Magazine, 4:123-124, 1869)

## ELECTRIC STORMS AND TORNADOES IN FRANCE ON AUG. 18 AND 19, 1890
**Hazen, H. A.; *Science,* 17:304-305, 1891.**

These facts show clearly that there were several violent storms on the 18th [August 1890] running in parallel lines, beginning toward the west early in the evening and occurring at points farther east later on; that is to say, the several appearances near Pire and Dreux were separate occurrences, and the violent storm did not go from one to the other, but each devastated its own narrow strip. It will be seen that this bears a most remarkable resemblance to the action of tornadoes in this country.

At Pire the trombe [tornado] was investigated by M. G. Jeannel. There was an apparent whirlwind, transported parallel to itself, and turning counter-clockwise, as shown by the fallen trees. The first thrown down were from the south-east, the next from the east, and so on to the north-west. The greater damage was on the right hand of the track. The velocity of gyration was great and that of translation relatively much less.

The roofs damaged were peculiar. On the right of the path those facing north were carried away, while those facing south were unharmed; on the left of the track just the reverse was true. During the whole time the lightning was continuous. The odor of ozone was noted at different places. At Reinou a woman tending a cow, grazing in the meadow, saw her enveloped in violet flames. These were so intense that the woman, from fright, covered her face with her handkerchief. A moment later the wind struck down every thing.

At Domagne Dr. Pettier suddenly heard an extraordinary indefinite roaring. He rushed toward the garden, where the firs were being plucked up. At the gate he felt a kind of pressure from above; he noticed an unusual smell of ozone; then he felt himself raised up, and this not by the wind, for it was calm, but as though by some invisible force. On many trees the foliage was scorched. About a mile west of Domagne, hail of the size of a walnut fell to a depth of over three inches, covering the ground.

At Dreux the report was by M. Bort. At 10 p.m. a great cumulo-nimbus thunder-cloud was seen to the south-south-west of the town. On its upper part a very brilliant plume of sparks was directed toward heaven. In this cloud the lightning was incessant and the thunder loud. After some hail had fallen, at about 10.25 p.m., a loud roar was heard, like that of a train entering a tunnel, and in less than a minute the storm reached the town. It blew off the tiles, plucked out the trees, and destroyed many houses. At the moment of the passage the sky was on fire, and some persons saw a cloud which reached the height of a house. Reaching the Blaise valley it plucked up many poplars, and left them lying generally from south-south-west to north-north-east. In the environs of Fontaine many trees were uprooted. At Brissard the hurricane made a passage through the western part of the village, destroying twenty houses. At another point most of the trees lay from south-west to north-east, but there were many, 220 yards from the first, that lay in an opposite direction.

Lightning strokes were very rare, because no traces were found upon trees, and no houses were fired. There was a remarkable exception, however, in the Vivien house, built solidly of brick, which had traces of electric discharges. Some window-panes were pierced by circular holes, and these holes had a sharp edge on the outside. On the inside the edge had suffered a beginning of fusion, which had rounded it off. The damage was reported at $300,000 in Dreux, and one person was killed. At the instant of the passage all the gas-lights were extinguished, and it is suggested that "this indicated a rarefaction of the air near the centre of the whirl." By the synoptic charts it appears that the passage of this trombe was coincident with the existence of a secondary barometric depression in the west of France, its path being recognized from Vendir to Ardennes. (Science, 17:304-305, 1891)

## ELECTRICITY AND TORNADOES
**Moore, H. D.; *American Meteorological Journal*, 2:517-518, 1886**

I read, yesterday, a copy of your journal (September number), from a friend of yours. It was the first I had seen, and I was much interested in its contents, especially the article on "Tornado Generation." Several years ago, a tornado passed near this place, which gave me an opportunity to pass over its path of destruction, and I saw some things from which I believed then, as I do yet, that there was some other force besides the wind which put in motion the objects in its way.

The tornado was preceded by a dark cloud from which there was considerable thunder and lightning, but there was very little, if any, near the tornado. After the storm had passed, the air was saturated with ozone to such a degree that even the small children noticed it, who compared it to the odor of burning brimstone or burning matches. Unfortunately, I once stood a few rods from a tree when it was struck by lightning, and I distinctly remember the peculiar odor which I attributed to ozone which had been formed by the passage of lightning through the air. As the odor which followed or accompanied the tornado was similar, I concluded there must have been electricity with it. I followed the path of destruction, which was not over ten rods wide, for a couple of miles, and at different places saw small hickory and white oak trees, from one to two inches in diameter, ruptured or burst open, with crevices on all sides as though they had been exploded with some explosive; the fine splinters standing out, brush-fashion, on all sides. These small trees grew on ground which had once been farmed, and were of slow growth and the toughest which grow. There was such a marked difference between these, and those which had been twisted by the wind, that I had no difficulty in distinguishing one from the other. They were ruptured by electricity. At other places I saw small roots which had been laid bare, split open without any other marks about them, and I believed them to have been split by electricity. Had they been split by any visible or hard object, it must have left some other marks upon them. This, again, must have been the work of electricity.

The above evidences of the force and other signs of electricity in tornadoes I have seen, while I have read and heard of many others. In an adjoining county, two years ago, a little village was almost totally destroyed by a tornado and several lives lost. Some of the killed were found to be terribly burned, while all who saw the tornado declare they saw the fire descend from the clouds. All the evidence points to electricity as the origin of the fire. Persons who have been carried from the ground in one of these storms, experienced a prickly sensation all over the body, which must be attributed to electricity. (American Meteorological Journal, 2:517-518, 1886)

# UNUSUAL WATERSPOUT STRUCTURES

## WATERSPOUT
**McCrone, J.; *Marine Observer,* 27:80-81, 1957.**

S. S. Clan Chattan. Captain J. McCrone. Port Said to Liverpool. Observer, Mr. K. Barr, 3rd Officer.

28th May, 1956, 0920 G. M. T. A most unusual waterspout was observed bearing due E. from the ship. It originated from a large Cb with rain falling, and stretched diagonally towards the horizon, as illustrated in the drawing on page 81. The waterspout lasted about 3 min before dissolving. There were 4/8 cloud at time of observation, consisting of $C_L2$, $C_M6$ and $C_H2$. Wind SW'ly, force 4, visibility excellent.

Position of ship: 42° 28'N., 9° 39'W.

Waterspout with horizontal segment

Note. This is a very unusual form of waterspout, of which the upper portion is is roughly horizontal. According to Wegener's theory of the formation of waterspouts, quoted in Part I of Mr. Gordon's article on waterspouts on page 50 of the January 1951 number of this journal, the upper end of a waterspout is a horizontal vortex within the Cb cloud, this vortex bending downwards to emerge from the cloud towards the sea. The article states that the horizontal part of the vortex, normally hidden by the surrounding cloud, has been seen once or twice, but gives no indication that it has ever been seen reaching far out of the cloud as in the present observation, which may therefore be a unique one. (Marine Observer, 27:80-81, 1957)

## FORKED WATERSPOUT
**Rippon, R. G.; *Marine Observer,* 38:62-63, 1968.**

m.v. Flintshire. Captain R. G. Rippon. Port Swettenham to Hong Kong. Observer Mr. P. M. Whitworth, 4th Officer.

19th May 1967. A waterspout which divided into two branches about half-way between the cloud base and the sea, as shown in the sketch, was seen at 0715 LMT about 10 miles off. The two parts joined together about 10 min later to form a single column, at the base of which great turbulence was observed. A shower was falling in close proximity. Air temperature 84°F, wet bulb 80.5°, sea 83.7°.

Position of ship: 14° 15'N, 112° 24'E.

Note. We wonder whether this is in reality what it appears to be. Some doubt has been expressed as to whether such a formation is dynamically possible and the idea most favoured is that there were in fact two separate spouts, the upper part of the distant spout being in the same line of sight as that of the nearer one. (Marine Observer, 38:62-63, 1968)

Forked waterspout in South China Sea

## LAND WATERSPOUTS
**Busk, H. G.; *Meteorological Magazine,* 61:289-291, 1927.**

I also feel that it is time that someone with scientific experience took up the cudgels on behalf of the many sea captains who write both to the Meteorological Magazine and to Nature iterating that water spouts are tubular. Their stories are generally followed by an editorial comment to the effect that water spouts are formed by the condensation of water vapour by rapid expansion in a vortex, and as such cannot be tubular, but consist of "solid" (if the term can be applied) mist.

The waterspout illustrated here occurred on the land, 150 miles from the nearest sea, at Naft-i-Safid, Bakhtiari Country, in Persia, on Monday, December 27th, 1915. Its distance from myself as observer was about two miles, and it was viewed through a pair of Zeiss glasses with a magnification of eight. As the pendant gradually dropped from the cloud, a hollow column of dust from the plain rose to meet it. The pendant and column eventually joined, the whole swaying about in a particularly beautiful manner. The height of column from the plain to the cloud was about 2,000 feet, and the motion of the column about fifteen miles per hour, but very irregular. I have no note on the direction of rotation of the column. The whole phenomenon lasted about fifteen minutes, and the pendant portion withdrew into the cloud first, leaving the lower portion rotating at the bottom. At the base of the column there was a fountain of large fragments thrown out, stones, sticks and anything that was loose. The whole column appeared like a translucent glass tube, but flexible, that is, there was a dark edge and a brighter middle, just as I have drawn it. A solid column of mist would have given an exactly opposite effect, namely, a dark middle and a bright edge. There is no doubt in my mind at any rate that the column was hollow, and that the lower part of it was composed of dust, which was of a browner colour than the cloud above it.

Such a "landspout" is, of course, not to be confused with the ordinary dust devil, which is quite common in any desert on a bright day.

I have frequently seen waterspouts at sea, and the tubular effect is always there. (Meteorological Magazine, 61:289-291, 1927)

## EMBRYO WATERSPOUT
**Kenyon, G. J.; *Marine Observer,* 32:174, 1962.**

s.s. Makalla. Captain G. J. Kenyon. Takoradi to Santa Isabel. Observers, Mr. H. R. Owen, 3rd Officer and Mr. P. L. Brown, Radio Officer.

29th November 1961. During a heavy rain squall at 1235 GMT, a circular patch of agitated water was sighted about 50 yd. away on the starboard bow. On closer inspection it was found to be approximately 60 ft. in diameter, the water within the circle being whipped round in an anticlockwise direction. At the centre, the disturbed water was apparently about 18 in. above the level of the surrounding sea. Air temp. 81°F, wet bulb 78°, sea 82°. Wind NE'ly, force 3.

Position of ship: 4° 00'N, 6° 25'E. (Marine Observer, 32:174, 1962)

## WATERSPOUTS
**Carpenter, G. D. Hale, and Brunt, D.; *Nature,* 110:414-415, 1922.**

Waterspouts on Lake Victoria are very commonly seen from Entebbe, but at a long distance away, and though I have worked on the lake shores for nearly four years it was only two days ago that I first saw one near enough to be of real interest.

I was in camp on the north end of Bagalla, the largest island of the Sese Archipelago. The camp lay about 300 yards from the shore of a small bay. At daybreak on June 30 there were very lowering black clouds and every indication of an immediate heavy storm. While looking out from the tent I suddenly saw that a waterspout was travelling obliquely towards us, and as it eventually came to within about 100 yards of the shore a very good view was obtained for about five minutes before it came to an end.

The pedicle arose from a well-marked circular area on the water, which was otherwise only faintly rippled by the preliminary puff of wind before the approaching storm.

This circular area was evidently very violently disturbed as a cloud of vapour, greatly agitated, rose from it for a little distance.

The pedicle was extremely narrow at its lower end, and not quite straight, being sinuous in outline. It broadened out gradually into a column which went up into the low cloud; the core of this column was much less dense than the periphery, and the violent upward spiral ascent of the water could be clearly seen.

So far I have described nothing unusual, but the following was quite new to me and seemed of great interest.

Surrounding the central core, but separated from it by a clear narrow space, was a sheath, the lower end of which faded away some distance above the water. The profile of this sheath was undulating, it being thicker in some places than others. A curious point is that this sheath seemed to pulsate rhythmically, but I could not say whether the appearance of pulsation might not have been an illusion caused by waves travelling up its outer surface.

This pulsation gave an uncanny suggestion of a live thing, which was aided by the violent spiral movement upwards in the central core, the clouds of vapour boiling round its base, and the movement of the whole across the water---indeed, we watched it spellbound until the pedicle dissolved away at the bottom, and the ascent of the part above brought the phenomenon to an end.

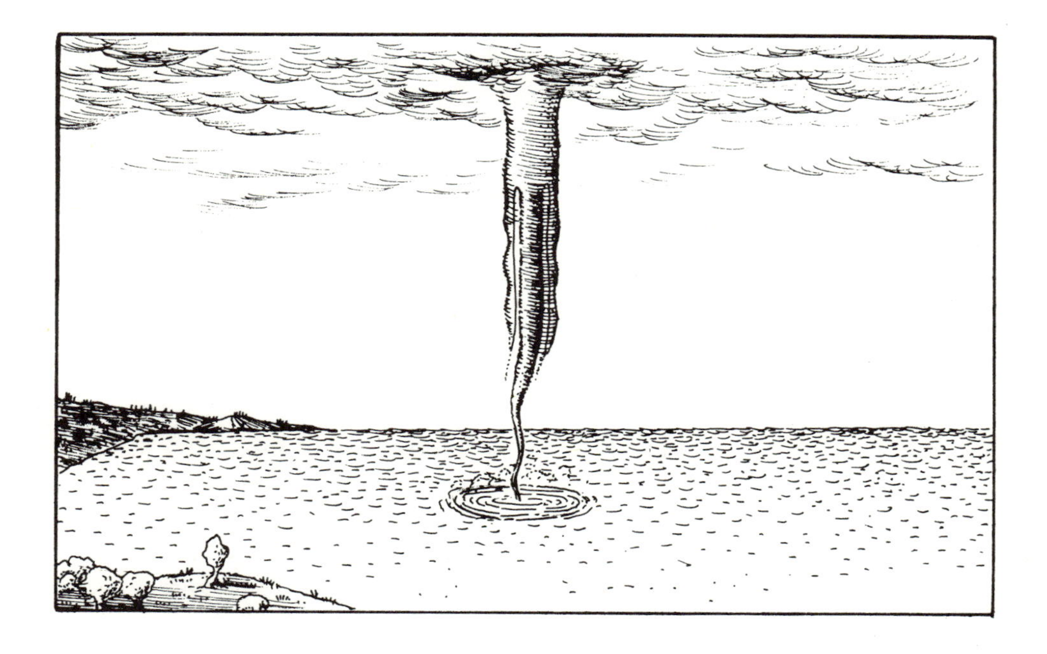

Double-walled waterspout

My wife watched with me, and is in entire agreement about the curious appearance of pulsation of the outer sheath.

Fig. 1 is a reproduction of a pencil drawing which may give some idea of what we saw. I cannot estimate the height to which the column rose. Its cessation was followed by violent rain and thunder. (G. D. Hale Carpenter)

Dr. Hale Carpenter's letter brings out one feature which has never, to my knowlege, been noted in a waterspout, namely, the sheath, separated from the main body of the whirl by a clear space. Wegener, in his book on "Wind- und Wasser-hosen in Europa," gives illustrations of a large number of waterspouts, but in no case is there mention of two trunks one within the other. The nearest approach to the phenomenon noted by Dr. Hale Carpenter is the not infrequent occurrence of waterspouts which show two clearly defined parts, an upper thick column with a lower whirl of much smaller thickness.

The accepted explanation of waterspouts is that they consist of whirls in rapid rotation with a discontinuity at the outer boundary. The rotation produces a rapid lowering of pressure within the whirl, and consequently a lowering of temperature, which may easily be sufficient to bring the air in the whirl down below its dew point. This is sufficient to explain the main features of the typical waterspout. The amount by which the temperature is lowered decreases outward from the "axis" of the whirl, while the difference between the air-temperature and dew point normally increases downward from the cloud level. The thickness of the visible column or zone of condensation therefore diminishes downward, giving the form of an inverted cone of irregular shape. Near the water the air is again near saturation, and the difference between air temperature and dew point is small, so that the base of the whirl is widened. It frequently happens that the portion at middle heights is not visible, on account of the relative dryness of the air.

F. J. W. Whipple (Meteorological Magazine, February 1922), writing on cloud pendants, shows that within a whirl of 20 metres diameter, rotating once in a second, with a lowering of pressure of 30 mb. at the centre, the maximum wind speed would be 70 metres per second, or 160 miles per hour, in agreement with winds estimated in tornadoes. A deficiency of 30 mb. pressure represents a suction sufficient to support 1 foot of water only. The solid appearance of a waterspout is therefore not to be ascribed to the lifting up into the air of a solid column of water, but is due partly to the condensation of water vapour within the whirl itself, and partly to water drops which are carried upward in spiral paths. This upward motion of water drops is a well-marked feature of most waterspouts.

The existence of an outer sheath, separated from the central core by a clear space, would appear to require a discontinuity of water content of the air, symmetrical about the axis of the whirl. It does not appear possible to explain it even as the effect of discontinuities of velocity within the whirl. No physical explanation of this clear space can be suggested.

It is usually suggested in text-books (for example, Humphrey's "Physics of the Air," p. 213) that waterspouts are formed at the boundaries of wind currents of different directions. But as such boundaries are of considerable extent, it is difficult to understand why single waterspouts ever come into existence. One would rather expect to find large families of waterspouts distributed over a considerable area. It is true that usually several are seen at the same time, but isolated cases are not infrequent.

The fact that Dr. Hale Carpenter, while standing within about one hundred yards of the waterspout he describes, apparently felt no wind from the whirl, testifies to the very limited diameter of the whirl in question. (D. Brunt) (Nature, 110:414-415, 1922)

## CLOUD PHENOMENON
**Haselfoot, F. E. B.; *Marine Observer,* 6:102, 1929.**

The following is an extract from the Meteorological Log of H. M. S. Herald, Captain F. E. B. Haselfoot, D. S. O., R. N. Observer, Lieutenant R. H. Kennedy, R. N.

"At 1600 hours on the 10th May, 1928, in Latitude 4° 10' N., Longitude 112° 14'E. Wind S. W. force 2. Sky three parts clouded with heavy Cumulus. Rain squalls round the horizon. Observed peculiar cloud formation to Eastward, Altitude about 3500'. It appeared as if a heavy rainfall was taking place from one cloud to another. It was quite apparent that no rain was falling into the sea.

"This rain was 'funnel'-shaped, and it is considered that a small waterspout had formed in the sky and was sucking rain from a heavy cumulus cloud to another cloud (small, fluffy and whitish) above it." (Marine Observer, 6:102, 1929)

## WATERSPOUTS BETWEEN CLOUDS
**Major, T. W.; *Marine Observer,* 6:127, 1929.**

The following is an extract from the Meteorological Report of S. S. Dryden, Captain T. W. Major, Liverpool to Montevideo. Observer, Mr. G. Oldfield, 2nd officer.

"14th June, 1928, at 1.55 p.m., a nimbus cloud bearing W. N. W. and moving in a N. N. E'ly direction was observed to be connected to a bank of cumulus cloud by two waterspouts, which did not reach sea, but merely stretched between the two cloud-banks. Position of ship at Noon, Latitude 6° 27' N., Longitude 27° 36' W." (Marine Observer, 6:127, 1929)

Waterspouts connecting clouds

# WHIRLWINDS AND DUST DEVILS

While they are rotary disturbances like tornadoes and waterspouts, whirlwinds and dust devils seem different phenomena altogether. They are considerably smaller in size and exist without menacing clouds overhead; they are in fact characteristic of warm, sunny days. Whirlwinds and dust devils, however, may be violent on occasion, sometimes appearing very suddenly with a peculiar detonation or rumbling sound. The "pranks" of whirlwinds and dust devils are legend, and we describe only a few of their capricious antics here.

A few very rare whirlwinds display the fire and smoke of some of the tornadoes and meteor-like phenomena described elsewhere. Such revolving masses of fire and smoke may all belong together regardless of their size or apparent origin.

## EXPLOSIVE ONSET OF WHIRLWINDS

### SINGULAR PHENOMENON
**Anonymous; London *Times,* July 5, 1842.**

Wednesday forenoon [June 29] a phenomenon of most rare and extraordinary character was observed in the immediate neighborhood of Cupar. About half past 12 o'clock, whilst the sky was clear, and the air, as it had been throughout the morning, perfectly calm, a girl employed in tramping clothes in a tub in the piece of ground above the town called the common, heard a loud and sharp report overhead, succeeded by a gust of wind of most extraordinary vehemence, and only of a few moments duration. On looking round, she observed the whole of the clothes, sheets, &c., lying within a line of certain breadth, stretching across the green, several hundred yards distant; another portion of the articles, however, consisting of a quantity of curtains, and a number of smaller articles, were carried upwards to an immense height, so as to be almost lost to the eye, and gradually disappeared altogether from sight in a south-eastern direction and have not yet been heard of. At the moment of the report which preceeded the wind, the cattle in the neighboring meadow were observed running about in an affrighted state, and for some time afterwards they continued cowering together in evident terror. The violence of the wind was such that a woman, who at the time was holding a blanket, found herself unable to retain it in fear of being carried along with it! It is remarkable that, while even the heaviest articles were being stripped off a belt, as it were, running across the green, and while the loops of several sheets which were pinned down and snapped, light articles lying loose on both sides of the holt were never moved from their position. (London Times, July 5, 1842)

### A CURIOUS PHENOMENON
**Anonymous; *Scientific American,* 43:25, 1880.**

The Plaindealer, of East Kent, Ontario, states that a curious and inexplicable phenomenon was witnessed recently by Mr. David Muckle and Mr. W. R. McKay, two citizens of that town. The gentlemen were in a field on a farm of the former, when they heard a sudden loud report, like that of a cannon. They turned just in time to see a cloud of stones flying upward from a spot in the field. Surprised beyond measure they examined the spot, which was circular and about 16 feet across, but there was no sign of an eruption nor anything to indicate the fall of a heavy body there. The ground was simply swept clean. They are quite certain that it was not caused by a meteorite, an eruption of the earth, or a whirlwind. (Scientific American, 43:25, 1880)

### A WHIRLWIND IN SURREY
**Anonymous; *Symons's Monthly Meteorological Magazine,* 29:90, 1894.**

Mr. Sidney H. Burchall writes from Reigate under date of June 21st: Whilst standing to-day in an open hay field, situated about a mile south of this town, I and others witnessed what may be described as a whirlwind, but of such unusual character that it may perhaps be worth recording. The field lies on high ground, exposed to the south, and at the time of my visit about one o'clock, several men were engaged in turning over a heavy crop of hay. The day had been a brilliant one, and was then very warm, and I could not perceive any movement of the trees which bounded the field on the east and west sides. Without warning an extraordinary whirlwind broke on the field within a few yards of where I stood, catching up the hay with violence. For some moments the air was filled with wisps of hay, whirling in rapid movements, and in circles gradually getting larger and larger as they rose in the air; but, what was so striking, rising to a great height. At the same time that this happened several rooks were passing. These appeared also to be caught by the tornado, and were likewise carried up to a great elevation; in fact, until I could no longer follow them. The passage of the storm was from west to east, but it seemed to travel very slowly, and was ultimately dispelled by the trees at the opposite end of the field. (Symons's Monthly Meteorological Magazine, 29: 90, 1894)

## CURIOUS ACTIONS OF WHIRLWINDS

### A REMARKABLE WHIRLWIND
**Capes, J. L.; *Nature,* 135:511, 1935.**

Dust-devils, or rotating columns of sand travelling rapidly across open spaces, are not uncommon objects to desert travellers. Their height and breadth is often very considerable and the evidence of the eddies causing them very great.

The smallest of the type I have seen was only about 5 ft. high, that is, the visible column of sand, and less than a foot in diameter. It passed so close to me that it was easy to see its narrow cycloidal path marked on the sand, which was deposited and lifted as the eddy travelled on at not less than 15 miles an hour, although the wind was actually very light.

I recently encountered a much more remarkable example while walking over a smooth surface of desert on a flat, calm day. Hearing a swishing sound behind me, I turned and observed a large revolving ring of sand less than a foot high approaching me slowly. It stopped a few feet away and the ring, containing sand and small pieces of vegetable debris in a sheet less than one inch thick, revolved rapidly around a circle of about 12 ft. diameter while the axis remained stationary. It then moved slowly around me after remaining in one spot for at least thirty seconds, and slowly died down. It would be interesting to know if others acquainted with the desert have come across similar examples of a broad, flat eddy.

The ancient superstition among desert tribes that these whirlwinds are spirits, called 'afrit' or 'ginni' (the 'genii' of the "Arabian Nights"), would seem to have a reasonable foundation in face of such an 'inquisitive' apparition. (Nature, 135:511, 1935)

## THE TALE OF — A GUST
**Godden, William; *Symons's Meteorological Magazine,* 46:54, 1911.**

I hope you will see your way to take a similar course with reference to a report from Bradford, which appeared in the daily papers recently. A gust of wind is said to have carried a girl to a height of twenty feet or so, whence she fell with fatal result. Onlookers seem to have experienced no inconvenience at all. Surely on the face of it a more preposterous story never appeared in print. What was the strength of the gale: and did nothing else of consequence happen owing to the wind at Bradford on the morning of 23rd February? (William Godden)

[Acting on this suggestion, we communicated with Mr. H. Lander, the rainfall observer at Lister Park, Bradford, who kindly sent us a copy of the Yorkshire Observer for February 25th, in which there was a fairly full report of the inquest on the school-girl who was undoubtedly killed by a fall from a great height in an extremely exposed playground during very gusty weather. One witness saw the girl enter the playground from the school at 8.40 a.m., and saw her carried in three minutes later. Another witness saw the girl in the air parallel with the balcony of the school 20 feet above the ground, her arms extended, and her skirts blown out like a balloon. He saw her fall with a crash. The jury found a verdict, "Died as the result of a fall caused by a sudden gust of wind."---(Ed. S. M. M.) (Symons's Monthly Meteorological Magazine, 46:54, 1911)

## A WATERSPOUT
**Gray, J.; *Nature,* 5:501, 1872.**

On Saturday last, April 16, whilst fishing in the river Elwy, at a point about two miles above the well-known Cefn caves, and five from St. Asaph by the river, I witnessed a very singular phenomenon. My attention was suddenly called up-stream by a remarkably strange hissing, bubbling sound, such as might be produced by plunging a mass of heated metal into water. On turning I beheld what I may call a diminutive waterspout in the centre of the stream, some forty paces from where I was standing. Its base, as well as I could observe, was a little more than two feet in diameter. The water curled up from the river in an unbroken cylindrical form to a height of about fifteen inches, rotating rapidly, then diverged as from a number of jets, being thrown off with considerable force to an additional elevation of six or seven feet, the spray falling all round as from an elaborately arranged fountain, covering a large area. It remained apparently in the same position for about forty seconds, then moved slowly in the direction of the right bank of the river, and was again drawn towards the centre, where it remained stationary as before for a few seconds. Again it moved in the former direction, gradually diminishing and losing force as it neared the bank, and finally collapsed in the shallow water. Strange to say, its course was perpendicular to the bank and not with the current.

At the time of the occurrence the river was still high, from the recent heavy rain, though the depth of water at the spot where I first observed it was not more than four feet. The current, of course, was stronger than usual, but presented a comparatively smooth surface. The day was fine and sunny, with a slight breeze from the S. E. The event occurred about 12. 15, and lasted seventy or eighty seconds, as well as I could judge. The atmosphere in the immediate vicinity seemed, from the way in which the spray was scattered, to be somewhat agitated; but my impression was that such agitation was the result of the phenomenon, rather than its cause. I had fished over the spot a few minutes previously, and examined it afterwards with great care, but saw nothing to account for the wonder. (Nature, 5:501, 1872)

## WHIRLWIND ON THE THAMES
**Hall, H.; *Royal Meteorological Society, Quarterly Journal,* 22:294-295, 1896.**

[On the Thames, September 20, 1896] Ten minutes or a quarter of an hour afterwards I was taking off my oilskins on deck and remarking the peculiar appearance of the departing squall, which looked like a row of webs suspended from sky to sea, when an exclamation from the helmsman called my attention to a new and very terrible danger. In two minutes we had every stitch of sail off the boat, although, being in a narrow channel, we were in imminent risk of drifting aground, and had leisure to observe the phenomenon with comparative equanimity. We were off Reculvers; the time was about 4 p. m. On our port bow was a circle (ten or twenty feet in diameter) of boiling steaming water, travelling down with the wind, though apparently in no very straight line, and revolving rapidly the while in "the way of the sun" upon its own centre. Perhaps the best idea I can give of it is by asking you to imagine a cauldron of the diameter mentioned immersed to the brim and full of water in violent ebullition, the water leaping up in places and covered with spray and spoondrift, which in the upper layers were so finely divided as to look like steam, the whole whirling round with extraordinary rapidity. There was no appearance of a waterspout---perhaps because it was not raining at the time---and the "steam" did not rise more than two or three feet above the surface of the surrounding (normal) waves. The phenomenon, that is to say the effect of the whirlwind, was visible only on the surface of the water. I cannot describe the weirdness and wickedness of its appearance, an effect principally due to the impression which it produced upon the mind that it was a living, sentient, and evil thing. A paid hand on board compared it to "a nest of white serpents twirling and twisting and gambolling," and spoke of the apparition as he. He said that in fifteen years' experience off the east coast as fisherman and yacht hand, he had never seen anything like him. He passed us at a distance of about 200 yards in about 2 fathoms of water. We saw him pass two other yachts and a barge about half a mile astern, which, like us, had lowered all sail upon his approach. He seemed to be making straight for an excursion steamer which was coming through the Horse Channel, and we expected to see the air filled with hats and parasols and "presents from Margate," but he passed astern of the steamer and we saw him no more, being too busily occupied in getting up sail to watch him further. The sky was dull (white), but the light was good, and the clouds (white) apparently high, but a quarter of an hour or so after the thing had passed there was a flash of lightning and a peal of thunder so simultaneous and startling that I instinctively glanced at the masthead, and, indeed, I think the discharge must have passed through the shrouds (steel) into the sea, which at the time was washing over the lee rail. I may add that at five o'clock we saw two mock suns, and that for half an hour or so after the moon rose the reflection of frequent flashes of lightning was visible between her and the horizon. (Royal Meteorological Society, Quarterly Journal, 22:294-295, 1896)

## A NARROW SQUALL
**Clark, J. Edmund; *Symons's Meteorological Magazine,* 49:52, 1914.**

On 19th July, 1913, a narrow strip of wind 1-1/2 ft. wide, drove through a hedge lifting newly cut hay 60 yards high, proceeded through a tree with great violence, blowing off the leaves of a hedge and tree, and upset a hay-cock in an adjoining field. It lasted some 10 minutes and was accompanied by great noise. I was informed by an old inhabitant such occurrences were frequent at Coquet Head in the Cheviots. I doubt they are all as narrow as this was. (Symons's Monthly Meteorological Magazine, 49:52, 1914)

## METEOROLOGICAL PHENOMENON
**Galton, Francis; *Nature,* 44:294, 1891.**

I have received in a letter from a friend residing in Boraston, Shropshire, the following account of a remarkably interesting meteorological phenomenon, which is well worth putting on record:---

"We had a curious sight from this house yesterday [July 26]. It was a dead calm, but in a field just below the garden, with only one hedge between us and it, the hay was whirled up high into the sky, a column connecting above and below, and in the course of the evening we found great patches of hay raining down all over the surrounding meadows and our garden. It kept falling quite four hours after the affair. There was not a breath of air stirring as far as we could see, except in that one spot." (Nature, 44:294, 1891)

## WHIRLWIND SUCKS UP CARP AND PIKE
**Anonymous; *Nature,* 125:728-729, 1930.**

May 15, 1586. Waterspout.---At Kestrzan in Bohemia a powerful whirlwind carried the water of two ponds into the air with all the carp and pike that they contained. On the following day there was a waterspout on Lake Constance near Meilen. (Nature, 125:728-729, 1930)

## NOTES OF LOCAL WHIRLWINDS IN NEW BRUNSWICK
**Kain, Samuel W.; *Monthly Weather Review,* 28:488-489, 1900.**

While Mr. Barber was standing by the side of a pool of water about six miles from Clarendon Station, Charlotte County, he heard in the distance a shrieking whistling sound; this continued to increase in intensity, and turning to seek a cause he noticed a whirlwind advancing from the hills, its course indicated by the swaying shrubs and a noise somewhat like that produced by a express train, but not so loud. It struck the pool about three feet from the shore and raised the water in a foaming mass of froth and spume to a height of 5 feet, and crossing threw the water upon the farther shore. Its path across the pool was about 15 feet wide. Mr. Barber was standing about one hundred feet from the path of the whirlwind. The sky was clear all day. In the morning there were a few light clouds, but after 2 p.m. the sky was cloudless.

The wind was northeast till noon, then shifted to southwest and south. It was light all day. The barometer was steady; at 8 a.m., 30.081; at 2 p.m., 30.129; at 8 p.m., 30.185. In St. John the highest temperature was 65°F., but at Clarendon the temperature was probably about 5° higher, the preceding days had been cold, and the change in temperature as considerable and rapid.

A very similar phenomenon was observed on Wednesday, June 14, at Grassy Lake, Kings County.

Dr. Colter, post office inspector, and Mr. Richard Magee, of the railway mail service, were fishing from a moored boat on the lake. It was a fine, clear day, and a good breeze was blowing when about 2 o'clock in the afternoon they heard a roaring in the woods, and with a rushing noise a few hundred feet from them the water of the lake commingled with reeds, lillies, and mud was torn up and hurled into the air, forming a waterspout apparently about 30 feet in diameter.

It lasted about two minutes, and for about that time the air seemed somewhat darkened. The violence of the wind drew their boat from its moorings in among the reeds, and it was fortunate that they were far enough from the path of the whirlwind to escape any more serious results. (Monthly Weather Review, 28: 488-489, 1900)

# STEAM DEVILS

## "STEAM DEVILS" OVER LAKE MICHIGAN DURING A JANUARY ARCTIC OUTBREAK

**Lyons, Walter A., and Pease, Steven R.; *Monthly Weather Review*, 100:235-237, 1972.**

January 1971 was one of the colder winter months of this century in Wisconsin, and, because Lake Michigan remained largely ice-free, there were many spectacular manifestations of the Great Lakes influence upon mesoscale meteorological conditions such as snow squalls. The upcoming (1972-73) International Field Year for the Great Lakes (IFYGL) will, among other things, be concerned with the various aspects of air-water interaction during winter from both the theoretical and observational viewpoints. In light of the recent interest in small-scale atmospheric vortexes, such as the waterspout studies of Golden (1971) and the dust devil investigations of Kaimal and Businger (1970), we would like to suggest an additional phenomenon for possible consideration by IFYGL investigators---the "steam devil".

Figure 1 [not reproduced] shows several well-developed steam devils a few miles off the Milwaukee shoreline of Lake Michigan. The period Jan. 30-31, 1971 was intensely cold in Milwaukee, the highest temperature reaching only -6°F with west-northwest winds frequently gusting to 40 mi/hr. Lake snow squalls were present in excelsis along the lee shores of all the Great Lakes. But most interesting was the view from the upwind shoreline. During both days, one could see the typical shallow, dense steam fog near the lake surface (water temperature 33°F) and the cumulus clouds building a few miles offshore. What was unusual were the numerous distinct fingers or columns of vapor swirling out of the steam fog layer directly into the overlying cumulus clouds. It is estimated that they were 50-200 m in diameter, traveled more or less with the wind, were nearly vertical, and showed a slow but distinct rotation (mostly cyclonic) of up to several revolutions per minute. The steam devils tended to be rather short lived, the longest surviving perhaps for 3 or 4 min.

An even more interesting view of the same phenomena was taken from a commercial airliner on Jan. 30, 1971...... Visible are small cumulus (which were forming further offshore on this day), plus the steam devils and a highly patterned effect on the surface steam fog. It definitely appears that there were quasi-hexogonal cells elongated along the surface wind direction, the largest steam devils being present at the vertexes of the hexagons. Robert W. Pease, of the University of California at Riverside, who took this picture, estimated the height of the tallest steam devil to be approximately 1, 500 ft by using photogrammetric techniques.

Earlier this same winter, time lapse films of the late under similar conditions, taken from atop a tall building in Milwaukee, showed a similar patterned effect to the steam fog. The steam devils were clearly seen to be areas of condensate rising from the surface directly to cloud base, although on that day no rotation was apparent.

The literature has several references to possibly related phenomena. Falconer et al. (1964) show a picture similar to ours over Lake Ontario. In a review of papers on sea smoke and steam fog, Saunders (1964) quotes several authors whose observations suggest similar structures. Lenshow (1965) shows a picture of what appears to be a distinct waterspout over Lake Michigan during a period of extreme cold-air advection.

The question arises as to what actually is the structure of this phenomenon. Are we dealing with a phenomenon dynamically related to and as vigorous as the water-spouts which inhabit the Florida Keys region, for example? It seems unlikely. While rotation is apparent, it is quite slow. Perhaps these are much more similar in nature and intensity to the dust devil over land. For this reason, we suggest calling them steam devils rather than water-spouts. However, it is hard to be certain until actual field measurements are made. Although steam fog is common-place over the lake during periods of cold advection, the fingerlike extensions to cloud level are rare and appear only when winds are quite strong (in excess of 25 mi/hr), though this is not to say they cannot occur without condensate making them visible. The one set of available time lapse films of this phenomenon suggests that steam devils are related to traveling subcloud convective plumes rising out of surface superadiabatic layers as described by Warner and Telford (1967). Although the observational difficulties would be formidable, it seems this phenomenon should merit attention during the IFYGL inasmuch as it may be one of the key factors in our understanding of the intense heat, moisture, and momentum transports that occur over the Great Lakes in winter. (Monthly Weather Review, 100:235-237, 1972)

# ELECTRICAL CHARACTER OF DUST DEVILS

## THE ELECTRIC FIELD OF A LARGE DUST DEVIL
**Freier, G. D.; *Journal of Geophysical Research,* 65:3504, 1960.**

At the time of obtaining measurements on the earth's electric field during an eclipse of the sun in the Sahara Desert at Kidal, French West Africa, an interesting record was obtained on the electric field associated with a large dust devil. This record is shown in Figure 1, where positive values of the field imply positive charge above

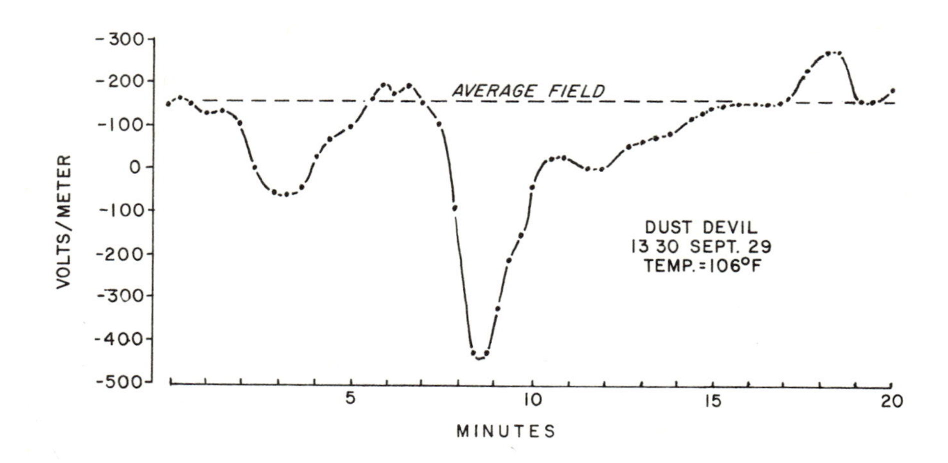


Electric field measured during passage of a dust devil

the earth. The dust devil was estimated to be between 100 and 200 meters high and about 8 meters in diameter, and its distance of closest approach to the electric field mill was about 30 meters. It was of sufficient strength to raise an army-type sleeping cot several meters into the air, and it was followed by truck for about 1 mile before it dissipated its energy. Its speed across country was about 4 m/sec.

The two negative excursions with a tendency to go positive between them suggests a sort of dipole structure for the dust devil with negative charge above (inverted thunderstorm). (Journal of Geophysical Research, 65:3504, 1960)

# WHIRLWINDS OF FIRE AND SMOKE

## FIERY WHIRLWIND
**Anonymous; *Monthly Weather Review,* 9:19, 1881.**

In northern Kentucky, southern Ohio and Indiana, violent wind and rain storms, causing great damage to crops and buildings were experienced on the 21st. Americus, Ga., 18th, at some distance from the town a small whirlwind, about 5 feet in diameter and sometimes 100 feet high, formed over a corn-field where it tore up the stalks by the roots and carried them with sand and other loose materials high into the air. The body of the whirling mass was of vaporous formation and perfectly black, the centre apparently illuminated by fire and emitting a strange "sulphurous vapor" that could be distinguished a distance of about 300 yards, burning and sickening all who approached close enough to breath it. Occasionally the cloud would divide into three minor ones, and as they came together again there would be a loud crash, accompanied by cracking sounds, when the whole mass would shoot upwards into the heavens. (Monthly Weather Review, 9:19, 1881)

## THE NEWBOTTLE WHIRLWIND OF NOV. 30th, 1872
**Beesley, T.; *Symons's Monthly Meteorological Magazine, 8:150,* 153, 1873.**

"May I ask you to insert the following account of a curious atmospheric phenomenon which occurred here to-day [Banbury, November 30, 1872]? About 12 o'clock we had a heavy storm of rain and hail, in the middle of which there was a very vivid flash of lightning, with almost instantaneous thunder of a very peculiar rattling sound. About five minutes afterwards, as I was leaving the house, my gardener called me to come quickly and see the ball of fire. I was unfortunately half a minute too late, but I have seen four persons who saw it from different points, and who all agree they heard a whizzing, roaring sound like a passing train, which attracted their attention, and then saw a huge revolving ball of fire travelling from six to ten feet off the ground. The smoke was whizzing round and rising high into the air, and a blast of wind accompanied it, carrying a cloud of branches along and destroying everything in its way. The havoc done is very considerable---large trees bodily uprooted, others broken off about ten feet from the ground, others have all their branches snapped off; in one place about 100 yards of a wall laid flat, and the remainder thrown over at intervals, as if the ball had rebounded, and some of the stones carried ten yards off. I rode this afternoon along the whole line of its journey, about two miles in length; the direction was first from S. W. to N. E., and near the end it turned N. W. Where it first began the breadth of ground travelled over was very narrow, but increased as it proceeded, till in the last field the debris covered a space quite 150 yards wide, and here it seems to have exhausted itself, as all the witnesses agree that the ball of fire seemed to vanish at this spot without any explosion. Here the ground had been cut in places as if by a cannon ball, but I could find no cause for this, and I saw no signs of fire on its route. One man, however, says there was a strong sulphurous smell after it had passed. About the time of this occurrence, my farm men at work about a mile in quite another direction, saw the water of a pond carried up into the air by whirlwind. The wind all day has been light from S. S. E. My pocket aneroid (made by Bryson, of Edinburgh) stood this morning at 28.45. At this moment it stands at 28.27, showing that the atmospheric disturbance has not yet begun to subside."

..........

William Marshall, gardener at Newbottle Manor, was returning from the stables to the house. He heard a noise like a long railway train crossing a bridge, and saw leaves and branches whirled into the air above the Spinney, and immediately afterwards "a dark ball, as big as a carriage," and sending up "a cloud of smoke," come out of the trees with a shower of branches, and roll "over and over," down the hill in the direction of the bridle road; the cloud of smoke at the same time whirling "round and round" with a "buzzing noise." He distinctly saw sparks of a red colour emitted from the ball about six feet from the ground, and this is confirmed by another man, William Jilson, of Astrop, who, from a field on the west, saw fire and ran away affrighted. Notwithstanding the natural tendency in uneducated observers to assume that "where there is smoke there is fire," I am disposed to credit these observations, although not supported by the other spectators, who, however, had not so good a view as these men had. It has been suggested that these sparks were produced by the friction of branches and twigs whirled round by the rotatory current; but I think it more likely that they were connected with electrical discharges consequent upon the friction of currents of air. Mr. Cartwright and Marshall went over the ground a few minutes after the occurrence; but they saw no trace of charred wood, nor anything to indicate combustion, and no smell of burning or of "sulphur" was perceptible. No other witness noticed the "rolling ball," but none other saw the cloud as it came out from amongst the trees loaded with leaves and branches. (Symons's Monthly Meteorological Magazine, 8:149-154, 1873)

# TEMPERATURE ANOMALIES

## GREAT HEAT IN A THUNDER SQUALL
**Parkinson, F. B.; *Symons's Meteorological Magazine,* 46:201, 1911.**

It may interest you to know that at 9 p.m., on the 20th September, a thunder cloud approached from the west, bringing with it a squall of wind that caused the temperature to rise in a few minutes to 110°F. By 9.45 p.m. it had fallen again to 67°, which I expect was about the temperature before the squall. I do not think my thermometer responded quick enough to register the highest point, but it is safe to say it rose 40° in five minutes. (Symons's Meteorological Magazine, 46:201, 1911

## THE REMARKABLE TEMPERATURE FLUCTUATIONS IN THE BLACK HILLS REGION, JANUARY 1943
**Hamenn, Ronald R.; *Monthly Weather Review,* 71:29-32, 1943.**

January 13, 1913. Rapid City. 8 a.m. temperature -17°F; 10 p.m. temperature 47°F; an increase of 64° in 14 hours.

January 22, 1943. Rapid City. Beginning at 7:32 a.m. the temperature rose 49° in two minutes---from -4° to 45°F. Frost appeared almost instantaneously on car windshields. (Monthly Weather Review, 71:29-32, 1943)

## HOT WEATHER AND BURNING WINDS
**Anonymous; *Scientific American,* 3:106, 1860.**

In every quarter of the south-west, the heat of the present summer appears to have been unprecedented. In Montgomery, Ala., the thermometer stood for several days at 103° in the shade. In Mississippi, Louisiana, and Missouri, it has ranged from 95° to 105° in the shade, and the people call it "the fiery term."

Several currents of intensely hot air have been experienced which appear to be similar to those which are common in Egypt, Persia, and some portions of India. A hot wind extending about 100 yards in width, lately passed through middle Georgia, and scorched up the cotton crops on a number of plantations. A hot wind also passed through a section of Kansask it burned up the vegetation in its track and several persons fell victims to its poisonous blast. It lasted for a very short period, during which the thermometer stood at 120°---far above blood heat. (Scientific American, 3:106, 1860.

# Chapter 4
# MYSTERIOUS NATURAL SOUNDS

## INTRODUCTION

When the Sourcebook Project began, the number and variety of mysterious natural sounds was not foreseen. Strange lights, mirages, and weather oddities were believed to constitute the majority of strange phenomena. The famous Barisal Guns were recognized, of course, due to the publicity given them by Sir George Darwin. Additional curious sounds have been discovered in our foray through the literature at a modest but surprising rate. Even now, after years of search, this chapter cannot be considered close to completion, for how many individuals take the time to write to scientific journals about ephemeral elusive sounds? For that matter, who, in these noisy latter days, can separate natural and man-made sounds? Almost all strange sounds reported below have been heard either in remote spots of the globe, where almost all sounds heard are natural, or in earlier and quieter days.

Modern technology for all of its masking of natural sounds has also led us to the knowledge of infrasound; that is, pressure waves at very low frequencies. As infrasound observations accumulate, puzzling data will doubtless be found. Much infrasound seems to originate with the auroras, with weather systems, and with seismic activity---just those phenomena that give rise to many of the curious observations reported here from all sensory channels.

# EXTRAORDINARY DETONATIONS

In this day of sonic booms and frequent blasting, detonations heard afar are written off as manmade with good reason. Eighty years ago the world was quieter and people tried to track down the causes of strange sounds. For example, E. Van den Broeck compiled several hundred pages of testimony on "mistpouffers" in the French scientific journal Ciel et Terre (Sky and Earth) in 1895 and 1896. Mistpouffers are dull, distant explosive sounds heard around the coast of Europe all the way to Iceland. Every country has its own name for them, and it turns out that they are also heard on North American and Asian coasts, too. Despite the universality of unexplained detonations and despite ample proof of their existence, they are a dead issue in the scientific world.

The most famous unexplained detonations are the Barisal Guns heard in the vicinity of the Ganges Delta. Waterguns, however, are not confined to salt water. Some of New York's finger lakes have their own private waterguns; so do lakes in Europe. Doubtless, many waterguns have never seen print in scientific journals and are thus excluded from this collection. We have to assume that they are as common on freshwater shores as salt.

"Land guns" are also ubiquitous. Residents in dozens of localities all over the world hear strange detonations year after year, often emanating from beneath their feet, sometimes apparently from the atmosphere overhead. Scientists ascribe these noises to seismic hot spots, where subterranean forces are slowly cracking rocks below the surface. Some but not all of these explosive sounds can be correlated with earth tremors, lending credence to the seismic explanation.

Those elusive detonations accompanied by "swishes" and sounds like escaping steam may tentatively be assigned to meteors. Here, we include only cases where there are no visible manifestations of meteors. Presumably clouds and distance may conceal some of the sound-producing events. There is always the possibility, though, that some of the detonations blamed on meteors, especially those in the interior of Australia, may actually be seismic.

## MISTPOUFFERS OR "FOG GUNS"

### SEISMIC AND OCEANIC NOISES
**Kain, Samuel W., et al; *Monthly Weather Review,* 26:153, 1898.**

Everybody who has been much upon our Charlotte County coast must remember that upon the still summer days, when the heat hovers upon the ocean, what seem to be gun or even cannon reports are heard at intervals coming from seaward. The residents always say, in answer to one's question: "Indians shooting porpoise off Grand Manan." This explanation I never believed; the sound of a gun report could not come so far, and, besides, the noise is of too deep and booming a character. I have often puzzled over the matter, and it is consequently with great pleasure that I find in Nature for October 31, 1895, a short article of Prof. G. H. Darwin, in which he calls attention to the occurrence of what is obviously the same phenomenon in the delta of the Ganges, upon the coast of Belgium, and in parts of Scotland, and in which he asks for experiences from other parts of the world. Two explanations are suggested by his correspondent, M. Van den Broeck, of Belgium, who called his attention to the phenomenon, one that the reports are of atmospheric origin, due to peculiar electrical discharges; the other that they are internal in the earth, due, perhaps, to shock of the internal liquid mass against the solid crust. The following number of Nature contains notes which suggest that the reports may accompany the formation of faults or may result from earthquakes too slight to be otherwise perceived, and later numbers of that journal contain numerous letters upon strange sounds heard in different parts of the world, with various explanations.

The discussion upon the subject by this society on December 3, 1895, has called out further information showing that others besides myself have noticed these or similar sounds in New Brunswick. The late Edward Jack, a keen observer of things in nature, wrote me under date December 13, 1895, "I have often noticed in Passamaquoddy Bay, when I was duck shooting in the early spring mornings, the noises of which you speak; they always seemed to come from the south side of the bay. They resembled more the resonance from the falling of some heavy body into the water than that of the firing of a gun, such as is produced by a cake of ice breaking away from a large sheet of it and toppling over into the sea. These noises were heard by me only in very calm spring mornings when there was no breath of air; * * * there was nothing subterranean in them." Capt. Charles Bishop, of the schooner Susie Prescott, has told Mr. S. W. Kain that he has heard these sounds 40 miles from land between Grand Manan, the Georges Banks, and Mount Desert Rock. They are reported also from the Kennebecasis. Mr. Keith A. Barber, of Torryburn Cove, wrote December 26, 1895, to this society: "I have heard sounds similar to those on the Kennebecasis in the warm days of summer. They seemed to come from a southeasterly direction." Mr. Arthur Lordly, a member of this society who resides in the summer at Riverside, has also told Mr. Kain that he has heard similar sounds, on clear warm days, on the Kennebecasis, from a southwest direction. No other reports of this occurrence in New Brunswick have reached me. The Scientific American (June 27, 1896, p. 403) has called attention to them and requested that observations be communicated to its columns, but apparently so far without result. (Monthly Weather Review, 26:153, 1898)

## MISTPOUFFERS IN THE PHILLIPPINES
**Anonymous; *Nature,* 85:451, 1911.**

The Rev. M. Saderra Maso, who has for many years studied the earthquakes of the Philippine Islands, is now turning his attention to the subterranean noises known in other countries under various names, such as mist-poeffeurs, marinas, brontidi, retumbos, &c. In the Philippines many terms are used, generally signifying merely rumbling or noise, while a few indicate that the noises are supposed to proceed from the sea or from mountains or clouds. Most of the places where they were observed lie along the coasts of inter-island seas or on enclosed bays; very few are situated on the open coast. The noises are heard most frequently at nightfall, during the night and in the early morning, especially in the hot months of March, April, and May, though in the towns of the Pangasinan province they are confined almost entirely to the rainy season. They are compared in 70 per cent. of the records to thunder. With rare exceptions, they seem to come from the mountains inland. The instances in which the noises show any connection with earthquakes are few, and observers usually distinguish between them and the low rumblings which occasionally precede earthquakes. It is a common opinion among the Filipinos that the noises are the effect of waves breaking on the beach or into caverns, and that they are intimately connected with changes in the weather, generally with impending typhoons. Father Saderra Maso is inclined to agree with this view in certain cases. The typhoons in the Philippines sometimes cause very heavy swells, which are propagated more than a thousand kilometres, and hence arrive days before the wind acquires any appreciable force. He suggests that special atmospheric conditions may be responsible for the great distances to which the sounds are heard, and that their apparent inland origin may be due to reflection, possibly from the cumulus clouds which crown the neighbouring mountains, while the direct sound-waves are shut off by walls of vegetation or inequalities in the ground. (Nature, 85:451, 1911)

## MISTPOUFFERS ON THE GULF OF MEXICO
**Cooper, W. S.; *Scientific American,* 75:123, 1896.**

On the evening of December 28, 1885, I was with a companion in a sailboat on the Gulf of Mexico about 20 miles S. E. of Cedar Keys, Florida. We were becalmed. The next morning the sky was cloudless. There was a light fog and no breeze. The atmosphere was bracing but not frosty. We were about ten miles out but in shallow water. Shortly after sunrise we heard reports as of a gun or distant cannon. They came at intervals of about five minutes. We were not certain as to the direction. My companion, who lived several miles further down the coast, said he had often heard the reports on still mornings. (Scientific American, 75:123, 1896)

## OCEANIC AND SEISMIC NOISES
**Anonymous; *Monthly Weather Review,* 26:216, 1898.**

The mysterious phenomenon known as "Barisal guns," or "Mist poeffers," forms the subject of a useful paper by Dr. A. Cancani, in the last Bollettino, Vol. III, No. 9, of the Italian Seismological Society. The observations on which his discussion is founded are collected from places in or near the inland province of Umbria, where the noises are known as "marina," it being the popular belief that they come from the sea. The sound is quite distinct and easily recognized; it is longer than that of a cannon shot, and though more prolonged and dull, it is not unlike distant thunder. It invariably seems to come from a distance, and from the neighborhood of the horizon, sometimes apparently from the ground, but generally through the air. The weather, when the "marina" is heard, is calm as a rule, but that it often precedes bad weather is shown by the common saying, "Cuando tuona la marina o acqualo vento o strina" (when the ocean thunders, expect rain or wind or heat). The interval between successive detonations is very variable, sometimes being only a few minutes or even seconds. They appear to be heard at all times of the day and year, the experience of observers differing widely as to the epochs when they are heard most frequently. With regard to the origin of the "marina," Dr. Cancani concludes that they can not be due to a stormy sea, because "mist-poeffers" are frequently observed when the sea is calm; not to gusts of wind in mountain gorges, for they are heard on mountain summits and in open plains. If their origin were atmospheric they would not be confined to special regions. Nor can they be connected with artificial noises, for they are heard by night as well as by day, and in countries where the use of explosives are unknown. There remains thus the hypothesis which Dr. Cancani considers the most probable, that of an endogenous origin. To the obvious objections that there should always be a center of maximum intensity, which is never to be found, and that they are so rarely accompanied by any perceptible tremor, he replies that, in a seismic series, noises are frequently heard without any shock being felt, and of which we are unable to determine the center. (Monthly Weather Review, 26:216, 1898)

### EARTH-SOUNDS IN THE EAST INDIES
**Anonymous; *Nature,* 134:769, 1934.**

Capt. P. Jansen, St. Helens Court, London, E.C.3, has sent us an interesting account of sounds heard by him near the mouths of rivers in the Dutch East Indies. Except in their higher pitch, they seem to resemble the barisal guns of the Ganges delta and the brontides of certain districts in Italy. On the roads of Sourabaya in Java, he says, two or three noises, as of foghorns of different notes, were heard at irregular intervals of a few seconds, each lasting for one or two seconds. In the hold of an empty ship, the noise was deafening. After continuing for one or two hours, the noises ceased as suddenly as they began. Capt. Jansen has heard the same noises, but less frequently, at the mouth of the Palembang River in Sumatra. At the mouths of some of the rivers of the Malay Peninsula, other noises were heard, like that of plucking the strings of a musical instrument all on the same note and at irregular intervals. Although barisal guns and brontides have for a long time been carefully studied, their origin is still obscure. They are heard frequently in seismic districts and also in countries free from earthquakes. Possibly they have more than one origin, but their frequent occurrence near the mouths of great rivers seems to connect them with the settling of the delta or of the underlying crust. (Nature, 134:769, 1934)

## THE BARISAL GUNS

### BARISAL GUNS
**Scott, G. B.; *Nature,* 53:197, 1896.**

I first heard the Barisal Guns in December 1871, on my way to Assam from Calcutta through the Sunderbans. The weather was clear and calm, no sign of any storms. All day the noises on board the steamer prevented other sounds from being heard; but when all was silent at night, and we were moored in one or other of the narrow channels in the neighbourhood of Barisal, Morelgunge and upwards, far from any villages or other habitations, with miles and miles of long grass jungle on every side, the only sounds the lap of the water or the splash of earth, falling into the water along the banks, then at intervals, irregularly, would be heard the dull muffled boom as of distant cannon. Sometimes a single report, at others two, three or more in succession; never near, always distant, but not always equally distant. Sometimes the reports would resemble cannon from two rather widely separated opposing forces, at others from different directions but apparently always from the southward, that is seaward. We were not very far from the sea when I first heard them, and on mentioning to an old lady on board that I heard distant cannon, she first told me of the mysterious sounds known as the "Barisal Guns." For the next two years I was in Upper Assam, above Goalpara, and do not remember ever hearing them there; but in 1874 I was working in the Goalpara district in the tract south of Dhubri, between the Brahmaputra and the Garo Hills; sometimes near the river, sometimes near the foot of the hills, at others between the two. I gradually worked down as far as Chilmari Chat (I think it is called), the landing-place for Tura, the headquarters of the Garo Hills district, and distant quite 300 miles from the mouths of the Brahmaputra and Ganges. The villages are few and far between and very small, firearms were scarce, and certainly there were no cannon in the neighbourhood, and fireworks were not known to the people. I think I am right in saying I heard the reports every night while south of Dhubri, and often during the day. The weather on the whole was fine. Short, sharp "nor'westers" occasionally burst on us of an evening, with much thunder and lightning; but the days were clear, and, as a rule, the sounds were heard more distinctly on clear days and nights.

I specially remember spending a quiet Sunday, in the month of May, with a friend at Chilmari, near the river-bank. We had both remarked the reports the night before and when near the hills previously. About 10 a.m. in the day, weather clear and calm, we were walking quietly up and down near the river-bank, discussing the sounds, when we heard the booming distinctly, about as loud as heavy cannon would sound on a quiet day about ten miles off, down the river. Shortly after we heard a heavy boom very much nearer, still south. Suddenly we heard two quick successive reports, more like horse pistol or musket (not rifle) shots close by. I thought they sounded in the air about 150 yards due west of us over the water. My friend thought they sounded north of us. We ran to the bank, and asked our boatmen, moored below, if they heard them, and if so in what direction. They pointed south! (Nature, 53:197, 1896)

## BARISAL GUNS
**Schurr, Henry S.; *Nature,* 61:127-128, 1899.**

Barisal Guns are heard over a wide range extending from the Twenty-four Pergunnahs through Khulna, Backergunge and Noakhali, and along the banks of the Megna to Naraingunge and Dacca. They are heard most clearly and frequently in the Backergunge district, from whose headquarters they take their name.

These Guns are heard most frequently from February to October, and seldom in November, December or January. One very noticeable feature is their absence during fine weather, and they are only heard just before, during, or immediately after heavy rain.

The direction from which they are heard is constant, and that is the south or south-east. I have heard them west of me when down in the extreme south of the district, but never north of me. On the other hand, I have been told by captains of river-going steamers that they have heard these reports to their north. These gentlemen, however, ply along waters outside the range of my observations, which lie on the mainland and its adjacent waters.

These Guns are always heard in triplets, i.e. three guns are always heard, one after the other, at regular intervals, and though several guns may be heard the number is always three or a multiple of three. Then the interval between the three is always constant, i.e. the interval between the first and the second is the same as the interval between the second and the third, and this interval is usually three seconds, though I have timed it up to ten seconds. The interval, however, between the triplets varies, and varies largely, from a few seconds up to hours and days. Sometimes only one series of triplets is heard in a day; at others, the triplets follow with great regularity, and I have counted as many as forty-five of them, one after the other, without a pause.

The report is exactly like the firing of big guns heard from a distance with this peculiar difference, that the report is always double, i.e. the report has (as it were) an echo. This echo is so immediate that I can best describe its interval by an illustration. Suppose a person standing near the Eden Gardens hard the 9 o'clock gun fired from Fort William, he would first hear the report of the gun and its immediate echo from the walls of the High Court. The Barisal Guns sound exactly like this, only as if heard from a distance of several miles, very much the same as the sound of the Fort gun heard at Barrackpore on a clear night in the cold weather.

The report varies little in intensity, and I cannot recollect that there was much difference in the sound, whether heard at Barisal itself or some 70 or 80 miles to the south at the extreme end of the district. The state of the atmosphere may affect it, but to no appreciable extent.

The Backergonian peasant is celebrated for the bombs he is in the habit of firing at his weddings and festivals, and many residents have asserted they can distinguish no difference between the reports of these festive bombs and the so-called "guns"; but to any one with a fairly acute sense of hearing, who listens attentively, the difference is very marked, and their assertions are completely refuted by the facts---

(1) that wedding bombs vary very noticeably in the intensity of their sound;

(2) are wanting in the very marked feature of the triplets, and

(3) are naturally confined to the wedding season, a very short season in each year, whilst the Barisal Guns are heard almost through the year, and very noticeably during the annual fast---the Roza---when, of course, there can be no festivals of any kind. (Nature, 61:127-128, 1899)

# LAKE GUNS

## REMARKABLE SOUNDS
**Smith, W. S.; *Nature,* 53:197-198, 1896.**

Lough Neagh is a sheet of water covering an area of upwards of 150 square miles, with very gradually receding shores, excepting at one or two spots. For many years after my settlement as minister here from England, I heard at intervals, when near the lake, cannon-like sounds; but not being acquainted with the geography of the distant shores, or the location of towns, or possible employments carried on, I passively concluded that the reports proceeded from quarrying operations, or, on fine summer days, from festive gatherings in Co. Derry, or Co. Tyrone. In time I came to understand that it was not from the opposite shores, but from the lake itself that the sounds proceeded. After questioning many of the local residents, I extended my inquiries to the fishermen, but they could assign no cause. A strange thing about the matter is that the people generally knew nothing of the phenomenon, and that it is shrouded in mystery. I have heard the sounds during the whole year . . . I have heard the reports probably twenty times during the present year, the last being on a Sunday afternoon a month since, when I heard two explosions; but with two exceptions they have all seemed to come from many miles away, from different directions at different times. They have come apparently from Toome Bay, from the middle of the lake, and from Langford Lodge Point, about nine miles distant. A fisherman thought they must be the result of confined air that reached the lake by means of springs that are believed to rise here and there in the bottom. But the lake is shallow, seldom more than 45 feet deep. The depression now covered by the lake having been caused, it is believed, by volcanic action when the trap-rock of Co. Antrim was erupted, there may possibly be subterranean passages, though I confess their occurrence does not seem very probable; while the sounds emanate, as stated, from various parts of the lake. I have as yet spoken to no one who observed any movement of the waters when explosions took place, nor have I spoken to any one who was close to the spot at the time. Rather every one seems to have heard them only in the distance, which is strange, as fishermen are on the lake during many months in the year, at all hours of the day and night.

Last winter the whole of the lake was frozen over, for the first time since 1814. One fine afternoon, when the air was still, I was skating in the neighbourhood of Shande's Castle, when these mystical guns boomed forth their reports every five or six minutes. On the last day of the skating, when thousands of people from Belfast and elsewhere were assembled in Antrim Bay, there were two fearful boomings, that startled every one near me. They seemed to think some dreadful catastrophe had occurred, as the sounds appeared to proceed from not more than half a mile away. I never before heard them so near. The ice in Antrim Bay remained as it was, but I afterwards learned that it was then breaking up six miles away, but with no alarming sounds. (Nature, 53:197-198, 1896)

## THE "GUNS" OF LAKE SENECA, N.Y.
**Anonymous; *Monthly Weather Review,* 31:336, 1903.**

In the Monthly Weather Review for September, 1897, page 393, we have given some account of the "barisal guns," the "mistpouffers," and similar phenomena whose origin is as yet not certainly understood. The following letter describes an analogous phenomenon in Seneca Lake, N.Y., and it may well be that the barisal guns have their origin in the escape of bubbles of gas just as do the "guns" of Seneca Lake.

Mr. Wm. A. Prosser, of Dresden, Yates County, N.Y., writes as follows, under date of August 18, 1903.

So far as I am personally concerned I know of no explosions of inflammable gas, and the newspaper stories are fabrications in this respect.

The "lake guns" are evidently caused by gas escaping from the sand at the bottom.

Long Point is situated about 15 miles south of Geneva, N.Y., and about 25 miles north of Watkins, N.Y., on the west side of Seneca Lake. Directly off the Point the water is very deep. Heavy currents pass either north or south at regular intervals. A heavy wind for a few hours will change the position of the extreme end of land (which extends 1-1/w miles eastward) several rods. When the swell is not too heavy you can always see the gas rising in bubble form, which, as a rule, makes very little noise, but larger eructations evidently produce these "lake guns." The sand would not stay in place were it not for the water holding it there at the extreme point. Large steamers can land there with but the aid of an ordinary gang plank. (Monthly Weather Review, 31:336, 1903)

## "GUNS" OF SENECA LAKE
**Ingalls, Albert G.; *Science*, 79:479-480, 1934.**

"Silencing the 'Guns' of Seneca Lake" was the title of a discussion which appeared in Science for April 13, (page 340), having been communicated by Professor Herman L. Fairchild, of the University of Rochester. In it Professor Fairchild offered an explanation of a mystery of sound which, as he states, "has hovered over Seneca Lake, in central New York, for more than a century." This is what is known by the local residents as the "lake guns," an occasional low, dull boom suggesting a distant muffled explosion.

The explanation offered in Professor Fairchild's communication was that Seneca Lake occupies the upper part of a glacially deepened valley which cuts the Oriskany Sandstone at a depth of about 1,450 feet below the surface of the lake, and that the "lake guns" are caused by volumes of natural gas which escape from that porous gas-bearing stratum, find their way upward through the 450 feet of superincumbent glacial drift and ascend rapidly through the 600 feet of lake water above it, exploding at the surface. He states that these "guns" have now suddenly ceased, "with no sound reported for the last summer," and that they are probably silenced forever. The interpretation given is that the many gas wells which have recently been drilled in the Wayne-Dundee field not far from Seneca Lake have diminished the gas pressure in the Oriskany, so that the gas no longer escapes into the lake.

This hypothesis is ingenious and at least plausible on the evidence presented, but the purpose of the present communication is to state that Professor Fairchild's obituary of the "guns" comes before their demise, for the writer heard these sounds as recently as last October.

During many years the lake "guns" were heard on innumerable occasions when the writer lived adjacent to Seneca Lake and, as a youth, spent month in and month out on it and in it, camping, swimming, sailing, boating and fishing. Each autumn at the present time he returns to the same scene for a vacation, and this is spent in a way which should bring to notice most of the audible lake sounds; that is, in working alone from morning till night laying stone masonry on a rock cabin situated actually in the lake waters, at the base of a cliff (which may incidentally be instrumental in concentrating sounds which proceed across the lake). This is at Glenora, close to the Dundee gas field, and it was from there that the "guns" were heard last October. The previous absence of reported observations of lake "guns" in 1934 may have been due to the fact that it is the fishermen who most frequently report them, and that last autumn the bass fishing was so poor that relatively few fished.

In itself the hypothesis of escaping gas bubbles is not new, having been commonly current among the native residents of Seneca's shores for many decades, but its association with the depletion of the nearby gas fields, credited by Professor Fairchild to Mr. A. M. Beebee, geologist of the Rochester Gas and Electric Corporation, is a new and ingenious hypothesis which probably will acquire standing if the lake "guns" actually become silent later on.

Interested readers will find in Davison's "Manual of Seismology" a discussion of phenomena of a similar description, under the section "brontides." "Guns" are evidently heard in several localities---Italy, the Philippines, Africa, Haiti, Belgium ---though their cause is unknown. Those heard at Seneca Lake seem to be most frequent in the autumn and in the daytime---in fact, the writer has never heard them after dark. Their direction is vague, and like the foot of a rainbow, they are always "somewhere else" when the observer moves to the locality from which they first seemed to come. No large bubbles or volumes of gas have actually been observed broaching on the surface of the lake, but is it not probable that a large single volume of gas, starting from the bottom of the waters 600 feet below, would be broken up into very many small bubbles during its upward passage through that amount of water? Theoretically, this should not occur, but actually, due to some form of instability, it possibly would.

Finally, is it not likely that gas from the Oriskany, seeping upward through the 450 feet of drift between that horizon and the top of the drift filling, would be stopped and held temporarily by the layers of fine compact lake-bottom clay which have accumulated there since the recession of the Wisconsin ice; and then, when a sufficient volume of gas had thus become pocketed under the clay (perhaps in a large "blister"), the latter would suddenly give way and cause the explosion, not at the surface, but at the bottom of the late. The sound would then be transmitted upward from the bottom by the water, emerging with the vague non-directional origin so often noted in connection with the phenomenon heard.

Since the above was written, Mr. E. R. Dobbin, of Geneva, N. Y., has stated in The Syracuse Herald (April 25) that at Kashong, near Dresden, N. Y., the Seneca Lake guns "were quite as distinct last summer as formerly, perhaps not so frequent. They have also been heard quite distinctly this spring....the Seneca Lake guns," he adds, "are still in existence." (Science, 79:479-480, 1934)

# UNEXPLAINED DETONATIONS: POSSIBLY METEORIC

## CURIOUS ELECTRICAL PHENOMENON

**Bourhill, H.; *Royal Meteorological Society, Quarterly Journal,* 30:55-56, 1904.**

In case it may prove of interest to some Fellows of the Society, I beg to relate an unusual occurrence which took place here early on the morning of June 14, 1903. At about 3.30 a.m. I was awakened by a report like the firing of a cannon, followed by a more or less long whizzing noise, with a second report a second or two after the first one. At first I thought a cannon must have been fired somewhere not very far off, only the sound seemed to be up in the air. Making up my mind to try and find out in the morning what caused the noise, I tried to go to sleep again. I had hardly turned over when a terrific explosion took place, as it appeared to me, just over the house I am living in. It was a sudden, terrifying report followed by a tearing, rending noise, giving one the idea that some large wooden structure was being torn asunder by some great force, and ending with a peculiar crackling sound.

My first idea was that the dynamite magazine here had exploded, and I jumped out of bed immediately to look around outside. I saw at once, by former experience, that the magazine had not exploded. My next idea was that it must have been a meteor, more especially as both I and another member of my family were sure that we had heard some hard particles falling on the iron roof of the house. But on examination in the morning, just after sunrise, all round the house and on the roof, no meteoric particles were to be found. I then came to the conclusion that the explosion must have been caused by some electrical disturbance or phenomenon; and on discussing the matter with the electrical engineer on this mine, he agreed there was no other way of accounting for it, stating at the same time that the electrical conditions of the atmosphere here were most unusual. As far as I have been able to ascertain at present, the explosion was heard at places 10 or 12 miles apart, and was noticed by about twenty people. There are miles of country here without a house, so that, taking that into consideration, it was heard by quite a number of persons. The night was perfectly calm and clear, without a cloud in the sky, and there was brilliant moonlight. I served through the siege of Kimberley, and so heard many 100-pounder shells explode, but the noise they made was small compared to the explosion I have been writing about (Royal Meteorological Society, Quarterly Journal, 30:55-56, 1904)

## BARISAL GUNS IN AUSTRALIA
## Cleland, J. Burton; *Nature,* 81:127, 1909.

In Nature of June 4, 1908 (vol. lxxviii., p. 101), under the title of "Barisal Guns in Western Australia," you published a note from me describing a peculiar, loud detonation heard by my companions and myself while on the Strelley River, in the north-west of Australia. In reading Captain Sturt's "Two Expeditions into the Interior of Southern Australia during the Years 1828, 1829, 1830, and 1831," I find that, when camped on the newly discovered Darling River, near what is now the town of Bourke, in New South Wales, in February, 1829, a very similar sound was heard by the explorers. Sturt's words are as follows:---"About 3 p.m. on the 7th Mr. Hume and I were occupied tracing the chart upon the ground. The day had been remarkably fine, not a cloud was there in the heavens, nor a breath of air to be felt. On a sudden we heard what seemed to be the report of a gun fired at the distance of between five and six miles. It was not the hollow sound of an earthly explosion, or the sharp cracking noise of falling timber, but in every way resembled a discharge of a heavy piece of ordnance. On this all were agreed, but no one was certain whence the sound proceeded. Both Mr. Hume and myself had been too attentive to our occupation to form a satisfactory opinion; but we both thought it came from the N.W. I sent one of the men immediately up a tree, but he could observe nothing unusual. The country around him appeared to be equally flat on all sides, and to be thickly wooded: whatever occasioned the report, it made a strong impression on all of us; and to this day, the singularity of such a sound, in such a situation, is a matter of mystery to me" (2nd edition, 1834, vol. i., p. 98) (Nature, 81:127, 1909)

## BARISAL GUNS IN WESTERN AUSTRALIA
**Cooke, W. E.; *Nature*, 78:390, 1908.**

I have just received the following note from Mr. H. L. Richardson, Hillspring Station, 100 miles northeast of Carnarvon, on our west coast:---"A peculiar incident happened here last evening (June 26) about an hour after sunset. In a south-easterly direction from here three reports took place high up in the air, and then a rushing noise like steam escaping, lasting for a few seconds, and gradually dying away. Mr. Loeffler, one of the owners of the station, was standing outside with me at the time. It was a beautifully clear evening, and there was nothing visible at all in that direction. The reports sounded like explosions of some combustible to which there was no resistance. (Nature, 78:390, 1908)

## AIR QUAKES
**Clark, Joseph; *English Mechanic*, 82:433, 1905.**

I have been interested in what I have read at page 401, "Scientific News," concerning what Mr. H. G. Fordham calls an "air quake," as it goes to explain what was heard here a few minutes past 3 p.m. on Nov. 18---namely, what was taken for thunder. From what I have ascertained from people who heard it the sound was as loud as thunder, but not exactly like thunder. It consisted of three reports---one loud, the next very loud, and the last more like a reverberation. People sitting indoors heard it as well as those out, and it was heard in a straight line at places three to four miles apart---from Ashcot in N.W. to Outleigh Wooton, S.E. ---but I have not heard of anyone hearing it either N. or S. of the S.E. to N.E. line. Street is lat. 50° 7', long. 2° 44'. (English Mechanic, 82:433, 1905)

# UNEXPLAINED DETONATIONS: POSSIBLY SEISMIC

## BARISAL GUNS
**Robinson, Charles H.; *Nature*, 53:487, 1896.**

On July 4, 1808, the expedition of Captains Lewis and Clark was at this place. Under that date we find the following entry in their journal: "Since our arrival at the Falls we have repeatedly heard a strange noise coming from the mountains in a direction a little to the north of west. It is heard at different periods of the day and night, sometimes when the air is perfectly still and without a cloud, and consists of one stroke only, or five or six discharges in quick succession. It is loud, and resembles precisely the sound of a six pound piece of ordnance at the distance of three miles. The Minnatarees frequently mentioned this noise like thunder, which they said the mountains made, but we paid no attention to it, believing it to be some superstition or falsehood perhaps. The watermen also of the party say that the Pawnees and Recaras give the same account of a noise heard in the Black Mountains [Black Hills] to the west of them."

The mountains towards which these noises were heard were the main range of the Rockies, and distant about eighty miles. In 1854, Mr. Doty, of Governor Stevens's party, heard similar noises. He was near enough to the mountains to be certain that the noises came from them. The locality where Mr. Doty heard them was where the direction observed by Lewis and Clark would strike the mountains.

Plenty of white men have been in this country for the last thirty years, or since 1866. I have made careful inquiry among pioneers, but cannot learn that the noises have been heard since Mr. Doty's report. (Nature, 53:487, 1896)

## SUBTERRANEAN SOUNDS
**Anonymous; *American Journal of Science,* 1:10:377, 1826.**

At the village of Babino-poglie in the centre of a valley in the Island of Meleda, in the Adriatic Sea, remarkable sounds were heard for the first time on the 20th March, 1822. They resembled the reports of cannon, and were loud enough to produce a shaking in the doors and windows of the village. They were at first attributed to the guns of some ships of war, at a distance, in the open sea, and then to the exercise of Turkish artillery, on the Ottoman frontiers. These discharges were repeated four, ten, and even a hundred times in a day, at all hours and in all weathers, and continued to prevail until the month of February, 1824, from which time there was an intermission of seven months. In September of the same year, the detonations recommenced, and continued, but more feeble and rare, to the middle of March, 1825, when they again ceased.

These noises have been accompanied by no luminous phenomena or meteors of any kind. Dr. Stulli; who furnishes the statement, conjectures that these remarkable detonations arise from sudden emissions of gas, elaborated in cavities beneath the islands, and which issuing through subterranean passages, strike the air with such force as to produce loud detonations. (American Journal of Science, 1:10:377, 1826)

## EXPLOSIVE NOISES AT FRANKLINVILLE, N.Y.
**Kales, J. W.; *Monthly Weather Review,* 25:393, 1897.**

About 9 o'clock a.m., October 10, 1896, the writer was driving on the Cuba road (see Chart VI) toward Franklinville, N.Y. At the point marked 1 on the map a loud explosive sound was heard which appeared to come from the center of East Hill. When point 2 was reached a similar sound was heard, also at point 3. These sounds succeeded each other at intervals of about five minutes. They closely resembled the sounds produced by coarse black powder used in blasting rocks in the construction of tunnels. The same sounds were heard at the same time by Mr. McStay at his residence, marked McStay. McStay attributed the sounds to the firing of cannon at Cuba, but there was no cannon at Cuba, 13 miles distant. East Hill lies between McStay's and Cuba. The dotted lines on the map all run through the center of East Hill. This hill is about 500 feet high, i.e., using the valley at Franklinville as a base. This region is covered with deep drift. The valley at Franklinville is filled with "till" 100 to 150 feet deep. The underlying rocks belong to the Chemung group, dip to the southwest, and are formed of thin lamellae of sandy shale and thick beds of sandstone. The surface of the soil is strewn with quartz and limestone pebbles, sandstone, and granite boulders. Many moraines extend along the hillsides, showing that this section was once covered with glaciers. On the summit of East Hill is a large sandstone boulder in which is a depression---a mortar---said to have been formerly used by the Indians for grinding corn. Single sounds, like those described, are heard in the hills about here, but so far as the writer knows no series of sounds have been so closely located as those of October 10, 1896, in East Hill; they appear to be due to breaking of the strata of underlying rocks. (Monthly Weather Review, 25:393, 1897)

## SEISMIC NOISES IN NORTH CAROLINA AND GEORGIA

**Hawkins, Barry C.; *Monthly Weather Review,* 25:393-394, 1897.**

Mention has been made in the Monthly Weather Review of certain sounds heard on Black Mountain, N.C., in 1876, and obviously caused by the slow falling or sliding and crushing of rocks. But I am going to describe a phenomenon which seems to be very similar to the famed "barisal guns," and located right in the United States. No account of these sounds has ever been published, and no scientist has ever taken the slightest interest in them, or paid any attention to them, so far as the writer knows.

In northern Georgia, in the extreme north of Rabun County, close to the North Carolina State line and thirty-fifth parallel of latitude, is Rabun Bald Mountain, forming one of the highest peaks on the very crest of the Blue Ridge. This mountain has the same bulky shape and long rambling ridges running for miles in all directions as are spoken of by Hugh Miller as characterising the gneissic mountains of Scotland. On the east side there is a small cliff over which a small stream falls in wet weather, and from the ranges to the east the peak appears in form exactly like a brace, viz, ⏞ . The entire mountain is of gneiss.

Now, on this mountain are heard mysterious sounds resembling distant cannon firing, and these sounds have been heard for many years, probably at least fifty; they have been heard in all kinds of weather and at various points on the mountain.

Numerous observers have noted the sounds, and two reliable gentlemen once spent a night on the summit. About 10 o'clock p.m., sounds were heard which were supposed to be cannon firing in Walhalla, S.C., in celebration of the presidential election, this being in November, 1884; but soon the sounds were found to issue from the ground and from a ridge to the southwest of the mountain. The explosive sounds continued till late in the night. At times they seemed to proceed from the ground immediately under the observers. In early days when bears were plentiful the pioneers said the sounds were caused by these animals rolling small boulders off the mountain sides in search of worms, snails, etc., but the bears have passed and the sounds still continue. Later the sounds were ascribed to "harnts" (haunts or ghosts); two men were murdered in "the sixties" and buried at some unknown point on the "Bald." Some have heard these sounds so near them in the woods that the sound was like that of a falling tree. But ordinarily the sound is like distant firing, as noted above. They are not heard at all times, people having spent the night on the peak and not heard them. (Monthly Weather Review, 25:393-394, 1897)

## THE BARISAL GUNS
**Wadsworth J.; *Weather,* 5:293, 1950.**

The letter from Mr. Cave on the Barisal Guns in your issue of April 1950, recalls an article which appeared in the June 1913 number of the Australian Monthly Weather Report and Meteorological Abstract. The writer of this article, S. H. Ebury, tells us that peculiar explosive sounds, known locally as "Hanley's guns", were heard from time to time in the districts lying between Daylesford and Maryborough in the Talbot county of Victoria. It was originally thought in Daylesford that the noises were merely those ordinarily caused by people shooting rabbits on an estate owned by Mr. Hanley, which lay eight miles to the north-west of the town; but this explanation was shown to be wrong when it was discovered that people living far to the north-west of Hanley's property also heard the noises in a north-westerly direction. Neither were the noises due to blasting in the mines of Moolort, for they did not cease when the mines were eventually closed down.

During the years 1912-14 Mr. Ebury listened for the noises at Sandon South, recording the time of occurrence and the direction from which the sound appeared to come, and he collected similar observations from residents of the Kooroocheang district. The Weather Bureau of Melbourne also gathered information from various localities and placed it at Mr. Ebury's disposal. It appears that the noises could be heard at any hour of the day or night, but that they were most frequent in the forenoon, the average time of occurrence during 1913 and 1914 being about 11 a.m. The chart which accompanied Ebury's article shows the direction from which the explosive sounds appeared to come in various parts of the area. It is seen that the majority of the observations indicate an origin of the sounds in the district known as Stony Rises, lying two miles to the north-west of Mount Kooroocheang. (Weather, 5:293, 1950)

## THE DETONATIONS AT COMRIE, SCOTLAND
**Anonymous; *Edinburgh New Philosophical Journal,* 31:119, 1841.**

About 3 p.m., I was walking, when I heard a sound somewhat resembling a peal of thunder at a great distance---or rather the echo which succeeds a louder thunder peal passing along through the clouds. I would have believed it was from thunder, had I not felt the motion of the ground under me, as if a heavy carriage had passed over it rapidly at a short distance from me. The sound preceded the movement of the ground about 3" or 4".

.... The shocks of 12th at 1 and 4 p.m. (which were very similar and which were attended with a considerable tremor of the earth), were accompanied with a noise resembling a mixture produced by the rush of the strong wind and the peal of distant thunder. It was different from any noise which I ever heard before. The shock of 3 p. m. (which was far more severe than any that had preceded it, and which was attended with greater tremor or heaving of the earth), was accompanied by a noise which at first resembled the murmurings of distant waters. This continued increasing in intensity for about 2", and then followed a very loud and terrific sound resembling that of a double shot for blasting rock immensely charged.

There was a very loud noise, like the emptying of a cart of stones. At Ardvoirlich, the first of these shocks is stated by those who heard it, "as particularly severe, and the noise was described as having a sort of hissing sound, and was compared to a large steam-vessel letting off steam.

In the great shock of the 23d there were two reports, with an interval of 4" or 5" between the first report and the commencement of the second, before any sensible vibration or concussion ensued. The nature of the noise usually resembles the report of a gun discharged among rocks, when the sound produced is deep and hollow. This marks the first explosion. Then follows the sharp rumble, as if through a cavity in the earth, and in the sharper shocks produces a jingle like the jarring of some metallic body in the earth. (Edinburgh New Philosophical Journal, 31:119, 1841)

## EARTHQUAKE IN CONNECTICUT
**Anonymous; *American Journal of Science,* 1:39:335-342, 1840.**

What we have at our disposition shall be devoted to a scene of local disturbance in Connecticut which has been observed ever since the settlement of the country. The region is around East Haddam, on the Connecticut river, a few miles below Middletown. The following memorandum was by request communicated to the senior editor of this Journal twenty five years ago, by the late Rev. Henry Chapman, and it has been kept on file with the expectation of making an investigation on the spot; but, as that which has been so long delayed may never be done, we are induced to give the fragment on the present occasion.

"In attempting to give an account of the circumstances attendant on subterranean noises, so frequently heard at East Haddam, perhaps it may be proper to mention the common opinion respecting them.

"East Haddam was called by the natives Morehemoodus, or place of noises, and a numerous tribe of cannibals resided there. They were famous for worshipping the evil spirit, to appease his wrath. Their account of the occasion of the noises is, 'that the Indian god was angry because the English god intruded upon him, and those were the expressions of his displeasure,' Hence it has been imagined that they originated after the arrival of the English in this country.

"About fifty years ago, an European by the name of Steele came into the place and boarded in the family of a Mr. Knowlton for a short period. He was a man of intelligence, and supposed to be in disguise. He told Mr. Knowlton in confidence, that he had discovered the place of a fossil which he called a carbuncle, and that he should be able to procure it in a few days. Accordingly, he soon after brought home a white round substance resembling a stone in the light, but became remarkably luminous in the dark. It was his practice to labor after his mineral in the night season. The night on which he procured it he secreted it in Mr. K.'s cellar, which was without windows, yet its illuminating power was so great that the house appeared to be on fire, and was seen at a great distance. The next morning he enclosed it in sheet lead, and departed for Europe, and has never since been heard of. It is rumored that he was murdered on his way by the ship's crew. He said that this substance was the cause of the noises---that a change of temperature collects the moistness of the atmosphere, which causes an explosion.

"These shocks are generally perceived in the neighboring towns, and sometimes at a great distance. They begin with a trembling of the earth, and a rumbling noise nearly resembling the discharge of very heavy cannon at a distance. Sometimes three or four follow each other in quick succession, and in this case the first is generally the most powerful.

"The particular place where these explosions originate, has not been ascertained. It appears to be near the northwest corner of the town. It was near this place that Steele found his fossil. The place where the ground was broken when the first one occurred which I mentioned above, was about three and a half miles from this place. There was no appearance of a deposit near where the ground was broken, but it has been observed that this place has been repeatedly struck with lightning.

. . . . . . . . . .

"The awful noises, of which Mr. Hosmer gave an account, in his historical minutes, and concerning which you desire further information, continue to the present time. [1729] The effects they produce, are various as the intermediate degrees between the roar of a cannon and the noise of a pistol. The concussions of the earth, made at the same time, are as much diversified as the sounds in the air. The shock they give to a dwelling house, is the same as the falling of logs on the floor. The smaller shocks produced no emotions of terror or fear in the minds of the inhabitants. They are spoken of as usual occurrences, and are called Moodus noises. But when they are so violent as to be heard in the adjacent towns, they are called earthquakes. (American Journal of Science, 1:39:335-342, 1840)

## SUBTERRANEAN SOUNDS HEARD IN THE WEST INDIES
**Templeton, E. C.; *Seismological Society of America, Bulletin,* 5:171-173, 1915.**

Father J. B. Julio, curate at Pilate, states that: "A hollow sound as of distant thunder was heard from time to time on the heights of Pilate between 4:00 and 9:00 a. m. on July 15th. The weather was calm and favorable to the distinct perception of this noise. To the observant this is merely the sound of waves which, rushing into the caverns of the sea of Borgne, produce the "gouffre," a noise which is echoed from mountain to mountain. But the people of Borgne maintain that it is produced in the sea by millions of little fishes."

. . . . . . . . . .

Notes from Canon L. M. Caze at Ganthier, Fonds-Verrette, describe: "Certain gaping pits of great depth near Fonds-Verrette, Bois Tombe and Ravine Contree from which strange sounds as though produced by explosions, escape periodically at the approach of the rainy season of the year, and sometimes also during the dry season when strong winds are blowing. The explosions are said to be followed by the escape of great jets of a white vapor from the pits and to be accompanied by a violent trembling of the ground within a radius of 500 or 600 meters. There is evidently some superstition connected with these pits, since it is claimed that objects cast into the pit one day have been found on the following day lying on the ground near the pit's mouth" (Seismological Society of America, Bulletin, 5:171-173, 1915)

## THE DEERFIELD (N.H.) PHENOMENA
**Anonymous; *Scientific American*, 2:2, 1846.**

During the last twelve years, certain curious, not to say alarming phenomena in the town of Deerfield, N. H., have excited the fears of the inhabitants, and we think should, ere this, have attracted the attention of the scientific. These are reports or explosions in the ground, apparently of a volcanic or gaseous nature. When first heard they were attributed to the blasting of rocks in Manchester, a new town some ten miles distant; but from the frequency of the reports at all hours in the night as well as the day, from the consideration that they were so loud, and were heard in all seasons, winter as well as summer, it was soon concluded that they had some other origin. The explosions, if they may be so called, commenced on a ridge of land running S. E. and N. W. some five miles in length, and principally on that portion called the South Road. They have, however, extended, and are now heard in a northerly direction. The sounds have become louder, and during the last fall and the present spring or summer, as many as twenty have been heard in one night. Many of them jar the houses and ground perceptibly, so much so, that a child whose balance is not steady, will roll from one side to the other. They are as loud as a heavy cannon fired near the house, with no reverberation, and little roll. Last fall some of the inhabitants were riding in a wagon when an explosion was heard, and they saw the stone wall, which was apparently quite compact, fall over on one side of the way, and a second after upon the other. The stone wall of an unfinished cellar also fell in. This can be attested by many witnesses. There is no regularity in these reports, as they are heard at intervals of a day, a week, and sometimes of months: but for the last year they have become very common, and are heard almost every week more or less. (Scientific American, 2:2, 1846)

## THE MOODUS NOISES
**Anonymous; *Science*, 6:834-835, 1897.**

A correspondent of the New York Sun states that the "famous and mysterious disburbances of the lower Connecticut valley, the 'Moodus noises,' are being heard again" after a silence of twelve years. The Indians knew of them before the coming of white men. For twenty years, up to 1729, the villagers thereabouts heard the noises almost continuously, 'shaking the houses and all there is in them.' They were again heard in 1852 and 1885. On the recent recurrence there was a sound like a clap of thunder, followed for some two hours by a roar like the echoes of a distant cataract. A day later there was a crashing sound like heavy muffled thunder, and a roar not unlike the wind in a tempest. The ground was shaken, causing houses to tremble and crockery to rattle, 'as though in an earthquake.' In view of the compressed condition of the rocks in the Monson quarries, described by Niles some years ago, these indications of local disturbance are of much interest and deserve special study from local observers. The region is one of deformed crystalline rocks, but all the disturbances that can be dated geologically are of great antiquity. The nearest comparatively modern disturbances are in the Cretaceous and Tertiary strata of the islands south of New England. (Science, 6:834-835, 1897)

# HUMS AND HISSES

Detonations intrude upon our sensory apparatus more readily than hums and hisses. Civilization's background of traffic noise, aircraft, and air-conditioners conceals natural hums and hisses very effectively. Yet, nature generates a variety of curious murmurs, sighs, swishes, and buzzes. Most of these sounds are readily attributed to insects, wind in the trees, and so on. Happily for this book, not all hums and hisses can be explained away so readily.

Swishes, hums, and cracklings have long been associated with auroras and meteors. In both cases, these sounds have been synchronous with the motions of the phenomena. The phenomena, however, are too far away (typically 50-100 miles) for the sounds to reach the observer instantaneously. Scientists regarded such observations as auditory illusions or the consequence of inexperience for many years, mainly because no acceptable physical mechanisms were available to explain the observations. Nevertheless, high quality observations kept accumulating. Many investigators now accept the reality of these sounds and ascribe them to: (1) electromagnetic radiation perceived as sound (some people can hear radar!); or (2) electrical discharge (brush discharge) arising from electrical effects created by the phenomena.

More mysterious are the hums and hisses not associated with any obvious physical event. Blowing sand, running water, seismic activity, insects, and other agents are blamed for sounds that seem to come from nowhere. Hums and hisses of this kind are generally perceived only where man-made noise is far-removed. A famous example in this category are the Yellowstone Lake Whispers that have been heard in Wyoming for generations.

Geophysical disturbances, such as storms and earth tremors, are known to generate very-long-wavelength sound (infrasound) that propagates hundreds of miles. Aurora, meteors, and winds gusting over mountains create infrasound, although the frequencies are too low to be detected by the ear. It is possible, though, that these sources of infrasound may, on occasion, produce higher frequencies that we perceive as hums and hisses that seem to come out of thin air.

# THE SUPPOSED SOUND OF THE AURORA

## AUDIBILITY OF THE AURORA AND LOW AURORAS
**Davies, F. T., and Currie, B. W.; *Nature*, 132:855-856, 1933.**

An extensive inquiry among traders, policemen, missionaries and Eskimos concerning the audibility of the aurora was made by members of the Canadian Polar party stationed near Chesterfield Inlet on the west coast of Hudson Bay. The majority of the white people who were questioned had more than twenty years' experience in the eastern Canadian arctic. The Eskimos were questioned only after we found that they were calling us "foolish white men", because we had said that the aurora was inaudible. Natives living in the region extending from Repulse Bay in the north to

Eskimo Point in the south, and from Baker Lake in the west to Southampton Island in the east were included in the inquiry.

All the white people insisted that they had heard rustling or swishing sounds accompanying brilliant auroral displays. These sounds were heard more frequently by them at Burrel on Hudson Straits, at Harrison on the east coast of Hudson Bay, and in the region between Chesterfield Inlet and the Churchill River than farther north. Bishop Turquetil, who has had more than twelve years' experience at both Reindeer Lake in northern Saskatchewan and Chesterfield Inlet, stated that sounds with aurora were more frequent at the former place. Nothing in their description of aurora at such times would suggest unusually low aurora, because forms common to normal displays were always mentioned.

The natives in this region call the aurora, akshanik, meaning that which moves rapidly. The same expression is used also by them for Chesterfield Inlet, because of the rapid flow of the tides in the Inlet. Stories involving the audibility of the aurora have been handed down from one to another for many generations. One that is repeated frequently explains that the sounds are made by spirits playing a game. A small group of the natives have also a story of the aurora coming low enough to kill some of their people. In this particular case the story may have originated from the effects of a destructive lightning flash, although lightning is so rare in this region that the natives often seek shelter with white men during even the mildest storms.

Very few natives from Baker Lake, Chesterfield Inlet, and north of these places, had heard the aurora, although all knew people who had heard it. Natives from Southampton Island knew of no one who had heard it there. Practically all natives from south of Chesterfield Inlet had heard the aurora, and described the sounds by blowing through rounded lips. According to them the aurora was heard more frequently during some winters than others. None had heard it during the winter of 1932-33. Nothing of a definite nature concerning low aurora could be found from them.

The collected testimony of both whites and natives indicates that the region of maximum audibility lies in the region of maximum auroral frequency. Unfortunately, this does not distinguish between an objective or a subjective explanation of the sounds, since the greater the number of aurora seen the more likely it is that conditions favourable to either effect may occur.

The extent to which the testimony of natives can be relied upon is debatable. As observers of unusual occurrences in their native habitat they are superior to the average white man. However, traditional accounts of the sounds may be faulty and have induced a greater susceptibility to a subjective effect.

The members of the party listened occasionally for auroral sounds during brilliant displays, but were unsuccessful except on the night of March 20, when J. Rea, assistant observer, heard sounds with a brilliant display. We were occupied with double station photographs at the time and listened carefully but unsuccessfully for the sounds. Our failure to hear the sounds may have been due to less sensitive hearing since conditions for detecting objective sounds were extremely good. During the day, a pilot balloon had been followed for 138 minutes, and the same clear, calm conditions prevailed during the night with a minimum temperature of -17°F. During the interval in which the sounds were heard, brilliant greenish-white flashes were darting overhead from south to north, and some of these may have come momentarily lower than usual. Height determinations have yet to be made from the double station photographs taken at this time, but the plates show no unusual displacements. Rea is of the opinion that the sounds did not occur simultaneously with the flashes.

One account which we heard of audible aurora indicates a subjective origin for the sound. In this case the aurora was heard in central Saskatchewan from the observation platform of a moving train. Here the noise of the train would have drowned all sounds coming from the aurora, and having the low intensity commonly reported.

On at least three occasions at Chesterfield an auroral condition prevailed which may explain the reports of individuals who claim that they have walked in the aurora. Following the active display with ray structures predominating, a continuous auroral glow covered the whole sky except for a small area in the north. The light from this glow, together with the light reflected from the snow, made it difficult not to feel that one was in an auroral mist or fog.

Continuous records were taken of atmospheric potential gradients from March to the end of August. No effect could be detected on the recording galvanometer during auroral storms. (Nature, 132:855-856, 1933)

## THE AURORA OF OCTOBER 15, 1926, IN NORWAY AND SOUNDS ASSOCIATED WITH IT
**Jelstrup, Hans S.; *Nature,* 119:45, 1927.**

Some curious phenomena accompanying the splendid aurora of Oct. 15, 1926, were observed by me. On the night in question I was working as observer of international determinations of longitude at the top of a hill named Voxenaasen in the neighbourhood of Oslo (approximate altitude, 470 metres). I was at work in a field observatory with a transit instrument registering star transits and chronometer beats for time determinations, when an initial aurora attracted my attention. My assistant was Mr. G. Jelstrup, electrotechnical student.

I was able, during intervals between my observations of time and polar stars, to observe the aurora, which was certainly one of the most splendid I had ever seen. But what is of preponderant interest is the following fact: When, with my assistant, at 19h 15m Greenwich Civil Time, I went out of the observatory to observe the aurora, the latter seemed to be at its maximum: Yellow-green and fan-shaped, it undulated above, from zenith downwards---and at the same time both of us noticed a very curious faint whistling sound distinctly undulatory, which seemed to follow exactly the vibrations of the aurora.

The sound was first noticed by me, and upon asking my assistant if he could hear anything, he answered that he noticed a curious increasing and decreasing whistling sound. We heard the sound during the ten minutes we were able to stay outside the observatory, before continuing our observations.

From 20h 1m to 20h 6m (Greenwich Civil Time) we registered on our radio-receiving set the rhythmic time-signals from the LY station (Bordeaux). We secured the whole series of tops---but at the same time the 'aurora statics' disturbed the pen of the registering instrument. The impulses thus registered are of varying strength, and each of them is of course exceptionally well determined in time, being 'received' at the same time as the scientific time signals. I therefore think that they may be of some interest. The maximum impulses of 'aurora statics' and their duration were:

| No. | Greenwich Civil Time. | Duration. |
|---|---|---|
| 1 | 20h 4m 28s.60 | 0s.08 |
| 2 | 20h 4m 29s.49 | 0s.10 |
| 3 | 20h 4m 39s.90 | 0s.25 |
| 4 | 20h 4m 40s.50 | 0s.25 |

As regards the intensity of these impulses, I find that in each case the vertical component was greater than 100 microvolt/metre.

When, after the reception of the time signals, we again went out of the observatory, the curious sound had absolutely ceased, and later in the night, when also the aurora had vanished, we noticed that the atmosphere was as if swept clean from statics and disturbances of our wave-length.

Concerning the curious sound, I would only remark that the weather was absolutely calm when it was heard. As regards our antennae system, it may be said that it consists of 5 strands of 40 metres each. Our receiver set is an aggregate, consisting of a three-circuits tuner, two high-frequency valves, one modulator, one heterodyne, four low-frequency valves, relay and chronograph. (Nature, 119:45, 1927)

## ON THE AUDIBILITY OF THE AURORA BOREALIS

**Garder, Clark M.; *Science*, 78:213-214, 1933.**

The proposition of the audibility of the aurora borealis has been the subject of considerable speculation and much doubt. Some scientists have claimed with much positiveness that the aurora emits no audible sounds and that the beams of light or electrical waves, such as they may choose to call them, do not come close enough to the earth's surface to be audible, even if any sound were emitted. In my own mind there can be no doubt left as to the audibility of certain types of aurora, for I have heard them under conditions when no other sound could have been interpreted as such, for no other sounds were present.

From the Eskimos I first learned that the aurora could be heard and, like most people, was rather skeptical about it, believing that their statements were based to a great extent on their superstitions. I was told by some of the older Eskimos that when the aurora displays become audible they are able to imitate the sound by whistling in such manner that the beams of light will be attracted or drawn down to them. This, of course, is purely superstition. However, it does bring out the fact that the Eskimos were frequently able to hear the aurora.

The following is my own personal experience which convinced me that the aurora borealis was actually audible. In the winter of 1925-1926 I was engaged in making a drive of reindeer across the mountain range bordering the Arctic coast north of Cape Prince of Wales on Bering Strait. One night during this drive found me traveling by starlight across the divide at the head of Nuluk River. This divide has an elevation of approximately two thousand feet. It was two o'clock in the morning when my native driver and I broke camp in order to overtake the reindeer herd ahead of us. As we climbed with our dog team to the summit of the divide we were both spellbound and astounded by the magnificent display of aurora, the most wonderful display I have ever witnessed during my eight years of life among the Eskimos.

Great beams of light shot up from the northern horizon as if a battery of gigantic searchlights were searching the arctic landscape. In front of these beams and throughout the whole length of the northern horizon great waves of iridescent light traveled from west to east like gigantic draperies before the stage of nature's amphitheater. Great folds or waves, ever changing in color, traveled one after another across the horizon and from behind them streamed the powerful beams of white light. These beams of light could be seen passing directly over our heads, and when one chanced to come over the divide it appeared to be not more than a hundred feet above the surface. The spectacle was so awe inspiring that the dog team was stopped and I sat upon the sled for more than an hour absorbing the marvelous beauty of this most unusual display. As we sat upon the sled and the great beams passed directly over our heads they emitted a distinctly audible sound which resembled the crackling of steam escaping from a small jet. Possibly the sound would bear a closer resemblance to the cracking sound produced by spraying fine jets of water on a very hot surface of metal. Each streamer or beam of light passed overhead with a rather accurate uniformity of duration. By count it was estimated to require six to eight seconds for a projected beam to pass, while the continuous beam would often emit the sound for a minute or more. This particular display was so brilliant that traces could easily be seen long after daylight. (Science, 78:213-214, 1933)

## AURORAL AUDIBILITY
**Silverman, S. M., and Tuan, T. F.; *Advances in Geophysics,* 16:193, 1973.**

A few reports have noted the presence of an odor at the time of an auroral display, described as either like a strong smell of sulfur or like ozone. Beals mentions a number of reports of odor during displays when they were accompanied by sounds. During the aurora of October 19, 1726 (new style), sounds and odor were noted at London, though whether these were simultaneous or not is not noted. In 1870, Rollier, the balloonist, descended on a mountain in Norway 1300 meters high, saw auroral rays across a thin mist, and heard a muttering. When the sound ceased he perceived a very strong smell of sulfur. This is the only account in which any sense of time sequence is given. In one instance an odor was noted without any mention of accompanying sounds. It was noted, however, that this had also been observed during other auroral displays. Trevelyan quotes the report of a person in the Faroes, "who stated that, when the color of the Aurora Borealis is dark red, and extends from west to east with a violent motion, he had experienced a small similar to that which is perceived when an electric machine is in action."

It appears, therefore, that an odor can accompany an auroral display, possibly even when sounds are not present. The simultaneous noting of both sound and odor would not be expected to be a common occurrence, even if both were always present, because of the masking effect which might be anticipated for either or both from the normal noise and odor environment. The few instances recorded are thus more significant than indicated by their numbers. (Advances in Geophysics, 16:193, 1973)

## OBSERVATIONS OF ACOUSTIC AURORA IN THE 1-16 HZ RANGE
**Procunier, R. W.; *Geophysical Journal,* 26:183, 1971.**

Abstract. Acoustic aurora have been heard by long-term residents of the arctic. They have also been recorded on microbarographs. Acoustic events associated with aurora are now reported in the near infrasonic range (1-16 Hz) at Barrow, Alaska. These observations were made with the aid of a resonant detector achieving a high signal-to-noise ratio similar to that used in extending the rocket-grenade technique to 109 km. Over 100 impulsive events of a quasi-repetitive nature were recorded on a patrol basis during January 1970. Acoustic events were correlated with disturbed magnetic conditions and optical aurora but uncorrelated with lower-frequency auroral microbarograph events at College or Inuvik.

It is hoped that these initial observations will persuade interested parties to a more complete study of this phenomena and encourage an explanation of the generation mechanism for auroral infrasonic waves. (Geophysical Journal, 26:183, 1971)

# HISSES ACCOMPANYING METEORS

## AN AUDIBLE METEOR?
**Huntley, E.; *Symons's Meteorological Magazine*. 41:190-191, 1906.**

Between half-past ten and a quarter to eleven on the night of August 7th last I was riding home on my bicycle, when I distinctly heard a faint but clear sound above my head like that made by a rocket as it passes through the air at a considerable distance. On looking up I saw a shooting-star still on the move, but leaving a streak of light behind it. It is difficult to describe, but when I first looked up there was a streak of light which got increasingly longer from one end until after quite an appreciable time the moving end suddenly stopped. The streak seemed to remain still for a while, and then quite gradually ceased to exist, the light diminishing from both ends and then going out.

I should like to know whether you think the sound heard had anything to do with what I saw, as the combination of the two things struck me as being very odd. I may add that the night was a fine one and very still. (Symons's Meteorological Magazine, 41:190-191, 1906)

## HISSING SOUNDS HEARD DURING THE FLIGHT OF FIREBALLS
**Khan, Mord. A. R.; *Nature*, 155:53, 1945.**

Many responsible eye-witnesses, in their descriptions of fireballs, have emphatically stated that they have occasionally heard a peculiar hissing sound simultaneously with the flight of a meteor. From personal observation, I can also testify to the validity of these statements. Fireball literature is full of such accounts. Three recent cases (connected with fireballs seen by a number of competent observers at Hyderabad, on October 13, 1936, on March 25, 1944, and on August 6, 1944, respectively) have placed the matter beyond any doubt whatever.

The obvious difficulty is about the simultaneity of the light and sound phenomena noticed by observers fifty to a hundred miles distant from the meteor. But it must be remembered that the fireball rushes through the upper atmosphere with parabolic speed (about 26 miles per second); its duration of visible flight is generally 6-8 seconds. Assuming its height to be roughly 75 miles, matter from a friable aerolite can issue in a regular stream along its entire path, into the lower atmosphere, with velocity large enough to bring it in the vicinity of an observer while the meteor is still in sight. For the height assumed, four or five seconds may suffice (even allowing for air resistance) for the matter from the meteor to reach the air in the neighbourhood of the observer, and thus give rise to sounds variously described as like the swish of a whip, the hissing noticed while a cutler sharpens a knife on a grindstone, or a hot iron being plunged into cold water.

A shower of fine sand beating against the leaves of trees was noticed immediately after the apparition of the fireball of October 13, 1936, described in detail in Science and Culture, Calcutta, 2, No. 5, 273 (1936). (Nature, 155:53, 1945)

# THE YELLOWSTONE LAKE WHISPERS

## OVERHEAD SOUNDS IN THE VICINITY OF YELLOWSTONE LAKE

**Linton, Edwin; *Science,* 22:244-246, 1893.**

While engaged in making certain investigations for the United States Fish Commission in the summer of 1890 my attention was called to an interesting phenomenon in the vicinity of Yellowstone Lake, of which I am pleasantly reminded by the following brief but vivid description in a recent report by Prof. S. A. Forbes.

Under his description of Shohone Lake, Professor Forbes, in a foot note, thus alludes to this phenomenon:

"Here we first heard, while out on the lake in the bright still morning, the mysterious aerial sound for which this region is noted. It put me in mind of the vibrating clang of a harp lightly and rapidly touched high up above the tree tops, or the sound of many telegraph wires swinging regularly and rapidly in the wind, or, more rarely, of faintly heard voices answering each other overhead. It begins softly in the remote distance, draws rapidly near with louder and louder throbs of sound, and dies away in the opposite distance; or it may seem to wander irregularly about, the whole passage lasting from a few seconds to half a minute or more. We heard it repeatedly and very distinctly here and at Yellowstone Lake, most frequently at the latter place. It is usually noticed on still bright mornings not long after sunrise, and it is louder at this time of day; but I heard it clearly, though faintly, once at noon when a stiff breeze was blowing. No scientific explanation of this really bewitching phenomenon has ever been published, although it has been several times referred to by travellers, who have ventured various crude guesses at its cause, varying from that commonest catch-all of the ignorant, "electricity," to the whistling of the wings of ducks and the noise of the "steamboat geyser." It seems to me to belong to the class of aerial echoes, but even on that supposition I cannot account for the origin of the sound." (A Preliminary Report on the Aquatic Invertebrate Fauna of the Yellowstone National Park, etc. Bulletin of the United States Fish Commission for 1891, p. 215. Published April 29, 1893).

In a paper which was read before the Academy of Science and Art of Pittsburg, Pa., March 18, 1892, entitled "Mount Sheridan and the Continental Divide," I recorded my recollections of this phenomenon and reproduce them here with no alteration. Although the style is, perhaps, somewhat lacking in seriousness, the descriptions were made from notes taken at the time and written out while the memory of the facts was still fresh. Indeed, even now, after a lapse of three years, I have a very distinct recollection of the sound, vivid enough at least to teach me how imperfect my description of it is. Words describe an echo very inadequately when one is in ignorance of the original sound, and especially so when he is in doubt as to whether the sound is the echo of a noise or the noise itself.

Following is the account of these overhead noises given in the paper alluded to above and published soon after by the academy:

Overhead Noises.---The last topic which I shall discuss in this somewhat desultory paper, is what I shall call overhead voices.

Lest I be thought to be indulging in some ill-advised or disordered fancy I shall first quote from Hayden's Report for 1872, on Montana, Idaho, Wyoming, and Utah. Mr. F. H. Bradley, p. 234, in that part of his narrative which relates their visit to Yellowstone Lake, says: "While getting breakfast. [This was near the outlet of the lake] we heard every few moments a curious sound, between a whistle and a hoarse whine, whose locality and character we could not at first determine, though we were inclined to refer it to water-fowl on the other side of the lake. As the sun got higher the sound increased in force, and it now became evident that gusts of wind were passing through the air above us, though the pines did not as yet indicate the least motion in the lower atmosphere. We started before the almost daily western winds, of which these gusts were evidently the forerunners, had begun to ruffle the lake."

With this warrant I shall proceed to describe as well as I can my impressions of these overhead noises, which appear to belong exclusively to the lake region of the Park.

The first time I heard them, or it, was on the 22d of July, about 8 a.m., on Shoshone Lake. Elwood Hofer, our guide, and I had started in our boat for the west end of the Lake. While engaged in making ready for a sounding on the northern shore, near where the lake grows narrow, I heard a strange echoing sound in the sky dying, away to the southward, which appeared to me to be like a sound that had already been echoing some seconds, before it had aroused my attention, so that I had missed the initial sound, and heard only the echo. I looked at Hofer curiously for an explanation. He asked me what I thought the sound was; I immediately gave it up and waited for him to tell me, never doubting that a satisfactory explanation would be forthcoming. For once this encyclopedia of mountain lore failed to come up to date. His reply was, that it was the most mysterious sound heard among the mountains. From the first this sound did not appear to me to be caused by wind blowing. Its velocity was rather that of sound. It had all the characters of an echo, but of what I am not even yet prepared to give an altogether satisfactory answer. I am afraid that my conclusions are about as satisfactory as those of the Irishman, who having been sent out from camp in the night to investigate a strange noise believed to be made by some wild beast, returned with the announcement that "it was nothing at all, only a noise just." Upon our return to camp I questioned both our guides and one of the packers, who had had much experience in the mountains. They agreed substantially in what they had to say about it. They had never heard it farther west than Shoshone Lake, nor farther east than Yellowstone Lake, and not at all north of these lakes. Hofer thought he had heard it once about 30 miles south of Yellowstone Lake. Dave Rhodes had heard it usually shortly after sunrise and up to perhaps half-past eight or nine o'clock. Hofer said that he had heard it in the middle of the day but usually not later than ten o'clock a.m. Neither of them remembered to have heard it before sunrise.

On the following morning we heard the sound very plainly. It appeared to begin directly overhead and to pass off across the sky, growing fainter and fainter towards the southwest. It appeared to be a rather indefinite, reverberating sound, characterized by a slight metallic resonance. It begins or is first perceived overhead, at least, nearly every one, in attempting to fix its location, turns his head to one side and glances upward. Each time that I heard the sound on Shoshone, it appeared to begin overhead, or as one of the men in the party expressed it "all over," and to move off to the southwest. We did not hear the sound while on Lewis or Heart Lake. The next time I heard the sound was on August 4th, when we were camped on the "Thumb" of Yellowstone Lake. Professor Forbes and I were out on the lake making soundings about 8 a.m. The sky was clear and the lake was quiet. The sun was beginning to shine with considerable power. The sound seemed loudest when overhead, and apparently passed off to the southward, or a little east of south. It had the same peculiar quality as that heard on Shoshone Lake, and is just as difficult to describe. There was the same slight hint of metallic resonance, and what one of the party called a kind of twisting sort of yow-yow vibration. There was a faint resemblance to the humming of telegraph wires, but the volume was not steady nor uniform. The time occupied by the sound was not noted, but estimated shortly afterward to be probably a half a minute. As I heard it at this time it seemed to begin at a distance, grow louder overhead where it filled the upper air, and suggested a medley of wind in the tops of pine trees, and in telegraph wires, the echo of bells after being repeated several times, the humming of a swarm of bees, and two or three other less definite sources of sound, making in all a composite which was not loud but easily recognized, and not at all likely to be mistaken for any other sound in these mountain solitudes, but which might easily escape notice if one were surrounded by noises. On August 8th, at 10.15 a.m., Professor Forbes and I heard the sound again while we were collecting in Bridge Bay at the northern end of the lake.

While on Shoshone Lake I ventured the suggestion that the sound might be produced beyond the divide east of us, and be reflected from some upper stratum of air of different density from that below. Hofer evidently considered himself responsible for an explanation of the origin of the sound, and frequently remarked that it reminded him of the noise made by the escaping steam of the so-called Steamboat Geyser, on the eastern shore of Yellowstone Lake, about 6 miles from the outlet. I passed between Steamboat Point and Stevenson's Island twice, but was not near enough either time to hear the escaping steam. Moreover, on each occasion the wind was blowing a lively breeze in the direction of Steamboat Point. On the afternoon of August 9th, at 3.20 p.m. while in a row-boat on the south eastern arm of Yellowstone Lake, near the entrance of the upper Yellowstone River, I heard a sound overhead, like rushing wind, or like some invisible but comparatively dense body moving very rapidly through the air, and not very far above our heads. It appeared to be travelling from east to west. It did not have the semi-metallic, vibrating, sky-filling, echoing resonance of the overhead noises that I had heard before, and was of rather shorter duration. It had, however, the same sound-like rapidity of the other. The sky was clear except for a few light fleecy and feathery clouds, and there was just enough wind blowing to ruffle the surface of the water. If this sound was produced by a current of air in motion overhead, it is difficult to understand why it did not give some account of itself, either in the clouds that were floating at different levels in the upper air, or among the pines which covered the slope that rose more than 1000 feet above our heads, or on the waters of the lake itself.

I am inclined to attribute the typical echoing noise to some initial sound, like that of escaping steam for example, from some place like Steamboat Geyser, and which is reflected by some upper stratum of air, that is differently heated from that below by the rays of the sun as they come over the high mountain ridges to the east of the lake. The sound may thus be reflected over the low divides west to Shoshone, and south to Heart Lake, or even farther in the direction of Jackson's Lake. I am not strenuous for this theory, and will be glad to hear a better explanation of this phenomenon. I have a dim recollection of some legend of phantom huntsmen, and a pack of ghostly but vocal hounds which haunt the sky of the Hartz Mountains. Can any one tell whether there is any natural phenomenon belonging to mountains or mountain lakes, which could give foundation to such legend?

The phenomenon has not yet been successfully explained, and I do not know that any similar phenomenon has been observed elsewhere.

It is to be hoped that some one will investigate the matter soon and give a scientific explanation of its cause. (Science, 22:245-246, 1893)

## MYSTERIOUS ACOUSTIC PHENOMENA IN YELLOWSTONE NATIONAL PARK

**Smith, Hugh M.; *Science*, 63:586-587, 1926.**

It is highly gratifying that the American people are making yearly increasing use of the Yellowstone National Park and that each season many thousands become for the first time acquainted with its beauties and wonders. Among future visitors there will be some who will experience the strange and bewildering musical sounds that for many years have been noted in certain parts of the park. As a partial contribution to the solution of the mystery, a personal observation on one manifestation of the weird phenomena may be of interest.

This subject has received scant mention in various published reports and articles, and by some people has been relegated to the category of yarns and myths which helped to make Jim Bridger famous. The best account seems to be that contained in the best work on the park, that of General Chittenden, from which the following quotation is taken (pages 288-289):

"A most singular and interesting acoustic phenomenon of this region, although rarely noticed by tourists, is the occurrence of strange and indefinable overhead sounds. They have long been noted by explorers, but only in the vicinity of Shoshone and Yellowstone Lakes. They seem to occur in the morning and to last only for a moment. They have an apparent motion through the air, the general direction noted by writers being from north to south. They resemble the ringing of telegraph wires or the humming of a swarm of bees, beginning softly in the distance, growing rapidly plainer until directly overhead, and then fading as rapidly in the opposite direction. Although this phenomenon has been made the subject of scientific study, no rational explanation of it has ever been advanced. Its weird character is in keeping with its strange surroundings. In other lands and times it would have been an object of superstitious reverence or dread and would have found a permanent place in the traditions of the people."

In the summer of 1919, during a visit to the park in connection with governmental fish-cultural and fish-planting work therein, the present writer, in company with Mr. A. H. Dinsmore, of St. Johnsbury, Vermont (a former superintendent of fish hatcheries in the park), made a camping trip to Lewis and Shoshone Lakes, employing pack horses, with helper, and having as a part of the equipment an Oldtown canoe transported by motor truck from Yellowstone Lake to Lewis Lake.

About eight o'clock on the morning of July 30, after having camped for two days and two nights on the shore of Shoshone Lake at its outlet at the southern end, we entered the canoe and pushed off from the shingly beach, headed for the northern end of the lake. The surface of the lake was glassy, the air was still, a faint haze overhung the water, the sky was cloudless, and the lake for a considerable distance out was in the shadow of heavily timbered hills. The canoe had barely gotten under way and was not more than twenty meters from the shore when there suddenly arose a musical sound of rare sweetness, rich timbre, and full volume, whose effect was increased by the noiseless surroundings. The sound appeared to come from directly overhead, and both of us at the same moment instinctively glanced upward; each afterward asserted that so great was his astonishment that he was almost prepared to see a pipe organ suspended in midair. The sound, by the most perfect gradation, increased in volume and pitch, reaching its climax a few seconds after the paddling of the canoe was involuntarily suspended; and then, rapidly growing fainter and diminishing in pitch, it seemed to pass away toward the south. The sound lasted ten to fifteen seconds and was subsequently adjudged to range in pitch approximately from a little below center C to a little above tenor C of the piano-forte, the tones blending in the most perfect chromatic scale.

It was at once realized that this was probably a manifestation of the strange phenomena referred to by General Chittenden, whose book had been read in camp, and that an opportunity was presented to investigate and perhaps elucidate the mystery. So favorable was the opportunity that in a short time we were reproducing the sound at will.

Following the dying away of the music and the short period in which we were held spellbound, paddling was resumed and, as the canoe gained sufficient headway, the music recurred in practically the same form as at first, and, as the paddling ceased and the momentum of the canoe fell off, the sound died away. It was then perceived that the sound was coincident with the motion of the canoe: when a certain speed was reached and maintained, the sound was produced and maintained; before that speed was attained there was no sound, and after that speed was lost the sound ceased.

A search for the cause of the sound disclosed the following situation: A jointed bamboo salmon rod with its butt-end touching the side of the canoe was projecting backward about half a meter beyond the gunwale; the waxed silk line was reeled in, but about a meter of line with the lure at the end was wrapped several times around the terminal joint; a lead sinker weighing one hundred grams, that had been attached in order to carry the lure into deep water frequented by the large trout with which the lake had been stocked, was dangling from the end of the rod about five centimeters below the surface of the water. As the canoe moved through the water, the short length of free line, held taut by the sinker, rapidly vibrated in conformity with the speed of the boat; the vibrations were transmitted through the bamboo rod to the canoe, whose thin, curved, rigid sides and bottom acted as a sounding board and gave out an augmented volume of sound that seemed to be concentrated or focused overhead. The combination of essential factors present in this case seems to have been a smooth water surface, a vibrating cord, a resonant body to which the vibrations were transferred, and still air, with perhaps other favorable atmospheric conditions. Later, in other parts of the lake, where the surface was disturbed by ripples or waves and there was slight to moderate movement of the air, attempts to reproduce the sound were futile, although the other factors were as in the original manifestation.

It is not to be inferred that the conditions and explanation herein given are held to account for all the mysterious musical sounds that have been heard in the park. On the contrary, owing to the rarity of the occasions when canoes had been used in park waters, it seems likely that the exact combination of factors noted must have been exceptional. All that is claimed for this manifestation is that it fulfilled the requirements of the mystery as described by the best observers and that its cause was ascertained. An account of the occurrence was promptly given to the superintendent of the park.

That acoustic conditions on Shoshone Lake that morning may have been peculiar is suggested by another observation as the canoe was skirting the eastern shore of the lake and the Shoshone Geyser Basin came within visual range. During several of the eruptions of the principal geyser, the splashing of the geyser water on the hard siliceous platform was distinctly heard. The distance in an air line, according to the official Geological Survey map, was approximately nine kilometers. (Science, 63:586-587, 1926)

## OVERHEAD SOUNDS OF THE YELLOWSTONE LAKE REGION

**Linton, Edwin; *Science*, 71:97-99, 1930.**

In Nature Notes from Yellowstone Park, Mr. L. S. Morris, ranger naturalist, describes aerial sounds which were heard by himself and companions over Grebe Lake, near the canyon of the Yellowstone in the park.

After reading Mr. Morris's narrative, I looked over a communication of my own to Science. I also reread the notes upon which that communication was based.

Since there are some items in these notes which were not included in the published account it seems to me, especially in view of recently awakened interest in these phenomena, to be worth while to make a record of them.

Following are all the references to overhead sounds which I find in the diary which I kept during the six weeks of our stay in the park. I copy from the diary without making any changes in the text, such comments as seem to be called for being enclosed in parentheses.

(1) July 23 (1890). Yesterday, when (Elwood) Hofer and I were on our way to the upper (western) end of the lake (Shoshone), I heard a strange noise, which I supposed was off to the southward and echoing among the mountains. (At the time, about 8 A. M., I was seated in our Osgood canvas boat, with the oars in my hands, but not rowing. I was probably ten or twelve feet from the shore, where Hofer was seated measuring off our dredge rope, which we were going to use for a sounding line. As I remember the situation we were from twenty to thirty feet apart.) Hofer asked me what I thought it was, and where it seemed to be. I told him the apparent direction and asked him what it was. He replied that it was the most mysterious sound that was heard in the mountains. Since then we have talked about the sound a good deal in camp, and this morning heard it again very plainly. (My recollection of this event is that it occurred just after we had had breakfast, and before we had separated for the day's work.) This time it appeared to be directly overhead, and to pass off across the sky, growing fainter and fainter toward the southwest. Hofer and Dave Rhodes, both of whom have had wide experience in the mountains, agree in their testimony in regard to it. They say they have never heard it anywhere out of the park, except to the south, about the forty-fourth parallel, some thirty miles south of here. He (Hofer) does not remember to have heard it farther west than Shosone Lake, or east than Yellowstone Lake---not to the north of these points. It is heard mostly in the morning, shortly after sunrise, and up to, perhaps, half past eight, or nine o'clock. Hofer says he has heard it in the middle of the day, but usually not later than 10 o'clock A. M., doesn't remember to have heard it before sunrise. The description given of the sound before I heard it and since agree with my observations with regard to it. (The meaning of this somewhat obscure sentence is that the descriptions which Hofer and Rhodes give of the sound as they have heard it on previous occasions and their description of the sound as they have heard it here on Shoshone Lake agree with my own observations.) When heard best it appears to be a rather indefinite, reverberating sound in the sky, with a slight metallic resonance, which begins, or at least is at first perceived, overhead; at least, nearly every one in attempting to locate it turns his head to one side and glances upward. (I remember that while I was having my first experience with this sound in the sky I noticed that Hofer was watching me very closely. Later I found that he had been observing my reactions. He told me that people invariably behaved that way when they were trying to locate the source of the sound. I had at first looked up, and then had tried to follow the diminishing sound toward the southwest.) The sound is as difficult to describe as an echo, which has been repeated several times in quick succession. Each time I have heard (it) here on Lake Shoshone it appeared to begin

to the southward, or, when first noticed, beginning overhead, or, as some one (Dave Rhodes) expressed it, "all over," and moving off toward the south.

(2) Camp on the "Thumb," Yellowstone Lake. August 4. While out on the lake this morning Professor Forbes and I heard again the strange noise which we heard several times on Shoshone Lake. It was about 8 A. M., morning still and clear, lake quiet, sun beginning to shine with considerable power (this mention of overhead sounds was preceded by an entry, which may, of course, have nothing to do with this phenomenon: "minimum temperature last night 33.5° F."); sound loudest almost overhead---seemed to pass to the southeast (so it stands in my diary. Since my other entries, where the apparent direction of these elusive sounds is recorded, indicate a direction west of south, which is in accord with my recollection of these events, I am inclined to think that this may be an example of those slips which Oliver Wendell Holmes cites in "Over the Teacups," where one writes north when he means south, and the like); sound of same nature as that heard on Shoshone (Lake)---very hard to describe---a certain metallic resonance---Professor Forbes calls it a kind of twisting, yow-yow vibration, resemblance to sound made by telegraph wires, but not a steady, uniform volume. The sound lasted probably half a minute, time not noted. As I have heard the sound here it seemed to begin at a distance, something like a mixture of wind in pine tops, in telegraph wires, the echo of bells, after being repeated several times, the humming of a swarm of bees and two or three other sources of sound, all making a not loud, but easily recognized sound, not at all likely to be mistaken for any other sound, but easily overlooked if one is surrounded by noises. The party on shore heard the same sound at the same time that we heard it on the lake. (On this occasion Professor Forbes and I were in our canvas boat, one hundred yards, more or less, from shore. Hofer was on the beach at the water's edge. The others were at the camp, which was some fifty feet or more back from the edge of the lake terrace, in a grove of pine-trees. I have a kodak picture of the scene which was taken from our boat as we were nearing shore on our return from this trip.) Hofer says he doesn't remember to have heard it when the sky was cloudy, has heard it when "quite considerable breeze" was blowing. Remembers that he was usually heard it when the sky is clear, or with few clouds, and the morning calm; as a rule in the morning, but has heard it as late as noon. Dave Rhodes thinks he has heard it only in the mornings, and when the sky is clear, or with light, fleecy clouds.

(3) August 8. Professor Forbes and I rowed up to Bridge Bay (northwest end of Yellowstone Lake), where we collected---back (to hotel) about 12:30 (P. M.).... We heard our mysterious sound again this morning, at 10 and 10:15, while out (on the lake) collecting. There is reason to believe that it is caused by the steamboat _Geyser_ (on east shore of the lake), and heard through some peculiar condition of the atmosphere at distances of several, perhaps thirty to fifty miles away. (Here I seem to have arrived at that state of mind where consolation is derived from belief in a theory.)

(4) August 9. 2:20 P. M., at head of southeast arm of (Yellowstone) lake. While in boat heard sound overhead, like rushing wind traveling very rapidly or like something rushing through the air, did not have the semimetallic sound, or like echo; seemed to travel from east to west; clear, except light, fleecy and feathery clouds, enough wind to ruffle the surface of the water. (On this occasion I was one of a party of five which had left the Lake Hotel on the afternoon of the eighth in two rowboats, with tent and camp outfit. My companions, who were engaged on the construction of a new building, were Mr. L. D. Boothe and Messrs. Couglin, Curl and Thomson. Mr. Curl and I were in one boat, and three others in the other. We were making our way from the east to the west side of the east arm of the lake, and were rowing slowly in the very shallow water of this part of the lake when the sound attracted our attention.)

The following note is included in Mr. Morris's article: "In the _Ranger Natural-_

ists' Manual for 1928 there appears a rather complete summary of the recorded observations of this weird phenomenon by Ranger Marguerite Arnold." (Science, 71:97-99, 1930)

# HUMMING SOUNDS HEARD ON LAND

## NOTES ON A METEOROLOGICAL PHENOMENON HEARD AT SESKIN, FEBRUARY 8th, 1928
**Grubb, L. G., and Grubb, I.; *Meteorological Magazine,* 63:66-67, 1928.**

A sound locally known as "wind in the mountains," was heard here all day. When first noticed before 8h. it sounded like the noise of a motor running, and distinct from the noise of the wind. It was persistent, unvarying except in the degree of loudness, and without anything like throbbing. In the afternoon its volume of sound was like that of the rush of a heavy train through a tunnel near by. (There is no railway tunnel within about twenty miles. It seemed to come from the Comeragh mountains, a few miles to the south-west. One man who has lived in the district all his life had thought it was the noise of the Millvale stream, but as that is a slow flowing, shallow stream more than a mile away such an explanation does not seem likely.

Locally it is considered a presage of storm and rain. It was heard also about a fortnight ago for a short time we understand.

About 16-1/2h. the SSW wind here was very light, the trees not moving but the sound was as described above, then suddenly a strong wind sprang up here which within two hours had reached force 8 or 9. The gale continued until after midnight but had lessened by 3h. Only 3.1 mm. fell of rain during the night. At 18-1/2h. and possibly later the sound from the direction of the mountains could be heard apart from the noise of the gale.

On February 1st, 1923, a similar occurrence was reported to you from here, but then the noise was much louder and lasted for a much shorter period. (Meteorological Magazine, 63:66-67, 1928)

## DESERT SOUNDS
**B., N. H.; *American Meteorological Society, Bulletin,* 12:40, 1931.**

The following account of a strange phenomenon occurring in the South American Desert as told the writer by Mr. Travers Ewell, a consulting engineer who has made frequent and extended trips into the desert regions of northern Chile and Peru.

"Shortly after darkness sets in the mysterious sound begins. At first it is a sort of hum, fairly high pitched, but weird, eerie, and almost beautiful. Soon, mingled with the hum at regular intervals, there is a deep bass boom, like the far off beating of a great drum. Sometimes the sound lasts for an hour or two, sometimes all night. I have heard it as often as three or four times a month. I have been told that the phenomenon is brought about by the movement of the sand under certain conditions of wind, temperature, and humidity producing the proper vibrations. I know that on the nights when the sound occurs, there is a moderate southwest wind blowing, sufficient to move a film of sand along the surface. But whatever the cause, the phenomenon is certainly very real and very astonishing." (American Meteorological Society, Bulletin, 12:40, 1931)

## SOUND TRANSMITTED THROUGH EARTH
**Durrant, Reginald G.; *Nature*, 107:140, 1921.**

In June, 1903, I was trekking towards the Victoria Falls. On the night before arrival we "outspanned" some twelve miles to the south, and on retiring to rest on the bare ground I became aware of a curious, rhythmic sound, quite distinct when my ear was pressed against the soil. I told my two brothers, who found they also could hear the pulsation, and one of them suggested that it must be due to the booming of the distant cataract.

To me the most interesting point is not that the sound was transmitted by the earth, but that it was transformed into rhythmic vibration---very different from the constant roar one hears when close to the Falls. Some process of interference would seem to occur and give rise to this result. (Nature, 107:140, 1921)

## REMARKABLE SOUNDS
**Tomlinson, C.; *Nature*, 53:78, 1895.**

In a book that was popular about fifty years ago, entitled "Journal of a Naturalist," the author says that the purely rural, little noticed, and, indeed, local occurrence, called by the country people "hummings in the air," was annually to be heard in fields near his dwelling. "About the middle of the day, perhaps from twelve o'clock till two, on a few calm sultry days in July, we occasionally hear, when in particular places, the humming of apparently a large swarm of bees. It is generally in some spacious open spot that this murmuring first attracts our attention. As we move onwards the sound becomes fainter, and by degrees is no longer audible." The sound is attributed to insects, although they are invisible.

A writer in the Edinburgh Philosophical Journal objects to this sound being attributed to insects, first because the fact is stated as being local and partial, heard only in one or two fields, at particular times of the year when the air is calm and sultry. He has often heard a similar humming in a thick wood, when the air is calm, and has diligently searched for insects, but in no case was able to detect them in numbers sufficient to account for the sound.

The same writer refers to remarkable sounds heard in a range of hills in Cheshire. When the wind is easterly, and nearly calm on the flats, a hollow moaning sound is heard, popularly termed the "soughing of the wind," which Sir Walter Scott, in his glossary to "Guy Mannering," interprets as a hollow blast or whisper. The explanation seems to be that a breeze, not perceptible in the flat country, sweeps from the summit of the hills, and acts the part of a blower on the sinuosities or hollows, which thus respond to the draught of air like enormous organpipes, and become for the time wind instruments on a gigantic scale. (Nature, 53:78, 1895)

## HUMMING IN THE AIR CAUSED BY INSECTS
**Bath, W. Harcourt; *Nature*, 34:547, 1886.**

In a letter to the Hon. Daines Barrington (letter lxxx.) the Rev. Gilbert White, the well-known author of the "Natural History of Selborne," mentions a strange humming sound in the air. He writes:---"There is a natural occurrence to be met with upon the highest parts of our downs in hot summer days which always amuses me much without giving me any satisfaction with respect to the cause of it: and that is a loud audible humming is of bees in the air, though not one insect is to be seen. This sound is to be heard distinctly the whole common through from the Honey Dells to my avenue gate. Any person would suppose that a large swarm of bees was in motion, and playing about over his head. This noise was heard last week on June 28."

It is singular that no explanation has been offered by any one or such a common phenomenon. I am convinced that the humming sound mentioned by Gilbert White was nothing more than the noise occasioned by the vibrations of millions of insects' wings in the air. In hot summer evenings in particular I have heard these peculiar humming sounds, and know them to be caused by immense hordes of gnats and midges which fill the air with their numbers. (Nature, 34:547, 1886)

# BELLS, MUSICAL NOTES, MELODY

A patient observer of nature soon discovers many sources of musical tones and bell-like notes---all above and beyond the common sounds of birds, trees creaking in the wind, and the like. A frozen lake, for example, is a whole orchestra of squeaks and throbbing notes. But here we include only those musical sounds that are difficult to identify or are strange due to their mode of production. Also excluded are musical sands and ringing rocks, which have been assigned to the companion handbook on geology.

Mysterious sounds near water are often created by fish which possess a surprising repertoire of vocal talents. Unlike bird songs, fish calls are not well-identified in the field guides. In this ignorance, some of the oceanic sounds reported below are likely the products of aquatic life. As for the rest, who knows? The wind, moving ice, electrical discharges, and other natural agents may be at work.

## SOUNDS FROM THE GREENLAND INTERIOR

### STRANGE SOUNDS FROM INLAND ICE, GREENLAND
**Dauvillier, A.; *Nature*, 133:836, 1934.**

During the month of August 1932, when setting up the French Expedition of the International Polar Year in Scoresby Sound, on the East Greenland coast, some of my colleagues and I heard four times the mysterious sound called by the late Prof. A. Wegener the "Ton der Dov-Bai". The sound was heard in the morning, generally at 11 a. m. (G. M. T.), and also during the afternoon. It was a powerful and deep musical note coming far from the south, lasting a few seconds. It resembled the roaring of a fog-horn. After that it was not heard during the course of the Polar Year.

A. Wegener and five of his companions heard it eight times in five different neighbouring places, both during the day and the polar night. It lasted sometimes a few minutes and Wegener ascribed it to the movements of inland ice. In fact, it seemed, in Scoresby Sound, to come from beyond Cape Brewster, precisely from the part of the coast where the inland ice flows into the sea from the large glaciers.

Is this vibrating sound really caused by the detachment of icebergs or is it similar to the 'desert song', that strange musical note produced by the sand? In fact, there is a close analogy between the fields of powdery dry snow of the inland ice and the fields of sand of the Arabian desert. (Nature, 133:836, 1934)

# DESERT SOUNDS

## WIND-SOUNDS IN THE DESERT
**Anonymous; *Popular Science Monthly,* 22:285, 1882.**

The travelers' tales of sounds like the ringing of bells, which they have heard in deserts and lonely places, are familiar. Some of them are too well substantiated to admit of serious dispute. Among them is that of the noises heard at the Gebel Nakus, in the Sinaitic Peninsula, which the Arabs say proceed from a convent of damned monks; the musical cliffs of the Orinoco, told of by Humboldt; and the sounds which the French savants Jollois and Devilliers declare they heard at sunrise at Karnak, Egypt, and described as comparable to the ancient fable of the vocal Memnon. The sounds are not always or exactly like the ringing of a bell; sometimes they resemble the music of a string, and may be generally described as of an intermediate character between the two classes. A characteristic of the sounds is, that no one can discern where they come from. M. Emile Sorel, fils, in order to determine their origin, has made some successful experiments in reproducing them artificially. Taking his gun into an open field, he placed it at an angle of 45$^{o}$ against the wind, when it gave forth a sound. Then moving it around, he caused it to utter the exact tone he sought. The sound could not be localized. Addressing a peasant, he asked him, "Do you hear my gun?" "Pardon, monsieur, it is the bells of ____." A similar answer was got from every one whose attention was called to the noise. It was believed to come from about two miles and a half to the windward. M. Sorel believes this experiment authorizes the hypothesis that the ringing is the result of the blowing of the wind over a slope at the foot of which is something that may act as a resonator. What is done on a small scale in a gun may be done on a large scale in nature, on the face of a mountain or a rock which is backed by a valley or a ravine, or which is itself elastic enough to give the resonant effect. The sounds are apparently not as readily given when the vibrating surfaces and media are moist. (Popular Science Monthly, 22:285, 1882)

# NATURAL MELODY

## NATURAL MELODY
**Paget, R. A. S.; *Nature,* 130:701, 1932.**

On the evening of October 8, I was at Angmering-on-Sea, and my host, Mr. Kenneth Barnes, called me to listen to the wind playing, and took me to the bathroom, which faced down wind. The wind was blowing hard and gustily, and was producing a most amazing effect---exactly as though a flageolet were being played by a human performer.

The melody was in E. major, with A# substituted for A♮, and it ranged over five semitones, of approximate frequency 1290, 1448, 1625, 1824, 1932. The melody did not slur up and down, as when the wind whistles through a cranny, but changed by sharply defined steps from note to note. The melody included runs, slow trills, turns and grace notes, and sounded so artificial that I felt bound to open the window and make sure that the tune was not being played by a human performer out of doors.

The sounds were traced to the overflow pipe of the bath, through which air was rushing in at the rosette (of six holes, each 9 mm. diameter, set in a circle) covering the inner end of the pipe where it joined the bath.

Next morning I examined the pipe. It was about 3 cm. in diameter, and about 3 ft. 5 in. long, in the form of an S-bend, of which the lower portion passed through the outer wall, and ended in an open mouth. The natural frequency of the pipe when blown into by mouth was about 161---that is, three octaves below the keynote of the scale previously indicated. Evidently the wind was playing on the 7th, 8th, 9th, 10th and 11th overtones of the pipe, and the melody was being produced by the rapid fluctuations of wind-pressure.

I wonder whether such an effect can ever have occurred in Nature? a broken bamboo stem, for example, partially obstructed at its windward end, and so 'shielded' by vegetation, soil, etc., as to produce a pressure difference between its open ends?

The effect of elaborate melodies thus produced without human intervention would be highly magical and suggestive. (Nature, 130:701, 1932)

## NATURAL MELODY
**Scrivenor, J. B.; *Nature*, 130:778, 1832.**

In his letter in Nature of November 5 on "Natural Melody" Sir Richard Paget inquires whether such an effect can have been produced in Nature, as, for example, by a broken bamboo stem. The following extract from Godinho de Eredia's "Report on the Golden Chersonese": 1597-1600 (English translation by Mr. J. V. Mills of the Malayan Civil Service) may be of interest, though it concerns speech, not melody:

"To conclude entirely with the Peninsula, I will relate a curious phenomenon which occurs at the mouth and entrance of the River Penagim [now called the River Linggi. J. B. S.] here there are dense thickets of Bamboos, and among them are two very tall stout Bamboos which are set in such a manner that one of them towers over the other; now it is an actual fact that by day and by night human voices are heard proceeding from these Bamboos; one of them says 'Suda', that is to say, 'Enough', and the other replies 'Bolon', which is as much as to say 'Not yet'.

"I always regarded this as a worthless fairy-tale, until Affonso Vicente, Ambassador to Achem, assured me that he personally heard these voices saying 'suda', 'bolon', when he went to this place on the Panagim for the sole purpose of observing this most curious occurrence in the year 1595." (Nature, 130:778, 1932)

## MUSICAL THUNDER
**Martyn, G. H.; *Nature,* 74:200, 1906.**

Early this morning a storm broke in this neighbourhood accompanied by heavy thunder. During the storm I noticed that two of the peals began with a musical note of distinct and definite pitch. The "musical" portion of the peal lasted for about two seconds in each case, and the frequency of the note was both times about 400 per second.

This sound closely resembled a foot-fall in a narrow alley between high walls, and was only heard in two consecutive peals, separated by an interval of about a minute, the first being much more definitely musical than the second. In each case the interval between the flash and the first sound of thunder was about five seconds.

As is well known, a peal of thunder from lightning near at hand frequently sounds like a quick succession of raps or a volley of guns. Can the successive raps have followed one another so rapidly in this case that they combined to form a note?

If so, and if this note was due to a special configuration of reflecting surfaces in the clouds, possibly to others in slightly different positions, considerably different frequencies may have been observed.

The fact that two peals only sounded in this manner separated by the short interval of about one minute, and that the second was not so decidedly musical as the first, seems to indicate that they were due to some rapidly changing source such as one might expect the reflecting surfaces of a cloud to be. I listened carefully to determine that the note had its origin outside and was not due to resonance within the room, and in the second peal it was certainly outside, and probably had the first had its origin within the room I should have observed it.

I should be very glad to hear if anyone has observed a similar phenomenon. (Nature, 74:200, 1906)

# MUSICAL SNOWFLAKES

## MUSICAL SNOWFLAKES
**Gibson, H. H.; *Symons's Meteorological Magazine,* 50:42, 1915.**

I read with great interest the importance which is attached to the want of information as to the depth of snow in this country.

I wonder if any of your readers have ever observed during a snow storm, when the air is absolutely calm, that the large flakes in certain conditions, when falling, cause a musical sound. I am doubtful if it is imagination or a fact that the dry flakes falling cause this. Perhaps some of those interested in the larger question of snow would give their experiences. (Symons's Meteorological Magazine, 50:42, 1915)

# OCEANIC MUSIC

## STRANGE NOISES HEARD AT SEA OFF GREY TOWN
**Dennehy, Charles; *Nature*, 2:25-26, 1870.**

In submitting the following to the notice of your readers, I am guided only by the desire of seeking a solution of what to me and to many others appears a very curious phenomenon. The facts related can be vouched for by numbers of the officers and crews of any of the R. M. Company's ships.

I must premise that this phenomenon only takes place with iron vessels, and then only when at anchor off the port of Grey Town. At least, I have never heard of its occurring elsewhere, and I have made many inquiries.

Grey Town is a small place, containing but few inhabitants, situated at the mouth of the river St. Juan, which separates Nicaragua from Costa Rica, and empties itself into the Atlantic, lat. 10° 54'N., and long. 83° 41'W. In this town there are no belfries or factories of any kind.

Owing to a shallow bar, vessels cannot enter the harbour or river, and are therefore obliged to anchor in from seven to eight fathoms of water, about two miles from the beach, the bottom consisting of a heavy dark sand and mud containing much vegetable matter brought down by the river. Now, while at anchor in this situation, we hear, commencing with a marvellous punctuality at about midnight, a peculiar metallic vibratory sound, of sufficient loudness to awaken a great majority of the ship's crew, however tired they may be after a hard day's work. This sound continues for about two hours with but one or two very short intervals. It was first noticed some few years ago in the iron-built vessels Wye, Tyne, Eider, and Danube. It has never been heard on board the coppered-wooden vessels Trent, Thames, Tamar, or Solent. These were steamers formerly employed on the branch of the Company's Intercolonial service, and when any of their officers or crew told of the wonderful music heard on board at Grey Town, it was generally treated as "a yarn" or hoax. Well, for the last two years the company's large Transatlantic ships have called at Grey Town, and remained there on such occasions for from five to six days. We have thus all had ample opportunity of hearing for ourselves. When first heard by the negro sailors they were more frightened than astonished, and they at once gave way to superstitious fears of ghosts and Obeihism. By English sailors it was considered to be caused by the trumpet fish, or what they called such (certainly not the Centriscus scolopax, which does not even exist here). They invented a fish to account for it. But if caused by any kind of fish, why only at one place, and why only at certain hours of the night? Everything on board is as still from two to four, as from twelve to two o'clock, yet the sound is heard between twelve and two, but not between two and four. The ship is undoubtedly one of the principal instruments in its production. She is in fact for the time being converted into a great musical sounding board.

It is by no means easy to describe this sound, and each listener gives a somewhat different account of it.

It is musical, metallic, with a certain cadence, and a one-two-three time tendency of beat. It is heard most distinctly over open hatchways, over the engine-room, through the coal-shoots, and close round the outside of the ship. It cannot be fixed at any one place, always appearing to recede from the observer. On applying the ear to the side of an open bunker, one fancies that it is proceeding from the very bottom of the hold.

Very different were the comparisons made by the different listeners. The blowing of a conch shell by fishermen at a distance, a shell held to the ear, an aeolian harp, the whirr or buzzing sound of wheel machinery in rapid motion, the vibration of a large bell when the first and louder part of the sound has ceased, the echo of chimes in the belfry, the ricocheting of a stone on ice, the wind blowing over telegraph wires, have all been assigned as bearing a more or less close resemblance; it is louder on the second than the first, and reaches its acme on the third night; calm weather and smooth water favour its development. The rippling of the water alongside and the breaking of the surf on the shore are heard quite distinct from it.

What is, then, this nocturnal music? Is it the result of a molecular change or vibration in the iron acted on by some galvanic agent peculiar to Grey Town? for bear in mind that it is heard nowhere else, not at Colon, some 250 miles distant on the same coast, not at Porto Bello, Carthagena, or St. Marta. The inhabitants on shore know nothing of it. If any of your numerous readers can assign a likely cause, will they be pleased to state by what means, if any, its accuracy may be tested? If required, I can forward a specimen of the mud and sand taken from the anchor. (Nature, 2:25-26, 1870)

## STRANGE NOISES HEARD AT SEA OFF GREY TOWN

**Kingsley, C.; *Nature*, 2:46, 1870.**

I am glad to see that the vexed question of the noise heard from under the sea in various parts of the Atlantic and Pacific has been re-opened by a gentleman so accurate and so little disposed to credulity as Mr. Dennehy. The fact that this noise has been heard at Grey Town only on board the iron steamers, not on board the wooden ones, is striking. Doubtless if any musical vibration was communicated to the water from below, such vibration would be passed on more freely to an iron ship than to a wooden one. But I can bring instances of a noise which seems identical with that heard at Grey Town being heard not only on board wooden ships, but from the shore.

I myself heard it from the shore, in the island of Monos, in the Northern Bocas of Trinidad. I heard it first about midnight, and then again in the morning about sunrise. In both cases the sea was calm. It was not to be explained by wind, surf, or caves. The different descriptions of the Grey Town noise which Mr. Dennehy gives, will each and all of them suit it tolerably. I likened it to a locomotive in the distance rattling as it blows off its steam. The natives told me that the noise was made by a fish, and a specimen of the fish was given me, which is not Centriscus scolopax, the snipe-fish, but the trumpet-fish, or Fistularia. I no more believe that it can make the noise than Mr. Dennehy believes (and he is quite right) that the Centriscus can make it.

This noise is said to be frequently heard at the Bocas, and at Point a Pierre, some twenty-five miles south; also outside the Gulf along the Spanish main as far as Barcelona. It was heard at Chagreasancas (just inside the Bocas) by M. Joseph, author of a clever little account of Trinidad, on board a schooner which was, of course, a wooden one, at anchor. "Immediately under the vessel," he says, "I heard a deep and not unpleasing sound, similar to those one might imagine to proceed from a thousand AEolian harps; this ceased, and deep and varying notes succeeded; these gradually swelled into an uninterrupted stream of singular sounds, like the booming of a number of Chinese gongs under water; to these sounds succeeded notes that had a faint resemblance to a wild chorus of a hundred human voices singing out of time in deep bass."

He had, he says, three specimens of the trumpet-fish, said to make the noise, either by "fastening the trumpet to the bottom of a vessel or a rock," or without adhering to any object. The whip-like appendage to the tail, which he describes, marks his specimens at once as Fistularias.

..........

Another instance of this sound being heard on board a wooden ship (and this time again in the Pacific) is given (in p. 304 of Mr. Griffith and Colonel Hamilton Smith's edition of Cuvier's Fishes, on no less an authority than that of Humboldt who (say the editors and authors of the Appendix) did not suspect the cause. "On the 20th of February, 1803, toward seven in the evening, the whole crew were astounded by an extraordinary noise, which resembled that of drums beating in the air. It was at first attributed to the breakers. Speedily it was heard in the vessel, and especially toward the poop. It was like a boiling, the noise of the air which escapes from fluid in a state of ebullition. They then began to fear that there was some leak in the vessel. It was heard unceasingly in all parts of the vessel, and finally, about nine o'clock, it ceased altogether. From the narration (says Cuvier) which we have extracted, and from what so many observers have reported touching various Sciaenoids, we may believe that it was a troop of some of these species which occasioned the noise in question."

For there is, without doubt, a great deal of evidence to show that certain Sciaenoids make some noise of this kind. The Umbrinas, or "maigres" of the Mediterranean and Atlantic are said to be audible at a depth of twenty fathoms, and to guide the fishermen to their whereabouts by their drumming. The fishermen of Rochelle are said to give the noise a peculiar term, "seiller," to hiss; and say that the males alone make it in spawning time; and that it is possible, by imitating it, to take them without bait. The "weak-fish" of New York (Labrus squetaquee of Dr. Mitchell) is said to make a drumming noise. But the best known "drum-fishes" are of the genus Pogonias, distinguished from Umbrina by numerous barbules under the lower jar, instead of a single one at the symphysis. M. Cuvier names them Poganias fusca, and mentions that "it emits a sound still more remarkable than that of the other Sciaenoids, and has been compared to the noise of several drums." The author of the Appendix states that these "drumfish" swim in troops in the shallow bays of Long Island; and according to Schoepf (who calls them Labrus chromis) assemble round the keels of ships at anchor, and then their noise is most sensible and continuous. Dr. Mitchell, however, only speaks of their drumming when taken out of the water. Species of the same genus, if not identical, are found as far south as the coast of Brazil; and it is to them, probably, that that noise is to be attributed which made the old Spanish discoverers report that at certain seasons the nymphs and Tritons assembled in the Gulf of Paria, and made the "Golfo Triste glad with nightly music." (Nature, 2:46, 1870)

## SOUNDS MADE BY FISHES IN THE EAST INDIES
**Anonymous; *Nature*, 135:426-427, 1935.**

In Nature of November 17 (p. 769), we quoted an interesting account of sounds heard in the East Indies by Capt. P. Jansen. We have received a letter from Dr. J. D. F. Hardenberg, of the Laboratory for Investigation of the Sea, Batavia, with reference to this note. He states that the comparison of these noises with the sounds made by foghorns is quite correct. They remind one also of the sounds made by motor traffic on a busy thoroughfare when heard at a distance of about a hundred yards. The noises, however, do not proceed from the earth, but are made by fishes of the genus Therapon, as described by Dr. Hardenberg in a recent paper (Zool. Anz., 108; 1934). The other sounds mentioned by Capt. Jansen have also been heard by Dr. Hardenberg, though less frequently, and once, when in the Java Sea, he heard sounds as if made by silver bells. Their origin is still unknown, but he supposes that they are also made by animals. (Nature, 135:426-427, 1935)

## MYSTERIOUS MUSIC ON THE GULF SHORE
**Anonymous; *Scientific American,* 3:51, 1860.**

The mystic music sometimes heard at the mouth of the Passagoula river, on a still night, is one of the wonders of our coast. It is not confined, however, to the Passagoula river, but has often been heard at other places. At the mouth of the Bayou Coq del Inde and other inlets opening into the Gulf along the coast of our own country, the curious listener, lying idle in his boat, with lifted oars, when every other sound is hushed, may sometimes hear its strains coming apparently from beneath the waters, like the soft notes of distant Eolian harps. We have always supposed that this phenomenon, whatever its origin might be, natural or supernatural, was peculiar to our own coast. It appears, however, from Sir Emerson Tenant's recent work on Ceylon, something very like it is known at Battialloa, in that island, and it is attributed to rather less poetical and mysterious origin---that it is a peculiar species of shellfish. They are said to be heard at night, and most distinctly when the moon is nearest the full. (Scientific American, 3:51, 1860)

# Chapter 5
# THE STRANGE PHENOMENA OF EARTHQUAKES

## INTRODUCTION

A sensory channel most humans find hard to ignore is that of physical motion; a category that includes shock, vibration, and gross physical movement. Short, sharp shocks are sensed kinesthetically like bumps on a highway. The frequent tilting effects of earthquakes, though, disturb one's sense of balance and induce disorientation. Whatever the mode of detection, the onset of physical motion of the earth's crust signals the existence of powerful local forces and thus the possibility of a variety of strange phenomena.

The physical motions of an earthquake need not occupy our attention here. Rather, it is the remarkable spectrum of parallel events occurring before, during, and after the quake proper that command our curiosity. The phenomena fall into two major divisions: (1) concurrent effects, such as "earthquake weather," that have elusive and/or questionable origins; and (2) concurrent phenomena, such as solar activity, that seem to trigger seismic activity via unknown or hard-to-explain mechanisms. The mystery in this chapter, then, is not in the earthquake itself but in the curious effects and correlates that surround the event in time and space.

# PHENOMENA ACCOMPANYING EARTHQUAKES

A major earthquake disrupts the balance of terrestrial forces from deep in the rocky mantle to high in the ionosphere. As the earth's surface vibrates---not unlike the skin of a drum---great sea waves (tsunamis) and atmospheric pressure waves spread out rapidly from the epicenter. The sea waves may travel 10,000 miles, while the air waves may penetrate into the charged layers of the upper atmosphere upsetting the electrical equilibrium. A truly great quake is felt globally.

Near the epicenters of big earthquakes, unusual phenomena abound---these are of course in addition to the physical motion of the earth itself. Loud groans, detonations, and grinding noises are emitted by the rent strata. Less common are flames issuing from the ground, moving lights, aurora-like beams, and sky glows. Most of these luminous effects are documented in Chapter 1, but a few observations especially those with odd electrical effects, are retained in this chapter. Both earthquake lights and electrical effects are usually attributed to the piezoelectric effect stimulated by earthquake forces. These same pressures also force up geysers of water, sand, and even coal. The powerful quakes near New Madrid, in the Mississippi Valley, in 1811 and 1812, produced a wide spectrum of such phenomena, as the first accounts below will prove.

The prediction of earthquakes is becoming more scientific. The scientists involved in these efforts have been intrigued by the ability of many animals to sense the coming of an earthquake long before humans hear the grating roars and feel the heaving ground. In Japan and China, fish, pheasants, and other animals have long been valued for their abilities to detect earthquake precursors. The ways in which animals sense the oncoming quake are not well-understood and likely involve the detection of low-level sounds and slight tiltings of the ground.

Some people still swear that "earthquake weather" exists, even though some of the greatest quakes were not preceeded by the hot, sultry, forboding calm that legend tells us presages a big convulsion. Still, there are startling correlations between earthquakes, heavy rainfall, fog, and other weather events. It is understandable how the dust raised by an earthquake can trigger rainfall, and also how heavy rain can "lubricate" strata and help release strains. Cause-and-effect relationships between quakes and other weather phenomena are harder to discern. In any case, big earthquakes release so much energy that almost all geophysical parameters are shaken a bit.

## ASSORTED PHENOMENA OF THE GREAT NEW MADRID QUAKE

### NEW MADRID EARTHQUAKE
**Shepard, Edward M.; *Journal of Geology*, 13:47, 1905.**

The first shock [at New Madrid, MO] came at 2 a.m., December 16, 1811, and was so severe that big houses and chimneys were shaken down, and at half-hour intervals light shocks were felt until 7 a.m., when a rumbling like distant thunder was heard, and in about an instant the earth began to totter and shake so that persons could neither stand nor walk. The earth was observed to roll in waves a few feet high, with visible depressions between. By and by these swells burst, throwing up large volumes of water, sand, and coal. Some was partly coated with what seemed to be sulphur. When the swells burst, fissures were left running in a northern and southern direction, and parallel for miles. Some were 5 miles long, 4-1/2 feet deep, and 10 feet wide. The rumbling appeared to come from the west and travel east. Similar shocks were heard at intervals until January 7, 1812, when another shock came as severe as the first. Then all except two families left, leaving behind them all their property, which proved to be a total loss, as adventurers came and carried off their goods in flat boats to Natchez and New Orleans, as well as all their stock which they could not slaughter. On February 17 there occurred another severe shock, having the same effect as the others, and forming fissures and lakes. As the fissures varied in size, the water, coal, and sand were thrown out to different heights of from 5 to 10 feet. Besides long and narrow fissures, there were others of an oval or circular form, making long and deep basins some 100 yards wide, and deep enough to retain water in dry seasons. The damaged and uptorn country embraced an area of 150 miles in circumference, including the old town of Little Prairie [now called Caruthersville], as the center, a large extent on each side of Whitewater, called Little River, also both sides of the St. Francis in Missouri and Arkansas. Reelfoot Lake, in Tennessee, sank 10 feet. (Journal of Geology, 13:47, 1905)

## THE NEW MADRID EARTHQUAKE
**Broadhurst, Garland C.; *American Geologist*, 30:78-79, 1902.**

Prof. J. W. Foster in his Physical Geography of the Mississippi valley furnishes us with much interesting information concerning the New Madrid earthquake, as follows:

The series of shocks began near the close of 1811 and continued until 1813. The telluric activity of which these events were a part extended over half the hemisphere and was manifested in a series of stupendous phenomena such as the elevation of Sabrina, one of the Azores, to the hight of 320 feet above the sea. The city of Caraccas with its 12,000 inhabitants was destroyed, an eruption of a volcano of St. Vincent, and fearful noises on the Llanois of Calabazo.

New Madrid seems to have been one of the foci of disturbance, and shocks were repeated almost every hour for months in succession. Mr. A. N. Dillard, of New Madrid, gave Mr. Foster much information. The weather had been warm and pleasant, the air was hazy like Indian summer. About midnight, while the French population were engaged in dancing the first shock came, and houses and fences were shaken down. The greatest consternation prevailed, and the people rushed out and Protestant and Catholic knelt side by side and offered up solemn supplications. The shocks were stated to continue for 20 to 30 months. In every instance the motion was from the west or southwest. Fissures would be formed 600 to 700 feet long and 20 to 30 feet wide through which water and sand spouted 40 feet high. There issued no burning flames but flashes such as would result from an explosion of gas, or from passing of electricity from cloud to cloud. Oak trees would be split in the center and for 40 feet up the trunk, one part standing on one side of a fissure, the other part on the other. Mr. Dillard says that his grandfather had received a boat-load of castings from Pittsburgh and had them stored in his cellar. During one of the shocks, the ground opened under the house and they were swallowed up and never afterwards seen. Near the St. Francis river there is a great deal of sunk land, caused by the earthquake of 1811. Here are large trees submerged ten or twenty feet beneath the water. Previous to the earthquake keelboats would come up the St. Francis river and pass into the Mississippi three miles below New Madrid. The bayou is now dry land. From one of the fissures there was ejected the cranium of an extinct species of ox. In Reelfoot lake the fisherman floats his canoe above the branching submerged tops of cypress trees. Reelfoot lake in Obion Co., Tennessee, nearly 20 miles long and seven broad owes its origin to the sinking of the ground during this period.

Timothy Flint who at one time resided at St. Charles, Mo., and published in 1826 a book of "Recollections" was in this region about 1820 and furnishes a very graphic account of the earthquake phenomena.

Flint says that a tract near Little Prairie became covered with water three or four feet deep and when the water disappeared, there remained a stratum of sand. Lakes twenty miles in extent were formed in the course of an hour while others were drained. The grave-yard of New Madrid was precipitated into the stream and most of the houses thrown down. The whole country between the mouth of the Ohio and the St. Francis including a front of 300 miles was convulsed to such a degree as to create lakes and islands and to cover a tract of many miles extent near Little Prairie with water three or four feet deep, and when the water disappeared a stratum of sand remained. Birds lost all power to fly and retreated to the neighborhood of man. A few persons sank in the chasms but were extricated. Trees were cut so as to fall across the chasms and on these the people would rest. Many persons perished with their boats upon the river. There were many shocks but two series of consussions were particularly terrible. There were two classes of shocks: those in which the motion was horizontal, and those in which it was perpendicular. The latter were attended with explosions and the terrible mixture of noises that preceded and accompanied the earthquakes in a louder degree, but were by no means so desolating and destructive as the other. In the interval of earthquakes, there was one evening---and that a brilliant and cloudless one---in which the western sky was a continual glare of vivid flashes of lightning and of repeated peals of subterranean thunder, seeming to proceed, as the flashes did, from below the horizon. This is said to be the same night on which the earthquake was so severe at Caraccas, and these flashes and that event were part of the same scene. (American Geologist, 30:78-79, 1902)

# CURIOUS EARTHQUAKE SOUNDS

## NOTES ON THE ASSAM EARTHQUAKE
**Kingdon-Ward, F.; *Nature,* 167:130, 1951.**

[Assam, August 15, 1950] I find it very difficult to recollect my emotions during the four or five minutes the shock lasted; but the first feeling of bewilderment---an incredulous astonishment that these solid-looking hills were in the grip of a force which shook them as a terrier shakes a rat---soon gave place to stark terror. Yet my wife and I, lying side by side on the sandbank, spoke quite calmly together, and to our two Sherpa boys who, having already been thrown down twice, were lying close to us.

The earthquake was now well under way, and it felt as though a powerful ram were hitting against the earth beneath us with the persistence of a kettledrum. I had exactly the sensation that a thin crust at the bottom of the basin, on which we lay, was breaking up like an ice floe, and that we were all going down together through an immense hole, into the interior of the earth. The din was terrible; but it was difficult to separate the noise made by the earthquake itself from the roar of the rock avalanches pouring down on all sides into the basin.

Gradually the crash of falling rocks became more distinct, the frightful hammer blows weakened, the vibration grew less, and presently we knew that the main shock was over. The end of the earthquake was, however, very clearly marked by a noise, or series of noises, which had nothing to do with falling rocks. From high up in the sky to the north-west (as it seemed) came a quick succession of short, sharp explosions---five or six---clear and loud, each quite distinct, like 'ack-ack' shells bursting. It was the 'cease-fire'. (Nature, 167:130-131, 1951)

## ON EARTHQUAKES IN CALIFORNIA FROM 1812 TO 1855
**Trask, J. B.; *American Journal of Science*, 2:22:110, 1856.**

From careful inquiry of the older residents, I can learn of but one shock that has proved in the slightest degree serious, causing the destruction of either life or property to any extent. This was the earthquake of September, 1812, which destroyed the Mission San Juan Capistrano, in Los Angeles county, and that of Viejo, in the valley of San Inez, in the county of Santa Barbara.

The following is the history of that event as I have obtained it from the native inhabitants, and older foreign residents on this coast:

The day was clear and uncommonly warm; it being Sunday the people had assembled at San Juan Capistrano for evening service. About half an hour after the opening of service, an unusually loud, but distant rushing sound was heard in the atmosphere to the east and over the water, which resembled the noise of strong wind, but as the sound approached no perceptible breeze accompanied it. The sea was smooth and the air calm. So distant and loud was this atmospheric sound that several left the building on account of it.

Immediately following the sound, the first and heaviest shock of the earthquake occurred, which was sufficiently severe to prostrate the Mission church almost in a body, burying in its ruins the most of those who remained behind, when the first indication of its approach was heard.

The shock was very sudden and almost without warning, save from the rushing sound above noted, and to its occurrence at that moment is to be attributed the loss of life that followed. American Journal of Science, 2:22:110, 1856)

## THE NEW MADRID EARTHQUAKE
**Fuller, Myron L.; *U.S. Geological Survey Bulletin* 494, 101-102, 1912.**

The noises accompanying the more important shocks are among the most noticeable features of the earthquake. Bradbury tells of being "awakened by a tremendous noise" at the time of the first shock and noted that "the sound, which was heard at the time of every shock, always preceded it at least a second and uniformly came from the same point and went off in the opposite direction." Audubon, speaking of one of the more severe shocks, describes the sound as like "the distant rumbling of a violent tornado." Bringier does not mention the subterranean noise, but describes the "roaring and whistling produced by the impetuosity of the air escaping from its confinement" in the alluvial materials. A similar roaring and hissing, like the escape of steam from a boiler, is also described by Hildreth. Eliza Bryan describes the sounds of the shock itself as "an awful noise resembling loud and distant thunder but more hoarse and vibrating." Flint compares the sound of the ordinary shocks to rembling like distant thunder, but mentions the fact that the vertical shocks were accompanied by "explosions and a terrible mixture of noises." Linn describes the phenomena as beginning with "distant rumbling sounds succeeded by discharges, as if a thousand pieces of artillery were suddenly exploded," and notes the hissing sounds accompanying the extrusion of the water.

In Tennessee, as described by Haywood, "in the time of the earthquake a murmuring noise, like that of fire disturbed by the blowing of a bellows, issued from the pores of the earth. A distant rumbling was heard almost without intermission and sometimes seemed to be in the air." Explosions like the discharge of a cannon a few miles distant were also recorded, and "when the shocks came on the stones on the surface of the earth were agitated by a tremulous motion like eggs in a frying pan, altogether made a noise similar to that of the wheels of a wagon in a pebbly road."

In the more distant localities we must again rely on the compilation of Mitchill. At Herculaneum, Mo., the noise was described as a roaring or rumbling resembling a blaze of fire acted upon by wind; at St. Louis before the shocks "sounds were heard like wind rushing through the trees but not resembling thunder," while more considerable noises accompanied the shocks. At Louisville the noise was likened to a carriage passing through the street; at Washington, D.C., the sound was very distinguishable, appearing to pass from southwest to northeast; at Richmond similar noises were heard, but they appeared to come from the east; at Charleston there was "a rumbling like distant thunder which increased in violence of sound just before the shock was felt;" while at Savannah the noise was compared to a rattling noise like "that of a carriage passing over a paved road."

From the above it would appear that earth noises were heard at most points where the earthquake was felt. In the region of marked disturbance there were the additional noises made by escaping air, water, crashing trees, and caving river banks. According to the best information the sound in the Mississippi Valley was a somewhat dull roar, rather than the rumbling sound of thunder with which it was compared at certain of the more remote localities. In reality, as has been stated to the writer in regard to the recent Jamaica earthquake, although suggesting many of the common noises, it was essentially unlike anything ever heard by the observers before. (U.S. Geological Survey Bulletin 494, 1912, 101-102)

# ELECTRICAL EFFECTS OF EARTHQUAKES

## EARTHQUAKE LIGHTS AND ELECTRICAL EFFECTS
**Anonymous; *Nature,* 6:89-90, 1872.**

With reference to the connection between electricity and earthquakes, the Pall Mall Gazette quotes from a Californian paper, the Inyo Independent, the following curious statements respecting the prevalence of electrical phenomena at the time of the recent earthquake in that State:---"A few days after the big shock, so called, at Cerro Gardo, very loud thunder was heard during a violent snowstorm. With the exception of the snow, the same thing occurred here, and perhaps at other places in the valley. This is remarkable, because almost unprecedented. Immediately following the great shock, men whose judgment and veracity are beyond question, while sitting on the ground near the Eclipse Mines, saw sheets of flame on the rocky sides of the Inyo Mountains, but half a mile distant. These flames, observed in several places, waved to and fro, apparently clear of the ground, like vast torches. They continued for only a few minutes. In this office, one day last week, while one of the proprietors was running a large number of sheets of flat-cap paper through a job press, these sheets, after leaving the press, were affected by the movements of the operator's hand, as a strong magnet would affect iron filings. When his hand was near them, the whole pile, or at least a hundred of them from the top, seemed to float in the air, like tissue paper in a slight breeze. The top sheet would rise at each end up to the hand when held four inches above it, and thus by attraction be moved entirely away from the others. At times during the night sparks of fire were repeatedly emitted from a woollen shawl on being touched by the hand. At the Kearsarge Mill, located at an altitude of nearly 80,000 feet above the sea, the following occurrence was noted by Harry Clawson and P. J. Joslyn:---The former, while sitting with his knees within three inches of a castiron stove, felt a peculiar numbing sensation, and, supposing his limbs were 'asleep,' essayed to rub them with his hand. As soon as his hand touched his knee he felt a shock, and immediately after, and for a couple of seconds, a stream of fire ran between both knees and the stove. (Nature, 6:89-90, 1872)

## EARTHQUAKE AT ANCHOR
**Anonymous; *Franklin Institute, Journal,* 23:308, 1937.**

The following notes, made on board H. M. S. Volage, while at anchor in Callao Roads, during the severe earthquake which occurred in March, 1828, are of considerable interest. We understand, that part of the chain cable of the Volage which was exposed to its effects, from being then in use, is now in the possession of that learned gentleman, Mr. Faraday.

March 30, 1828. The morning clear, and a light breeze from the southward. At 7h. 28m. a black thin cloud passed over the ship, with very heavy distant thunder. At the same moment we felt the shock of a severe earthquake. I should think it continued seventy or eighty seconds. The ship trembled violently, and the only thing I can compare it to is, the ship being placed on trucks, and driven with rapidity over coarse paved ground. The ship was moored with two chain-cables, and on weighing the anchors a few days after, we found 56 links of the best bower cable much injured; the iron had the appearance of being melted, and nearly one-sixth of the link was destroyed. This piece was 30 fathoms from the anchor, and 20 fathoms from the ship. The bottom was soft mud, in which the cable was buried. During the earthquake the water alongside was full of little bubbles; the breaking of them sounded like red-hot iron put into water. The city of Lima suffered considerably, and a number of lives were lost. This severe shock was felt for nearly one hundred miles north and south of Lima. (Franklin Institute, Journal, 23:308, 1837)

# HOW EARTHQUAKES MAY AFFECT THE ATMOSPHERE AND IONOSPHERE

## SEISMIC AIR WAVES FROM THE GREAT 1964 ALASKAN EARTHQUAKE

**Bolt, Bruce A.; *Nature,* 202:1095-1096, 1064.**

Exceptional atmospheric waves, resembling those from large nuclear explosions, have been recorded at Berkeley after the Alaskan earthquake of March 28, 1964. This severe earthquake occurred 3,130 km away from Berkeley near the head of Prince William Sound. A preliminary determination made by the U.S. Coast and Geodetic Survey gives an origin time of 03 h 36 min 13 sec G.M.T.; the geographic latitude and longitude are 61.05°N., 147.50°W. Displacements of the adjacent sea-bed generated a damaging tsunami which was observed at numerous places around the Pacific. Changes in sea-level indicate that the mainland in the disturbed region along the coast rose about 7 ft., while a portion of Kodiak Island subsided by about 5 ft.

..........

In addition to the seismic waves recorded by pendulum seismographs, two distinct wave trains were recorded by a microbarograph in the seismic vault. The instrument is a Davies Marion barovariograph designed to respond to pressure fluctuations with periods down to 10 sec. An electric flow sensor is placed in an air-line between the atmosphere and a thermally insulated 40-1. flask. The sensor detects relative variations in the resistance of two thermistors placed symmetrically in the air-line about a heated thermistor. The signal is recorded continuously on a paper chart which moves at 6 mm/min. Sensitivity in March 1964 was set such that 1 mm of pen deflexion corresponded to a pressure variation of 0.0048 millibars.

Fortunately the air pressure was extremely steady in the whole region at the time. The synoptic weather charts show prevailing stable conditions from Alaska to San Francisco. The path of propagation of an atmospheric wave between these places lies along a broad ridge of 1016 millibars pressure. At Berkeley, the pressure was steady at 1016 millibars with a light wind having an average velocity of a few knots from the north.

The microbarograph commenced to respond to short-period pressure variations 14 min after the earthquake. The onset of this wave train is a sharp compression which corresponds to a pressure variation of 0.021 millibars; the first variations have periods near 1 min. The arrival time corresponds closely to the arrival of the large amplitude part of the seismic surface-wave train. These air-pressure fluctuations are evidently coupled to the Rayleigh waves, which had a frequency spectrum rich in harmonics with periods of 10 sec 2 min. They are clearly visible on the barogram for an interval of 4 h; the most conspicuous period is about 15 sec.

The most interesting feature on the barogram is a large pressure pulse which arrives as a compression with an emersio onset about 2 h 39 min after the earthquake. The maximum pressure variation in the pulse is about one-tenth of a millibar. Dispersion of the pulse is well developed in a manner which resembles the dispersion in acoustic waves generated by nuclear explosions.

Suppose that the source of this pulse was in the neighbourhood of the earthquake epicentre. Then the first rarefaction maximum has a group velocity of 317 m/sec and an apparent period of 3 min. At 06 h 31 min, several pressure variations with periods about 1 min appear; the group velocity of the first rarefaction in this train is 298 m/sec. These values are in reasonable agreement with the empirical curves for normal dispersion constructed by Donn and Ewing. These authors used acoustic waves recorded on barograms from a number of stations after atmospheric nuclear explosions. The pressure variation also suggests the occurrence of inverse dispersion of the pulse. If variations with periods below 4 min are filtered from the record, three pressure fluctuations with period increasing from 9 to 12 min are apparent. More detailed measurement of the dispersion is being undertaken by Fourier analysis of the pulse.

The pressure pulse may be explained as a seismic air wave which was generated by sudden vertical displacement or tilting of the Earth's surface near the epicentre of the earthquake. Such movements account for the tsunami as well as the observed changes in sea-level. Computation shows that the approximate amplitude of the recorded atmospheric wave with period 3 min and velocity 317 m/sec is 0.75 m. This is of the order of the noted changes in ground elevation. The seismic air wave would have a more extensive source than the corresponding waves from nuclear explosions. Although an extended source complicates the wave form there is the possibility that correlation between similar recordings made at stations distributed around the epicentre may provide information on the direction of tilting of the displaced region.

The tsunami arrived off San Francisco about 2.5 h after the seismic air wave. A tide gauge operated by the U.S. Coast and Geodetic Survey near the entrance of San Francisco Bay responded to the first surge of the tsunami 5 h 9 min after the earthquake. (The mareogram shows a long train of oscillations of predominantly 40-min period. These are seismic seiches of the water of San Francisco Bay.)

Pressure pulses propagating in the atmosphere after such natural events as the Krakatoa volcanic eruption and the Siberian meteoritic impact have been examined previously. A few observations of acoustic waves generated by displacement in the meizoseismal area of an earthquake have also been discussed. For example, Benioff and Gutenberg identified pressure variations with period 0.5-2 sec on barographs written 670 km from an earthquake epicentre. In 1926, Thomson reported sounds heard out to 270 km from a New Zealand earthquake; F. J. Whipple remarked that he could find no earlier references to sounds at such great distances. An intense dispersive seismic air wave of the type generated by the Alaskan earthquake does not appear to have been previously reported. It is important to know how widely it was recorded by barographs.

The present observation suggests the possibility of an additional method of prediction of a tsunami. After the detection of a large Pacific earthquake the recording of a seismic air wave at a time appropriate to the epicentral distance would provide evidence that Earth movement favourable to seismic sea-wave generation may have occurred. (Nature, 202:1095-1096, 1964)

## OBSERVATION OF IONOSPHERIC DISTURBANCES FOLLOWING THE ALASKA EARTHQUAKE

**Leonard, R. S., and Barnes, R. A., Jr.; *Journal of Geophysical Research,* 70:1250-1251, 1965.**

The Alaska earthquake, which occurred at 5.36 p.m. local time on Friday March 27, 1964 (0336, March 28, 1964 UT), initiated a large traveling ionospheric disturbance. This disturbance undoubtedly affected many different types of ionospheric sensors; however, the data reported here are limited to ionosondes operating at College, Alaska; Adak, Alaska; Palo Alto, California; and Maui, Hawaii.

The ionosonde data show indications of the perturbation in two different forms. The form depends on the distance between the observing station and the epicenter of the earthquake.

The ionograms from the two Alaska sites show the propagating disturbance as a perturbation of the F-layer trace. The less affected of these two sites was the closer station at College, Alaska, where the disturbance appeared as an abnormal spread in the trace near $f_0F_2$ and $f_xF_2$ that lasted for at least 10 minutes. The traveling disturbance produced a remarkably perturbed ionosphere over the sounder located at Adak, Alaska. Line drawings made from the Adak ionograms are shown in Figure 1 [not reproduced]. An attempt was made to draw only the ordinary component; however, the ionograms starting at 0355 UT and continuing for approximately 45 minutes were quite disturbed, and the identification of the ordinary and extraordinary modes is not always clear. An abnormal increase in critical frequency is apparent on the sounding made at 0355 UT.

The disturbance is indicated at the more remote stations (Palo Alto, California, and Maui, Hawaii) by an abnormal increase, possibly preceded by a slight decrease, in the critical frequency. Examples of these disturbances can be seen in the plots shown in Figure 2 of the critical frequencies as a function of time. The days preceding and following the day of the earthquake are included to indicate typical variations in the critical frequency. Unfortunately, the ionosondes are usually programmed to operate for a 15-second sampling interval once during each 15-minute period; these observations are shown as points on Figure 2 [not reproduced] with smoothed lines drawn to indicate probable variations between the observations. The disturbances in the critical frequency appear to be similar to the ionospheric disturbances that have been associated with nuclear explosions.

Because the sounder data are taken every 15 minutes, only limits can be established on the arrival time of the disturbance. The arrival time for each outlying station has been taken as occurring sometime between the first sounding showing an abnormal increase in critical frequency and the preceding sounding. The arrival times for the Alaska stations were taken as occurring between the first disturbed ionogram and the preceding sounding. These times are shown in Figure 3, where time of arrival is plotted versus the distance between the station and the epicenter of the earthquake. The velocity for the remote observations, approximately 400 m/s, is in reasonably good agreement with the observations of Anastassiades et al. [1962] for the traveling disturbances launched by nuclear explosions. The observations made at Adak do not fit this pattern, since the mean velocity is much higher--- approximately 2 km/sec. This rather high mean velocity may indicate that the wave did not originate at the epicenter of the earthquake.

Pressure disturbances at ground level were observed at the University of California seismologic station at Berkeley, California, with a mean travel velocity from the epicenter of roughly 300 m/sec [Bolt, 1964]. Although the mean velocity of this ground level disturbance is somewhat lower than the mean velocity of the ionospheric disturbance, they may be similar phenomena. A possible initiating mechanism could be either the large tidal actions (as much as 80 feet at some locations) and/or large in-phase ground movements. These in-phase ground movements, deduced from seismic data, indicate that an area of the ground perhaps as large as 150 km in diameter moved in-phase with vertical excursions of many inches. Smaller areas within this in-phase area moved much greater distances vertically, but perhaps out-of-phase with the much larger area disturbance. (Journal of Geophysical Research, 70:1250-1253, 1965)

# THE MYSTERIOUS RELATION BETWEEN EARTHQUAKES AND WEATHER

## EARTHQUAKE WEATHER
**Proctor, Richard A.; *Living Age,* 160:307, 1884.**

Anybody who has ever lived for any length of time at a stretch in a region where earthquakes are common objects of the country and the seaside, knows perfectly well what earthquake weather in the colloquial sense is really like. You are sitting in the piazza, about afternoon teatime let us say, and talking about nothing in particular with the usual sickly, tropical languor, when gradually a short of faintness comes over the air, the sky begins to assume a lurid look, the street dogs leave off howling hideously in concert for half a minute, and even the grim vultures perched upon the housetops forget their obtrusive personal differences in a common sense of general uneasiness. There is an ominous hush in the air, with a corresponding lull in the conversation for a few seconds, and then somebody says with a yawn, "It feels to me very much like earthquake weather." Next minute you notice the piazza gently raised from its underpropping woodwork by some unseen power, observe the teapot quietly deposited in the hostess's lap, and are conscious of a rapid but graceful oscillating movement, as though the ship of state were pitching bodily and quickly in a long Atlantic swell. Almost before you have had time to feel surprised at the suddenness of the interruption (for the earth never stops to apologize) it is all over; and you pick up the teapot with a smile, continuing the conversation with the greatest attainable politeness, as if nothing at all unusual had happened meanwhile. With earthquakes, as with most other things and persons, familiarity breeds contempt. (Living Age, 160: 307, 1884)

## PHENOMENA OF THE MESSINA EARTHQUAKE
**Anonymous; *Nature,* 79:255, 1908.**

The most disastrous earthquake in Europe for many years was experienced in Calabria and the district of Messina, in Sicily, on Monday, December 28. The shock occurred at 5.20 a.m., and was followed by a great seawave, which appears to have destroyed Messina and Reggio, and also the greater part of the villages on each side of the Straits of Messina. Reports from Catanzaro state that the first intimation of the disturbance was a prolonged, thunderous noise followed by a vivid flash of lightning, and at the same time by a series of violent shocks which seemed interminable. Heavy torrential rain then fell, and continued to fall during Tuesday. According to reports from Times correspondents, so complete has been the destruction of Messina that it is almost impossible to obtain any connected account of the character of the earthquake. The centre of the disturbance seems to have been in the Straits, and it is greatly feared that the whold conformation of the neighbouring coast-line has been changed. (Nature, 79:255, 1908)

## EARTHQUAKE PRECEDED BY WINDS AND DUST STORM
**Anonymous; *Nature,* 38:485, 1888.**

A telegram from the city of Mexico states that on the night of the 6th instant there occurred the heaviest shocks of earthquake ever recorded in the city. The houses swayed, the walls cracked, and people rushed into the streets to pray. There was for a few moments much apprehension. The phenomenon was preceded by high winds and dust-storms. (Nature, 38:485, 1888)

## EARTHQUAKE AND HEAVY RAIN
**Anonymous; *Nature,* 60:535, 1899.**

A succession of earthquake shocks occurred on Monday night, September 25, in the district of Darjeeling, involving great loss of life and damage to property. No details as to the exact times of the shocks have been received. The earthquake was accompanied by a remarkable rainfall, and was followed by extensive landslips. It is reported that in twenty-four hours over 20 inches of rain fell, and in all 28 inches fell in thirty-eight hours. (Nature, 60:535, 1899)

## EARTHQUAKES AND ELECTRICITY
**Parnell, Arthur; *Journal of Science,* 20:7, 1884.**

To the above lists it may perhaps be interesting to add a few meteorological statistics, gathered from the records of earthquake incidents, collected by the writer for analysis, up to the present date. Most of the manifestations named are probably closely connected with the action of terrestrial electricity. It is to be understood that either shortly before, or during, or shortly after the occurrence of shocks, these additional phenomena were among the attendant circumstances. The number of separate earthquake cases from which they are gleaned amounts to 490.

| | No. of cases. |
|---|---|
| Thunder, detonations, and rumblings | 156 |
| Isolated rushes or currents of wind, or hissing sounds, giving the idea of an escape of force | 31 |
| Waves or commotions of the sea | 28 |
| Aurorae | 23 |
| Meteors | 73 |
| Ignes Fatui | 2 |
| Lightning flashes in the atmosphere (exclusive of thunderstorms) | 15 |
| Flames seen to issue from fissures | 10 |
| Magnetic disturbances . | 22 |
| Tempests, hail, and rain (exclusive of thunderstorms | 62 |
| Whirlwinds | 7 |
| Snowstorms | 8 |

(Journal of Science, 20:7, 1884)

# THE SENSITIVITY OF ANIMALS TO IMPENDING EARTHQUAKES

## HOW ANIMALS ACT DURING EARTHQUAKES

**Anonymous; *American Review of Reviews*, 37:104, 1908.**

The news dispatches, it will be remembered, announced that on the eve of the great earthquake at Karatagh, in Central Asia, on October 20, all the dogs of the region set up a howling, horses stampeded, and cattle bellowed with fright. This report is in singular confirmation of some general principles as to the conduct of animals during earthquakes laid down in an article in a recent number of the Dutch review, Vragen van den Dag.

The writer of this article reminds us of the frequent contention that some animals are able to feel in advance certain conditions of the weather or other natural phenomena, and that they are thus, in this respect, better endowed than man.

Whether this power has been lost to man in the process of civilization, or that it was never possessed by him at all, we would not undertake to affirm. Although animals are not to be wholly regarded as weather prophets, still by a close observation of their behavior under particular circumstances of the kind, something may be gained in this line of human knowledge.

In connection with the fearful catastrophies of recent date in Italy, California, and elsewhere, which, like so many others of like nature, will long retain a hold on human memory, attention has again been called to the fact that many animals give intimations of such great disturbances in advance by certain particular and often unusual conduct. It is particularly such animals as have their abode under ground that often indicate, days before the event, that something unusual in nature is about to occur, by coming out of their hiding places under ground into the open.

Aelian mentions that, in the year 373 before Christ, five days before the destruction of Helike, all the mice, weasels, snakes, and many other like creatures, were observed going in great masses along the roads leading from that place. Something similar was noticed also, later, though not to so marked an extent as in the case mentioned by Aelian. This leaving of their subterranean abodes by underground creatures on such occasions might possibly be explained by the emission of various malodorous and noxious gases during these disturbances of the earth.

But not only do animals living under ground furnish indications that something out of the ordinary is about to happen. The larger animals on the surface, such as cows, horses, asses, sheep, and many birds, even, seem to get premonitions of particular natural phenomena and events.

Thus it is related that in 1805, during an earthquake, the cattle at Naples and its neighborhood set up a continuous bellowing some time before the event, at the same time trying to support themselves more firmly by planting the forefeet widely apart; the sheep kept up a continuous bleating, and hens and other fowl expressed their restlessness by making a terrible racket. Even the dogs gave many indications of uneasiness at the time. The actions of animals observed during the great earthquake of 1783 seem to have been most remarkable. Thus the howling of the dogs at Messina became so unendurable that men were sent out with cudgels to kill them. Their noise was most marked during the progress of the earthquake, while it was difficult to pacify the animals in the vicinity for some time even after the cessation of the shocks. Dogs and horses ran about meanwhile with hanging heads, or stood with outstretched legs, as if aware of the need of planting themselves firmly. Horses that were ridden at the time stopped and stood still without orders, trembling so at the same time that no rider could remain in the saddle. Scophus tells the story of a cat during an earthquake at Locris which set up a most dismal caterwauling at the approach of each new shock, meanwhile constantly jumping from one point to another. The roosters kept up a continual crowing, both before and during the earthquake. In the fields Scophus observed hares so under the influence of the terrestrial disturbance that they made no attempt to escape and seemed in no way disturbed by his presence. A flock of sheep could not be kept on the right road, notwithstanding the efforts of shepherd and dogs, but fled in affrightened haste to the mountains. During the same year of 1783, fear had taken such possession of the peasants of Calabria that they were seen to flee from their huts the moment dogs began to howl, asses to bray, or cows to bellow. Birds, also, seem to have premonitions of the coming of such catastrophies. During the earthquake at Quintero, in Chile, in November, 1822, the gulls uttered all sorts of unusual cries during the whole of the preceding night, and were in constant restless motion during the quake. On February 20, 1835, the day before the earthquake at Concepcion, in Chile, at ten in the morning, great flocks of sea-birds, mostly gulls, were seen to pass over the city landward, a phenomenon not to be explained by any stormy condition of the weather. It was fully an hour and a half after their passage, at 11.40 of the forenoon, before the earthquake came, one so disastrous that nearly the entire city was reduced to ruins. Even the fish in the sea seem to be disturbed at the approach of an earthquake. Thus during the one of 1783, quantities of fish were caught at Messina, of a kind that usually keeps hidden in its secret abodes at the ocean's bottom. And Alexander von Humboldt, the famous traveler and naturalist, tells of having observed the crocodiles of the Orinoco leaving the water and fleeing to the forest during an earthquake. (American Review of Reviews, 37:104, 1908)

## EFFECTS OF EARTHQUAKES ON ANIMALS
**Anonymous; *Nature,* 38:500, 1888.**

In the last issue of the Transactions of the Seismological Society of Japan, Prof. Milne discusses the effects of earthquakes on animals. The records of most great earthquakes refer to the consternation of dogs, horses, cattle, and other domestic animals. Fish also are frequently affected. In the London earthquake of 1749, roach and other fish in a canal showed evident signs of confusion and fright; and sometimes after an earthquake fish rise to the surface dead and dying. During the Tokio earthquake of 1880, cats inside a house ran about trying to escape, foxes barked, and horses tried to kick down the boards confining them to their stables. There can, therefore, be no doubt that animals know something unusual and terrifying is taking place. More interesting than these are the observations showing that animals are agitated just before an earthquake. Ponies have been known to prance about their stalls, pheasants to scream, and frogs to cease croaking suddenly a little time before a shock, as if aware of its coming. The Japanese say that moles show their agitation by burrowing. Geese, pigs, and dogs appear more sensitive in this respect than other animals. After the great Calabrian earthquake it is said that the neighing of a horse, the braying of an ass, or the cackle of a goose was sufficient to cause the inhabitants to fly from their houses in expectation of a shock. Many birds are said to show their uneasiness before an earthquake by hiding their heads under their wings and behaving in an unusual manner. At the time of the Calabrian shock little fish like sand-eels (Cirricelli), which are usually buried in the sand, came to the top and were caught in multitudes. In South America certain quadrupeds, such as dogs, cats, and jerboas, are believed by the people to give warning of coming danger by their restlessness; sometime immense flocks of sea-birds fly inland before an earthquake, as if alarmed by the commencement of some sub-oceanic disturbance. Before the shock of 1835 in Chili all the dogs are said to have escaped from the city of Talcahuano. The explanation offered by Prof. Milne of this apparent prescience is that some animals are sensitive to the small tremors which precede nearly all earthquakes. He has himself felt them some seconds before the actual earthquake came. The alarm of intelligent animals would then be the result of their own experience, which has taught them that small tremors are premonitory of movements more alarming. Signs of alarm days before an earthquake are probably accidental; but sometimes in volcanic districts gases have emanated from the ground prior to earthquakes, and have poisoned animals. In one case large numbers of fish were killed in this way in the Tiber, and at Follonica, on the morning of April 6, 1874, "the streets and roads were covered with dead rats and mice. In fact, it seemed as if it had rained rats. The only explanation of the phenomenon was that these animals had been destroyed by emanations of carbon dioxide." (Nature, 38:500, 1888)

## BIRDS AND EARTHQUAKES
**Anonymous; *Nature,* 132:964, 1933.**

The Long Beach earthquake of March 10, 1933, began at 5.55 p.m. with the most severe of a succession of shocks which continued for twenty hours. At this time, about sunset, a flock of a hundred Brewer blackbirds (Euphagus cyanocephalus) had retired to roost in some medium-sized trees. M. P. Skinner records that although no preliminary shocks were felt by human beings, these birds became uneasy before the severe shock (Condor, 35,200; 1033). During the shock the birds began to leave the roost, and rose slowly in ascending spirals above the trees to a height of about 140 ft. They then descended slowly and settled noisily in the roost; thereafter throughout the minor shocks they showed no sign of disturbance. At their usual time near dawn, meadow-larks and mocking-birds began to sing and kept up their morning songs in spite of the tremors that were occurring almost every minute. (Nature, 132:964, 1933)

## EARTHQUAKES AND PHEASANTS
**Anonymous; *Nature,* 112:20-21, 1923.**

Pheasants, it has long been known, are peculiarly sensitive to the effects of slight tremors, and in many earthquake countries they are supposed to give notice of a coming shock. Prof. Sekiya's attempt, nearly forty years ago, to study the behaviour of pheasants before and during earthquakes was unsuccessful, probably because the birds were not under natural conditions. Recently, Prof. Omori (Bull. Imp. Earthq. Inves. Com. of Japan, vol. II, 1923, pp. 1-5) has been able to observe those living in a neighbouring park, usually within a distance of a hundred yards, and during the quiet hours of the night. In three years he recorded 22 cases of the disturbance of pheasants. On seven occasions the birds crowed before the tremor was felt, on five at the same time, and on five afterwards. In four cases they crowed while no tremor was felt, though slight movements were recorded by the seismographs; and in only one case did an earthquake occur without the accompaniment of pheasant-crowing. Thus, in half the cases observed, the movement was noticed by pheasants more readily than by a trained observer under good conditions. (Nature, 112:20-21, 1923)

## SENSITIVITY OF FISH TO EARTHQUAKES
**Anonymous: *Nature,* 132:817, 1933.**

Two Japanese seismologists, Dr. Shinkishi Hatai and Dr. Noboru Abe, observed that catfish (Siluridae) in natural conditions showed signs of restlessness about six hours before earthquake disturbances were registered on their recording apparatus. Since catfish are, ordinarily, placid unresponsive creatures, experiments were made to test the seeming responsiveness (Science Service, Washington, D.C.). Catfish placed in an aquarium were tested three times a day by tapping on the supporting table. When no earthquake was impending, the fish moved lazily or not at all; but about six hours before a shock the fish jumped when the table was tapped, and sometimes swam about agitatedly for a time before settling down upon the bottom again. Several months' testing showed that in a period when 178 earthquakes of all degrees of severity had been recorded, the fish had correctly predicted 80 per cent of the shocks. They showed no discrimination in their movements between slight local shocks and more serious distant shocks. The experimenters think that the catfish are made sensitive through electrical changes in the earth, since it was only when the aquarium was electrically earthed, through the drainpipe, that they responded to a coming earthquake. (Nature, 132:817, 1933)

# COLLECTION OF HAIRS AFTER EARTHQUAKES

## NOTE ON EARTHQUAKES IN CHINA
**Macgowan, D. J.; *Nature,* 34:18, 1886.**

The tremors that are experienced in Chehkiang, Kiangsu, and coterminous regions to the west, are sometimes followed by the appearance on the ground of substances that in Chinese books are styled "white hairs." When I first called attention to records of that kind that are found in local gazetteers, I suggested that they might be crystals precipitated by gaseous emissions, such as were once reported as occurring after an earthquake in the south-west of the United States; from later descriptions of these "horsetail-like" substances I incline to the opinion that they are organic, perhaps mycillium.

In the summer of 1878 the vernacular press gave an account of the occurrence of the phenomena at Wusoh, a city on the grand canal, thirty miles north of Suchau. "At noon, June 12th of that year, shocks of an earthquake were experienced, which lasted several minutes (Sin. 'for the space of time taken in swallowing half a bowl of rice'); the motion was so great that sitting or standing was difficult, but no harm was done. Two days later at night there was a severe shock, after which, within and without the walls of the city, white hairs resembling a silvery beard, about three inches in length, were found, which boys pulled out of the ground, gathering handfulls in a short space of time." My list of Chinese earthquakes for the past two thousand years having been destroyed by fire I am unable to indicate the regions in which earthquakes were followed by the emission of "hairs," but my impression is that all, or nearly all, are alluvial valleys. (Nature, 34:18, 1886)

## COLLECTION OF HAIRS AFTER EARTHQUAKES IN CHINA
**Dyer, W. T.; *Nature,* 34:56-57, 1886.**

In 1852, during one of the late Mr. Fortune's visits to China, he experienced the shock of an earthquake at Shanghai. He gives the following curious account in "A Residence among the Chinese" (pp. 4, 5), of the subsequent search for the hairs:---

"Groups of Chinese were seen in the gardens, roadsides, and fields engaged in gathering hairs which are said to make their appearance on the surface of the ground after an earthquake takes place. This proceeding attracted a great deal of attention from some of the foreign residents in Shanghai, and the Chinese were closely examined upon the subject. Most of them fully believed that these hairs made their appearance only after an earthquake had occurred, but could give no satisfactory explanation of the phenomenon, while some, more wise than their neighbours, did not hesitate to affirm that they belonged to some huge subterraneous animal whose slightest shake was sufficient to move the world.

"I must confess, at the risk of being laughed at, that I was one of those who took an interest in this curious subject, and that I joined several groups who were searching for these hairs. In the course of my travels I have ever found it unwise to laugh at what I conceived to be the prejudices of a people simply because I could not understand them. In this instance, however, I must confess the results were not worth the trouble I took. The hairs, such as I picked up, and such as were shown me by the Chinese, had certainly been produced above the earth and not below it. In some instances they might readily be traced to horses, dogs, and cats, while in others they were evidently of vegetable origin. The northeastern part of China produces a very valuable tree known by the name of the hemp-palm [Chamaerops Fortunei, see Kew Report, 1880, p. 31], from the quantity of fibrous bracts it produces just under its blossoms. Many of these fibres were shown to me by the Chinese as a portion of the hairs in question; and when I pointed out the source from which such had come, and which it was impossible to dispute, my friends laughed, and, with true Chinese politeness, acknowledged I was right, and yet I have no doubt they still held their former opinions concerning the origin of such hairs. The whole matter simply resolves itself into this: if the hairs pointed out to me were the true ones, then such things may be gathered not only after earthquakes, but at any other time. But if, after all, these were not the real things, and if some vegetable (I shall not say animal) production was formed, owing to the peculiar condition of the atmosphere and from other causes, I can only say that such production did not come under my observation." (Nature, 34:56-57, 1886)

## A DROLL PHENOMENA
**Anonymous; *Scientific American*, 4:139, 1848.**

The Singapore Free Press, East Indies, of September 14, gives an account of a strange phenomena that occured at Chantibun in the eastern part of the kingdom of Siam on the 13th of May last. First a violent shock of earthquake was felt accompanied with tremendous noise and subterranean roaring. The doors and partitions of the houses were cracked and strained. But more extraordinary than all, it says: "During the shock, there spontaneously came out of the ground a species of human hairs in almost every place---in the bazaars, in the roads, in the fields and the most solid places. These hairs, which are pretty long, stand upright and adhere strongly to the ground. When they are burned they twist like human hairs, and have a burned smell which makes it to be believed that they are really hairs; they all appeared in the twinkling of an eye during the earthquake.---The river of Chantibun was all rippling, and bubbles rose to the surface, so that the water was quite white. It is thought that these hairs may have been produced by electricity. The mountains of Chantibun run nearly from north to south and are united to the system which separates Camboja from Siam."

The hairs were probably some interior bituminous substance melted and blown through the pores of the earth into fine strings and congealed, resembling-hairs, as every one knows is the case with rosin. &c. (Scientific American, 4:139, 1848)

# EARTHQUAKES AND POSSIBLE TRIGGERING FORCES

In the absence of external triggering forces, earthquakes would presumably be distributed uniformly in time. However, lengthy records of earthquakes are emphatic that earthquakes occur more often during some time intervals than others. For over 200 years scientists have been correlating earthquakes with every imaginable natural phenomenon in hopes of discovering triggering forces. The search has been less than successful. In fact, in the case of earthquakes and the earth's wobble there seems to be doubt about which is cause and which is effect.

Earthquakes definitely seem associated with solar activity (sunspots and flares), the earth's wobble (measured by latitude variations) and geomagnetism. It is also known that the active sun projects solar plasma toward the earth and that the interaction of the plasma with the earth may affect the earth's rotation. Changes in the earth's rate of rotation, in turn, may give rise to earthquake triggering forces in the mantle. Thus, a cause-and-effect relationship does exist here. But is the earth's 14-month Chandler wobble stimulated by solar activity or by earthquake activity?

It is also conceivable that tidal stresses created by the moon initiate quakes. (Seismometers left on the moon indicate that the earth may also trigger moonquakes, which is only fair.) But how could Uranus, at its great distance, or the even more remote pulsars stimulate earthquakes? Nevertheless, such correlations seem to exist, although some scientists have rejected them for statistical reasons. It may be, though, that the enthusiasm of the detractors is inspired by the lack of known physical mechanisms that might reach from Uranus and pulsar to earth.

## CORRELATION OF EARTHQUAKES WITH THE EARTH'S WOBBLE, SOLAR ACTIVITY, AND GEOMAGNETISM

### CLUSTERING AND PERIODICITY OF EARTHQUAKES
**Davison, Charles; *Nature,* 120:587-588, 1927.**

Every few years we pass through a season of marked earthquake-frequency, or, as Mallet called it, a 'period of paroxysmal energy.' In the mere fact of clustering, there is nothing remarkable; it is its intensity, rather than its existence, that is worthy of notice. In Italy, for example, as Mercalli pointed out, 209 destructive earthquakes occurred between the years 1601 and 1881, and 182 of them in 103 years. In other words, the average frequency of great earthquakes during one of the cluster-years was eleven times that during one of the remaining years.

The first seismologist who examined the clustering of earthquakes, the first, indeed, who possessed the necessary materials, was Robert Mallet. In 1858 he drew curves representing the frequency of earthquakes in each year from 1500 to

1850. The intervals between successive maxima in these curves varied widely. Though he could trace no law in their variation---it should be remembered that he counted weak shocks as well as strong---their average duration was found to lie between five and ten years, the shorter intervals being those of fewer, and usually weaker, earthquakes.

Fifty years later, John Milne went close to the heart of the problem by suggesting that variations of seismic activity in distant regions may be synchronous. Considering the earthquakes from 1899 to 1907, he noticed that, during the last six years, the annual numbers of 'large earthquakes' on the eastern and western sides of the Pacific rose and fell together. In the following year he found that the destructive earthquakes of Italy and Japan during the last three centuries occurred in periods during the last three centuries occurred in periods of activity, the intervals between them ranging from five to twenty years, and that of the eighteen years of maximum frequency in Italy, fourteen agreed with years of similar activity in Japan. "These coincidences suggest," he says, "that a relief of seismic strain in one part of the world either brings about a relief in some other part or that relief is governed by some general internal or external agency."

At this time, Milne's catalogue of more than 4000 destructive earthquakes was still unpublished, though it must have been approaching completion. It is on the fuller portion of this great work---that relating to the northern hemisphere from 1750 (and for some purposes from 1701) to 1899---that the present inquiry is based. My object at first was merely to ascertain if the cluster-years for different intensities were the same or in any way related, and this work led to the recognition of several periods in the recurrence of cluster-years. I hope to give fuller details of these periods in a later paper.

To each earthquake Milne assigned an intensity according to a scale of three degrees. Shocks of intensity I. were strong enough to crack walls or chimneys or to shatter old buildings. By those of intensity II., buildings were unroofed or shattered and some were thrown down. Earthquakes of intensity III. destroyed towns and isolated districts. In order to smooth away minor inequalities, I have taken three-yearly means of the annual numbers of earthquakes of each intensity. The corresponding curves show definite years of clustering, which are given in the following table:

| | Int. III. | Int. II. | Int. I. | Sunspot Minimum. |
|---|---|---|---|---|
| A | 1754-56 | 1755-56 | 1755-56 | 1755 |
| B | 1765-66 | 1767 | 1766-67 | 1766 |
| | | | 1771-73 | 1775 |
| C | 1785-90 | 1784-86 | 1785-86 | 1785 |
| | | 1789-91 | 1791-92 | |
| | | | 1799-1800 | 1798 |
| D | 1810-11 | 1811 | 1811-13 | 1810 |
| E | 1822-23 | 1818-20 | 1821-22 | 1823 |
| F | 1828-29 | 1827 | 1827-29 | -- |
| | 1834-36 | 1832 | -- | 1833 |
| G | 1840-41 | 1840-42 | 1840-42 | 1843 |
| H | 1845-47 | 1846-48 | 1846-47 | -- |
| J | 1852-57 | 1852-56 | 1852-57 | 1856 |
| | | 1859-60 | 1861-63 | -- |
| K | 1868-72 | 1869-72 | 1867 | 1867 |
| | | | 1873-74 | 1878 |
| | | | 1880-81 | |
| L | 1884-86 | 1884 | 1885-86 | 1889 |
| M | 1892-93 | 1893-94 | 1895 | |

It will be seen that, in all three classes, there are twelve clusters at approximately the same times. They are denoted by the letters A-M. There are also six other clusters, two common to two classes, and four of earthquakes of intensity I. only. Of the latter, two may not be real exceptions, for in the other classes there are traces, too slight to be otherwise noticed, of clusters about the years 1773 and 1798. In any case, the existence of at least four of these additional clusters seems to imply that they are not consequences of the great shocks of intensity III., but rather that the clusters of all classes are the effects of some common cause or causes.

More important, however, than the coincidences of the clusters are the intervals between the years of maximum frequency. Some of the intervals are so suggestive of an eleven-year period that it was only natural that the cluster-years should be compared with the turning epochs of sunspot frequency. The last column of the table shows that eight clusters in all three classes (A-E, G-K) agree closely with the years of low sunspot frequency, and that this correspondence also holds for three other clusters (1771-73, 1799-1800, and 1832 or 1834-36) which appear in only one or two classes. The only divergences are for the last two years of sunspot minima.

In order to test more closely the existence of an eleven-year periodicity, I counted the numbers of earthquakes of each intensity in the years 1755, 1766, . . . ., 1887; 1756, 1767, . . . ., 1888; and so on, and took five-yearly means of these eleven sums, with the following results, the first subsequent year of least sunspot frequency being 1765:

| Intensity | First Maximum Epoch | Amplitude |
|---|---|---|
| III. | 1764 | 0.16 |
| II. | 1762-63 | 0.12 |
| I. | 1763-64 | 0.09 |

Thus, in destructive earthquakes of every intensity, the maximum-epoch of the eleven-year period occurs shortly before the epoch of least sunspot frequency; and the amplitude increases with the intensity of the shock. It may be added that the epochs are almost exactly the same for the same three classes for the separate intervals 1701-99 and 1801-99, and for all three intensities together for the interval 1501-1698, and for each season of the interval 1701-1898.

It is interesting to notice that the same periodicity holds in widely separated regions. Taking the three intensities together, the maximum occurs in 1764 in Europe, in 1763-64 in Asia, in 1764 in Italy, in 1764-65 in China, and in 1763-64 in the island groups of the western Pacific. Even in the slight earthquakes of Great Britain, the same period is present with its maximum in 1763-64. Thus, by a somewhat different line of evidence, Milne's remarkable generalisation seems to be confirmed.

In addition to the eleven-year period of destructive earthquakes, there are other clearly marked periods of 19, 22, and 33 years, the maximum epochs of which (1754, 1760, and 1757) agree closely in widely distant regions of the northern hemisphere. Some of these maxima seem to be responsible for clusters in the above table, especially for those of the years 1791, 1847-48, 1880-81, and about 1887. To the occurrence of their minima about the years 1798, 1833, and 1878, the absence or slightness of clusters is probably due.

During the present century, the maxima of the 11, 33 and 19 year periods occurred in 1918, 1922 and 1925; those of the 11, 22 and 33 year periods are due in 1951, 1958 and 1955; and those of the 11, 22, 33 and 19 year periods in 1984, 1980, 1988 and 1982. The times of unusually frequent earthquakes are thus, 1918-25, 1951-58 and, especially, 1980-88. The first of these intervals is notable for its very numerous earthquakes, some of which, such as the Chinese earthquake of

1920 and the Japanese earthquake of 1923, were of great violence. (Nature, 120: 587-588, 1927)

## SOLAR ACTIVITY AS A TRIGGERING MECHANISM FOR EARTHQUAKES

**Simpson, John F.; *Earth and Planetary Science Letters,* 3:417, 1967.**

Abstract. Solar activity as indicated by sunspots, radio noise and geomagnetic indices, plays a significant but by no means exclusive role in the triggering of earthquakes. Maximum quake frequency occurs at times of moderately high and fluctuating solar activity. Terrestrial solar flare effects which are the actual coupling mechanisms which trigger quakes appear to be either abrupt accelerations in the earth's angular velocity or surges of telluric currents in the earth's crust. The graphs presented in this paper permit probabilistic forecasting of earthquakes, and when used in conjunction with local indicators may provide a significant tool for specific earthquake prediction. (Earth and Planetary Science Letters, 3:417, 1967)

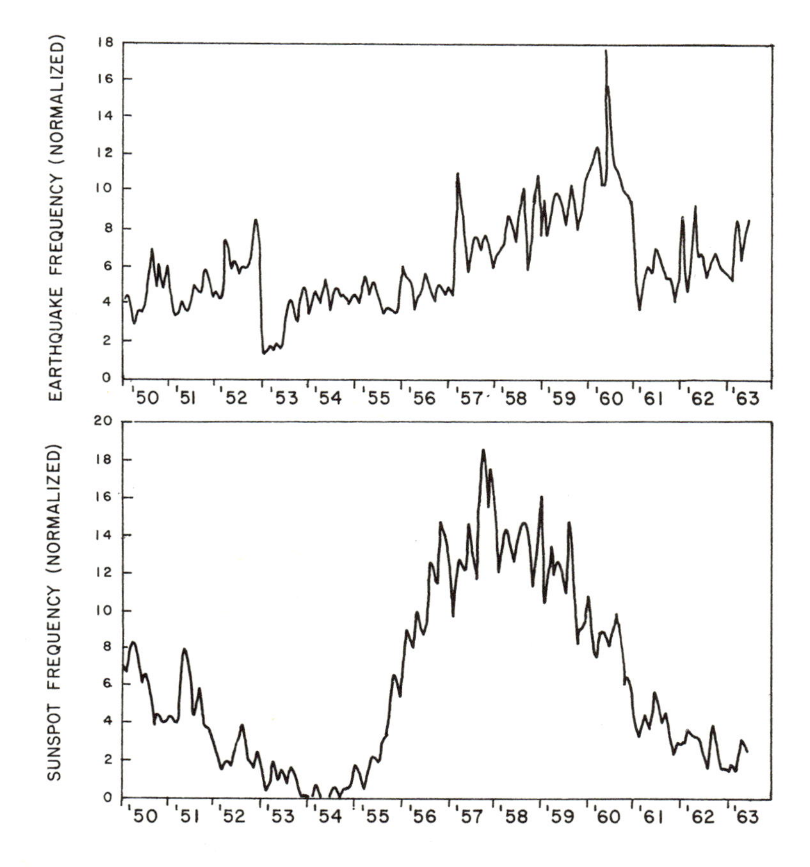


Frequencies of earthquakes (top) and sunspots (bottom) from 1950 to 1963.

## SEISMIC ACTIVITY, POLAR TIDES AND THE CHANDLER WOBBLE
**Pines, David, and Shaham, Jacob; *Nature,* 245:77, 1973.**

Abstract. Earthquakes may be triggered by changes in the elastic energy which are neither too fast (~days) nor too slow ($\sim 10^6$ yr). Such a point of view provides a qualitative understanding of suggested correlations between polar motion and the annual seismic energy release. The Chandler wobble and earthquakes may have a common excitation source---the release of elastic energy stored in the Earth---and we show that a resonant coupling between seismic activity and polar motion has the right sign and strength to explain the pumping of the Chandler wobble as a consequence of in-phase release of elastic energy. (Nature, 245:77, 1973)

## CHANDLER WOBBLE, EARTHQUAKES, ROTATION, AND GEOMAGNETIC CHANGES
**Press, Frank, and Briggs, Peter; *Nature,* 256:270, 1975.**

Abstract. A pattern recognition algorithm can be applied to the seismicity of major earthquake belts, to the amplitudes of Chandler Wobble, to changes in the rotational velocity of the Earth, and to the drift of the eccentric geomagnetic dipole for the years 1901-64. The patterns which emerged suggest that all of these diverse phenomena are related. (Nature, 256:270, 1975)

# ARE NOCTURNAL EARTHQUAKES MORE FREQUENT?

## NOCTURNAL EARTHQUAKES
**Anonymous; *Nature,* 239:131, 1972.**

The suggestion that seismic activity fluctuates with the seasons or the time of day has been made on numerous occasions in the past; but as Bullen (Introduction to the Theory of Seismology, Cambridge University Press; 1963) pointed out some years ago, claims for such periodicities have never been substantiated. Not, that is, until last year when Shimshoni (Geophys. J., 24, 97; 1971) produced evidence which seemed to show that seismic activity is significantly higher during the night than during the day. At first sight such a correlation appears inherently unlikely, except that it is never possible to ignore completely the idea of trigger forces. In other words, although there can be no suggestion that minor forces, such as, for example, temperature effects, tidal effects and abnormal changes in barometric pressure, produce the elastic strain which is ultimately released in earthquakes, it is just conceivable that minor forces act as the "last straw" (Bullen) leading to the actual release of strain.

But whatever the cause may be, a correlation is a correlation; and that is just what Shimshoni seemed to have from a statistical analysis of the 15,325 seismic events reported by the US National Oceanic and Atmospheric Administration for 1968 to 1970. If seismic events occur randomly with local time throughout the twenty-four hour period, the number of events within each hourly period over the three years should be 638.5 (that is, 15,325/24). In practice, however, $X^2$ for the randomness hypothesis turned out to be 80.1137, the probability of obtaining a figure as high as this for a truly random process being less than $10^{-7}$. Shimshoni thus concluded that seismic events do not occur randomly throughout the twenty-four hours, and went on to show that within the daily cycle there is a peak of seismic activity around midnight and a minimum around noon.

Three authors or groups of authors have now descended upon Shimshoni to point out that, although there may indeed be a correlation of the sort he demonstrates, it represents a data deficiency rather than a real variation in seismicity. Davies (Geophys. J., 28, 305; 1972), for example, notes that the number of earthquakes recorded over any given period of time will reflect not only the number of earthquakes actually occurring over that period, but also the detection capability of the recording network, and that it is the latter which varies between night and day.

To see how this effect works, he considers an earthquake occurring on the equator at the equinox. The location of such an event will be determined from observations made within epicentral distances of 100°; and so if the shock occurs at midday it will be observed (P waves) at most of the relevant stations in daylight, and if it occurs at midnight it will mostly be observed in darkness. Seismic background noise---most of it probably local cultural noise---correlates strongly with light and dark, however, the detection threshold at many stations being significantly higher during the day. According to Davies, it is possible to observe at many stations a dawn noise level rise of 10 to 20 db.

The result of all this is that more of the earthquakes occurring at night will actually be detected---a circumstance which is bound to bias the data in any published list of earthquakes. On the other hand, this situation will not obtain for events occurring far from the equator. At high latitudes, P waves from midnight earthquakes will be observed at many stations during daylight. Under these circumstances no day/night correlation would be expected, nor is it found. Events occurring at middle and lower latitude will, however, be sufficient to give an overall bias.

Flinn et al. (Geophys. J., 28, 307; 1972) raise the same basic objection as Davies, but they also present data to make explicit the point that it is the detection rate of the low magnitude earthquakes that will increase as the detection threshold lowers at night. Moreover, this will have a particularly significant effect in giving the Shimshoni correlation simply becuase most events are of low magnitude anyway (only 4 per cent have body wave magnitudes greater than 5.5). In a very satisfying series of graphs of earthquake frequency against the time of occurrence, Flinn and his colleagues are able to show that the diurnal effect is definitely not present for events in the body wave magnitude range 5.5 to 6.5, is possibly present in the range 4.5 to 5.5, and is certainly present below 4.5.

Knopoff and Gardner (Geophys. J., 28, 311; 1972) again raise the question of bias introduced by the higher daytime noise, but they also add two other possible sources of bias. First, they claim that there is bound to be a bias against reporting small shocks because of a "finite and variable density of mesh of seismic stations." In addition, they propose a bias caused by aftershocks. The argument here seems to be that if one of the relatively few earthquakes of magnitude 8 or higher happens to occur at night the large number of associated aftershocks will give a false impression of nocturnal seismicity. Knopoff and Gardner then go on to describe an elaborate technique for the identification of aftershocks and their elimination from the record. When the three sources of bias are removed from the original record (actually for the period March 24, 1963, to December 31, 1968, rather than 1968 to 1970), no day/night correlation is evident. The conclusion then is that Shimshoni's correlation must be produced by one or more of the three biases.

As Shimshoni (Geophys. J., 28, 315; 1972) points out in his reply to the critics, the elimination of aftershocks by Knopoff and Gardner is rather "drastic", not least because the process will also remove many events not related to the principal shock. On the other hand, he accepts the basic point about variation in detection capability. But this is not necessarily the last word on the matter. The question still remains as to whether the detection threshold effect can explain all of the day/night correlation. (Nature, 239:131, 1972)

## MORE EARTHQUAKES OCCUR AT NIGHT
**Anonymous; *Nature*, 99:392, 1917.**

The statistics of Italian earthquakes during the twenty years 1891-1910 are examined by Dr. A. Cavasino in a recent paper (Boll. Soc. Sismol. Ital. vol. xx., for 1916, pp. 3-31). The total number of earthquakes recorded is 5922, giving an average annual number of 296. The Calabrian earthquake of 1905 was followed by a train of 396 after-shocks, the Messina earthquake of 1908 by 1227. It is generally supposed that the greater frequency of earthquakes at night is apparent and due to the quiet of the midnight hours. Dr. Cavasino shows that this is not the case for Italian earthquakes by considering only shocks of such intensity that they could not escape notice at any hour of the day. Of these, 865 occurred during the twelve night hours, and 638 during the day hours. (Nature, 99:392, 1917)

## THE DIURNAL PERIODICITY OF EARTHQUAKES
**Davison, Charles; *Journal of Geology*, 42:449, 1934.**

Abstract. Though non-instrumental records of earthquakes give an apparent nocturnal maximum, it is shown that, for several regions in which earthquakes are weak or moderately strong, there is a real diurnal period, with its maximum about midnight. The instrumental records obtained in Japan and Italy and at various seismological observatories are examined, and it is shown that the maximum epoch of the diurnal period usually falls about noon or midnight, and that the noon maximum of the diurnal period is associated, as a rule, with a summer maximum of the annual period, and the midnight maximum of the former with a winter maximum of the latter. It is suggested that the noon and summer maxima occur in earthquakes caused by an elevation of the crust, and the midnight and winter maxima in those caused by a depression of the crust. It is noticed that the midnight and winter maxima prevail in regions in which the earthquakes are of slight or moderate intensity, and the noon and summer maxima in those visited by the most destructive shocks. In the after-shocks of great earthquakes, the maxima epoch are suddenly reversed, usually from near noon to near midnight, and the duration of the reversal varies from about a week to a year or more. (Journal of Geology, 42:449, 1934)

# EARTHQUAKES AND THE MOON'S PHASES

## LUNAR PERIODICITY OF EARTHQUAKES
**Anonymous; *Nature,* 142:81, 1938.**

In a recent paper (Union Geod. Geoph. Intern., Publ., Ser. A, fasc. 15, 244-257; 1937), H. T. Stetson has studied the frequency of earthquakes in connexion with the hour angle of the moon. During the years 1918-29, 2,569 earthquakes were recorded at stations more than 80$^{o}$ from the origin. Arranging these according to the hour-angle of the moon referred to the meridian of the epicentre of each earthquake, he finds two maxima of frequency, at 7 and 18 hours. For the smaller district consisting of the Philippines and Japan, the maxima occurred at 6 and 21 hours. Again, for 113 deep-focus earthquakes, with depths of 100 km. or more below the surface, the maxima were at about 5 and 16 hours, and the curve representing the means of the numbers of earthquakes for lunar-hourly intervals corresponds closely with the curve of the east and west component of the lunar tidal force. (Nature, 142:81, 1938)

## THE LUNAR PERIODICITY OF EARTHQUAKES
**Anonymous; *Nature,* 121:920, 1928.**

During the year 1927 the seismographs at the Hawaiian Volcano Observatory recorded 1149 local earthquakes (Volcano Letter, No. 170, Mar. 29). Among them, the regular recurrence of times of increased seismic activity was too noticeable to be set aside as accidental. These epochs of greater activity occur at intervals of about two weeks and near the times of the first and last quarter phases of the moon. According to Perrey's first law, similar epochs coincide nearly with the times of new and full moon, but it should be remembered that Perrey's law relates to ordinary earthquakes, while most of the Hawaiian tremors are of volcanic origin. (Nature, 121: 920, 1928)

## TIDAL FREQUENCY OF EARTHQUAKES
**Anonymous; *Nature,* 127:908, 1931.**

The relation between tidal phases and the frequency of earthquakes is discussed in a recent paper by S. Yamaguti (Bull. Earthq. Res. Inst., vol. 8, pp. 393-408; 1930). In earlier investigations on the subject, the origins of the earthquakes considered were unknown. Mr. Yamaguti confines his study to the after-shocks of the Kwanto earthquake of 1923, the Tango earthquake of 1927, etc., in which the epicentres were distributed within well-defined areas. In all of these, he finds that the frequency is greatest at, or a little before, the time of low-water at a neighbouring station;

while a secondary maximum appears before high-water on the Kwanto curve and a little after high-water on the other curves. There is, however, an interesting difference between the Tango after-shocks that originated beneath the sea-bed and on land on the northern side of the Yamada fault. The former have two maxima, 1-1/2 hours before and after high-water; the latter a principal maximum about low-water and a secondary maximum about high-water. (Nature, 127:908, 1931)

# OTHER PECULIAR CORRELATIONS OF EARTHQUAKES AND NATURAL PHENOMENA

## EARTHQUAKES, FISHERIES, AND FLOWER FALL
**Anonymous; *Nature,* 130:28, 1932.**

Prof. T. Terada has shown that there exists a curious relation between the numbers of earthquakes in the Idu peninsula and the numbers of fishes caught near the northern end of Sagami Bay (Proc. Imp. Acad. Tokyo, vol. 8, pp. 83-86; 1932). During the spring of 1930, swarms of earthquakes occurred in the neighbourhood of Ito on the east coast of the peninsula (Nature, vol. 126, pp. 326, 971). It was found that the epochs of abundant catches of horse mackerel (Caranx) at the Sigedera fishing ground coincided very nearly with those of the earthquakes. This result led Prof. Terada to compare the numbers of fishes caught in the six years 1924-29 with the numbers of felt and unfelt earthquakes in and near the Idu peninsula. For the year 1928, the parallelism of the two curves was very close, though in other years it was less conspicuous. During 1928, the curve representing the numbers of immature tunny (Thynnus) caught shows a remarkable similarity with the horse mackerel and earthquake curves. In another paper (Bull. Earthq. Ves. Inst., vol. 10, pp. 29-35; 1932) Prof. Terada points out that, though the daily numbers fluctuate, the time-distribution curve of the Ito earthquakes resembles on the whole the probability curve, and he shows that the daily number of falls of camelia flowers follows a similar statistical distribution. (Nature, 130:28, 1932)

## EARTHQUAKES AND RAINFALL
**Anonymous; *Nature,* 136:958-959, 1935.**

Mr. J. F. Brennan, Government meteorologist in Jamaica, has recently made an interesting comparison between the mean monthly frequency of earthquakes and the mean monthly rainfall in Jamaica (Earth. Notes, East. Sect. Seis. Soc. America, 7, 25-26; 1935). During the years 1908-34, 480 earthquakes were recorded. The curve representing the mean monthly frequency shows two maxima, in February (60 earthquakes) and July (46), both months of minimum rainfall, and two minima in May (25) and October (31), months of maximum rainfall. The author suggests that the greater frequency of earthquakes during the dry seasons may be due to the withdrawal of water from underground water-courses, which may thus facilitate the fall of large masses of rock. (Nature, 136:958-959, 1935)

## GREAT EARTHQUAKES AND THE ASTRONOMICAL POSITIONS OF URANUS
**Tomaschek, R.; *Nature*, 184:177-178, 1959.**

Thanks to the excellent collection of uniform data of earthquakes given by Gutenberg and Richter, it is now easily possible to study statistically the influence of different factors on earthquakes. In the course of a study of tidal effects on earthquakes, the astronomical positions of the planets have also been taken into account and a remarkable correlation between the positions of Uranus and the moment of great earthquakes has been established for a certain period. Gutenberg and Richter's data of all earthquakes equal or greater than magnitude 7-3/4 have been used. The investigations will be published in detail later, but here attention is directed to the results concerning the position of Uranus.

A total of 134 earthquakes has been investigated. In this a fairly significant amount of cases has been found, where Uranus was very near its upper or lower transit of the meridian of the epicentre in the time of great earthquakes. Closer investigation showed that this occurred especially during the years 1904, where Gutenberg-Richter's data start, and also 1905 and 1906.

. . . . . . . . . .

The correlation cannot be explained by a tidal effect, since the statistical investigation for all the great earthquakes (M > 7-3/4) during 1904-50 in regard to the absolute and relative positions of the Sun and the Moon give no indication of a significant deviation from chance distribution. The tidal forces of the planets are extremely small compared with those of the Sun and the Moon. That the accumulated stresses within the Earth's crust are released at times which, at least for a period of several years, are strongly correlated with certain positions of Uranus may, therefore, not be a relationship of cause and effect in the usual mechanical sense. (Nature, 184: 177-178, 1959)

## APPARENT RELATION BETWEEN CHINESE EARTHQUAKES AND CALIFORNIA TREE GROWTHS
**DeLury, Ralph E.; *Popular Astronomy*, 27:560-561, 1919.**

From an examination of the records of Chinese earthquakes, H. H. Turner has recently deduced a period of 240 years (Monthly Notices, May, 1919). It shows a similar periodicity to the Nile Floods, with a change of phase. Turner makes the suggestion that a variation of the Earth's radius could account for the earthquakes and possibly at the same time for the fluctuations of the moon's longitude by changing the speed of rotation of the earth. The measurements of the tree-growths in California from The Climatic Factor, by Ellsworth Huntingdon, reveal a striking similarity to the curve of the Chinese earthquakes, thus: (E, for earthquakes, and F for tree-growth in mm. per decade for two decades, from pp 323-4 The Climatic Factor):

| Date | 0 | | 240 | | 480 | | 720 | | 960 | | 1200 | | 1440 | | Total | Mean |
|---|---|---|---|---|---|---|---|---|---|---|---|---|---|---|---|---|
| | E | F | E | F | E | F | E | F | E | F | E | F | E | F | E | F |
| 10 | 2 | 7.1 | 5 | 7.0 | 1 | 6.4 | 3 | 6.1 | 1 | 6.6 | 11 | 6.7 | 5 | 6.9 | 28 | 6.68 |
| 30 | 0 | 7.2 | 7 | 7.0 | 1 | 6.5 | 1 | 6.3 | 2 | 6.9 | 3 | 6.7 | 22 | 6.4 | 36 | 6.71 |
| 50 | 1 | 7.2 | 28 | 6.8 | 5 | 6.4 | 4 | 6.2 | 17 | 7.5 | 3 | 6.6 | 38 | 6.5 | 96 | 6.74 |
| 70 | 1 | 7.2 | 11 | 6.8 | 5 | 6.4 | 30 | 6.4 | 8 | 7.2 | 2 | 6.6 | 47 | 6.8 | 104 | 6.78 |
| 90 | 5 | 6.7 | 15 | 6.8 | 3 | 6.6 | 5 | 6.4 | 12 | 6.8 | 7 | 6.5 | 14 | 7.0 | 51 | 6.67 |
| 110 | 16 | 6.7 | 11 | 6.7 | 3 | 6.7 | 6 | 6.4 | 14 | 6.7 | 35 | 7.0 | 20 | 7.2 | 105 | 6.76 |
| 130 | 15 | 6.8 | 11 | 6.7 | 2 | 6.7 | 6 | 6.4 | 15 | 6.8 | 33 | 7.6 | 27 | 7.2 | 109 | 6.88 |
| 150 | 13 | 6.8 | 10 | 6.6 | 3 | 6.2 | 8 | 6.3 | 1 | 6.7 | 36 | 7.6 | 28 | 7.1 | 99 | 6.76 |
| 170 | 11 | 6.7 | 6 | 6.3 | 9 | 6.1 | 4 | 6.7 | 10 | 6.7 | 26 | 7.3 | 24 | 7.4 | 90 | 6.74 |
| 190 | 2 | 7.0 | 2 | 6.4 | 2 | 6.2 | 0 | 6.9 | 6 | 6.4 | 4 | 7.2 | 40 | 7.3 | 56 | 6.75 |
| 210 | 0 | 7.2 | 1 | 6.3 | 6 | 6.3 | 6 | 6.7 | 12 | 6.4 | 7 | 7.0 | 12 | 7.3 | 44 | 6.74 |
| 230 | 5 | 7.1 | 3 | 6.4 | 4 | 6.2 | 2 | 6.5 | 10 | 6.5 | 14 | 6.9 | 1 | 7.4 | 39 | 6.71 |

It seems reasonable to suggest that the meteorological variations produced simultaneous changes in the tree-growth and in the Chinese earthquakes. (Popular Astronomy, 27:560-561, 1919)

## PERIODICITIES IN SEISMIC RESPONSE CAUSED BY PULSAR CP1133
**Sadeh, Dror, and Leidav, Meir; *Nature*, 240:136-138, 1972.**

A search for periodicities in seismic output which correspond to pulsars' frequencies was first made by Wiggins and Press. They analysed only 20.5 h of seismic data to see if a line split predicted by Dyson was present, and give an upper limit to the amplitude of the ground motion of $10^{-2}$ pm or $10^{-9}$ cm.

We repeated their experiment with a much larger integration time, using a vertical seismometer (Rocard, Ecole Normal Superior, Paris) placed in a seismic station 15 km north of Eilat, Israel (29° 40' N latitude, 35° E longitude). The seismometer is sensitive between 0.1 Hz to 10 Hz, and is connected through an amplifier which gives 0.5 V per 1 mm of vertical movement at 1 Hz to a correlator which digitizes the analogue signal and performs an on-line autocorrelation. This technique suppresses the non-periodic noise and enhances any periodic signal.

The time range of the correlator, 0.1 s/channel, was chosen to detect the periods of known pulsars, 0.2 s-10 s. After 13,107 s of correlation, the 100 channels are printed, the memory erased and the analysis starts again. To obtain the power spectrum, a separate Fourier analysis is performed for each 13,107 s of data. The first ten channels of autocorrelation were dropped because they contain most of the noise. This does not change the period, but gives a better signal to noise ratio.

Periodicity of Data. After the first few days of data reduction, it became apparent that the power spectra showed a persistent peak which corresponds to a period of 0.53-0.60 s. (The uncertainty in the period results from the time-uncertainty for each channel.) Fig. 1 shows an example of an autocorrelation and its power spectrum where this peak shows best. After several weeks, we noted changes in the height of this power spectrum peak during the day, and to see the exact periodicity of these changes a Fourier analysis on the amplitude of the power spectrum which corresponds to the 0.53-0.60 s period was performed. This analysis shows an approximate 24 h period (Fig. 2). To distinguish between the sidereal and solar period, the times and date when these amplitudes showed a maximum were determined. [Figures not reproduced]

..........

Gravitational or Electromagnetic Waves? Can the effects described above be due to gravitational waves from the pulsar CP1133? Weber, who pioneered a search for gravitational waves, calculated the response of the Earth to gravitational radiation from pulsars. He uses an example of the interaction of a gravitational wave with four point masses. The gravitational wave causes the masses, if they are aligned perpendicular to the direction of the wave, to vibrate at the frequency of the wave. If the masses are aligned in the direction of the wave, they do not move with respect to each other. This implies that if gravitational waves are emitted from a specific direction in the sky, all points on Earth which lie on a plane perpendicular to this direction will be excited, or a particular point on Earth will feel maximum vibrations twice a day as it crosses this plane. This will happen twice in a sidereal day or twice in 23 h 56 min.

The results shown in Fig. 3 are consistent with these predictions; if there was a source of gravitational waves at the position of CP1133 their effect on a seismometer in Eilat will be such as to give the maxima on the two lines plotted in Fig. 3.

Two more pieces of evidence support the suggestion that we are detecting this pulsar. Its period, as measured in the radiofrequency region, is 1.1879s. If the radio waves are emitted once for each rotation of a rod-type mass, the period of the gravitational wave will be half of the radio period 0.594 s. The period observed, 0.53-0.60 s, is consistent with this period. Secondly, CP1133 is the second nearest pulsar to Earth (80 pc) and thus would have the greatest effects. The nearest pulsar, CP0950, which has a 0.253 s period in the radio frequency or 0.126 in gravitational waves, could not be detected by our equipment because of the 0.1 s channel width. In order to detect CP0950, a narrower channel width is required.

On the other hand, the amount of Earth's displacement due to gravitational waves from pulsars as calculated by Weber and Dyson is of the order of $10^{-17}$ cm as compared with 2 x $10^{-9}$ cm found in this experiment. Thus, without an amplification mechanism in the Earth's crust and if pulsars have the properties used in the calculation, the effect found here is not due to gravitational waves. Several arguments can be raised as to why the effect cannot be due to the electromagnetic radiation from pulsars. We do not know of any other mechanism that can cause the effect described here. (Nature, 248:136-138, 1972)

# MISCELLANEOUS EARTH TREMORS AND VIBRATIONS

Adjustments in the earth's crust cause most earthquakes, however, some difficult-to-ignore tremors and vibrations are attributable to other forces, some natural, some man-made. The non-earthquake tremors and vibrations are often created when fluids and gases in the earth and atmosphere somehow couple their energies to the solid ground. Typical examples producing curious effects are heavy thunder and vibrating dams. Of course, warfare, city traffic, and heavy machinery also shake the earth, but these sources are excluded as commonplace and scientifically tractable.

## VIBRATIONS OF THE ICE-CAP OF POLAR SEAS
**Fakidov, Ibrahim; *Nature,* 134:537-538, 1934.**

In the course of the Polar expedition on the S. S. Cheluskin in the Chukehi Sea, 1933-34, I noted a very interesting phenomenon, concerning which I have been unable to find any descriptions in the literature available to me.

The solid ice-cap of the sea, which represents, as it were, an immense elastic plate on a liquid foundation, is in a state of perpetual vibration. Though I had no special seismic apparatus at my disposal, I was nevertheless able, by means of very primitive hand-made instruments, to detect and roughly to measure these vibrations. They proved for the most part to be caused by the wind, and the direction of the greatest amplitudes tallied with the direction of the prevailing wind. A few cases of considerable vibrations running in a determined direction were observed on a windless day. Some hours later (up to eight hours) the wind blew in that same direction; evidently it did not keep pace with the sound oscillations it had created in the ice. We call those vibrations 'wind vibrations', but we admit that they may consist in a periodic warping of the ice-plate.

Besides these strictly directed 'wind vibrations' there may be observed 'disturbance vibrations' in the ice, spreading equally from the centre---the spot of the breaking up of the ice in different directions. A systematic investigation of the 'wind vibrations' would evidently greatly assist arctic synoptics. The study of the 'disturbance vibrations' will obviously permit periodicities in the dynamics of the ice-covered sea to be determined.

At present it is proposed to take observations of both types of vibrations by means of special ice seismographs, which will be set up on the shore-ice. There could be created, by means of blastings, artificial vibrations in the ice, and this method could be used also for the determination of the limits of the ice-fields, this too being of considerable practical importance in ice-navigation. The data obtained from seismological investigations of the ice would be of great help in the prognosis of ice conditions. (Nature, 134:537-538, 1934)

## EARTH TREMOR DUE TO THUNDER NOTES
**Humphreys, W. J.; *Monthly Weather Review,* 45:515, 1917.**

Mr. Douglas F. Manning, Alexandria Bay, N.Y., sends the following report under date of October 28, 1917:

A peculiar effect of thunder was felt here last night (Oct. 27, 1917), between the hours of 10 and 11 p.m. The day had been ideal with a light south wind, mild temperature, and a few alto-cumuli moving lazily from the west; in fact, it was an "Indian Summer" type of day.

Toward evening my aneroid began to fall rapidly and the clouds increased, and by 8 o'clock a rain was falling. At about 10 [p.m.] I noticed a flash of lightning, and this was followed in a short interval by a deep, prolonged rumble, causing windows and doors to rattle, chinaware to jar, and a distinct earth tremor was felt; in fact many thought it was one. The lightning increased in intensity and frequency and the same marked earth tremors followed each flash at short intervals, and it seemed as if a series of earthquakes were taking place, so strong was the concussion produced. The storm gradually passed over accompanied by a tremendous but brief downpour of rain mixed with small hail, and by 11 o'clock all was still again.

To-day one hears many stories of the storm and its peculiar behavior, all making note of the trembling effect produced.

This instructive letter is published for the benefit of others interested in these problems.

Since "musical" notes of very low pitch and great volume are occasionally produced by a series of sequent or pulsating lightning discharges, it seems probable that the shaking described by Mr. Manning was owing in great measure to the resonance response of rooms to thunder notes of this character. (Monthly Weather Review, 45:515, 1917)

## MIMIC EARTHQUAKE NEAR AKRON, O.
**Claypole, E. W.; *American Naturalist,* 22:242-243, 1888.**

A district of country lying about five miles south of Akron, O., was on the night between Thursday, February 9th, and Friday, February 10th, the scene of a commotion that well simulated an earthquake on a small scale. About nine o'clock in the evening a smart shock disturbed the inhabitants and caused much consternation, which was intensified when between two and three the following morning a severer one, accompanied by a loud noise, as of an explosion, awoke the sleepers by shaking the houses and cracking the walls of some of them. When daylight came several long clefts in the ground were discovered, furnishing evidence of some subterranean disturbance during the night.

Similar phenomena occurred almost in the same spot in 1882 and 1883. At that time a cleft from two hundred to three hundred feet long was formed, which crossed a road, marking its course with a furrow, such as that made by a plough. This crack was not more than an inch or two in width, but was sounded with a stick to the depth of several feet (some say fifteen or twenty). It passed under a house, cracking the cellar-wall. The noise accompanying it was likened by some of those who heard it to a cannon fired in the cellar. The explosion of Friday morning last (February 10) was heard by several persons in Akron, at a distance of about five miles.

The writer visited the spot last summer, at the request of a gentleman who had leased several farms, with the intention of drilling for gas. On making inquiry of one of the oldest residents, he learned that an earlier event of the same kind took place about twenty-five years ago, but could get no details.

The phenomena pointed, not to seismic causes, but to subterranean explosions, presumably of gas. The ground is clay and gravel, moraine matter, probably not less than a hundred feet in depth.

Another account differs somewhat in the details: "After the explosion in 1882, then the fissures, some of them nearly half a mile long, radiating to the top of a rise of ground. Mr. Thornton dug a hole nine (9) feet deep at the point where the fissures crossed it formed a centre, and at that depth found the cleft in the earth as pronounced as it was at the surface."

So far as it is possible to determine it, the cause of the commotion is due to the presence of a certain, perhaps a small, amount of natural gas, which, in ordinary circumstances, escapes unnoticed. But when the ground is frozen (and all these explosions have occurred in the winter) the gas is unable to ooze through the soil and accumulates below the frozen crust, until its elasticity becomes sufficient to burst it; hence the explosion, the shock, and the cracks in the ground.

At the outburst in 1883 several of these cracks could be seen radiating from a central point in a field and extending to different distances. One of the spectators says these clefts divided the field into half or quarter-acre pieces. Similar results followed the late explosion. I have not seen the place since, but learn that the fissures very soon closed, or were filled up in consequence of a rapid thaw.

"One peculiar feature is that while former disturbances rent greater fissures and were accompanied by much greater damage to property than the recent ones, yet so far as is known, none of the former explosions was heard or felt in this city."

It would be interesting to inquire if any similar events are on record elsewhere in regions yielding natural gas. Perhaps this note may be the means of calling out such cases if they exist.

It may at first sight appear as if so powerful explosions and shocks must indicate natural gas in considerable quantity. But when we reflect that the pressure in a well of small yield rapidly rises to a high figure when the bore-hole is closed, we see that such an inference is not safe. No other indications of gas are yet known in the immediate neighborhood, as would probably be the case if a large supply were accessible. (American Naturalist, 22:242-243, 1888)

## UNIDENTIFIED EARTH TREMORS IN DOMINICA, WEST INDIES

**Robson, G. R., and Barr, K. G.; *Nature*, 188:306, 1960.**

Tremors the origin of which has not been identified were recorded by our seismograph station at Roseau in the volcanic island of Dominica during September 1959, and have been recorded at irregular intervals ever since.

Each tremor consists of an approximately sinusoidal ground motion of 0.3-sec. period. The first swing is always down, the first and eighth swings are maxima and there are about twelve swings in all. The ground amplitude of these maxima varies from about 0.1u, which is the point at which the tremors become masked by the microseismic background, to 1.0u. They are quite unlike the tremors produced by near earthquakes or by explosions on land or at sea. Factories, quarries and other possible artificial sources for the tremors in Dominica have been investigated and eliminated.

The tremors tend to occur in bursts, numbering from two or three to several hundred tremors, which may start at any hour of the day or night and continue for periods up to twelve hours. The average period between bursts has been seven days, but there is no evidence of periodicity.

We have operated three additional Willmore-Watts seismographs in Dominica for six weeks in January and February 1960 and have found that the tremors can be recorded only within a few kilometres of the town of Roseau. Because of long quiet periods between bursts of activity, useful recordings were obtained only of a burst of five tremors at one station 1 km. from the Roseau Station and at the Roseau Station itself. There is a possible correlation between the differences in arrival times and the ratios of the maximum amplitudes of the tremors at the two stations, which is consistent with each tremor having originated at a point source and having obeyed an inverse-square law of attenuation. Assuming such an attenuation law the amplitudes at the two stations indicate that the sources lie close to a circle of about 5 km. diameter, the northern part of which passes through the town of Roseau. The differences in arrival times at the stations indicate a seismic velocity between 150 m./sec. and 740 m./sec. depending on the locations of the sources on the circle.

The town of Roseau is situated on a delta of sand and gravel built by the Roseau River on the steeply shelving east coast of the island of Dominica. In such materials the velocity of longitudinal waves is about 750 m./sec. We have estimated the energy release at the source as about $10^{11}$ ergs for the largest tremor recorded. The island contains several active thermal areas and ancient centres of eruption, but none is near the apparent sources of the tremors. We can find no evidence that these are related in any way to the volcanic activity of the island; but we cannot put forward an alternative explanation for their origin. We should be grateful to hear from others who have had experience of similar tremors. (Nature, 188:306, 1960)

## ON VIBRATING DAMS

**Loomis, Elias; *American Journal of Science,* 1:45:363-365, 1843.**

Sometime in the winter of 1841-2, my opinion was asked respecting a remarkable phenomenon noticed at Cuyahoga Falls, a village on the Cuyahoga River about eight miles from Hudson. The phenomenon consisted in the vibrations of the doors and windows, and other movable objects belonging to the buildings in the village. They were noticed at certain stages of the water, and at times ceased entirely. They were generally ascribed to a certain dam in the river and various conjectures were formed as to the mode of their production. The subject was new to me, and I did not at first form any distinct idea of the phenomenon itself or its cause. Some weeks afterwards, I visited the locality, and although the water was at too low a stage to exhibit the phenomenon in question, I succeeded in obtaining a pretty good description of the facts and formed my opinion as to its cause. I published a notice of it in the Ohio Observer which led to some discussion, and brought to light several similar cases elsewhere. As I have not succeeded in finding any notice of this subject in such books as I have had the opportunity of consulting, I have thought it desirable that the facts should be placed on record. I propose therefore to communicate such information as I have been able to collect, and shall conclude with some speculations as to the cause of the phenomena.

Dam at Cuyahoga Falls, Ohio. This dam is a portion of an arc of a circle, the convexity of course being turned up stream. It is formed of hewn oak timbers one foot square, piled upon each other in tiers, all morticed firmly together, so as to form as it were one huge plank two feet thick; twelve and a half feet in breadth, and ninety feet in length, measured not between the banks, but between the points of support. Its curvature is described with a radius of a hundred and twenty feet; that is, the arc is about one eighth of the circumference of a circle. There is an embankment of earth upon the upper side, which until recently was left in an unfinished state. The bank did not rise to the top of the dam, and sloped off very abruptly. This dam was erected in the summer of 1840. During the winter of 1840-1 there was noticed considerable rattling of the windows of the neighboring houses; a phenomenon different from what had ever been noticed before; but during the winter of 1841-2, which was a very open and wet winter, the vibrations were more remarkable and became a matter of general complaint. The doors and windows of most of the houses in the village would shake for days together violently as with the ague, and to such a degree as seriously to disturb sleep. This phenomenon was apparently somewhat capricious. After continuing for a time, perhaps an hour or a day or longer, the vibrations would suddenly cease, and after some interruption might be as suddenly resumed. The rattling of vibrating objects would frequently cease, while the vibrations could still be felt. A window, when apparently at rest, if put in motion by the hand, would continue to rattle. The buildings themselves (stone as well as wood) would vibrate with the doors and windows. This might be felt and also seen; as for example, a slender branch of a grape vine trained up against the side of a stone building, was seen to vibrate in exact time with the doors and windows.

The vibrations ceased entirely when water ceased to pour over the dam; they were also inconsiderable when the depth of water was eighteen or twenty four inches. A depth of five or six inches produced the greatest effect. The number of vibrations per second was thought to be about constant, but no accurate experiments on this point were ever made. From the best estimate I could obtain, they amounted to twelve or fifteen per second. A heavy log resting against the dam materially impaired the effect. The vibrations were seldom noticed in summer, as the water rarely ran six inches over the dam; I however formed a plan during the summer of 1842, for a series of experiments during the ensuing winter and spring, to determine particularly the number of vibrations per second. I had several methods in mind to be tried if others should fail, but the one with which I expected most success was with a monochord; being simply a cat-gut, one end of which passed over a pulley and was stretched by a variable weight. The number of vibrations was too small for a musical sound, but by holding a small slip of paper near the string when vibrated, a succession of rattles is produced, which I hoped might be tuned to unison with the rattling of the windows. The vibrations of the string could be easily determined by the principles of acoustics. I was however never allowed the opportunity of testing my methods. In my first communication to the Ohio Observer, I had stated as a test of my theory, "if this dam were filled up to the top with earth it would probably cease vibrating." To my great regret the experiment was immediately tried. During the season of 1842 a large amount of rock, estimated at two hundred and fifty tons, was deposited upon the embankment. This raised the bank fully up to the level of the dam for a depth of six feet or more. Since that time the river has passed through every stage of elevation, known in ten years; from that in which you might walk with dry shoes over the entire length of the dam, to a depth of six feet on the brink of the earth, which happened June 5, 1843, the greatest rise known since 1832. The result is that the vibrations have entirely ceased. (American Journal of Science, 1:45:363-365, 1843)

# GEOGRAPHICAL ORGANIZATION OF EARTHQUAKES

Today, it is well known that the most earthquakes occur along the fringes of the tectonic plates that form the continental-oceanic mosaic. Although the concept of tectonic plates is now well-established, scientists noted long ago that earthquakes tended to occur along certain lines. Many were tempted to organize these lines into geometric global patterns. This geometric synthesis was closely related to efforts to explain the distribution of continents and oceans in terms of regular polyhedrons and other mathematical idealizations. Some of these fascinating explorations are documented in the companion handbook on geology.

In addition to the linear distribution of earthquakes, certain myserious seismic "hot spots" puzzle seismologists. Many of these areas of anomalous activity occur in eastern North America far removed from the boundaries of tectonic plates.

## EARTHQUAKES
**Kingsmill, Thomas W.; *Nature*, 35:319-320, 1887.**

In Nature of October 14 (p. 570) you published a letter from Prof. O'Reilly regarding the great earthquake of Carolina, and drawing attention to the tendency of earthquake lines to assume the direction of great circles. So far his observations were identical with a theory I had myself elaborated, and which I embodied in a paper written at the beginning of the year 1884, now in the hands of the Committee of the Geological Society of London, but never presented to the Society. So long ago as that period I had drawn attention to what I pointed out as the two principal earthquake great circles---one, the Japan and Rocky Mountain system, with one of its poles in 170° W. long., 25° S. lat.; the other, the Himalayic, with its north pole approximately in 45° N. lat., 160° W. long. The former has been frequently described, and Scrope ("Volcanoes," p. 303) suggested a theory to explain its occurrence. The latter is little less remarkable, and is at the moment even more interesting, as, with the exception of the Carolina earthquake, all the great earthquakes and volcanic eruptions of the last five years may be referred to it. I may instance the cases of Krakatao, Kashmir, the Caucasus, Spain, Cotopaxi, New Zealand, and the recent Mediterranean disturbance, all of which occurred within a few degrees of the line or actually on it. Now it is remarkable that this line is marked through a considerable portion of its course by the presence of disturbed Miocene rocks, so much so that I have felt justified in calling it the Miocene line. (Nature, 35:319-320, 1887)

## THE $E=MC^2$ OF SOLID EARTH THEORY
**Anonymous; *Science News*, 93:301-302, 1968.**

Earthquakes by and large never move straight up and down. Instead, they strike up at earth's crust at an angle. What started the hypothesis was doctoral candidate George E. Rouse's observation that the deep quake zones around the planet---called

Benioff zones---all seem to lie at surprisingly similar angles of about 60 degrees. Wondering why this should be so, he decided to see what would happen if he projected the Benioff zones into imaginary planes passing all the way through the globe.

Using a $1.50 toy globe, Rouse began his first circle at a point of relatively mild seismic activity in Chile. To his surprise, the completed circle also passed through a very active zone in Turkey, the site of a recent quake in the Pyrenees Mountains between France and Spain and another seismically active spot in Venezuela. Intrigued, he drew 15 more circles, beginning at different deep quake zones, and a striking conclusion appeared: Not only was the plane of every circle tangent to the outer core of the earth, but where the planes emerged, they coincided with the Benioff zones of other quake areas, and with volcanic and other seismic areas.

With 16 circles, in fact, Rouse found he had included most of the major seismic features of the globe; five more brought the total to over 90 percent. With the original 16 circles, there are 19 points on the globe at which three circles intersect. "All but five of these trisections occur at zones of major seismic or volcanic activity," says Dr. Ramon E. Bisque, a professor of geochemistry who is now helping Rouse with his research. "Most of the others are in ocean areas that could be or have been active." (Science News, 93:301-302, 1968)

## NEW MAPS PINPOINT EARTHQUAKES IN EASTERN U.S.
**Anonymous; *Department of the Interior News Release*, January 29, 1975.**

According to James F. Devine, geophysicist, U.S. Geological Survey National Center, Reston, Va., and coauthor of the report, "The sources of earthquakes in the East are not nearly as well known or understood as are the sources of earthquakes in the Western U.S. Among the obvious reasons for the lack of knowledge are the facts that eastern earthquakes occur less frequently, are generally smaller, and are thought to occur at greater average depths than they do along the West Coast. Also, eastern earthquakes appear to produce only vibratory ground motion and not the actual differential ground displacement of the Earth's crust that is so evident along the West Coast. Apparently as a result of all these factors, earthquakes have produced little evidence of surface fault movement east of the Rockies in at least recent geologic time. As a result, we usually have only secondary geologic effects, such as sand boils and ground slumping to aid us in our seismic studies."

"On the other hand, major earthquakes in the East have been felt over a much greater distance than similar-sized earthquakes in California," the USGS scientist said, "For example, the three large earthquakes near New Madrid, Mo., in 1811-1812 were felt in Boston, Mass., and New Orleans, La., and the 1886 Charleston, S.C., earthquake was felt in Des Moines, Iowa, and Providence, R.I."

"Because similar major earthquakes could occur again," Devine said, "This study was originally designed on behalf of the Atomic Energy Commission to develop evidence for the existence of tectonic regions or structural provinces that would be useful in defining seismic hazards related to engineered structures. As with many studies, this one did not produce the results envisioned at the outset, but it did result, we think, in assembling seismic and structural data useful in evaluating the distribution and origin of earthquakes in the eastern part of the United States, and in delineating areas of relatively high seismic activity. Furthermore, the new seismotectonic map provides us with the first preliminary framework to begin a more systematic study of the sources of these earthquakes and to better define earthquake hazards in the East," he concluded.

The report uses the Modified Mercalli intensity scale for earthquakes to allow tabulation and rough comparisons of earthquakes that occurred before scientific earthquake monitoring equipment was available. This scale is based on the subjective reaction and observations of the effects of earthquakes. In contrast, the Richter Scale, more commonly used to report the size of recent U. S. earthquakes, is based on instrumentally recorded measurements of the amplitude of seismic waves and can be related approximately to the amount of energy released.

Some highlights from the report:

An average of about 6 earthquakes of Modified Mercalli (MM) intensity IV-V or greater---strong enough to be generally felt and to cause slight damage---have occurred each year in the Eastern United States. The largest earthquakes documented in the East were the intensity MM XII New Madrid earthquakes of 1811-1812.

In some areas of the Eastern United States, earthquake activity appears to be associated with known tectonic structures, such as in the Mississippi embayment trough and in the Appalachian Mountains. In many other areas, seismicity, usually of low level, is so scattered that no localized faults or other structures are indicated. In a few places, such as central Virginia, central New England, and most notably in the vicinity of Charleston, S. C., the sources of seismic activity of intermediate or even high levels remain a geologic mystery and cannot, in the light of present information, be considered localized along known structures. (U. S. Geological Survey News Release, January 29, 1975)

# Chapter 6
# PHENOMENA OF THE HYDROSPHERE

## INTRODUCTION

Weather is a phenomenon of the atmosphere, but seven tenths of the earth is covered by a thin coating of water possessing "weather" of an entirely different sort. The most accessible of these oceanic phenomena occur on the ocean surfaces: the waves, the more ponderous water movements called tides, and localized disturbances. What transpires below the surface is still largely a mystery, we can conjecture from surface clues and a few subsurface measurements that the subsurface movements of water are very complicated.

Lack of knowledge of the subsurface realm focusses our attention on events occurring in the interface between ocean and atmosphere. Centuries of sea lore plus data from a fleet of research vessels and thousands of shore instruments have discerned tidal regularities (and irregularities) and charted wave patterns on many seas. It is, however, an empirical business, with exceptions frequently sending scientists back to their computers to modify general laws for local conditions. When such situations prevail, as they do for both atmospheric and oceanic phenomena, we must conclude that we do not yet understand all of the nuances of these two fluid media.

# SOLITARY WAVES AND SMALL WAVE GROUPS

Many mariners tell of "waves from nowhere" that suddenly broke over their ships, sweeping men and equipment into the sea. Great solitary waves are rare; and they pose problems of explanation. If a solitary wave is merely larger than the general run of waves, it can be written off as a chance addition of two smaller waves. Single, giant waves preceded and followed by calm water are not explained so readily. Distant storms and submarine disturbances are, of course, possible progenitors. Such sources, though, habitually generate wave trains with very long wavelengths. Tsunamis or earthquake-generated waves, for example, are barely noticed in the deep ocean and rise to catastrophic heights only as they approach shallow waters. Solitaries in constricted areas, such as the English Channel, may be the piling up of a storm surge, but a single giant in the open, calm ocean is harder to comprehend.

Small groups of waves come and go as mysteriously as the solitary giants. We may suppose that, like ripples in a pond, they are caused by some distant disturbance possibly thousands of miles away. Again, this is surmise, perhaps not a reasonable one at that. There are no reasonable alternatives however.

## GIANT SOLITARY WAVES

### SOME RECORDED EXTREMES OF METEOROLOGICAL ELEMENTS
**Hennessy, J.; *Marine Observer*, 10:17, 1933.**

Admiral FitzRoy the first Director of the Meteorological Office has left on record that he personally measured seas that were 60 feet in height. Waves exceeding this height are exceptional but there are a few cases on record of which the following are instances:---

The highest seas yet recorded, reaching to a height of 80 feet were measured by Sir Bertram Hayes and his officers in the R. M. S. Majestic during a North Westerly gale of hurricane force, in Latitude 48° 30' N., Longitude 21° 50' W., on December 29th, 1922. Dr. Vaughan Cornish, the British authority on water waves, in a letter to the Editor of Lloyd's List remarked that he had never met with any other account of waves of such a great height taken under conditions so favourable for observation.

The R. M. S. Olympic when bound from Southampton to New York on February 27th, 1925, was struck by a 70 feet wave, during a heavy gale, which badly damaged her navigation bridge, 75 feet above sea level. The R. M. S. Aquitania bound from New York to Southampton encountered the same gale during which a heavy sea swept her bridge carrying away a locker and smashing a chronometer and three sextants. Aquitania's bridge is 70 feet above sea level.

In 1875 the late Captain Kiddle of the R. M. S. Celtic from reliable measurements determined a height of 70 feet for several waves in Mid Atlantic.

Solitary waves of an exceptional height have been met with unexpectedly in mid ocean which cannot be accounted for by the wind then blowing. These waves are commonly referred to as tidal waves but this is a misnomer: they are in all probability due to submarine seismic disturbances.

In 1881 a solitary wave struck the barque Rosina in the North Atlantic sweeping overboard all hands then on deck. In the same year this unfortunate ship encountered another such wave while both watches were engaged shortening sail. On this occasion the whole crew were carried away with the exception of a sick seaman lying in his bunk. He was eventually rescued by a passing steamer.

In 1882 the well known four masted barque Loch Torridon when off the Cape of Good Hope encountered an unexpected solitary wave which washed overboard the Captain, Second Mate and the whole watch on deck.

In 1893 the barque Johann Wilhelm when 300 miles east of Charleston S. C. had a similar experience to the Rosina. A solitary wave struck the vessel and throwing her on her beam ends washed overboard all hands with the exception of one man who was later rescued by the S. S. Electrician.

The S. S. Rheinland was nearly lost some years ago in the North Atlantic when a solitary wave submerged the vessel for a short interval. Her commander, Captain Randle, stated that only the funnel and masts were distinguishable. Every boat was washed away, one man lost overboard, and several of the officers and crew received broken limbs. (Marine Observer, 10:17, 1933)

## ABNORMAL ATLANTIC WAVES
**Stromeyer, C. E.; *Nature,* 51:437-438, 1895.**

In 1887 (Nature, vol. xxxvii. p. 151) you kindly published a few remarks from me on the possible volcanic origin of the two waves which swept the decks of the steamers Umbria and Faraday, and which, from the data then available, seemed to have originated at a point in the Atlantic known as the "Faraday Reef," and marked with a cross (+) on the accompanying chart. I am now able to send you the details of a few similar cases which I have collected since then. The exact positions of the vessels, and the directions from which these solitary waves seemed to come, being also marked on the chart. In the case of H. M. S. Orontes the ship's course was not stated, and on account of darkness and other causes the directions from which some of the other waves came are not to be depended upon. None of these encounters would have been reported had they not caused much damage---masts and funnels going by the board, and bulwarks, deckhouses, and lifeboats being smashed; but many seafaring men can recall solitary and abnormally high waves having struck their vessels, although the sea was otherwise quiet. Amongst the strange results which these blows have produced, may be mentioned that the magnetism of the steamship Energia was thus suddenly altered sufficiently to introduce an error of $18^{o}$ into the compass readings. The full details about this and a few other vessels have not been obtained.

The following are short summaries of each case:

S. S. Faraday.---The wave was visible like a line of high land on the horizon about five minutes before it struck the vessel.

S. S. Westernland.---A huge wave rose to a great height just in advance of the ship. No other similar waves were met with. About noon the wind had changed from S. W. to W. N. W.

S. S. Germanic.---Wind W. N. W. with terrific squalls. Shipped a tremendous sea over bow.

S. S. Umbria.---The disturbance came from N. W. and consisted of two waves. The first one was broken, the second one green. The wind had previously changed from S. W. to N. W.

H. M. S. Orontes.---While steaming in smooth water a huge wave broke over the vessel forward.

S. Festina Lente.---A steep sea fell on board from both sides.

S. S. Manhattan. The sea was high, but fairly true until a mountainous wave broke on board from N. W.

S. S. Diamond.---Lying to, awaiting daylight to enter port. The wave was heard some time before it was seen, and then seemed to be about 40 feet high. The vessel never rose to it, but was literally submerged for a time. An examination of the chart will show that with the exception of the Westernland each wave may have originated at a common centre, and might therefore be due to subaqueous volcanic activity. However, as the solitary waves which strike the west coast of South America and the so-called Death Waves on the west coast of Ireland are said to be regarded as precursors of storms, it is possible that these solitary Atlantic waves may be due to a similar cause; but even then it is inexplicable how a number of comparatively small and regular waves can be converted into one abnormal one, or how the reported changes of wind and consequent confused sea could produce such a wave. It will be noticed that the dates for the Festina Lente, Manhattan, and Diamond are very close together, and therefore there is a possibility that they were struck by the same wave. (Nature, 51:437-438, 1895)

[Faraday Reef not marked on original chart in Nature.]

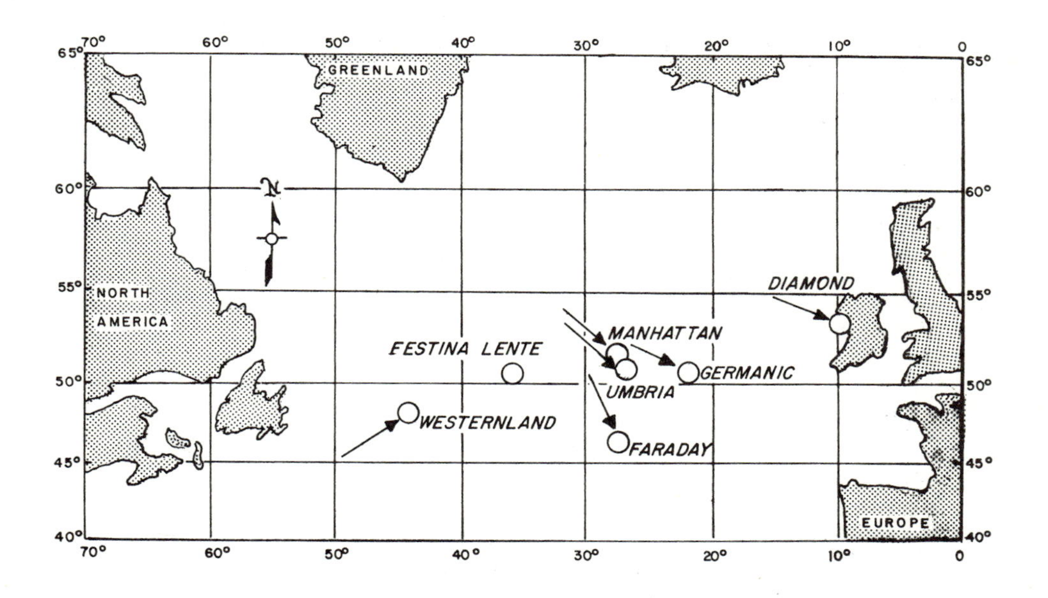


Possible point of origin of abnormal Atlantic waves

| Ship's name | Local time | Date | Latitude North ° | ' | Longitude West ° | ' | Speed Knots | Ship's course | Wave's course |
|---|---|---|---|---|---|---|---|---|---|
| Faraday | 6.45 a.m. | 14/2/84 | 46 | 11 | 27 | 53 | 6 | N. 72°E | Port beam |
| Westernland | 2.45 a.m. | 27/11/86 | 47 | 59 | 43 | 57 | 7 | S. 60°W. | Bow |
| Germanic | 9.40 a.m. | 5/5/87 | 50 | 36 | 22 | 8 | 4 | N. 68°W. | Bow |
| Umbria | 4.40 a.m. | 26/7/87 | 50 | 50 | 27 | 8 | 16 | S. nrly W. | 3 pts on strbd bow |
| H. M. S. Orontes | 5 p.m. | 18/2/91 | 36 | 12 | 32 | 50 | 9 | ? | Bow |
| Festina Lente | noon | 16/11/94 | 50 | 12 | 35 | 23 | ? | S. E. byS. | ? |
| Manhattan | 2 a.m. | 17/11/94 | 51 | 26 | 27 | 31 | ? | S. 86°W. | N. W. |
| Diamond | 10 p.m. | 21/11/94 | 53 | 9 | 9 | 52 | Lying to | W. N. W. | W. N. W. |

## EXTRAORDINARY WAVE IN THE CHANNEL
**Anonymous; *Symons's Monthly Meteorological Magazine,* 18:42, 1883.**

The Weymouth and Channel Islands Steam Packet Company's mail steamer Aquila left Weymouth at midnight on Friday for Guernsey and Jersey. On her passage across Channel, the weather was calm and clear, and the sea was smooth. When about one hour out, 1 a.m., 31st March, 1883, the steamer was struck violently by mountainous seas, which sent her on her beam ends and swept her decks from stem to stern. The water immediately flooded the cabins and engine room, entering through the skylights, the thick glass of which was smashed. As the decks became clear of water, the bulwarks were found to be broken in several places, one of the paddle boxes was considerably damaged, the iron rail on the bridge was completely twisted, the pump was broken and rendered useless, the skylight of the ladies' cabin was completely gone, and the saloon skylight was smashed to atoms. The cabins were baled out with buckets, while tarpaulins were placed over the skylights for protection. Five minutes after the waves had struck the steamer, the sea became perfectly calm. Several of the crew were knocked about, but no one was seriously injured. (Symons's Monthly Meteorological Magazine, 18:42, 1883)

## VALPARISO SEA SURGE
**Cornell, James, and Surowiecki, John; *The Pulse of the Planet,* 11, 1972.**

Valparaiso Sea Surge, Valparaiso, Chile, July 25-26, 1968. Suddenly and without warning, tidal waves 15 to 18 feet tall rolled into the Chilean coastline near the capital city of Valparaiso on July 25 and 26. The huge waves were apparently caused by a sea surge far offshore. Although much coastal property and many ships were damaged, there were no personal injuries reported.

The question that needed answering, however, was what caused the sea surge? Some reports indicated the waves were associated with seismic phenomena; other theories attributed the waves to a storm far out at sea. Today, however, their origin remains a mystery. (The Pulse of the Planet, 1972, 11)

## THE HAZARD OF GIANT WAVES
**James, Richard W.; *Mariners Weather Log,* 10:115-117, 1966.**

On April 12, 1966 the Italian liner Michelangelo was steaming westward at 16 kt. some 600 mi. southeast of Newfoundland. Generated by whole gale winds, waves of 25 to 30 ft. in height pounded the ship as she headed for New York. Suddenly a huge wave towered high above the others, smashed into the Michelangelo, inundating the forward half of the ship. Steel superstructure gave way, bridge windows 31 ft. above the waterline were crushed in, furniture in the public lounge flew through the air, heavy steel flaring on the bow was torn off, and the bulkhead under the bridge was bent back at least 10 ft. Three people were killed and 12 were injured. (For a description of the storm that generated this wave, and an account of the damages and casualties suffered by other ships caught in it, see Rough Log, North Atlantic Ocean, page 138.)

What causes these monster waves that occasionally strike ships? How do they form, and when can you expect them? Can they be predicted?

Unfortunately, although something is known of the nature and behavior of these oversized waves, the answer to the last question is "no." The exact time and place an oversized wave will develop cannot be forecast. The probability that one will occur however, and its approximate height can be calculated.

Under the influence of a strong wind, the ocean surface becomes complex, with mounds and hollows of water in chaotic motion. Careful observation of a number of waves passing a ship will disclose a variety of wave heights, with high waves followed by lower waves---or the reverse---in a random fashion. The average of such visually observed waves, called the significant wave height, occurs most frequently, but much higher and much lower waves also occur.

Ocean waves are the cumulative result of a large number of superimposed, simple sine wave surfaces of varying lengths and heights as shown in figure 3. Where the sine waves reinforce each other the waves are high; where they do not align a trough exists; and where there is slight misalignment a low wave is produced. Occasionally all the crests align perfectly and an oversized wave results.

If we make certain assumptions as to the mathematical model describing these combinations, it is possible to estimate the height and frequency of these oversized waves. On the average, of 20 waves passing a ship, one will be more than 1.73 times the significant height. If 1,000 waves pass the ship there is one chance in 20 that a wave 2.22 times the significant height may occur.

Since the significant waves encountered by the Michelangelo were around 30 ft. high, there was a chance of encountering a wave of 52 to 66 ft. With reported wave periods of 12 seconds it would have taken over 3 hours to encounter 1,000 waves. The probability would have been 1 in 20 that a monster wave of 66 ft. would form during that time.

Why aren't more of these huge waves observed? First, because in moderate seas, with wave heights of about 5 ft., little notice is paid to an occasional 8- to 10-ft. wave. It is only when the seas are high that oversized waves attract attention. Secondly, these waves are generally unstable, peaking and breaking before traveling any distance; nevertheless many ships have encountered such waves, usually with disastrous results. It should be noted that during any heavy weather the possibility exists of running into a giant wave. (Mariners Weather Log, 10: 115-117, 1966)

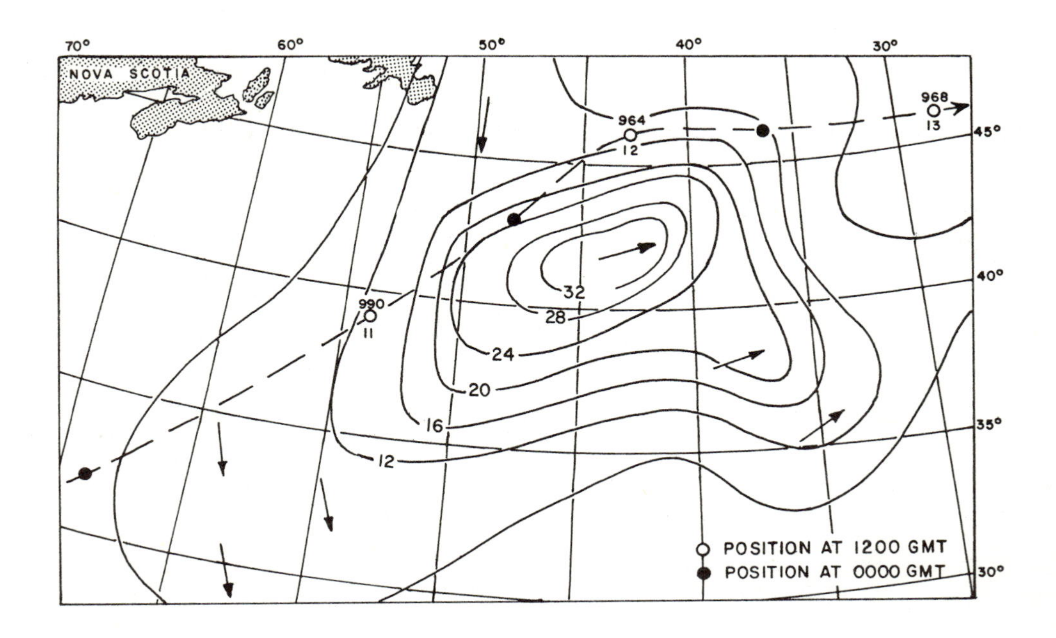


Wave height map for North Atlantic

# ANOMALOUS GROUPS OF WAVES

## SUDDEN ONSET OF SWELL
**Wortley, B. T.; *Marine Observer*, 32:116-117, 1962.**

S. S. City of Lucknow. Captain B. T. Wortley. Djibouti to Suez. Observer, Mr. D. Oldfield, Junior 2nd Officer.

27th July 1961. At 1930 GMT the wind veered from 330° to 060° and increased to force 4; the air temp. dropped from 89°F to 86° and a shower was seen near the ship. A sudden swell, 5 ft. in height, came in from 300° with a period of 5 sec. It arrived in four or five lines at a time, then died down for about 30 sec. to 1 min., after which interval it again moved in. This cycle kept repeating, though with decreasing height of waves, until 2120. Throughout the day the wind had been light and variable.

Position of ship: 18° 25'N, 39° 58'E.

Note. We are advised by the National Institute of Oceanography that there is no standard explanation for this phenomenon. From a meteorological point of view it was associated with the atmospheric division between dry northern hemisphere air and the more humid southern hemisphere air. (Marine Observer, 32:116-117, 1962)

## LARGE WAVES
**Wright, J. A.; *Marine Observer*, 14:7, 1937.**

The following is an extract from the Meteorological Record of S. S. Madura, Captain J. A. Wright, Marseilles to Gibraltar. Observer, Mr. A. W. Clarke, 2nd Officer.

"28th February 1936, 0240 G. M. T., proceeding down the Spanish coast, from Marseilles towards the Gibraltar Straits, seven miles from the coast. Prevailing weather, strong west winds, moderate sea, moderate short swell, fine and clear. The vessel, whilst under the lee of Cape de Gata, completely sheltered from the wind and experiencing only a moderate short swell, was met on the starboard bow from a W. N. W. direction by four large waves in succession, which caused a violent pitching and the shipping of a great deal of water; the fourth and final large wave rolled right over the forecastle head and completely filled the forward well deck. Upon righting again, conditions were as before, spraying forward and pitching easily. When Cape de Gata bore 270°, half an hour later, the full force of the westerly wind was again felt.

Position of ship, Latitude 36° 40'N., Longitude 2° 00'W." (Marine Observer, 14:7, 1937)

## ABNORMAL SWELL
**Davies, L. M.; *Marine Observer*, 27:203, 1957.**

S. S. Grelmarion. Captain L. M. Davies. Galveston to United Kingdom. Observer, Mr. R. Dixon, Chief Officer.

8th October, 1956. At 1235 G. M. T. while on a course of 043°, speed 12 kt, a small area of extremely turbulent sea was encountered, with steep swell and white-crested waves 12-15 ft in height running from a N'ly direction. As waves struck the vessel she was momentarily thrown off course and one large sea was shipped amidships. The phenomenon consisted of some four swells which approached the vessel and passed away on the quarter. Wind 045°, force 4-5. Sea and swell moderate, direction 045°.

Position of ship: 31° 10'N., 79° 01'W.

Note. At 1200 G. M. T. on 6th October a storm was centred 300 miles NE. of Newfoundland. The National Institute of Oceanography state that the winds recorded in this storm would not lead one to expect any swell at the ship's position on the 8th, but long waves of 15-20 sec period could just cover the distance in the time if the storm could generate them. (Marine Observer, 27:203, 1957)

## WAVES IN THE INDIAN OCEAN
**Davies, R. H. A.; *Marine Observer*, 35:60, 1965.**

R. R. S. Discovery. Captain R. H. A. Davies. Aden to Mauritius.

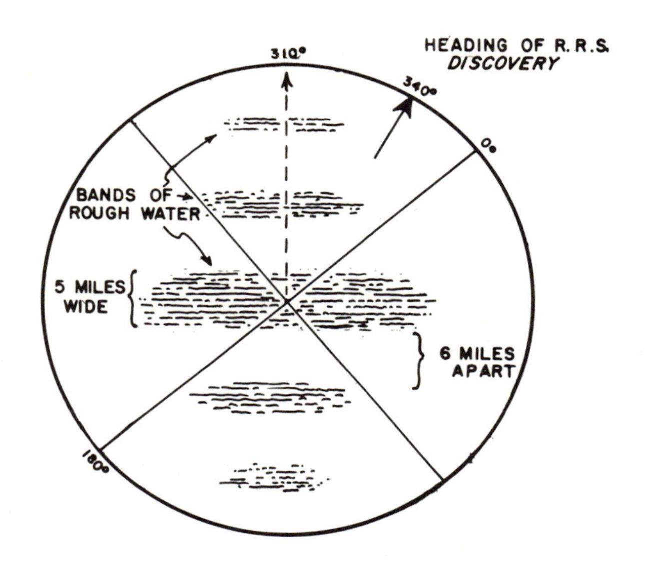


Radar display observed by R.R.S. Discovery

28th March 1964, 1930-2000 GMT. Bands of breaking waves of height 2 ft and period 1 sec, about 1/2 mile in width and 6 miles apart were observed by radar, heading in a 310° direction. When the bands passed the ship, which was hove-to, the wind did not change in direction or increase in force. The swell died down completely. There were five bands in all, the first and last being comparatively weak while the middle one was the strongest. They had the appearance of waves breaking on a flat sandy shore. Wind ENE, force 3. Sea temp. 85°F.

Position of ship: 11° 30'S, 58° 00'E.

29th March. Similar waves, moving from 300°, were observed from 0530-062 GMT. Wind NW, force 3. Sea temp. 84.9°.

Position of ship: 11° 30'S, 58° 00'E.

Note. This phenomenon appears to be a resonance between wind and current. The sea surface temperature was extremely high (85°F) and probably only a very shallow layer of the sea was involved in the waves. (Marine Observer, 35:60, 1965)

# THE BEWILDERING VARIETY OF TIDES

For most people living on the sea coast, tides come and go with clockwork regularity twice a day. Some places, though, experience only one tide per day; in others, the tides follow the sun rather than the moon. Elsewhere, the situation becomes even more complex.

Basically, tides are the sloshings of the oceans in their relatively shallow basins (2 miles deep on the average, but thousands of miles across) under the combined influences of the moon and sun. Obviously, the shape and bottom contour of the coastline have strong effects, too. Geophysicists are generally satisfied that tides, regardless of their idiosyncracies, can be explained by appealing to the fluctuating attractions of the moon and sun and the natural frequencies of oscillation of the various ocean basins. Despite this assurance, the reader will probably agree that some of the tidal phenomena described below are very strange.

Tidal bores, too, can be described in terms of the unrushing tide and the constricting influences of river estuaries. This explanation cannot detract from the eeriness of the foaming walls of advancing water seen on the Amazon and other properly configured rivers around the world.

## TIDES OUT OF THE MOON'S CONTROL

### VARIETY IN TIDES
**Marmer, H. A.; *Smithsonian Institution Annual Report,* 1934, 181-191, 1935.**

The usual explanation of the tide in textbooks of physical geography or astronomy makes of it a simple phenomenon. It is shown that the gravitational attraction of sun and moon on the earth gives rise to forces which move the waters of the sea relative to the solid earth. It is further shown that these forces have different periods, but that the predominant ones have a period of about half a day; therefore there are two high waters and two low waters in a day. And finally it is shown that the moon plays the leading role in bringing about the tide, for the tide-producing power of a heavenly body varies directly as its mass and inversely as the cube of its distance from the earth.

On the basis of these general considerations, numerous features of the tide can be explained. Thus, at the times of new and full moon, when the tide-producing forces of sun and moon are in the same phase, the tides are greater than usual, while at the times of the moon's quadratures the tides are less than usual. In the same way, when the moon is in perigee, or nearest the earth, greater tides result than when the moon is in apogee or farthest from the earth. To be sure, it is known that the tides at different places vary in time, in range, and in other features. But these differences are ascribed to modifications brought about by local hydrographic features. And thus the whole phenomenon is seemingly reduced to simple terms.

To the navigator who is familiar with the Seven Seas, however, this very much oversimplifies the subject. For he finds that the tides at different places vary not only in time and in range, but also in character of rise and fall. Quite apart from differences in time and in range, which may be regarded as merely differences in degree, it is found that tides present striking differences in kind. There is, in fact, an almost bewildering variety in tides.

Take for example the actual records of the rise and fall of the tide at three such well-known places as Norfolk, Va., Pensacola, Fla., and San Francisco, Calif., for the last 4 days of May 1931. These are shown in figure 1, the horizontal line associated with each tide curve representing the undisturbed or mean level of the sea. At Norfolk, it is seen, there are two high and two low waters in a day, morning and afternoon tides differing but little, and the high waters rising approximately the same distance above sea level as the low waters fall below it. At Pensacola, on the same days, there were but one high and one low water each day. And at San Francisco, while there were two high and two low waters each day, the morning tides differed very considerably from the afternoon tides.

. . . . . . . . . . .

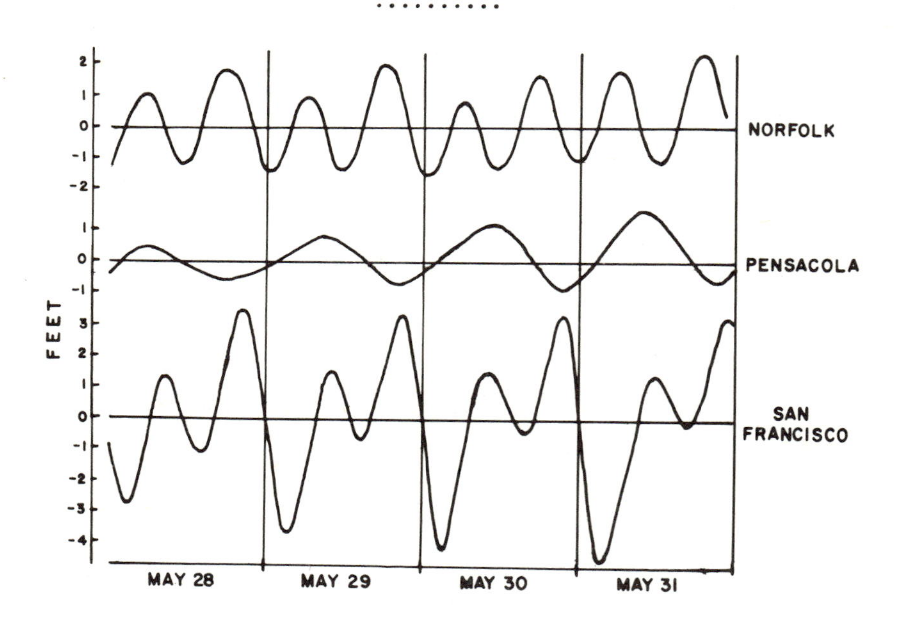


Various kinds of tides around U.S. coasts

Throughout the world, with but rare exception, the sovereignty of the moon over the tide is clearly exhibited by the retardation in the times of high and low water by about 50 minutes each day. Thus in figure 1 it is seen that the first high water of May 28 at Norfolk occurred at 6 o'clock and that each day thereafter it came about an hour leater. The other high water and also the low waters are seen to have occurred approximately an hour later each day. And at Seattle, with a totally different kind of tide, a similar retardation in the times of high and low water is seen to have occurred. This merely confirms the old adage that "the tide follows the moon." For the transit of the moon over any place occurs each day later by 50 minutes, on the average.

There are some places, however, where the tide appears to follow the sun rather than the moon. That is, instead of coming later each day by about 50 minutes, the tide comes to high and low water at about the same time day after day. Thus at Tahiti in the Society Islands, it has been known for many years that high water generally comes about noon and midnight and low water about 6 a. m. and 6 p. m. In fact, it appears that the natives use the same word for midnight as for

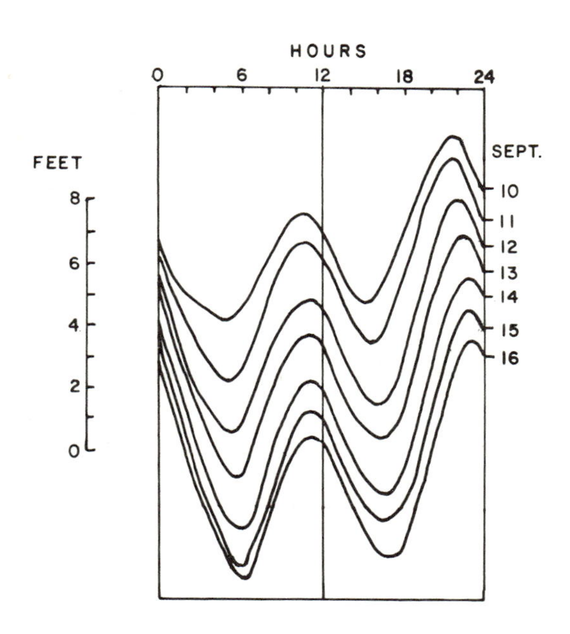


Sun-controlled tide at Tuesday Island in the Pacific

high water. At Tahiti, therefore, the tide is solar rather than lunar.

The range of the tide at Tahiti is small, less than a foot on the average. A better example of the solar type of tide has recently come to light at Tuesday Island, a small island in Torres Strait, lying about 15 miles northwesterly from the northern point of the Australian mainland. Here the range of the tide averages nearly 5 feet. The peculiar behavior of the tide here with regard to time is clearly brought out if the tide curves for a number of days are arranged in column, as in figure 5, which represents the tide curves for each day of the week beginning September 10, 1925.

It will be noted that the high and low waters in this figure fall practically in a vertical line; which means that instead of coming later each day by about 50 minutes, which is the state of affairs at most places in the world, the tide here comes about the same time day after day. That this is not a general feature of the tides in the South Pacific Ocean is evident from a comparison with the tide curves for Apia, Samoa, for the same week, which are shown in figure 6. It will be noted that here there is a distinct shift to the right in regard to the times of high and low water in following down the curves.

The mathematical process of harmonic analysis permits the tide at any place to be resolved into its simple constituent tides. At Apia the principal lunar constituent has a range of 2.5 feet, while the principal solar constituent has a range of 0.6 foot. Hence the tide here follows the moon. At Tuesday Island, the principal lunar and solar constituents both have the same range of 3.1 feet. Hence here the tide is no longer predominantly lunar but as much solar as lunar.

The answer to the question as to why the tide at some places is governed by the sun rather than the moon is again found in the physical characteristics of the various oceanic basins and seas. Where the conditions are such as to restrict the response to the lunar tide-producing forces but not the response to the solar tide-producing forces, the latter become the more prominent and give rise to solar tides.

An interesting form of tide is found at Jolo, in the Sulu Archipelago, P.I. Here the high waters follow the moon but the low waters appear to follow the sun. The tide curves for the week beginning September 10, 1925, at Jolo are shown in figure 7. The tide here is complicated by the fact that for part of the month there are two high and two low waters in a day, while at other times there is but one high and one low water. It is seen, however, that the high waters exhibit the distinct shift to the right characteristic of lunar tides, while the low waters lie almost perpendicularly under each other. (Smithsonian Institution Annual Report, 1934, 181-191)

# UNDERGROUND TIDES

## UNDERGROUND TIDES
**Belknap, Morris B.; *Nature,* 20:603, 1879.**

Mention is made in Nature, vol. xx, p. 401 of a spring in the Dux coal mines, Bohemia, exhibiting ebb and flow similar to tides. May not this be due to a subterranean syphon, acting precisely as a Field's flushing tank would in a house drainage system. Mention of such springs is made in Silliman's "Principles of Physics." We have in our neighbourhood [Louisville] not far from the Mammoth Cave a surface pool about 50 feet in diameter exhibiting this apparent tidal action. The pool is situated in the cavernous limestone country that forms such a large portion of our state, and is only a few hundred yards from Green River, whose peculiarly tinted waters it closely resembles. There is but little doubt that the river furnishes it with water until a level is reached, bringing one of the numerous underground conduits in the limestone into action, when the pool ebbs. (Nature, 20:603, 1879)

## TIDES IN SUBTERRANEAN WATERS
**Anonymous; *Nature,* 20:401, 1879.**

In one of the recently submerged coal mines in the Dux district (Bohemia) a remarkable phenomenon has occurred which Dr. Braumuller, an eminent Austrian mining engineer, describes. It appears that in the subterranean waters of the "Fortschritt" mine regular tides have been observed for the last six months. Both the Berlin and the Vienna Academies of Sciences, are devoting considerable attention to the strange phenomenon. A satisfactory explanation has, however, not yet been arrived at by either. (Nature, 20:401, 1879)

# UNUSUAL TIDAL BORES AND RIVER PHENOMENA

## THE "POROROCA" OR BORE, OF THE AMAZON
**Branner, John C.; *Science*, 4:448-490, 1884,**

While travelling upon the Amazon in 1881, I was fortunate in having an opportunity to observe some of the effects of a remarkable phenomenon which occurs at the northern embouchure of that river, in connection with the spring-tides. It is known to the Indians and Brazilians as the pororoca, and is, I believe, generally supposed to be identical with the 'bore' of the Hugli branch of the Ganges, of the Brahmapootra, and of the Indus. I regret very much, that like Condamine, who passed through this part of the country about 1740, I could not observe this phenomenon in actual operation; but the gentleman whose guest I was at the time, and upon whose boat I was a passenger, was fairly horrified at my suggesting such a thing, while his boatmen united in a fervent 'God forbid that we should ever see the pororoca!' and ever afterwards doubted my sanity. I venture, however, to give some of the results of my own observations, in order that those who in the future visit this region, concerning which so little is known, may be able to see, and establish as far as possible, the rate of destruction and building-up here being carried on.

. . . . . . . . . .

As I shortly afterwards met and conversed with a man who had seen the pororoca, I shall first give his description of it, and then speak of its effects as observed by myself. This man was a soldier in the Brazilian army, and, on the occasion referred to, was going with a few other soldiers from the colony to Macapa in a small open boat. Arriving at the mouth of the Araguary, they went down with the tide, and anchored just inside the bar which crosses the mouth of this stream, to await the turning of the tide, which would enable them to pass the shallows, and then carry them up the Amazon. Shortly after the tide had stopped running out, they saw something coming toward them from the ocean in a long white line, which grew bigger and whiter as it approached. Then there was a sound like the rumbling of distant thunder, which grew louder and louder as the white line came nearer, until it seemed as if the whole ocean had risen up, and was coming, charging and thundering down on them, boiling over the edge of this pile of water like an endless cataract, from four to seven metres high, that spread out across the whole eastern horizon. This was the pororoca! When they saw it coming, the crew became utterly demoralized, and fell to crying and praying in the bottom of the boat, expecting that it would certainly be dashed to pieces, and they themselves drowned. The pilot, however, had the presence of mind to heave anchor before the wall of waters struck them; and, when it did strike, they were first pitched violently forward, and then lifted, and left rolling and tossing like a cork on the sea it left behind, the boat nearly filled with water. But their trouble was not yet ended; for, before they had emptied the boat, two other such seas came down on them at short intervals, tossing them in the same manner, and finally leaving them within a stone's throw of the river-bank, where another such wave would have dashed them upon the shore. They had been anchored near the middle of the stream before the waves struck them, and the stream at this place is several miles wide.

But no description of this disturbance of the water can impress one so vividly as the signs of devastation seen upon the land. The silent story of the uprooted trees that lie matted and tangled and twisted together upon the shore, sometimes half buried in the sand, as if they had been nothing more than so many strings or bits of paper, is deeply impressive. Forests so dense that I do not know how to convey an adequate idea of their density and gloom, are uprooted, torn, and swept away like chaff; and, after the full force of the waves is broken, they sweep on inland, leaving the debris with which they are loaded, heaped and strewn through the forests. The most powerful roots of the largest trees cannot withstand the pororoca, for the ground itself is torn up to great depths in many places, and carried away by the flood to make bars, add to old islands, or build up new ones. Before seeing these evidences of its devastation, I had heard what I considered very extravagant stories of the destructive power of the pororoca; but, after seeing them, doubt was no longer possible. The lower or northern ends of the islands of Bailique and Porquinhos seemed to feel the force of the waves at the time of my visit more than any of the other islands on the south-east side of the river; while on the northern side the forest was wrecked, and the banks washed out far above Ilha Nova.

The explanation of this phenomenon, as given by Condamine, appears to be the correct one; that is, that it is due to the incoming tides meeting resistance, in the form of immense sand-bars in some places, and narrow channels in others. (Science, 4:488-490, 1884)

# CURIOUS PHENOMENA OF NATURAL WATER SURFACES

If wind-generated waves could be subtracted, the surfaces of the oceans would still be marked by curious rips, rollings, slicks, calms, and sundry disturbances. Many of these phenomena are probably manifestations of subsurface movements of water. Internal ocean waves that intersect the surface and submarine upwellings can turn a calm surface into patches and long patterns of choppy sea that contrast in color and temperature with the surrounding "normal" ocean. Submarine volcanoes and quakes can spew forth mud, bubbles, geysers, and dead fish. Huge shoals of fish and mile-long windrows of floating plants also festoon the ocean surfaces. This section presents a unique group of phenomena that are not so much unexplained as they are strange and unexpected (at least by landlubbers).

## FOAM STRIPS ON INLAND SEAS

### BANDS OF FOAM ON THE DEAD SEA
**Durward, J.; *Meteorological Magazine,* 67:41, 1932.**

The Dead Sea seen from the air during a gale on February 7th, 1932, presented a striking appearance. The height of the aircraft was about 4,000 feet, and at this height the wind was south-westerly 65-70 m.p.h. The wind in the deep depression near the Dead Sea was therefore probably S.-SSW., force 6-8. The surface of the sea was "moderate" on the sea disturbance scale, but the striking feature was the parallel bands of white foam which ran the whole length of the sea. These bands were orientated roughly south by west to north by east---i.e., slightly across the sea's major axis. The impression was that the lines of foam were interference bands due to two sets of waves or to one set of waves and their reflections from one shore.

As the sea was obscured at times by cloud and rain, it was not possible to count the number of bands accurately, but it is estimated that there were about 20 between the east and the west shore. (J. Durward)

[The Dead Sea has a length of 47 miles and is 9-1/2 miles across at its widest part, so that the bands of foam referred to would be somewhat less than a half-mile apart. According to the Encyclopaedia Britannica a line of white foam extends along the axis of the lake almost every morning. This was supposed by Blanckenhorn to mark the line of a fissure under the bed of the lake (which lies along the line of a great geological fault), but has also been explained as a consequence of the current of the Jordan.---Ed. M.M.] (Meteorological Magazine, 67:41, 1932)

## FOAM ON THE DEAD SEA
**Anonymous; *Nature,* 143:468, 1939.**

A peculiar phenomenon often to be seen on the northern waters of the Dead Sea is that of arcs of foam, more or less semicircular, spreading out fanwise from certain points of the west and east shores. The arcs seem to spread from the shores in early morning and often meet and even cross during the forenoon. These lines of foam often bear reeds and other vegetation debris that have reached the Dead Sea by inflowing rivers, and, at the seasons of migration, the foam may attract flocks of birds searching for food. Dr. D. Ashbel has recorded some of his observations on this phenomenon and offers an explanation (Geog. Rev., Jan. 1938). The arcs originate from springs on the two shores, and they make discontinuities between bodies of water of different salinity and density. When the outflow of the springs is strong, the arc is frequently not smooth but zig-zag. This explanation contradicts the earlier one of Blanckenhorn that the lines of foam originate from warm water arising along lines of fault on the sea floor. That suggestion does not explain lack of replacement of the lines as each moves forward, which Dr. Ashbel thinks is due to the gentle winds that frequently blow from the land on to the sea during the night and early morning. These winds do not ruffle the water and so mixing does not occur. Gusty west winds such as often occur in the afternoon cause mixing and so the lines of discontinuity disappear. (Nature, 143:468, 1939)

## FOAM STRIPS ON INLAND WATERS
**Richardson, W. E., and Ovey, C. D.; *Weather,* 5:361, 1950.**

The occurrence of 'foam strips' on inland waters provides a phenomenon for which I can find no serious explanation in meteorological literature. A recent visit to the Malham Tarn Field Centre (Council for the Promotion of Field Studies) revived my interest in the matter, for I have seldom seen it so well formed as there. Long ribbons of foam stretched almost from one side of the lake to the other, and this rich development would appear to be accounted for by the open nature of the surrounding landscape. Although I have not investigated the issue properly, the appearance of the strips seems to correspond roughly with the occurrence of Beaufort wind-forces of 5 or 6. It seems significant that the Beaufort Scale stipulates 'crested wavelets form on inland waters' at force 5, for the bubbles necessary for the strips are formed at this force.

Taking for an analogy the suggested mode of formation of the 'stone strips' seen on certain hill slopes, would it not be possible to interpret the foam strips on lakes as resultants from a similar cause? Thus, stone strips are believed to develop as part of a solifluction process, whereby the finer soil slips down the hillside leaving the irregular stones to 'ooze' sideways until they make contact with other stones, likewise displaced, from adjacent areas. At the point of contact the strip develops.

Returning to the lake, foam is produced by a fairly strong wind, and bubbles which make up this foam will act as obstructions similar to the boulders within the solifluction process on the hillside. That these bubbles should be deflected one way or another from the path of the wind would appear to be clear, and the result will be a zone of accumulation in the form of a foam strip.

Finally, I have observed 'calm strips' on the water of a local canal (i.e. at

Thorne) when it has been ruffled by winds of force 4 and when the direction of those winds have been aligned with that of the canal. The considerable quantity of surface oil on this canal appears to act in the same way as the bubbles in the foam strips. It may be presumed that the lower wind speeds necessary for this phenomenon are accounted for by the fact that the oil is always present on the canal irrespective of wind force, whereas the bubbles of the foam strip must be developed first.

Perhaps this explanation lacks attention to some vital aspect of physical behaviour which I have overlooked. I would be pleased therefore if I could hear other readers' views or of some publication which would assist me to understand the matter more clearly. (W. E. Richardson)

During the recent course at Malham Tarn the same phenomenon as that observed by Mr. Richardson was seen. Winds greater than about force 6 appeared to destroy the 'foam strip' effect, and a close examination of the crest-bubbles showed that those formed between the strips burst at once while those within them persisted. It seems likely that the strips may be due to a spiral cellular motion in the surface layers similar to the parallel development of cumulus clouds running down wind. The suggestion is offered that converging flow on the surface in this form of motion would concentrate algae and other microscopic planktonic organisms along lines of convergence. Water would descend along these strips but, because the surface tension would possibly not be broken until the wind exceed force 6 on the Beaufort Scale, microscopic organisms would concentrate forming a scum favourable for bubble persistence. In support of this it could be added that Malham Tarn, I believe, is particularly rich in planktonic organisms.

It is suggested that a close observation on this phenomenon should be kept at the Tarn noting the occurrence of it and the wind force. It may bear some relation to the effect described by A. K. Totton in the August issue of Weather where however the temperature in the calm plankton-rich lanes was observed to be higher than that of the surrounding water, indicating vertical water movements of a different nature. The physical processes involved in solifluction seem only to be partially analogous i.e. in respect of the main cellular flow system. (C. D. Ovey) (Weather, 5:361, 1950)

# STREAKS, SLICKS, AND CALMS

## STREAKS ON NATURAL WATER SURFACES
**Stommel, H.; *Weather*, 6:72-74, 1951.**

I was very much interested to see Mr. A. K. Totton's photographs of markings on the water (in the August 1950 issue of Weather) because recently I have been trying to collect as much information about such phenomena as possible. I enclose two photographs which may interest the readers of Weather; one, showing 'venous streaks' on the surface of the Great Salt Lake of Utah, was kindly given to me by Prof. A. J. Eardley of the University of Utah; the other, a picture of the Banana River in Florida, shows a remarkable development of 'parallel streaks', which was taken by Mr. Alfred Woodcock, of Woods Hole.

Most natural bodies of water have a streaky appearance which results from a natural process, although artificial streaks (due to the passage of a ship, for example) are also frequently observed. There has been quite a difference of opinion in the literature as to the nature of the process which produces streaks. I am personally inclined to admit that under certain circumstances any one of the processes suggested may account for them; to attribute all streaks to the same mechanism would be a great mistake. I should like, therefore, to review briefly what little material there is in the literature about these streaks on the surface of natural bodies of water.

The streaks themselves may consist of accumulations of floating objects, such as Gulf-weed, pine needles or bubbles of foam, all of which are visible because of their colour. Others are local concentrations of an oil film (presumably of planktonic origin) which damp the tiny wind-raised capillary waves. These are visible because of the marked difference in reflectivity of sky light between ruffled and slick areas.

Single Streaks. Single long streaks are often observed in harbours. They may stream out from a single continuous source of contamination, and run downwind with the surface water, or follow tidal or other currents. For convenience they may be called 'trail-streaks'. Single long streaks may also develop from an initially localized concentration of surface contamination in regions of strong horizontal shear by a simple stretching of elements of water in the shear zone. These 'shear-streaks' occur on the edges of strong tidal currents and in rivers. A third kind of commonly observed single streak is seen to remain parallel to a coast-line a hundred feet or so offshore, most frequently with an onshore wind. The cuase of these 'shore-streaks' is not known, but I venture to suggest that they are due to the horizontal convergence and sinking of surface water which is driven toward the shore by the wind, sinking somewhere off-shore, where the water is reasonably deep, rather than at the vanishingly small depths at the very water's edge.

The kinds of single streaks mentioned above are described in order to avoid confusion with other kinds which are really more interesting. I am referring to the large numbers of streaks often arranged in geometric patterns over large areas of a lake or ocean. Although it may be dangerous to draw a sharp distinction, these multiple streaks usually appear either in long parallel lines or with a vein-like pattern. I find it convenient to call them 'parallel' and 'venous' streaks respectively.

Parallel and venous streaks are plentiful in nature. Windrows of Gulfweed in the deep ocean are conspicuous examples of parallel streaks. Unfortunately there are no good photographs of these lines of weed. I have tried to photograph them on several occasions, as have others at this institution, and so far we have had no success. Similar streaks are observed on lakes, and have been carefully studied.

Air- and Water-Cell Theories. A large proportion of the parallel and venous streaks are aligned up- and down-wind. Two possible mechanisms have been suggested. One is the air-cell theory, which maintains that there is a cellular structure of the wind, which sweeps the surface water into lines. Langmuir on the other hand has reported that cellular motions exist in the water and that the streaks are located along lines of surface convergence and sinking water, a mechanism which may be called the water-cell theory. He has suggested that the cellular motion extends down to the thermocline, and indeed that it plays an important role in the formation of the thermocline. If this idea be true it certainly deserves further investigation, because the lack of an adequate theory of the thermocline is one of the most glaring lacunae in oceanography and limnology. Langmuir believes the water-cells are mechanically driven by the wind, although in just what manner is obscure. Woodcock indicated the presence of films of colder water on the surfaces of lakes and ocean waters and suggested that the water-cells are thermally driven convection-cells similar to those observed in the laboratory. I do not think the results of these laboratory studies are directly applicable to lake and ocean processes without further investigation.

Ewing has observed other parallel streaks whose orientation bears no obvious relation to the direction of the wind. He attributes them to alternate contractions and dilatations of the surface film due to progressive internal waves travelling along at some distance beneath the surface.

Cinematographic Studies. During June and July of last year I made a number of cinematographic studies of streaks on small ponds on Cape Cod, Massachusetts. They were of the venous and parallel types, and consisted of oil-slicks and lines of foam. The steadier the wind direction, the more nearly parallel they seemed. Upon showing the pictures at some 80 times their normal speed interesting features, which otherwise escaped the eye, were visible: the structure of the wind was so gusty and turbulent that it was obvious that no permanent structure in the air maintained the streaks; these were so quickly re-oriented after a shift of wind (only 1 or 2 minutes was necessary) that deep motions such as those reported by Langmuir did not seem likely. Thermal convection seemed also unlikely because the streaks occurred at times of intense heating of the ponds and with a very stable epilimnion ($0.1^{o}$ C. $m^{-1}$). Although surface convergence into the lines was confirmed, using Woodcock's ballasted-bottle technique, cellular motions were not observed at depths of one foot or deeper. The indications were that these streaks were due to a process confined to the top few inches at most, and did not play a role in the formation of the deep thermocline, which in this instance was at a depth of about thirty feet. It seems, therefore, that they were due to yet another process, perhaps associated with differences of wind-stress in and out of streaks, but which I must admit I do not understand. It would be misleading to conclude that this classification of various types of surface marking is complete. For example, Woodcock and Wyman have published photographs of mysterious bands on the surface of the ocean which may be due to some still undiscovered process. (Weather, 6:72-74, 1951)

## PATCH OF CALM WATER
**Newman, J. B.; *Marine Observer*, 34:174-175, 1964.**

m.v. Mahout. Captain J. B. Newman. Port Said to Wilmington. Observers, Mr. A. P. Spriggings, Chief Officer, and Mr. D. D. Pease, 4th Officer.

2nd October 1963. At 0700 GMT the vessel passed through a patch of calm water approx. 2,000 yd. wide and several miles long, extending in a line lying $350^{o}$-$170^{o}$. The wind at the time was force 5 and all the surrounding sea was covered with breaking wave crests. There was no sign of oil on the surface of the water. Sea temp. $68^{o}$F.

Position of ship: $36^{o}$ 21'N, $4^{o}$ 42'W.

Note. Dr. L. H. N. Cooper of the Marine Biological Association of the United Kingdom has given an extensive note on a calm patch in the North Atlantic in The Marine Observer of July 1961, page 118. Dr. Cooper has kindly passed the above report to M. Cousteau, the famous underwater explorer, because these phenomena are probably associated with the structure of the sea bottom. This calm patch was probably associated with descending cold Atlantic water at the easterly approaches to the Straits of Gibraltar. (Marine Observer, 34:174-175, 1964)

## ANOMALOUS PATCH OF BLUE WATER
**Cooke, A. H.; *Marine Observer*, 31:181, 1961.**

s.s. Pacific Envoy. Captain A. H. Cooke. Panama to London. Observers, Mr. G. H. Deere, 2nd Officer, Mr. T. A. Tate, 3rd Officer, Mr. D. L. Smith, Radio Officer and members of the crew.

23rd December 1960. At 1815 GMT a patch of brilliant blue water was observed some 50 by 100 yd. in extent and oval in shape. The surrounding sea was dark grey in colour and had been so, for the past 24 hours. Sea temp. 82°F. Wind E'N, force 5. Rough sea and short E'ly swell.

Position of ship: 20° 40'N, 64° 51'W.

Note. Dr. T. J. Hart, of the National Institute of Oceanography, comments: "Blue is the desert colour of the sea, and unless we are to assume that the surrounding dark grey sea was uniformly rich in plankton, and the blue patch barren, it is hard to see how it could have any biological explanation.

"The trade wind weather is usually fine, and plankton scanty in the Caribbean, except in winter close in to the Venezuelan coast, so far as we know." (Marine Observer, 31:181, 1961)

## HONEYCOMB APPEARANCE OF WATER
**Shaw, J.; *Nature,* 43:30, 1890.**

This afternoon, [Tynron, England] while ascending a mountain pathway adown which water was trickling, after the torrents of rain that fell in the morning had ceased, I observed an appearance of the surface of running water so exactly like the hexagons of the bees' cells that I looked at it carefully for some time. Little air-bells of water seemed to issue from under the withered leaves lying in the tract, which rushed towards the hexagons, occupying an irregular space about four inches by five. As soon as these air-bells arrived at the hexagons, they arranged themselves into new cells, making up, apparently, for the loss occasioned by the continual bursting here and there of the cell-walls. No sooner had these cell-walls burst, than others closed in and took their places. The worst-formed hexagons were those at the under or lower side of the surface---the part of the surface farthest down the hill; here they were larger, and more like circles. By an ingenious mechanical theory, Darwin accounts for the hexagonal structures of the cells of the hive-bee so as to supersede the necessity of supposing that the hive-bee constructed its comb as if it were a mathematician. But here the blind forces of Nature, under peculiar conditions, had presented an appearance, on running water less than half an inch in depth, so entirely like the surface of a honeycomb, that it would be a startling result could it be reproduced in a laboratory. (Nature, 43:30, 1890)

## CALM PATCH
**Batt, Edward H., and Cooper, L. H. N.; *Weather,* 16:86-87, 1961.**

An extraordinary experience happened to me during a passage from Gibraltar to the United Kingdom, and I should be grateful for an explanation. I am a master of a small motor yacht of a little over 60 tons, and on 16 September 1958, I was on passage from Gibraltar to England. I rounded Cape St. Vincent at about 0830 and ran into a pretty stiff north-west wind about Force 4; course was set, 310°, for Cape Espichel with the intention of putting into Lisbon, but after steaming for about four hours the wind from the north-west began to increase and a bad sea was now running with much salt water on board, and I was thinking of running all the way back to the shelter of Cape St. Vincent if the bad weather continued to increase.

However, by reducing speed from 10 knots to about 6, the going was made more comfortable, and I kept on course until we were abeam Sines, which was 12 miles distant. We were still having a pretty horrible time when there appeared ahead quite a calm patch with no broken water and a little more than half an hour's steaming brought us up to it.

I was quite dumbfounded at what I was about to experience. This calm patch was at least a cable wide and stretched away in a north-easterly direction for about 30 miles. I was able to steam along it for three hours in almost a complete calm at full speed, while on either side of me was a steep sea (Fig. I). The lane was absolutely dead straight and the demarcation line of where the broken water flattened itself out into this marine autobahn had to be seen to be believed. Needless to say, I kept in it and steamed right into Setubal without even any spray on deck.

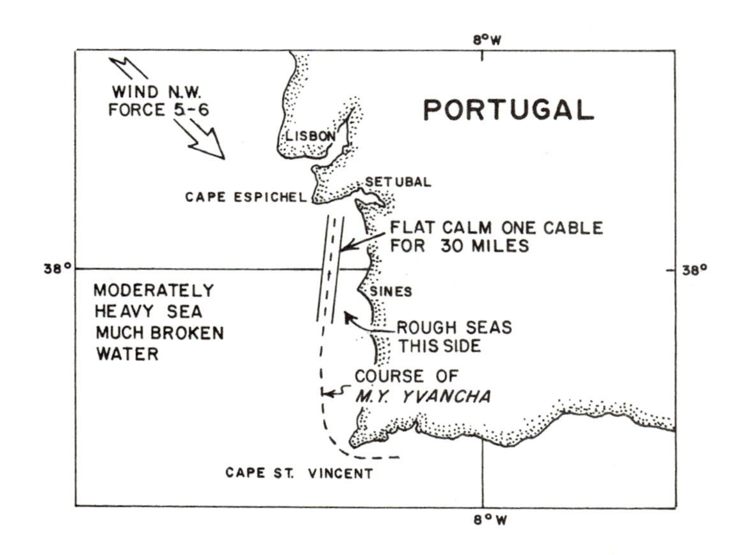


Peculiar calm patch of water off Portugal

The points to remember are: (i) the wind was north-west and blowing 5 to 6 on land, (ii) we were in deep water, (iii) we were 12 miles from the nearest point of land when we came upon the entrance (12 miles off Cape Sines).

I have spent over 30 years at sea and have made the passage to the Mediterranean on many occasions, but I have never experienced such a phenomenon as this before. It was most definitely not an oil slick; there was no sign of any oil and the lane was dead straight for 30 miles with clearly defined sides on both port and starboard sides. Had it come from an oil leakage I am sure it would have been curved from the source as it was blown away by the wind.

Probably there is a local explanation, but I was unable to enquire as I remained at anchor at Setubal for the night and sailed early next day. I shall look forward with interest to any explanation, if such is possible. Ryde, Isle of Wight (Weather, 16: 86-87, 1961)

# STRANGE SURFACE DISTURBANCES ON BODIES OF WATER

## JETS OF WATER RISE FROM SEA
**Evans, M.D.; *Marine Observer*, 31:183, 1961.**

S. S. Malmo. Captain M. D. Evans. Falmouth to Benghazi. Observers, the Master, Mr. J. B. Drinkall, 2nd Officer, Mr. A. B. Smith, 3rd Officer, Mr. O. Murphy, Radio Officer and Mr. M. B. Smith, Seaman.

4th December 1960. At 0830 GMT, as the ship was steaming through a calm to slight sea at 12 kt. on a course of 124° towards Benghazi, a strange-looking column of what seemed to be white Cu. cloud appeared to rise vertically from near the horizon, about 45° on the starboard bow, and vanished a few seconds later. The officer on watch believed his eyes to be deceiving him due to the sun's glare, but within a few minutes all the observers named above saw the column reappear. On examination with binoculars it was seen to be a jet of water rising into the air at regular intervals of about 2 min. 20 sec. Each spurt lasted for about 7 sec. and then disappeared. They were visible until the vessel left them far astern. The radar was switched on but no echo appeared on the screen; however, by sextant altitude and calculation the jets were found to be 494 ft. high; they resembled an underwater explosion but no such noise was heard.

Position of ship: 32° 08'N, 19° 32'E.

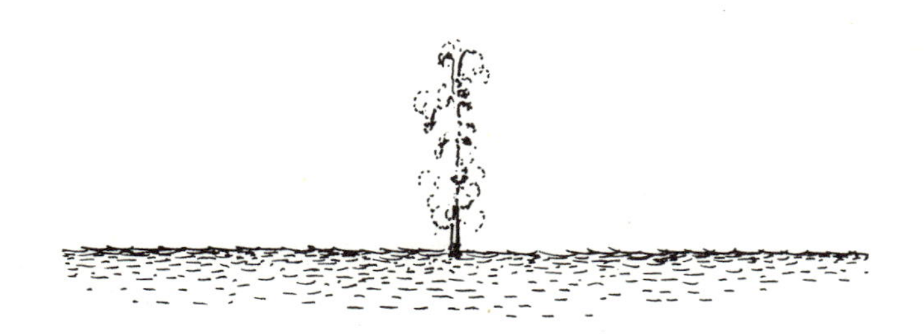

Jets of water rising from sea

Note. An examination of synoptic charts has revealed few indications of the nature of this phenomenon. Conditions appeared anticyclonic and stable although minor fronts existed in the central Mediterranean at the time which could have given rise to phenomena similar to waterspouts. (Marine Observer, 31:183, 1961)

## DISTURBANCE IN OCEAN
**Anonymous; *Nature,* 51:468, 1895.**

It is interesting to note, in connection with the subject of Mr. Stromeyer's letter published last week, that Reuter's correspondent at San Francisco reports that vessels arriving there announce the occurrence, on the 2nd inst. of an earthquake in the bed of the Pacific Ocean. The disturbance was accompanied by a loud roar, coming, apparently, from the sea, which became covered with a mass of white foam, and subsequently rose in numerous geyser-like columns. (Nature, 51:468, 1895)

## SUBMARINE DISTURBANCE
**Roberts, H. W.; *Marine Observer,* 32:59, 1962.**

S. S. Oronsay. Captain R. W. Roberts, O. B. E., D. S. C. Aden to Colombo. Observers, Mr. D. Hughes, 1st Officer, Mr. A. N. Perkins, Senior 3rd Officer, the Quartermaster and the look-out.

9th April 1961. At 2005 GMT a disturbance resembling an air bubble breaking surface was seen approximately 1/4 mile from the ship, causing discoloured water and dead or stunned fish to appear on the surface over an area of about 1/2 mile. No groundings were obtained on the echo sounding machine on any scale. Air temp. 82°F, sea 85°; wind NE, force 2. Sea calm.

Position of ship: 10° 12'N, 61° 57'F. (Marine Observer, 32:59, 1962)

## PATCH OF VIOLENTLY DISTURBED WATER
**Large, H. C.; *Marine Observer,* 28:69, 1958.**

M. V. Eumaeus. Captain H. C. Large. Pusan to Hong Kong.

7th May, 1957. At 0650 G. M. T. a patch of violently disturbed water was seen, which was about 40 sq. ft in area. It closely resembled a squall or shoal of jumping fish, but no fish were to be seen. Similar patches were sighted at 0710, all about the same size. The echo sounder gave 54 fm. Wind NE'N., force 5.

Position of ship: 27° 43'N., 123° 20'E. (Marine Observer, 28:69, 1958)

## SPECTACULAR CURRENT RIP
**Roberts, T.; *Marine Observer,* 34:118-119, 1964.**

m. v. St. John. Captain T. Roberts. Bahia (Salvador) to Las Palmas. Observer, Mr. K. B. Whitting, 3rd Officer.

29th August 1963. At 1630 GMT, the vessel passed across a 500 ft. wide band of agitated water which was darker than the smooth surrounding sea. It stretched from E to W as far as the eye could see. There were white breaking waves all along both edges of the band of disturbed water. The sea temp. was taken and found to be 76°F, as compared with 82° at noon and 81° at 1800. Wind ESE, force 3.

Position of ship: 0° 06'S, 30° 48'W.

Note. This observation was forwarded to Dr. L. H. N. Cooper of the Marine Biological Association of the United Kingdom at Plymouth who sought the opinion of Mr. William G. Metcalf of the Woods Hole Oceanographic Institution, Massachusetts, who had recently been working in the area. Mr. Metcalf comments:

"I have been studying the current system in the equatorial region and am delighted to learn of the very interesting phenomenon described so well.

In the location described, the Equatorial Undercurrent is now known to flow strongly to the east just below the sea surface. The temperatures measured before and after crossing the area of agitation (82°F and 81° respectively) are normal surface temperatures for the region. The 76° temperature measured in the turbulent strip is typical of water generally found at about 35 fathoms in this area. At that depth and temperature, the salinity can be expected to be more than a half of a part per thousand greater than is generally found at the surface. Correspondingly, the density, which increases with a temperature drop and a salinity rise, would be markedly greater than the usual surface density.

It has been my feeling that in the usual case, the surface layer which normally has a westerly drift in this area, will frequently slow down during periods of calm and even more to the east through frictional coupling with the swift flowing undercurrent.

However, in this instance we seem to have a much more violent and spectacular manifestation of the undercurrent whereby the surface layer is actually thrust aside and the undercurrent itself has reached the surface. The magnitude of the turbulent forces required to maintain this phenomenon is difficult to estimate, but it is very impressive that these forces are apparently able to thrust heavy water up through the light surface layer and maintain this condition along a narrow strip extending from horizon to horizon. It is not surprising that the sea surface in the strip was considerably agitated in view of the powerful force which must have been at work. (Marine Observer, 34:118-119, 1964)

## STRANGE LAKE WAVE
**Stewart, Charles; *Knowledge,* 4:395, 1883.**

On reading in Knowledge, No. 111, the letter (1043) from Port Louis, it occurred to me that the following extracts from a letter written in December, 1792, by the well-known William Cruch, of Edinburgh, and published in the "Edinburgh Fugitive Pieces" (pc. 119 and 121), might interest your readers:---

"In 1782, at the time of the dreadful earthquakes in Calabria, the mercury in the barometer in Scotland sunk within the tenth of an inch of the bottom of the scale; the waters in many of the lochs or lakes in the Highlands were much agitated."

"Upon the 12th of September, 1784, a very extraordinary phenomenon was observed at Loch Tay. The air was perfectly calm, not a breath of wind stirring. About six o'clock in the morning, the water at the east end of the loch ebbed about 300 feet and left the channel dry. It gradually accumulated and rolled on about 300 feet further to the westward, when it met a similar wave rolling in a contrary direction. When the waves met, they rose to a perpendicular height of five or six feet, producing a white foam upon the top. The water then took a lateral direction southward, rushing to the shore, and rising upon it four feet beyond the highest watermark. It then returned, and continued to ebb and flow every seven minutes for two hours, the waves gradually diminishing every time they reached the shore, until the whole was quiescent. During the whole of that week, at a later hour in the morning, there was the same appearance, but not with such violence."

I need scarcely add that Loch Tay is about fourteen miles long and one mile broad at the widest part, having the village of Killin at the west end and that of Kenmore at the east. I could give traditions of other interesting phenomena, showing some connection with earthquakes, but must not occupy your valuable space. (Knowledge, 4:395, 1883)

## BORE AT SEA
**Anonymous; *Science,* 5:61, 1885.**

The commander of the British steamship Bulgarian reports that on Dec. 29, in latitude 49° north, longitude 34° 30' west, at two P. M., while the sea was smooth and the wind moderate from south and west, he ran through a regular bore. The water boiled and seethed. The surface of the bore was about two feet above the general level of the ocean, and its extent about six miles long and from three to five miles wide, moving to the north-east. This is a very unusual phenomenon for such a place. (Science, 5:61, 1885)

# STRATIFIED WAVES

## PECULIAR STRATIFIED SHAPE OF TYPHOON WAVES
**Gherzi, E.; *Nature,* 175:310, 1955.**

While M. V. Tancred of the Barber Line was riding out the typhoon of early October 1954 in Kobe Bay, we were able to observe a unique shape of typhoon wave which, so far as we know, has not been described. The vessel was hove-to with two anchors down facing the incoming gale (133 km./hr.) and the waves with crests up to 10 and 15 ft. were passing neatly along her side. From the rail we were struck by the peculiar appearance of the wave-slopes facing the wind.

On many of these there were a number of well-defined steps, carved so to say into the water just like the steps of a ladder, starting from the trough of the wave up to about half its height. Although the waves were moving quickly the steps remained steadily extending parallel to each other for one or two metres in length. There were at times as many as twenty of these nicely successive steps cut into the body of the wave. We tried to photograph them; but the very poor visibility and the fast motion of the waves resulted only in a blurred print.

Were these steps carved in the wave slopes by high harmonics of the period of the typhoon squalls? These higher harmonics have been registered by sensitive modern microbarographs, and they might have something to do with the quite peculiar screaming of the wind, often reported by seamen during the squalls of tropical storms. Strangely enough, the peculiar whistling has never been noticed in the squall line of extratropical storms. (Nature, 175:310, 1955)

Stratified typhoon wave

# BLOWING WELLS

## BLOWING WELLS
**Hulburt, Ray G.; *Scientific American*, 95:115, 1906.**

In your issue of July 7, 1906, page 4, you publish a short article on "The Vagaries of Wells." In Water-Supply and Irrigation Papers of the U. S. Geological Survey, No. 29, Wells and Windmills in Nebraska, Washington, 1899, our State Geologist, E. H. Barbour, describes certain wells, in substance, as follows: One class of wells found throughout a large part of the State, especially south of the Platte, deserves particular notice. These wells are known as "blowing," "roaring," "breathing," "singing," or "wea ther" wells. These wells are held in doubt elsewhe re, but the fact of their existence is established. In some communities such wells are distinguished at a distance because of the mound of earth heaped up to check the wind. The attention of the writer was first called to this matter by inquiries for explanation of and remedy for the freezing of pipes in wells at a depth of 30, 50, 60, 80, and even 120 feet below the surface. Reports have come in from about twenty counties. The information is derived from land owners, farmers, well diggers, ministers, principals of schools, civil engineers, and students whose fathers own such wells. These accounts agree with personal observations. There are periods when these wells blow out for consecutive days, and an equal period when they are reversed. This is tested with the flames of candles and by dropping paper. chaff, feathers, etc., into the casing to see it blown out by some force, or drawn in. It is further stated that blowing often indicates high or low conditions of barometer, and that some wells blow most audibly when the wind is from the northwest, whereupon water rises to a higher level in the well than before; but when conditions are reversed, air is drawn in. Many observers notice a reverse of the current according as it is morning or evening, and according as the temperature is high or low. During the progress of a low barometer area over one of these regions, the wind is expelled with a noise audible for several rods. Upon the following of a high-barometer area, the current is reversed. Steam rises from the curbing, melting the snow. After the current is reversed, the thawed circle freezes again. The pipes are often thawed out when the well blows. The periods of most pronounced exhalation or inhalation are coincident with exceptionally low and exceptionally high barometer areas.

He then explains the geology of the country, and draws the same conclusion as did M. Grosseteste, and continues: "The wind may be the cause in some places. At times the friction of the wind is sufficient to drive the water of the Platte across its bed, leaving the north side dry while the south side is flooded. Equilibrium is disturbed. There must be readjustment. In the vicinity water rises in wells, at a distance there is a wave of transmitted energy which can but affect every portion of the underflow of the Platte. This may show itself in a rise of water and displacement of air, and a rise over a wide area might expel a large volume of air." (Scientific American, 95:115, 1906)

# Chapter 7
# FALLING MATERIAL

## INTRODUCTION

Anomalous rain, snow, and hail---colored and plain---are treated in Chapter 3. Beyond these nearly normal forms of precipitation are those falling materials that do not belong aloft at all: large ice chunks, living animals, non-meteoric stones, and many other nominally terrestrial materials. Charles Fort made much of falling materials, even though most of them can be explained rationally by appealing to recognized meteorological mechanisms; i.e., waterspouts, whirlwinds, etc. Fort had a point; the minority of falling material not succumbing to conventional explanations requires truly revolutionary explanations. Such is the claim for residues of unexplaineds in all areas of science---and this claim is perfectly valid.

Where would all the anomalous falling material come from? Fort, in his customary tongue-in-cheek way, postulated that near space was full of debris that occasionally fell to earth and was replenished by unspecified levitating mechanisms. A half century later, we have found that space is not as empty as we had supposed. In addition, terrestrial catastrophies (meteor strikes, for example) are now recognized as possible ways in which terrestrial (and lunar) materials can be ejected into near space. Other mechanisms may exist. Perhaps violent storms create conduits between earth and space. We do know that some violent thunderstorms and tornadoes reach far into the ionosphere. The atmosphere and magnetosphere, once believed to shield the earth's surface from space "weather", seem to have holes in them permitting a modicum of two-way traffic.

# FISH, FROGS, AND OTHER LIVING CREATURES

"Of course a waterspout or whirlwind deposited them," is the common response to news of a fish fall. No one can deny the possibility, even likelihood, of such a meteorological explanation after seeing the tornado of "The Wizard of Oz." But the venerable waterspout and whirlwind must possess some unusual refinements if it is to account for all of the facts concerning animal falls.

To begin, there exist many well-documented fish, frog, and animal falls---no one seriously denies that they do take place. The stranger aspects of these falls appear only after reviewing many reports. First, the transporting mechanism (whatever it may be) prefers to select only a single species of fish or frog or whatever animal is on the menu for that day. Second, size selection is also carefully controlled in many instances. Third, no debris, such as sand or plant material is dropped along with the animals. Fourth, even though saltwater species are dropped, there are no records of the accompanying rainfall being salty. All in all, the the mechanism involved is rather fastidious in what it transports. The waterspout or whirlwind theory is easiest to swallow when the fish that fall commonly shoal on the surface in large numbers in nearby waters. It is much harder to fit the facts when the fish are from deep waters, when the fish are dead and dry (sometimes headless), and when animals fall in immense numbers.

A final feature of animal falls hints ever so slightly that some falls (certainly not all) may come from very high altitudes. This feature is the "footprint" or pattern of the fall, which is usually rather small (a few hundred feet long) and highly elliptical. Bunched objects in outer space or in the very high atmosphere that enter the atmosphere land in just this kind of pattern.

## FISH

### RAINING FISH
**Lawrence, E. N.; *Weather*, 10:345-346, 1955.**

'Pouring cats and dogs' is a very common expression though a demand for substantiating evidence would challenge the most ardent investigator. But need we resort to Norse mythology for spectacular meteorological metaphor when we might as easily say 'raining sprats and whiting' and be within the bounds of scientific plausibility? For many such 'rains' or 'showers' have been noted over the last two thousand years. One of the earliest references to an occurrence of this kind was made by Athenaeus during the second to third century A. D. In his Deipnosophistae he remarks:

'I know also that it has rained Fishes. At all events Phoenias, in the second book of his Eresian Magistrates, says that in the Chersonesos it once rained Fishes uninterruptedly for three days, and Phylarchus in his Fourth book, says the people had often seen it raining Fish.'

The phenomenon has by no means confined itself to warm climes; Halsted's History of Kent states that 'about Easter 1666, in a pasture field in the parish of Stansted, which is a considerable distance from the sea and from any branch of it, a place where are indeed no fish ponds, there was found about a bushel of fish supposed to have rained down from the cloud, there having been at that time a great tempest of thunder, hail and wind, etc. These fish were about the size of a man's little finger; some were small whiting, others like sprats, and some rather like smelts. Several of these fish were shown publicly at Maidstone and Dartford. A report comes from still further north: in Reid's Law of Storms it is related that on 9 March 1830, after a day of heavy rain, the inhabitants of the island of Islay, Argyllshire were surprised to find strewn over their fields a number of small, perfectly fresh herrings, some even exhibiting signs of life. Resembling very small herrings, too, were those fish which showered down at eight o'clock one evening in 1833 on the edge of Lake Gwynant, according to the Caernarvon Herald.

Fish have rained in many lands. An American newspaper stated in 1835 that after a night of heavy rain a great number, mostly sun-perch and from two to three inches long, were found swimming in the gutters of Jefferson Street, Louisville. Following similar conditions at Jelapur, India, a fish was discovered 'about one cubit in length and weighing more than six pounds', and when hail fell during a heavy storm at Essen in 1896, one of the stones as large as a hen's egg, contained a Crucian carp about four centimetres long.

Altogether, more than fifty 'rains of fishes' are said to have been reported in various parts of the world. They are almost invariably accompanied by violent thunderstorms and heavy rain, usually over a restricted area and following long, fairly straight landtracks. In each case, young aquatic life appears to have been lifted from the sea by the force of intense convection, as in a waterspout or tornado, and carried up into the atmosphere, where it is sometimes able, for a time at least, to survive. The scarcity of incidents during the twentieth century could be ascribed to a decline in fish populations, but it may well be also a reflection of climatic change, for the cold middle-latitude winters of the seventeenth to nineteenth centuries favoured the development of very strong surface temperature gradients along the warmer continental-coastal areas. (Weather, 10:345-346, 1955)

## A RAIN OF SMALL FISH

**Anonymous; *Monthly Weather Review*, 29:263, 1901.**

Mr. J. W. Gardner, voluntary observer at Tillers Ferry, S. C., reports that during a heavy local rain about June 27 there fell hundreds of little fish (cat, perch, trout, etc.) that were afterwards found swiming in the pools between the cotton rows in a field belonging to Mr. Charles Raley.

It is a well-known fact that in such rains all sorts of foreign objects, whether sticks or stones, frogs or fish, or even debris of destroyed houses and crops, occur occasionally not only in America but in Europe and elsewhere. It is very rare that we are able to trace these objects back to their sources, but there can be no reasonable doubt that they were carried up from the ground by violent winds, such as attend thunderstorms and tornadoes. Light objects, such as sheets of paper, have been identified as falling at points twenty or fifty miles distant from their starting point, but it is hardly likely that heavier objects, such as fish, could be carried so far without coming to the ground. (Monthly Weather Review, 29:263, 1901)

## A SHOWER OF SAND EELS
**Meek, A.; *Nature*, 102:46, 1918.**

About 3 o'clock on the afternoon of Saturday, August 24 last, the allotment-holders of a small area in Hendon, a southern suburb of Sunderland, were sheltering in their sheds during a heavy thunder-shower, when they observed that small fish were being rained to the ground. The fish were precipitated on three adjoining roads and on the allotment-gardens enclosed by the roads; the rain swept them from the roads into the gutters and from the roofs of the sheds into the spouts.

The phenomenon was recorded in the local newspapers, the fish being described as "sile." I was away at the time, but, seeing the account, I wrote to Dr. Harrison, and thanks to him, and especially to Mr. H. S. Wallace, I obtained a sample of the fish, and I was able yesterday (September 5) to visit the place in the company of the latter gentleman.

From those who saw the occurrence we derived full information, which left no doubt as to the genuineness of what had been stated, and this we were able to put to the test, for a further sample was obtained from a rain-barrel which could have got its supply only from the spout of the shed to which it was connected. The precipitation of the fish, we were told, lasted about ten minutes, and the area involved Commercial Road, Canon Cocker Street, the portion of Ashley Street lying between these streets, and the adjoining gardens. The area measured approximately 60 yards by 30 yards, and was thus about one-third of an acre. It is not easy to say how many fish fell, but from the accounts it may be gathered they were numerous; there were apparently several hundreds.

There can be no question, therefore, that at the time stated a large number of small fish were showered over about one-third of an acre during a heavy rain accompanied by thunder; we were informed that no lightning was observed, and that the wind was variable.

All the examples which came into my hands from different parts of the ground and from the rainbarrel prove to be the lesser sand-eel (Ammodytes tobianus). They all, moreover, are about 3 in. in length, or 7.5 cm. to 7.9 cm. They are not "sile," a name usually given to the very small young of the herring. But the sand-eels are sea-fish, and it is evident that the sand-eels showered to the ground at Hendon were derived from the sea.

On sandy beaches around our coasts the lesser sand-eel is very common. As its name implies, it burrows into the sand, but in the bays it may often be seen not far from the surface swimming about in immense shoals---shoals which are characterised by the members being all about the same size.

The place where the sand-eels in question were deposited lies about one-quarter of a mile from the seashore, but it is probable that the minimum distance of transport was at least half a mile.

The only explanation which appears to satisfy the conditions, therefore, is that a shoal of sand-eels was drawn up by a waterspout which formed in the bay to the south-east of Sunderland, and was carried by an easterly breeze to Hendon, where the fish were released and deposited. It is significant that the area of deposition was so restricted, and that no other area was affected. The origin and the deposition were therefore local.

We were informed that the fish were all dead, and, indeed, stiff and hard, when picked up immediately after the occurrence. This serves to detract from the possibilities of distribution being influenced by such an occurrence, but it is possible that other species would be able to withstand such an aerial method of dispersion. It is more than probable that the vortical movement of a waterspout would transport plankton. This was naturally not observed in this case, and the small creatures, including eggs and young stages, would likely be carried over a wider area. (Nature, 102:46, 1918)

## SHOWER OF FISH AT HENDON, SUNDERLAND, AUGUST 24, 1918
**Anonymous; *Royal Meteorological Society, Quarterly Journal,* 44:270, 1918.**

On Saturday, August 24, 1918, there occurred at Hendon, near Sunderland, the somewhat unusual phenomenon of a shower of fish. According to a report in the Sunderland Daily Echo of August 26, this shower took place about 3 p. m., coinciding with the passage of a thunderstorm at that hour with rain of only a few minutes' duration. In the rear of the storm the ground was "strewn with thousands of little fishes." The fall was apparently confined within a very circumscribed area just outside the town of Hendon and about one-quarter of a mile from the sea coast. The fish were of the variety known locally as "sile," eel-like things, two to three inches long, which are found in very considerable numbers at this time of the year, both in the river and the sea. The fish were quite stiff when found, and many of them were broken by striking the ground.

Mr. T. W. Backhouse of West Hendon House has kindly supplied some further information bearing on this shower of fish. He was at his house at the time and did not remark anything unusual in the weather. The shower of fishes took place about three-quarters of a mile to E. S. E. At West Hendon there was a shower of a thundery nature, accompanied by considerable darkness, though thunder was not heard. The rainfall measured in the shower amounted to 0.04 inch, wind West to South-west, not strong. Mr. Backhouse has heard the fish described as sand-eels, and that they exist in considerable numbers near the shore. Two men who were sheltering in their allotment huts at the time of the shower picked up handfuls of fish.

The phenomenon is probably attributable to whirlwind or waterspout action in the neighbourhood of the coast or out at sea. (Royal Meteorological Society, Quarterly Journal, 44:270, 1918)

## SHOWER OF FISH
**Anonymous; *Annual Register,* 101:14-15, 1859.**

If any one has entertained doubts as to the possibility of this phenomenon, his hesitation will be put to rest by a well-certified occurrence at Mountain Ash, Glamorganshire. At 11 a. m. of the 9th of February, during a heavy rain, a stiff gale blowing from the south, a very large number of small fish were precipitated upon the fields and housetops at that place. The phenomenon was witnessed by a great number of persons: the Rev. Mr. Roberts, curate of St. Peter's, Carmarthen, and the Rev. John Griffith, the Vicar of Aberdare and Rural Dean, made inquiries on the spot, in order to preserve the facts of this curious occurrence. The following is the testimony of John Lewis, a sawyer, who was the principal witness:---

"On Wednesday, February 9, I was getting out a piece of timber, for the purpose of setting it for the saw, when I was startled by something falling all over me ---down my neck, on my head, and on my back. On putting my hand down my neck I was surprised to find they were little fish. By this time I saw the whole ground covered with them. I took off my hat, the brim of which was full of them. They were jumping all about. They covered the ground in a long strip of about 80 yards by 12, as we measured afterwards. That shed (pointing to a very large workshop) was covered with them, and the shoots were quite full of them. My mates and I might have gathered bucketsful of them, scraping with our hands. We did gather a great many, about a bucketful, and threw them into the rain pool, where some of them now are. There were two showers, with an interval of about ten minutes, and each shower lasted about two minutes or thereabouts. The time was 11 a.m. The morning up-train to Aberdare was just then passing. It was not blowing very hard, but uncommon wet, just about the same wind as there is to-day (blowing rather stiff, and it came from this quarter (pointing to the S. of W.). They came down with the rain in 'a body like.'"

Mr. Griffith collected 18 or 20 living specimens of the unexpected visitants and transmitted them to Professor Owen. The three largest were four inches long. Some, which died after capture, were fully five inches in length. (Annual Register, 101:14-15, 1859)

## DO FISH FALL FROM THE SKY?
**Bajkov, A. D.; *Science,* 109:402, 1949.**

In view of the prevailing skepticism about rains of fish, my own observations of this phenomenon may interest the readers of Science.

A rainfall of fish occurred on October 23, 1947 in Marksville, Louisiana, while I was conducting biological investigations for the Department of Wild Life and Fisheries. In the morning of that day, between seven and eight o'clock, fish ranging from two to nine inches in length fell on the streets and in yards, mystifying the citizens of that southern town. I was in the restaurant with my wife having breakfast, when the waitress informed us that fish were falling from the sky. We went immediately to collect some of the fish. The people in town were excited. The director of the Marksville Bank, J. M. Barham, said he had discovered upon arising from bed that fish had fallen by hundreds in his yard, and in the adjacent yard of Mrs. J. W. Joffrion. The cashier of the same bank, J. E. Gremillion, and two merchants, E. A. Blanchard and J. M. Brouillette, were struck by falling fish as they walked toward their places of business about 7:45 a.m. There were spots on Main Street, in the vicinity of the bank (a half block from the restaurant) averaging one fish per square yard. Automobiles and trucks were running over them. Fish also fell on the roofs of houses.

They were freshwater fish native to local waters, and belonging to the following species: Large mouth black bass (Micropterus salmoides), goggle-eye (Chaenobryttus coronarius), two species of sunfish (Lepomis), several species of minnows and hickory shad (Pomolobus mediocris). The latter species were the most common. I personally collected from Main Street and several yards on Monroe Street, a large jar of perfect specimens, and preserved them in Formalin, in order to distribute them among various museums. A local citizen who was struck by the fish told me that the fish were frozen; however, the specimens I collected, although cold, were not frozen. There is at least one record, in 1896 at Essen, Germany, of frozen fish falling from the sky. The largest fish in my collection was a large-mouth black bass 9-1/4 inches long. The largest falling fish on record was reported from India and weighed over six pounds.

The fish that fell in Marksville were absolutely fresh, and were fit for human consumption. The area in which they fell was approximately 1,000 feet long and about 75 or 80 feet wide, extending in a north-southerly direction, and was covered unevenly by fish. The actual falling of the fish occurred in somewhat short intervals, during foggy and comparatively calm weather. The velocity of the wind on the ground did not exceed eight miles per hour. The New Orleans weather bureau had no report of any large tornado, or updrift, in the vicinity of Marksville at that time. However, James Nelson Gowanloen, chief biologist for the Louisiana Department of Wild Life and Fisheries, and I had noticed the presence of numerous small tornadoes, or "devil dusters" the day before the "rain of fish" in Marksville. Fish rains have nearly always been described as being accompanied by violent thunderstorms and heavy rains. This, however, was not the case in Marksville.

Certainly occurrences of this nature are rare, and are not always reported, but nevertheless they are well known. The first mention of the phenomenon was made by Athanaseus in his De pluvia piscium nearly two thousand years ago, and E. W. Gudger, in his four collective articles, reports 78 cases of falling fish from the sky. There is no reason for anyone to devaluate the scientific evidence. Many people have never seen tornadoes, but they do not doubt them, and they accept the fact that wind can lift and carry heavy objects. Why can't fish be lifted with water and carried by the whirlwind? (Science, 109:402, 1949)

# FROGS AND TOADS

## SHOWER OF FROGS
**Anonymous; *Arcana of Science,* 217, 1830.**

As two gentlemen were sitting conversing on a causeway pillar near Bushmills, they were very much surprised by an unusually heavy shower of frogs, half formed, falling in all directions; some of which are preserved in spirits of wine, and are now exhibited to the curious by the two resident apothecaries in Bushmills.

Mr. Loudon also observes, when at Rouen, in September last, "we were assured by an English family resident there, that during a very heavy thunder shower, accompanied by violent wind, and almost midnight darkness, an innumerable multitude of young frogs fell on and around the house. The roof, the window-sills, and the gravel walks were covered with them. They were very small, but perfectly formed, all dead, and the next day being excessively hot, they were dried up to so many points or pills, about the size of the heads of pins, The most obvious way of accounting for this phenomenon is by supposing the water and frogs of some adjoining ponds to have been taken up by the wind in a sort of whirl or tornado. (Arcana of Science, 217, 1830)

## EXTRAORDINARY PHENOMENON AT DERBY
**Anonymous; *The Athenaeum*, 542, July 17, 1841.**

On Thursday week, during a heavy thunder-storm, the rain poured down in torrents mixed with half-melted ice, and, incredible as it may appear, hundreds of small fishes and frogs in great abundance descended with the torrents of rain. The fish were from half an inch to two inches long, and a few considerably larger, one weighing three ounces; some of the fish have very hard pointed spikes on their backs, and are commonly called suttle-backs. Many were picked up alive. The frogs were from the size of a horse-bean to that of a garden-bean; numbers of them came down alive, and jumped away as fast as they could, but the bulk were killed by the fall on the hard pavement. We have seen some alive to-day, which appear to enjoy themselves, in a glass with water and leaves in it. (The Athenaeum, 542, 1841)

## TOADS FALLING IN A SHOWER
**Winter, W.; *Zoologist*, 18:7146, 1859.**

I was out insect-catching by the side of the river Waveney, about a quarter-past 9 on Friday night, when a thunder-storm came on. I ran for shelter to the buildings at Aldeby Hall. The rain came down in torrents. Just before I was clear of the fens I observed some small toads on my arms, and several fell in my net, and on the ground and paths there were thousands. I am quite sure there were none in my net before I started, as I took a Leucania pudorina out of it. I believe they fell with the rain out of the clouds. Can you enlighten me on the subject? Two other persons have told me that they met with the same occurrence some distance from the spot in which I was situated. (Zoologist, 18:7146, 1859)

## A SHOWER OF FROGS
**Boyden, C. J.; *Meteorological Magazine*, 74:184-185, 1939.**

An account of a shower of frogs at Trowbridge, Wilts in the afternoon of June 16th appeared in The Times of the 17th. Mr. E. Ettles, superintendent of the municipal swimming pool stated that about 4.30 p.m. he was caught in a heavy shower of rain and, while hurrying to shelter, heard behind him a sound as of the falling of lumps of mud. Turning, he was amazed to see hundreds of tiny frogs falling on the concrete path around the bath. Later, many more were found to have fallen on the grass nearby.

A trough of low pressure was moving eastwards on that day, rain was reported from many places in the area and Torquay reported a line-squall in the afternoon. It is possible that the squall occurred also at Trowbridge, and that this was strong enough to have forced the frogs from the water, although Mr. Norman's theory given below may be the more probable explanation.

Showers of frogs are not uncommon and have been reported from time to time. Several subsequent letters have appeared in The Times citing instances at home and abroad.

Showers of fish have been reported from very early days, notably in India during stormy weather. Accounts and suggested explanations are given by Dr. S. L. Hora in the Journal of the Asiatic Society of Bengal 1933 and by J. R. Norman in the Natural History Magazine 1928. Mr. Norman suggests that in the case of frogs it is possible that numbers of tadpoles may undergo metamorphosis simultaneously, hide if the weather is at all dry and come out into the open with the first rain so suddenly that they appear to have fallen from the sky. (Meteorological Magazine, 74:184-185, 1939)

# SHELLFISH

## FALL OF MUSSELS
**Anonymous; *Nature*, 47:278, 1893.**

Das Wetter of December last contains an account of a heavy thunderstorm which occurred at Paderborn on August 9, 1892, in which a number of living pond mussels were mixed with rain. The observer who is in connection with the Berlin Meteorological Office sent a detailed account of the strange occurrence, and a specimen was forwarded to the Museum at Berlin, which stated that it was the Anodonta anatina (L.). A yellowish cloud attracted the attention of several people, both from its colour and the rapidity of its motion, when suddenly it burst, a torrential rain fell with a rattling sound, and immediately afterwards the pavement was found to be covered with hundreds of the mussels. Further details will be published in the reports of the Berlin Office, but the only possible explanation seems to be that the water of a river in the neighbourhood was drawn up by a passing tornado, and afterwards deposited its living burden at the place in question. (Nature, 47:278, 1893)

# TURTLES

## REMARKABLE HAIL
**Anonymous; *Nature*, 125:728, 1930.**

During a severe hailstorm at Vicksburg (U. S. A.) a remarkably large hailstone was found to have a solid nucleus, consisting of a piece of alabaster, from 1/2 to 3/4 inch in length. During the same storm, at Borina, 8 miles east of Vicksburg, a gopher turtle, 6 in. by 8 in., and entirely encased in ice, fell with the hail. These hailstorms occurred on the south side of a region of cold northerly winds; they were apparently accompanied by local whirls which carried heavy objects from the earth's surface up to the clouds, where they were encased by successive layers of snow and ice. (Nature, 125:728, 1930)

# INSECTS

## THE WONDERFUL SHOWER AT BATH

**Arnold, John, et al; *Symons's Monthly Meteorological Magazine,* 6:59, 1871.**

A most violent storm of rain, hail, and lightning visited Bath on Saturday night. The rain descended in torrents, causing the Avon to overflow its banks in the lower districts, especially at Salford, where whole tracts of land were laid under water. The storm was accompanied by a similar phenomenon to that of the previous Sunday; myriads of small annelidae enclosed in patches of gelatinous substance, falling with the rain and covering the ground. These have been microscopically examined, and show, under a powerful lens, animals with barrel-formed bodies, the motion of the viscera in which is perfectly visible, with locust-shaped heads bearing long antennae, and with pectoral and caudal fin like feet. They are each an inch and a half long, and may be seen by the curious at Mr. R. Butler's, The Derby and Midland Tavern, where scientific men, on inspecting them, pronounce them to be marine insects, probably caught up into the clouds by a waterspout in the Bristol Channel. (Symons's Monthly Meteorological Magazine, 6:59, 1871)

## BLACK WORMS FALL FROM SKY

**Anonymous; *Nature,* 6:356, 1872.**

A letter from Bucharest, given in the Levant Times, reports a curious atmospheric phenomenon which occurred there on the 25th of July, at a quarter past nine in the evening. During the day the heat was stifling, and the sky cloudless. Towards nine o'clock a small cloud appeared on the horizon, and a quarter of an hour afterwards rain began to fall, when, to the horror of everybody, it was found to consist of black worms of the size of an ordinary fly. All the streets were strewn with these curious animals. It is to be hoped that some were preserved, and will be examined by a competent naturalist. (Nature, 6:356, 1872)

# REPTILES

## A SNAKE RAIN

**Anonymous; *Scientific American,* 36:86, 1877.**

The Kentucky meat shower, which attracted so much attention recently, has now been supplemented by a rain of live snakes in Memphis, Tenn. Thousands of little reptiles, ranging from a foot to eighteen inches in length, were distributed all over the southern part of the city. They probably were carried aloft by a hurricane and wafted through the atmosphere for a long distance; but in what locality snakes exist in such abundance is yet a mystery. (Scientific American, 36:86, 1877)

# HAY, "ANGEL HAIR," AND OTHER ORGANIC SUBSTANCES

Tornadoes and whirlwinds operate like vacuum cleaners. They suck up loose materials, carry them for miles, and then deposit them on amazed countryfolk, This method of transportation is analogous to that of fish by waterspouts. Very likely there is much truth in both hypotheses. However, the remarks made in the section on fish falls about peculiar patterns of object selection and dispersal also apply here.

Hay is probably the most common material transported by whirlwinds. Indeed, few farmers have not seen small whirlwinds capriciously depositing their cut hay on telephone wires on hot summer afternoons. With such a well-recognized method of levitation and deposition, the emphasis in this section must be on stranger materials and mechanisms.

Ordinary wind is the transporting mechanism in the case of air-borne cobwebs, which most investigations prove are merely the products of bona fide spiders. The reason for introducing this seemingly tractable problem involves the question of "angel hair." Angel hair consists of wispy strands of material sometimes reported in the vicinities of UFO sightings. Scientific periodicals rarely deal with the angel hair situation, but a few curious "cobweb" phenomena have been recorded. Particularly puzzling is the apparent disappearance of some cobwebs when held in the hand. This is reminiscent of angel hair stories.

Finally, we have "manna from heaven." Like many tales of falling material, manna turns out to be a purely earthbound phenomenon. Humans, it seems, have a strong urge to assign the mysterious appearance of something to heavenly sources.

## HAY

### HAY TRANSPORTED BY WHIRLWIND
**Fraser, H. Malcom; *Meteorological Magazine,* 55:177,1920.**

A somewhat curious phenomenon occurred here about 5.30 p.m. on Saturday, July 31st. It rained hay over the whole of Stone, and to my own knowledge as far as Moddershall, over two miles away. A good many of the wisps which fell were larger than dinner plates, all were soaked through being in a cloud, and everyone agrees that the hay seemed to fall from a great height. The hay was accompanied by a whirling wind which knocked over an umpire on the cricket ground.

I have tried to give you certain facts, and avoid the many rumours which are flying round the town. (Meteorological Magazine, 55:177, 1920)

## REMARKABLE SHOWER OF HAY
**Moore, J. W.; *Symons's Monthly Meteorological Magazine*, 10:111-112, 1875.**

Sir,---You may care to insert in the Meteorological Magazine the enclosed cutting from the Dublin Daily Express, of Wednesday, July 28th. The fall of grass or hay was observed at Dean's Grange, Monkstown, more than a mile south of the place where Dr. Benson noticed it. My rainfall for July has been 2.751 inches on 18 days; max. in 24 hours was .506 on the 14th, very different from your great fall. ---Yours very truly, (J. W. Moore)

"Sir,---In connection with the great and wide-spread meteorological disturbances which have occurred of late in these and other countries, the following will not be devoid of interest to some of your readers who may not have seen the unusual phenomenon which I am about to describe.

"About half-past nine o'clock this morning, as I was standing at a window facing the east, in Monkstown, my attention was called by the Rev. T. Power to a number of dark flocculent bodies floating slowly down through the air from a great height, appearing as if falling from a very heavy, dark cloud, which hung over the house. On going to the hall door, a vast number of these bodies were discerned falling on all sides, near and far, to as great a distance as their size permitted them to be seen---over an area whose radius was probably between a quarter and half a mile. Presently, several of these flocks fell close to my feet. On picking up one, it was found to be a small portion of new hay, of which I enclose a sample. When it fell it was as wet as if a very heavy dew had been deposited on it. The duration of this phenomenon, from the time my attention was first called to it till it had entirely ceased, was about five minutes, but how long it may have been proceeding before this I cannot tell.

"The flocks of hay, from their loose, open structure, made a large appearance for their weight, but the average weight of the larger flocks was probably not more than one or two ounces, and, from that, all sizes were perceptible down to a single blade. There was no rain from the dark cloud overhead, though it had a particularly threatening appearance at the moment of observation. The air was very calm, with a gentle under-current from S. E. The clouds were moving in an upper-current from S. S. W.

"My friend, Dr. J. W. Moore, has kindly supplied me with the following particulars as to the state of the weather in Fitzwilliam Square, at 9 o'clock a. m., half an hour before the above observations were made:---

"'The barometer, corrected and reduced to 32 deg. at sea level, read 30.342 in. The air was tolerably warm and dry, the dry-bulb temperature being 60 deg., and the wet-bulb temperature 54.3 deg.; the dew-point 49.4 deg., and the percentage of humidity 69. A slight S. E. breeze was blowing, while rather heavy cirro-cumuli and cumuli were floating in an upper current from S. S. W.

"'The coincidence of a hot sun and two air currents probably caused the development of a whirlwind some distance to the south of Monkstown. By it the hay was raised into the air, to fall, as already described, over Monkstown and the adjoining district. (Symons's Monthly Meteorological Magazine, 10:111-112, 1875)

# LEAVES

## EXTRAORDINARY FLIGHT OF LEAVES
**Shaw, James; *Nature,* 42:637, 1890.**

The pastoral farm of Dalgonar is situated near the source of the Skarr Water, in the parish of Penpont, Dumfriesshire. The ridge of hills on the farm as per Ordnance Survey is 1580 feet above sea-level. There are only five trees on the farm---two ash and three larch. An extraordinary occurrence presented itself to the eyes of Mr. Wright, my informant, at the end of October 1889, on this farm, which has been narrated to me in a letter received from him, as follows:---

"I was struck by a strange appearance in the atmosphere, which I at first mistook for a flock of birds, but as I saw them falling to the earth my curiosity was quickened. Fixing my eyes on one of the larger of them, and running about 100 yards up the hill until directly underneath, I awaited its arrival, when I found it to be an oak leaf. Looking upwards the air was thick with them, and as they descended in an almost vertical direction, oscillating, and glittering in the sunshine, the spectacle was as beautiful as rare. The wind was from the north, blowing a very gentle breeze, and there were occasional showers of rain.

"On examination of the hills after the leaves had fallen, it was found that they covered a tract of about a mile wide and two miles long. The leaves were wholly those of the oak. No oak trees grow in clumps together nearer than eight miles. The aged shepherd, who has been on the farm since 1826, never witnessed a similar occurrence." (Nature, 42:637, 1890)

# SPIDER WEBS AND OR ANGEL HAIR

## A RAIN OF SPIDER WEBS
**Anonymous; *Scientific American,* 45:337, 1881.**

In the latter part of October the good people of Milwaukee (Wis.) and the neighboring towns were astronished by a general fall of spider webs. The webs seemed to come from "over the lake," and appeared to fall from a great height. The strands were from two feet to several rods in length. At Green Bay the fall was the same, coming from the direction of the bay, only the webs varied from sixty feet in length to mere specks, and were seen as far up in the air as the power of the eye could reach. At Vesburg and Fort Howard, Sheboygan, and Ozaukee, the fall was similarly observed, in some places being so thick as to annoy the eye. In all instances the webs were strong in texture and very white.

Curiously there is no mention, in any of the reports that we have seen, of the presence of spiders in this general shower of webs. It is to be hoped that some competent observer---that is, some one who has made a study of spiders and their habits---was at hand and will report more specifically the conditions of this interesting phenomenon.

Quite a number of notable gossamer showers have been reported in different parts of the world. White describes several in his history of Selborne. In one of them the fall continued nearly a whole day, the webs coming from such a height that from the top of the highest hill near by they were seen descending from a region still above the range of distinct vision.

Darwin describes a similar shower observed by him from the deck of the Beagle, off the mouth of La Plata River, when the vessel was sixty miles from land. He was probably the first to notice that each web of the gossamer carried a Lilliputian aeronaut. He watched the spiders on their arrival and saw many of them put forth a new web and float away. (Scientific American, 45:337, 1881)

## COBWEBS OR FLYING SAUCERS?
**Bishop, P. R.; *Weather*, 4:121-122, 1949.**

By a curious coincidence, I had just finished reading the letter to the Editors in the September issue of Weather, entitled "Cobwebs in the Rigging", when a colleague of mine started to describe a phenomenon he had observed the day before and which he had taken to be an actual manifestation of the illusive "Flying Saucers" which raised so much excitement some time ago. After talking the matter over we were both satisfied that what he had seen was a particularly fine example of a large mass of "cobwebs" described in the very interesting letter from Messrs. Ovey and Browning.

As his description may be of interest to your readers, coming so shortly after the above letter, I pass it on to you for publication.

"Sunday, September 26, 1948. Port Hope, Ontario. This day was warm and the sky cloudless. We had had dinner in the garden and I was lying on my back on the lawn, my head just in the shade of the house, when I was startled to see an object resembling a star moving rapidly across the sky. The time was 2 o'clock Eastern Standard Time.

At first it was easy to imagine that recent reports of 'Flying Saucers' had not been exaggerated.

More of these objects came sailing into view over the ridge of house, only to disappear when nearly overhead. With field glasses I was able to see that each was approximately spherical, the centre being rather brighter than the edges. The glasses also showed quite a number at such heights that they were invisible to the naked eye.

With only a gull flying in the sky for comparison, I should estimate the elevation of the lower objects to be about 300 ft. and the higher ones 2,000 ft; the size was about one foot in diameter and the speed about 50 m.p.h., in a direction SW to NE.

Also visible every now and then were long threads, apparently from spiders. Some of these were seen to reflect the light over a length of three or four yards, but any one piece may of course have been longer. Each was more or less horizontal, moving at right angles to its length. In one case an elongated tangled mass of these gave the appearance of a frayed silken cord. These threads appeared only in the lower levels.

It is reasonably certain that these objects were balls of spiders' threads, possibly with thistledown entangled in them, but the way in which they caught the rays of the sun and shone so brightly was very striking. P. L. Lewis"

This may really be the cause of the "Flying Saucer" scare. Port Hope is some 60 miles east of Toronto and a SW wind would waft these webs from the Middle West where the saucers were so often reported. The play of light would also explain why aeroplanes sent up to investigate could never find them. No one else seems to have seen them on this occasion, but perhaps Mr. Lewis was the only one to be taking horizontal, post-prandial repose at that time. (Weather, 4:121-122, 1949)

## SPIDERS' FILAMENTS
**Pape, R. H.; *Marine Observer*, 33:187-188, 1963.**

m.v. Roxburgh Castle. Captain R. H. Pape.

The following is the text of a letter received from the Master dated 10th October 1962:

At 2000 GMT while the Roxburgh Castle was moored to her berth (Section 24) in Montreal, I was walking round outside my accommodation and noticed fine white filaments of unknown kind hanging around stanchions and topping lift wires of derricks.

Calling the attention of the Chief Officer, I pulled one of these strands from a stanchion and found it to be quite tough and resilient. I stretched it but it would not break easily (as, for instance, a cobweb would have done) and after keeping it in my hand for 3 or 4 minutes it disappeared completely; in other words it just vanished into nothing.

Looking up we could see small cocoons of the material floating down from the sky but as far as we could ascertain there was nothing either above or at street level to account for this extraordinary occurrence.

Unfortunately I could not manage to preserve samples of the filaments as the disappearance took place so quickly.

I would be very glad to know what explanation, if any, can be given to account for the phenomenon.

Note. Mr. D. J. Clark, of the Natural History Museum, comments as follows:

"Spiders are, I think, responsible for the phenomena you describe. The majority of these particular spiders belong to the family Linyphiidae, and mature in the autumn. In the autumn, on fine, warm and sunny days, especially with a fairly heavy early morning dew, the spiders begin to disperse and migrate in order to colonise new areas where the food supply is greater. The method they use is known as 'ballooning'. As the sun dries off the dew, upward air currents are created. The spider runs to the top of a plant, fence etc. and lifting the tip of its abdomen emits a globule of liquid silk. This silk is drawn out in a thread by the air currents and hardens as a result of this drawing out, not simply by contact with the air. When the thread is long enough to support the spider, it lets go of its support and flies away. The spiders sometimes are carried many miles. Eventually, they come down to earth and on landing they free the 'parachute'. This again floats away and becomes entangled with other threads, sometimes quite thick bands are thus formed, and when this again settles down it is very conspicuous. The single thread is very fine and difficult to see unless the light is reflected from it, and when entangled together with other threads it is easy to see and quite tough and resilient.

I cannot explain the disappearance of these strands when held in the hand. It may be that the threads of the strand you describe were not so entangled and when handled broke up into individual threads thus becoming very inconspicuous. Spider silk cannot melt because heat does not affect it, it is on the whole less soluble than true silk." (Marine Observer, 33:187-188, 1963)

## SHOWER OF GOSSAMER AT SELBOURNE
**Anonymous; *Nature,* 126:457, 1930.**

Gilbert White records ("Natural History of Selborne") that before daybreak "I found the stubbles and clover grounds matted all over with a thick coat of cobweb. . . . When the dogs attempted to hunt, their eyes were so blinded and hoodwinked that they could not proceed, but were obliged to lie down and scrape the encumbrance from their faces with their forefeet. . . . About nine. an appearance very unusual began to demand our attention---a shower of cobwebs falling from very elevated regions, and continuing, without any interruption till the close of the day. These webs were not single filmy threads, floating in the air in all directions, but perfect flakes, or rags: some near an inch broad, and five or six long, which fell with the degree of velocity, that they were considerably heavier than the atmosphere. On every side, as the observer turned his eyes, he might behold a continual succession of fresh flakes falling into his sight, and twinkling like stars, as they turned their sides towards the sun. How far this wonderful shower extended, it would be difficult to say; but we know that it reached Bradley, Selborne and Alresford, three places which lie in a sort of triangle the shortest of whose sides is about eight miles in extent." The gossamer descended even on the highest part of the downs. (Nature, 126:457, 1930)

# MANNA

## FALL OF MANNA
**Anonymous; *Nature,* 43:255, 1891.**

The director of the central dispensary at Bagdad has sent to La Nature a specimen of an edible substance which fell during an abundant shower in the neighbourhood of Merdin and Diarbekir (Turkey in Asia) in August 1890. The rain which accompanied the substance fell over a surface of about ten kilometres in circumference. The inhabitants collected the "manna," and made it into bread, which is said to have been very good and to have been easily digested. The specimen sent to La Nature is composed of small spherules; yellowish on the outside, it is white within. Botanists who have examined it say that it belongs to the family of lichens known as Lecanora esculenta. According to Decaisne, this lichen, which has been found in Algeria, is most frequently met with on the most arid mountains of Tartary, where it lies among pebbles from which it can be distinguished only by experienced observers. It is also found in the desert of the Kirghizes. The traveller Parrot brought to Europe specimens of a quantity which had fallen in several districts of Persia at the beginning of 1828. He was assured that the ground was covered with the substance to the height of two decimetres, that animals ate it eagerly, and that it was collected by the people. In such cases it is supposed to have been caught up by a waterspout, and carried along by the wind. (Nature, 43:255, 1891)

## THE ORIGIN OF MANNA
**Timothy, B.; *Nature,* 55:440, 1897.**

The note in Nature, p. 349, concerning the "manna," reminds me of a passage in Daniele Bartoli's "Asia."

Speaking of the island of Ormuz---which is described as one of the places in the world worst supplied in even commonest necessities of life, and scarcely having any water---the historian tells us that "not even thorns and briars could grow on its barren soil; no animals or birds (sic) are seen there all the year round, but every morning a dew falls which congeals into grains, has a very sweet taste, and is called 'manna.'"

Now, tamarisks affect sandy soils or brackish shores; and as T. mannifera grows in Arabia, it may be that the exudations from the plants were blown from Oman, on the eastern shore of Arabia, across the Persian Gulf; or, perhaps, from the nearer coast of Persia. This would seem to confirm the belief that manna is the product of the tamarisk, and not of a lichen. (Nature, 55:440, 1897)

## THE BIBLICAL MANNA
**Anonymous; *Nature,* 124:1003-1004, 1929.**

There has always existed among scientific workers a wide divergence of opinion as to the true nature and origin of the manna, believed to have fallen from heaven to provide food for the Israelites in the Sinai desert during the Exodus from Egypt. Some authors considered the manna to be a desert lichen, Lecanora esculenta Nees, while others connected it with desert shrubs of the genus Tamarix and considered it to be either a physiological secretion of the plant, or its sap flowing from the wounds caused by insects. In order to solve this problem, the Hebrew University in Jerusalem organised in 1927 a small expedition to the Sinai Peninsula, and the leaders of that expedition, Dr. F. S. Bodenheimer and Dr. O. Theodor, have just published a very interesting account of their investigations.

The expedition visited some classical localities where manna was recorded. In the course of investigations, it was established beyond doubt that the appearance of manna is a phenomenon well known in other countries under the name of 'honey-dew', which is a sweet excretion of plant-lice (Aphidae) and scale-insects (Coccidae). Two scale insects mainly responsible for the production of manna were found, namely, Trabutina mannipara, Ehrenb., occurring in the lowlands, and Najacoccus serpentinus var. minor Green, which replaces the former in the mountains. Two other Hemipterous insects, Euscelis decoratus Haupt and Opsius jucundus Leth., also produce manna, but to a lesser extent. All these insects live on Tamarix nilotica var. mannifera Ehrenb.; no manna was observed on other species of Tamarix, a fact probably due to some physiological peculiarities of the former. The authors observed the actual excretion by the insects of drops of clear sweet fluid, and proved by experiments that the fluid is ingested by the insects from the vessels of the phloem. When in an experiment a twig bearing the insects was placed in water, and the bark was cut below the insects, the production of manna continued in a normal manner, but it stopped as soon as the flow of carbohydrate solution from the leaves was interrupted by cutting off the bark above the insects. The dry desert climate of Sinai causes the syrup-like fluid excretion to crystallise, and the whitish grains thus produced, which cover the branches or fall to the ground underneath them, constitute the true manna of the Bible.

A chemical analysis of the manna demonstrated the presence of cane sugar, glucose, fructose, and saccharose; pectines were also found, but there was no trace of proteins.

Detailed descriptions of the manna insects are given in the report, which includes very good photographs of various stages of the production of the manna. Notes on the course of the expedition and on the fauna of the Peninsula of Sinai in general provide very interesting reading on that still practically unexplored country. (Nature, 124: 1003-1004, 1929)

# MISCELLANEOUS OBJECTS AND MATERIAL

## LUMINOUS PHENOMENON
**Wilson, J. G.; *Marine Observer*, 24:205, 1954.**

S.S. Caxton. Captain J. G. Wilson. London to Newfoundland. Observer, Mr. R. L. Goodfellow, 3rd Officer.

11th November, 1953, 2330 G.M.T. This evening when the lookouts were relieving each other on the forecastle head, they noticed that an apparent phosphorescent substance, giving off a greenish glow, adhered to those parts of their clothing which were exposed to the wind, also to the apron of the forecastle head, the jackstaff and halyards. It could be readily brushed off or transferred to other objects. Wind N'ly force 3. Weather slightly cloudy, good visibility.

Position of ship: 52° 54'N, 46° 42'W.

Note. The cause of this phenomenon cannot be assigned with certainty. In the January, 1954, number of this journal an observation of St. Elmo's fire, made by the same observer in the same ship, was published. On that occasion it was clearly stated that the minute balls of light on the cap of the lookout man could not be brushed off and this is what would be expected with St. Elmo's fire. In the present instance the luminosity was clearly stated to be transferable, which seems to rule out an electrical phenomenon. An alternative explanation is that spray containing minute phosphorescent organisms was blown on to the foredeck. The possibility of this cannot be fully judged as the logbook contains no information about the state of the sea at the time, but the recorded wind of force 3 does not appear to be strong enough. Also, while phosphorescence is not unknown in November in the North Atlantic, it is considerably less likely than during the summer months. (Marine Observer, 24: 205, 1954)

## A SHOWER OF HAZEL NUTS
**Pim, Arthur; *Symons's Monthly Meteorological Magazine,* 2:59, 1867.**

Sir, ---I enclose you two extracts from one of our Dublin papers relative to some berries, which are reported to have fallen in large quantities in some parts of Dublin on the night of Thursday, 9th May. I have been given two of these berries; they are in the form of a very small orange, about half an inch in diameter, black in colour, and, when cut across, seem as if made of some hard dark brown wood. They also possess a slight aromatic odour.

Various speculations have been given forth as to their origin, but none of them seem to be worth much. If you think the extracts herein, worthy of a place in your Magazine you can insert them. ---Yours very truly (Arthur Pim)

Sir, ---I have daily expecting to see some notice of the strange phenomenon which took place during the tremendous rain-fall of Thursday night. None have appeared in any of the journals, I hope, through your columns, the public may learn to what cause we are to attribute the shower of aromatic smelling berries which fell over Dublin (and, possibly, other parts) on Thursday night.

Both on the north and south sides of the river these berries fell in great quantities and with great force, some being larger than the ordinary Spanish nut.

Numbers of these strange visitors were picked up in Capel-street, in Dame-street, and Bishop-street, and I am informed that so violent was the force with which they descended that even the police, protected by unusually strong head covering, were obliged to seek shelter from the aerial fusilade!---Yours truly, (Resticus Expectans)

Inquirer, who has sent us some "small balls," which he says "fell in large quantities on Thursday night," locality not specified, is informed that they are simply hazel nuts, preserved in a bog for centuries. How they came to descend on him we cannot say. (Symons's Monthly Meteorological Magazine, 2:59, 1867)

## "PURPLE PATCHES"
**Pedder, A.; *Nature,* 55:33, 1896.**

I should be very glad if I could obtain information as to the cause and nature of certain "purple patches" which I have noticed from time to time for many years past, but have been unable to get explained. The patches in question occur during, or immediately after, rain, on the pavement or roadway; dashes of vivid purple, or rather violet, varying in size from small splashes or drops to patches as large as the palm of one's hand, but most commonly they are about the size of a shilling. When quite fresh, sometimes a little clot is observable in the centre of the splash. Sometimes I find one patch completely isolated, sometimes two or three in close proximity; sometimes, again, numerous little drops scattered over a certain space; once I counted twenty or thirty tiny dashes in about ten yards of pavement. When quite wet the violet colour can be rubbed up with a handkerchief or paper, which it stains as with "aniline purple" dye, as it does the pavement, and when once dry it is quite inerasible, and lasts till it is worn away by exposure, or the feet of passersby. I observe it to occur chiefly during warm rain after a dry or cold spell; never during dry weather, whether in summer or winter. During the past hot summer there was none to be found, but directly the weather changed in July, I saw it in various localities. This was also the case in the long cold winter of 1895, when on the breaking up of the frost there was also a complete absence during the following summer, till the drought gave, and then again I found this appearance recur. I naturally observe it most in Bath, where I live; but it is not at all confined to one place or situation. I have found good specimens at such widely different places as the doorway of a hotel at Oban; the Castle Hill, Edinburgh; railway platform at Morecambe; doorstep at Windermere; in streets and roads at Cambridge, Bude, Penzance, St. Ives, Clevedon; once in a London street (Pall Mall East), and once some was found in a cold water bath.

I have from time to time made inquiries from various people who I though would know, but have not been fortunate enough to meet any scientific person who has observed it. But one learned professor to whom I described the "patches," suggested whether "purple bacteria" would prove a solution to the mystery, and recommended me to inquire through the medium of your columns. I should be much obliged if some one would enlighten me, or mention some authority to whom I could refer. (Nature, 55:33, 1896)

## A MARVELOUS "RAINFALL" OF SEEDS
**Wallace, R. Hedger; *Notes and Queries,* 8:12:228, 1897.**

Some days ago the province of Macerata, in Italy, was the scene of an extraordinary phenomenon. Half an hour before sunset an immense number of small blood-coloured clouds covered the sky. About an hour later a cyclone storm burst, and immediately the air became filled with myriads of small seeds. The seeds fell over town and country, covering the ground to a depth of about half an inch. The next day the whole of the scientists of Macerata were abroad in order to find some explanation. Prof. Cardinali, a celebrated Italian naturalist, stated that the seeds were of the genus Cercis, commonly called Judas Tree, and that they belonged to an order of Leguminosae found only in Central Africa or the Antilles. It was found, upon examination, that a great number of the seeds were actually in the first stage of germination. (Notes and Queries, 8:12:228, 1897)

## EXTRAORDINARY PHENOMENON IN FIFESHIRE
**Anonymous; *Symons's Monthly Meteorological Magazine,* 14:136, 1879.**

A discovery, startling and unusual, was made on Sunday on the Lomond Hills, near Falkland, which has since given rise to considerable speculation. The hills were found in several places to be covered with seaweed, or some substance as nearly resembling it as possible, and it was also seen hanging from the trees and shrubberies in the district. In some places the weed lay in a pretty thick coating, so that quantities of it could be collected from the grass. A heavy hailstorm, accompanied by T and L, passed over the district on Saturday evening (Aug. 30th), and it is thought that at a late hour a waterspout had burst over the hills. Many people have collected pieces of the weed for preservation. (Symons's Monthly Meteorological Magazine, 14:136, 1879)

## SHOWER OF GRAIN
**Anonymous; *American Journal of Science,* 1:41:40, 1841.**

Col. Sykes communicated the contents of a letter from India, from Capt. Aston, on the subject of a recent singular shower of grain. He stated that 60 or 70 years ago, a fall of fish had occurred during a storm in the Madras Presidency. This fact is recorded by Major Harriott, in his "Struggles through Life," as having taken place while the troops were on the line of march, and some of the fish falling upon the hats of the European troops, they were collected and made into a curry for the general. This fact was probably for fifty years regarded as a traveller's tale, but within the last ten years, so many other instances have been witnessed, and publicly attested, that the story is no longer doubted. The shower of grain above mentioned, took place March 24, 1840, at Rajket in Kattywar, during one of those thunder storms to which that month is subject, and it was found that the grain had not only fallen upon the town, but also upon a considerable extent of country round the town. Capt. A. collected a quantity of the seed, and transmitted it to Col. Sykes. The natives flocked to Capt. A. to ask for his opinion of this phenomenon: for not only did the raining of grain upon them from heaven, excite terror, but the omen was aggravated by the fact that the seed was not one of the cultivated grains of the country, but was entirely unknown to them. The genus and species was not immediately recognized by some botanists to whom it was shown, but it was thought to be either a Spartium, or a Vicia. A similar force to that which elevates fish into the air, no doubt operated on this occasion, and this new fact corroborates the phenomena, the effects of which had been previously witnessed. (American Journal of Science, 1:41:40, 1841)

# GELATINOUS MASSES OR "PWDRE SER"

"Star jelly" is still another name for pwdre ser, but nothing can be as descriptive as "rot of the stars." The basic phenomenon has been the same since written history began. A meteor is seen to land nearby, investigation reveals a jelly-like mass in the approximate location, thus we have "star jelly" or, more scientifically, a "gelatinous meteor." The incongruity is that common meteorites are hard lumps or rock and metal well-suited to surviving the fiery plunge through the earth's atmosphere from outer space. Gelatine would be consumed in a few seconds. Thus, the co-location of meteor impact and gelatinous mass must be coincidental.

Actually, several of the accounts that follow indicate that there is really little doubt that the observer did perceive gelatinous masses to fall and that there was also a luminous display of some kind. In addition, the testimony of legends and folk lore is overwhelmingly in favor of the reality of gelatinous meteors. Obviously, this does not mean that all gelatinous masses found in meadow and field fell from the sky, but neither can we deny the possibility of gelatinous meteors just because terrestrial nature does produce lumps of gelatin.

Scientists who have tried to explain pwdre ser always opt for the terrestrial interpretation; that is, some form of plant or animal life or possibly some half-digested matter disgorged by an animal. It seems that a common characteristic of pwdre ser is its smell and general rottenness. But if rotten fish can fall from the sky, why not offensive gelatin? Pwdre ser often has one other interesting feature: it seems to evaporate away rapidly, removing all evidence of the unusual phenomenon. Genuine animal matter is not so fleeting and emphemeral.

## PWDRE SER
**Hughes, T. McKenny; *Nature,* 83:492-494, 1910.**

In my boyhood I often lived on the coast of Pembrokeshire. Wandering about with my gun I was familiar with most natural objects which occurred there. One, however, which I often came across there, and have seen elsewhere since, greatly roused my curiosity, but I have not yet met with a satisfactory explanation of it.

On the short, close grass of the hilly ground, I frequently saw a mass of white, translucent jelly lying on the turf, as if it had been dropped there. These masses were about as large as a man's fist. It was very like a mass of frog's spawn without the eggs in it. I thought it might have been the gelatinous portion of the food disgorged by the great fish-eating birds, of which there were plenty about, as kingfishers eject pellets made up of the bones of the fish they eat, or that possibly there might be some pathological explanation connecting it with the sheep, large flocks of which grazed the short herbage. But the shepherds and owners of the sheep would have known if such an explanation were admissible. They called it 'pwdre ser," the rot of the stars.

Years afterwards I was in Westmorland, on the Geological Survey, and again not unfrequently saw the "pwdre ser." But I now got an addition to my story. Isaac Hindson, of Kirkby Lonsdale, a man whose scientific knowledge and genial personality made him a welcome companion to those who had to carry on geological research in his district, tole me that he had once seen a luminous body fall, and, on going up to the place, found only a mass of white jelly. He did not say that it was luminous. I have never seen it luminous, but that may be because when it was light enough to see the lump of jelly, it would probably be too light to detect luminosity in it.

Then, in my novel reading, I found that the same thing was known in Scotland, and the same origin assigned to it, for Walter Scott, in "The Talisman," puts these words in the mouth of the hermit:---"Seek a fallen star and thou shalt only light on some foul jelly, which in shooting through the horizon, has assumed for a moment an appearance of splendour." I think that I remember seeing it used elsewhere as an illustration of disappointed hopes, which were "as when a man seeing a meteor fall, runs up and finds but a mass of putrid jelly," but I have lost the reference to this passage.

Thus it appeared that in Wales, in the Lake District, and in Scotland, there existed a belief that something which fell from the sky as a luminous body lay on the ground as a lump of white jelly.

I asked Huxley what it could be, and he said that the only thing like it that he knew was a nostoc. I turned to Sachs for the description of a nostoc, and found that it "consists, when mature, of a large number of moniliform threads interwoven among one another and imbedded in a glutinous jelly, and thus united into colonies of a specifically defined form.... The gelatinous envelope of the new filament is developed, and the originally microscopic substance attains or even exceeds the size of a walnut by continuous increase of the jelly and divisions of the cells."

All the nostocs, however, that I have had pointed out to me have been of a green or purplish or brown-green colour, whereas the "pwdre ser" was always white, translucent in the upper part, and transparent in the lower part, which appeared to occur among the roots of the grass, as if it grew there. Moreover, the mass was much larger than a walnut, in fact, would generally about fill a half-pin mug.

The only reference I can find from which it would appear that the writer was describing a nostoc is the passage in Dryden and Lee (1678).

"The shooting stars end all in purple jellies." In the following note, appended to this passage, it is clear that the writer thought that the jelly-like matter found where shooting-stars had seemed to fall, was white.

Note.---"It is a common idea that falling stars, as they are called, are converted into a sort of jelly. Among the rest, I had often the opportunity to see the seeming shooting of the stars from place to place, and sometimes they appeared as if falling to the ground, where I once or twice found a white jelly-like matter among the grass, which I imagined to be distilled from them; and thence foolishly conjectured that the stars themselves must certainly consist of a like substance."

Poets and divines carry the record of this curious belief far back into the seventeenth century.

Suckling (1541) says:---

"As he whose quicker eye doth trace
A false star shot to a mark't place
Do's run apace,
And, thinking it to catch,
A jelly up do snatch."

Jeremy Taylor (1649):---

"It is weaknesse of the organ that makes us hold our hand between the sun and us, and yet stand staring upon a meteor or an inflamed gelly."

Henry More (1656):---

"That the Starres eat....that those falling Starres, as some call them, which are found on the earth in the form of a trembling gelly, are their excrement."

Dryden (1679):---

"When I had taken up what I supposed a fallen star I found I had been cozened with a jelly."

William Somerville (1740):---

"Swift as the Shooting Star that gilds the night
With rapid transient Blaze, she runs, she flies;
Sudden she stops nor longer can endure
The painful course, but drooping sinks away,
And like that falling Meteor, there she lyes
A jelly cold on earth."

Several old writers, however, while agreeing as to the mode of occurrence of the "pwdre ser," and recognising the widespread belief that it was something which fell from the sky and was somehow connected with falling stars, have tried to find some more commonplace and probable explanation of the phenomenon, and most of them refer it to the stuff disgorged by birds that had fed on frogs or worms.

Merrett (1667)*, for instance, in his work on meteors and wandering lights, says:---

"Sequuntur Meteora, ignita, viz. Ignis fatuus, the Walking fire, or Jack of the Lantern, Castor and Pollux, Helena, Ignis lambens. Draco, Stella cadens: Est sub-stantia quaedam alba et glutinosa plurimis in locis conspicua quam nostrates 'Star-faln' nuncupant, creduntq; multi originem suam debere stellae cadenti hujusq; materiam esse. Sed Regiae Societati palam ostendi solummodo oriri ex intestinis ranarum a corvis in unum locum congestis, quod aliis etiam ejusdem societatis viri praestantissimi postea confirmarunt."

The Rev. John Morton, of Emmanuel College (1712), is, however, the only one who, so far as I can ascertain, ever tried any experiments with the view of finding out what it really was. He set some of it on the fire, and when he had driven off all the watery part, there was left a film like isinglass, and something like the skins and vessels of animal bodies. He records many observations as to its time and mode of occurrence; for instance, he says that "in 1699-1700 there was no star-gelly found about Oxenden till a wet week in the end of February, when the shepherds brought me above thirty several lumps." This and other observations suggest that it is a growth dependent upon the weather, &c. On the other hand, he says that he saw a wounded gull disgorge a heap of half-digested earth-worms much resembling star-jelly, and that Sir William Craven saw a bittern do the same in similar circumstances.

The Hon. Robert Boyle, 1744, explaining how clammy and viscous bodies, such as white of egg, are reduced to a thin and fluid substance, says:---

"And I remember, I have seen a good quantity of that jelly, that is sometimes found on the ground, and by the vulgar called a star-shoot, as if it remained upon the extinction of a falling star, which being brought to an eminent physician of my acquaintance, he lightly digested it in a well-stopt glass for a long time, and by that alone resolved it into a permanent liquor, which he extols as a specifick to be outwardly applied against Wens."

Pennant seems to have supposed that its origin was that suggested by Morton, for in his description of the winter mew he says:---"This kind (i.e. the Coddy Moddy or Winter Mew) frequents, during winter, the moist meadows in the inland parts of England remote from the Sea. The gelatinous substance, known by the name of star shot, or star gelly, owes its origin to this bird or some of the kind, being nothing but the half digested remains of earth-worms, on which these birds feed and often discharge from their stomachs."

I have found it commonly near the sea, but have never seen any trace of earth-worms or other similar food in it.

Here, then, we have a well-known substance which may be of different origin in different cases, respecting the general appearance of which, however, almost all

---

*Merrett, "Christophorus, Pinax: Rerum naturalium Britannicarum, continens vegetabilia, Animalia, e. Fossilia in hac Insula reperta inchoatus," ed. 2der Lond., 1667, p. 219.

accounts agree. The variety of names under which it is known point to its common and widespread occurrence, e.g. pwdre ser, star-slough, star shoot, star shot, star-gelly or jelly, star-fall'n.

We have in every name, and in every notice in literature, a recognition of the universal belief that it has something to do with meteors, yet there does not appear to be any evidence that anybody ever saw any luminosity in the jelly. Nor has anybody seen it disgorged by birds, except in the case of those two wounded birds where some half-digested gelatinous mass was thrown up. Nor has anyone watched its growth like nostoc from the ground.

In 1908 I was with my wife and one of my boys on Ingleborough, where we found the "pwdre ser" lying on the short grass, close to the stream a little way above Gaping Ghyl Hole. For the first time I felt grateful to the inconsiderate tourist who left broken bottles about, for I was able to pack the jelly in the bottom of one, tie a cover on, and carry it down from the fell. I sent it, with the sod on which it appeared to have grown, to my colleague, Mr. E. A. Newell Arber, with a brief sketch of my story and the reason why I thought it of interest. Mr. Arber reported that it was no nostoc, and said that he had sent it over to Mr. Brookes, in the Botany School, who reported that it was a mass of bacteria.

That is the end of my story, but I confess I am not satisfied. The jelly seemed to me to grow out from among the roots of the grass, and the part still tangled in the grass was not only translucent but quite transparent.

What is it, and what is the cause of its having a meteoric origin assigned to it? Has anyone ever seen it luminous?

Should anyone come upon it I should be very grateful if they would send it, and the sod on which it is found, to the Botany School at Cambridge, with a label indicating what the parcel contains, so that it may be attended to before decay has perhaps obscured important features. (Nature, 83:492-494, 1910)

## PWDRE SER
### Adams, Ellen M., and Schlesinger, Frank; *Nature*, 84:105-106, 1910.

The following letter, which I received last winter, may possibly throw some light on the questions raised by Prof. Hughes in his paper on "Pwdre Ser" in Nature of June 23:---

"Allegheny, December 4, 1909---Dear Professor Schlesinger,---

"Referring to the falling meteor of which my husband made mention at your lecture last evening, the facts are as follows. One evening some years since my father, Mr. Joel Powers, while walking on Lawrence St., Lowell, Massachusetts, saw a brilliant shooting star or meteor flash downward through the atmosphere, striking the earth quite near him. He found it upon investigation to be a jelly-like mass, and almost intolerably offensive in smell. I have often heard my father allude to this event, which greatly interested him, he being a close observer and an extensive reader. Respectfully yours, Ellen M. Adams."

While I am of the opinion that the mass found by Mr. Powers had no connection with the meteor that he saw, it may be well to put this piece of evidence on record in view of Prof. Hughes's paper. (Nature, 84:105-106, 1910)

## ACCOUNT OF A GELATINOUS METEOR
**Graves, Rufus; *American Journal of Science,* 1:2:335-337, 1819.**

On the evening of the thirteenth day of August, 1819, between the hours of eight and nine o'clock, was seen in the atmosphere, at Amherst, Massachusetts, a falling meteor or fire ball, of the size, as represented by an intelligent spectator, of a man's hat, or a large blown bladder, of a brilliant white light resembling burnished silver.

The position of this spectator being in a direct line of the street where the luminous ball appeared, and at the distance of not more than five hundred yards, with the sight bounded by the buildings, there could be no deception relative to the direction that it took. Its altitude, at its first discovery, was two or three times the height of the houses; it fell slowly in a perpendicular direction, emitting great light, till it appeared to strike the earth in front of the buildings, and was instantly extinguished, with a heavy explosion. At the same instant, as appeared from the report, and from the ringing of the church bell, an unusually white light was seen a few minutes afterwards, by two ladies in a chamber of Mr. Erastus Dewey. While they were sitting with two candles burning in the room, a bright luminous circular spot suddenly appeared on the side wall of the chamber near the upper floor in front of them, of the size of a two feet stand-table leaf. This spectrum descended slowly with a tremulous motion nearly to the lower floor and disappeared.

In critically examining the chamber where the foregoing phenomenon was observed, it appeared that the light must have entered through the east front window in a diagonal direction, and impinged on the north wall of the chamber back of the ladies, and thence reflected to the south wall in front of them, forming the circular spectrum, with the corresponding tremulous motion of the meteor, and descending with it in the same direction, according to the fixed laws of incidence and reflection.

Early on the ensuing morning, was discovered in the door yard of the above mentioned Erastus Dewey, at about twenty feet from the front of the house, a substance unlike anything before observed by anyone who saw it. The situation in which it was found, being exactly in the direction in which the luminous body was first seen, and in the only position to have thrown its light into the chamber, (as before remarked,) leaves no reasonable doubt that the substance found was the residuum of the meteoric body.

This substance when first seen by the writer was entire, no part of it having been removed. It was in a circular form, resembling a sauce or sallad dish bottom upwards, about eight inches in diameter, and something more than one in thickness, of a bright buff colour, with a fine nap upon it similar to that on milled cloth, which seemed to defend it from the action of the air. On removing the villous coat, a buff coloured pulpy substance of the consistence of good soft soap, of an offensive, suffocating smell appeared; and on a near approach to it, or when immediately over it, the smell became almost insupportable, producing nausea and dizziness. A few minutes exposure to the atmosphere changed the buff into a livid colour resembling venous blood. It was observed to attract moisture very readily from the air. A half-pint tumbler was nearly half filled with the substance. It soon began to liquify and form a mucilaginous substance of the consistence, colour, and feeling of starch when prepared for domestic use. The tumbler was then set in a safe place, where it remained undisturbed for two or three days; and when examined afterwards, the substance was found to have all evaporated, except a small dark coloured residuum, adhering to the bottom and sides of the glass, which, when rubbed between the fingers, produced a fine ash-coloured powder without taste or smell; the whole of which might have been included in a lady's thimble.

The place where the substance was first found was examined, and nothing was to be seen but a thin membranous substance adhering to the ground similar to that found on the glass.

This singular substance was submitted to the action of acids. With the muriatic and nitric acids, both concentrated and diluted, no chemical action was observed, and the matter remained unchanged. With the concentrated sulphuric acid a violent effervescence ensued, a gaseous body was evolved, and nearly the whole substance dissolved. There being no chemical apparatus at hand, the evolving gass was not preserved, or its properties examined. (American Journal of Science, 1:2:335-337, 1819)

## LUMINOUS BODY DESCENDS
**Powell, Baden; *Reports of the British Association*, 94, 1855.**

"1844, Oct. 8th, near Coblentz, a German gentleman (a friend of Mr. Greg's), accompanied by another person, late in the evening, after dark, walking in a dry ploughed field, saw a luminous body descend straight down close to them (not 20 yards off), and heard it distinctly strike the ground with a noise; they marked the spot, and returning early the next morning as nearly as possible where it seemed to fall, they found a gelatinous mass of a greyish colour so viscid as 'to tremble all over' when poked with a stick. It had no appearance of being organic. They, however, took no further care to preserve it." (Reports of the British Association, 94, 1855)

## ACCOUNT OF AN EXTRAORDINARY METEORIC PHENOMENON
**Acharius, E.; *North American Review*, 3:320-322, 1816.**

Having received intelligence from several persons, although differently and variously related, of an extraordinary and probably hitherto unseen Phenomenon, which was observed in the air last month, at and about the village of Biskopsberga, near the Town of Skeninge, which accounts not being testified to me by eye-witnesses, nor agreeing as to facts and circumstances, I resolved to proceed to the place myself for obtaining an exact and detailed account thereof; and as I find so many singular circumstances attending this Meteor, which deserve to be known, I have thought it my duty to communicate them to the Royal Academy, and thereby to save so remarkable an occurrence from oblivion, which, although difficult to explain, still affords an additional proof of those many wonderful operations of Nature, which take place in our Atmosphere.

On the 16th of last May, being a very warm day, and during a gale of wind from south-west, and a cloudless sky, at about 4 o'clock, p. m. the sun became dim, and lost his brightness to that degree, that he could be looked at without inconvenience to the naked eye, being of a dark-red, or almost bright colour, without brilliancy. At the same time there appeared at the western horizon, from where the wind blew, to arise gradually, and in quick succession, a great number of balls, or spherical bodies, to the naked eye of a size of the crown of a hat, and of a dark brown colour. The nearer these bodies, which occupied a considerable though irregular breadth of the visible heaven, approached towards the sun, the darker they appeared, and in

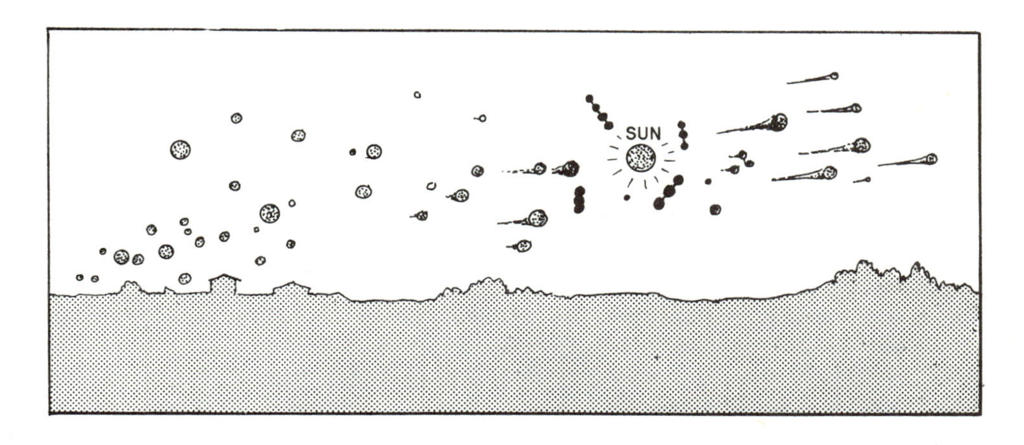


Strange spherical bodies move across sky

the vicinity of the sun, became entirely black. At this elevation their course seemed to lessen, and a great many of them remained, as it were, stationary; but they soon resumed their former, and an accelerated motion, and passed in the same direction with great velocity and almost horizontally. During this course some disappeared, others fell down, but the most part of them continued their progress almost in a straight line, till they were lost sight of at the eastern horizon. The phenomenon lasted uninterruptedly, upwards of two hours, during which time millions of similar bodies continually rose in the west, one after the other irregularly, and continued their career exactly in the same manner. No report, noise, nor any whistling or buzzing in the air was perceived. As these balls slackened their course on passing by the sun, several were linked together, three, six, or eight of them in a line, joined like chain-shot by a thin and straight bar; but on continuing again a more rapid course, they separated, and each having a tail after it, apparently of three or four fathoms length, wider at its base where it adhered to the ball, and gradually decreasing, till it terminated in a fine point. During the course, these tails which had the same black colour as the balls, disappeared by degrees.

It fortunately happened, that some of these balls fell at a short distance, or but a few feet from Mr. Secretary K. G. Wettermark, who had then for a long while been attentively looking at the phenomenon, in the aforesaid village. On the descent of these bodies, the black colour seemed gradually to disappear the nearer they approached the earth, and they vanished almost entirely till within a few fathoms distance from the ground, when they again were visible with several changing colours, and in this particular exactly resembling those air-bubbles, which children use to produce from soapsuds by means of a reed. When the spot, where such a ball had fallen, was immediately after examined, nothing was to be seen, but a scarcely perceptible film or pellicle, as thin and fine as a cobweb, which was still changing colours, but soon entirely dried up and vanished. As somewhat singular, it may be observed, that the size of these balls, to the sight, underwent no particular change; for they appeared of the same dimension, at their rise from the western horizon, as well as on their passing by the sun, and during the whole of their course to the eastern part of the heavens, where they disappeared.

Such have been the real circumstances attending this phenomenon, to which all the people in the village can testify. I have drawn up this report from the accounts of none but eye witnesses, and have compared them one with the other; and I cannot doubt the truth of the incidents, having been related to me in a manner agreeing in particulars and details. The labourers of Peter Manson, a farmer, being at work in the field, were the first who observed the phenomenon, and as it continued so long, all the people in the village were gradually observing it; it therefore could not be an illusion, possibly affecting one or other individual.

I leave it to the genius of more skilful and able men, to unfold the causes of this occurrence; but as an hypothesis may be hazarded, without being censured, it may be supposed, that perhaps a strong gust of wind, (coup de vent) from some mountainous or woody tracts, or regions at a distance, had loosened, collected and carried along with it, some probably vegetable substances of a jelly-like nature, which, in passing through the air, having incorporated some additional matter by a chemical union therewith, formed themselves into thin globular masses, or by the effect of the air and wind, were formed into bubbles, which became perceptible to the eye by the sun's light. But, why did the sun lose his brightness? and how could this unnumerable quantity of such a soapy and jellied substance, be generated or produced in one place? (North American Review, 3:320-322, 1816)

## THE KENTUCKY SHOWER OF FLESH

**Anonymous; *Scientific American Supplement*, 2:426, 1876.**

In 1537, while Paracelsus was engaged in the production of his "elixir of life," he came across a very strange-looking vegetable mass, to which he gave the name of "nostoc."

The want of rapid transportation, combined with the perishable nature of the substances fallen, have hitherto prevented a complete and exhaustive examination. The specimens of the "Kentucky shower," however, reached this city well-preserved in glycerine, and it has been comparatively easy to identify the substance and to fix its status. The "Kentucky wonder" is nothing more or less than the "Nostoc" of the old alchemist. The Nostoc belongs to the confervae; it consists of translucent, gelatinous bodies, joined together by thread-like tubes or seed-bearers. There are about fifty species of this singular plant classified; two or three kinds have even been found in a fossil state. Like other confervae, the Nostoc propagates by self-division as well as by seeds or spores. When these spores work their way out of the gelatinous envelope they may be wafted by the winds here and there, and they may be carried great distances.

Wherever they may fall, and find congenial soil, namely, dampness or recent rain, they will thrive and spread very rapidly, and many cases are recorded where they have covered miles of ground, in a very few hours, with long strings of Nostoc.

On account of this rapidity of growth, people almost everywhere faithfully believe the Nostoc to fall from the clouds, and ascribe to it many mysterious virtues. The plant is not confined to any special locality or to any climate; sown by the whirlwind, carried by a current of air, in need of moisture only for existence and support, it thrives everywhere. Icebergs afloat in mid-ocean have been found covered with it. In New Zealand it is found in large masses of quaking jelly, several feet in circumference, and covering miles of damp soil; and in our own country it may be found in damp woods, on meadows, and on marshy or even gravelly bottoms.

All the Nostocs are composed of a semi-liquid cellulose and vegetable proteine. The edible Nostoc is highly valued in China, where it forms an essential ingredient of the edible bird-nest soup. The flesh that was supposed to have fallen from the clouds in Kentucky is the flesh-colored Nostoc (N. carneum of the botanist); the flavor of it approaches frog or spring-chicken legs, and it is greedily devoured by almost all domestic animals.

Such supposed "showers" are not rare, and are entirely in harmony with natural laws. In the East Indies, the same Nostoc is used as an application in ulcers and scrofulous disease, while every nation in the East considers it nourishing and palatable, and uses it even for food when dried by sun heat. (Scientific American Supplement, 2:426, 1876)

# ICE CHUNKS AND HYDROMETEORS

Hail several inches in diameter, though unusual, has been recorded. (See Chapter 3) Large, solitary chunks of ice weighing several pounds also fall from the sky. Unlike large hailstones, the solitary chunks cannot be explained away by extrapolating accepted modes of hail formation to larger sizes. A standard theory, however, does exist: the chunks simply fall from aircraft flying overhead. A corollary is that cases recorded before the Wright Brothers must be dismissed as old wives' tales.

Doubtless some ice chunks do fall from aircraft, but a second possibility exists; namely, the chunks are truly meteoric (or cometary) and come from outer space, somehow surviving the searing descent through the atmosphere. In this radical context, it is curious that some hydrometeors fall during intense thunder and lightning, much like the legendary thunderstones.

## ICE FALL FROM A CLEAR SKY IN FORT PIERCE, FLA.
**Pielke, Roger A.; *Weatherwise,* 28:156-160, 1975.**

A little before 1900 EST on 13 December 1973, a chunk of ice, roughly 10 inches in diameter, fell from a clear sky through the roof of the home of Mr. and Mrs. Alex Cappar who live south of Fort Pierce, Florida. The ice fall occurred almost simultaneously with a Titan 3C rocket launch from Kennedy Space Center which took place at 1857. The incident was reported by the Fort Pierce News Tribune which contacted Dr. Neil Frank of the National Hurricane Center in order to ascertain its cause. The speculation was that it fell from a plane. In order to investigate the ice fall more quantitatively, it was decided to retrieve and examine the ice.

The damage to the Cappar house is shown in figure 1 [photos not reproduced] where the ice penetrated the ceiling of their den and smashed into the floor, break- a coffee cup and wristwatch, and littering Mr. and Mrs. Cappar with ice and roofing debris. If the ice had struck either one of them, it would have undoubtedly killed them.

Ice falls from a clear sky have been reported before. Snowden Flora in Hailstorms of the United States reported hail falls from a clear sky, but there were always thunderstorms in the near vicinity. His assumption was that the hail was spewed out of the thunderstorms by their strong updrafts.

Carte reported on an ice fall from a clear sky over the Brits district of the Transvaal in the Union of South Africa. He examined the internal structure of the ice, and finding traces of tea, coffee and detergent, determined the ice came from a passenger aircraft landing at Jan Smits Airport. The ice apparently formed from the contents of a galley sink on the aircraft which were released during flight. A defective electric heater supposedly caused a build-up of ice at the discharge point.

The ice which fell in Florida was collected by Mr. Cappar and stored in his home freezer. The total remaining ice fall, illustrated in figure 2, consisted of numerous small ice pieces and one large ice chunk.

Ice Structure. The large ice piece, which weighed about 5 kg, was photographed and sectioned twice while sections from two of the smaller pieces were also photographed. The external photographs show that the large ice chunk is wedge-shaped, with distinct flutings, or grooves on the two tapering sides of the wedge at about 70° to the sharp edge of the wedge. Figure 3 illustrates the ice chunk and the flutings.

A section was taken parallel to the sharp edge of the wedge, bisecting the wedge angle. Photographs of this section, illustrated in figure 4, were taken with bright-field illumination to show the air bubbles. The sharp end of the wedge is at the bottom right of the picture. The structure is remarkably lumpy and irregular, with the exception of the regular, parallel features at the bottom of the photograph, which extend from the sharp edge at the bottom towards the upper left at an angle of 70° or so, undoubtedly reflecting the fluting mentioned above.

A section was made perpendicular to the one described above and at right angles to the thin edge of the wedge. This section, illustrated in figure 5, with the sharp edge of the wedge at the top, shows features parallel to the sides of the wedge and the hint of growth layering. This section now begins to make the ice look like an accretion that grew with the thin end into the wind.

Thinner slices were magnified and photographed from the same section as illustrated in figure 5. Two sets of photographs from these sections are shown in figures 6 and 7 where the left-hand figure was made with bright-field illumination and the right-hand figure was made with crossed polaroids. The photograph in figure 6 illustrates the irregular alternations in the bubble structure while both photographs show the small, fairly uniform crystal structure. The air bubbles are in grain boundaries and triple junctions and are in nearly equilibrium configurations. This sort of structure does result from soaking snow in water for a considerable length of time and freezing the slush. It is not found in hailstone, for instance, but is common in old snowfields in the mountains, and can be made in the laboratory with ease by crushing some ice, making a ball of the material, soaking it with water and then freezing the mixture.

One of the small pieces of ice was about one inch thick with a tabular shaped end. Sections at right angles to each other and perpendicular to the plane of the tabular end were made and are illustrated in figure 8. The accretional nature of the piece is absolutely clear from the growth layering, the linear features, and the shape. It appears to be accretion at high temperature. On the basis of the ice structure, accreted and then frozen slush appears to be the only reasonable hypothesis for explaining the origin of the ice. It is also possible that the accretion was of wet snow directly, but then one would need to postulate some melting, to fill up most of the pores with liquid, and then refreezing. This is a more complicated origin.

Chemical Composition. A chemical study was performed on the ice sample in an attempt to learn of its origin. Every precaution was exercised in preparing the sample for analysis; however, due to possible contamination from the fall through the roof and the handling afterwards, the results should be viewed with reservation.

A qualitative analysis was performed on the sample using an x-ray spectrometer in series with a scanning electron microscope. A small section of the sample was placed on a 0.45 um millipore filter and sublimated down to a residue. This was put into the instrumentation to obtain a relative scan of the elements present. Results showed the typical elements to be present: silicon, phosphorous, sodium, chlorine, sulfur, calcium, and titanium. Trace amounts of silver, selenium, gallium, germanium, and mercury also were present.

Quantitative analyses were performed on the sample for a few common constituents by established techniques. A section of the sample was melted and then filtered to remove the debris that had been collected upon impaction through the roof. Results showed nothing unusual with respect to the nitrogen ($NO_2$, $NO_3$, $NH_4$) or phosphate (O-$PO_4$, total $PO_4$) groups and most of the common ions. A high sodium concentration of 63 ppm and sulfate concentration of 12 ppm were the only unusual results. The pH of 7.17 was high, but could be explained due to the time elapsed between collection and analysis.

A small portion of the remaining sample has been stored frozen for future heavy metal analyses. Until they are performed, the origin of the sample will remain an unknown. The unexpected elements of silver, selenium, gallium, germanium, and mercury detected with the x-ray spectrometer in series with a scanning electron microscope and the somewhat high sodium and sulfate concentrations do not in any fashion present a clear picture of the sample's origin.

Conclusion. The chemical and physical analyses of the ice pieces, although quantitatively describing the material, do not permit a definitive determination of the origin of the ice. The coincidence between a Kennedy Space Center rocket firing and the ice fall may suggest that the ice was dropped off during launch, but the trajectory of all rocket launches, of which we are aware, are to the southeast away from the Cape, and therefore well away from the Cappar home. A second, more plausible explanation, may be that slush was collected off an airport runway and froze on an aircraft on take-off from a northern airport, and then was dropped as the plane lowered to make a landing in south or central Florida. There was, in fact, a storm system producing snow on this date in the northern Midwest. Several airports in the immediate vicinity of Fort Pierce were contacted to determine if planes from northern airports were making landing approaches near the time of the ice fall, but none was reported.

Since Fort Pierce is under a major air route of flights into the south Florida area, it is reasonable to suspect one of these aircraft dropped the ice. Unfortunately, it was impossible to check into all commercial, private, and military air traffic over the Fort Pierce area at the time of the ice fall.

Our conclusion, therefore, is necessarily tentative and somewhat speculative. The physical appearance of the ice suggests it was frozen onto an aircraft during take-off, then dropped as the plane descended to land. The chemical composition of the ice, particularly the silver, selenium, gallium, germanium, and mercury, is somewhat unusual, and we have no explanation for this. The high sodium content could have resulted from salting of the runways from which the airplane would have taken off.

Finally, we encourage any reader who has additional suggestions to solve this riddle to send a correspondence to Weatherwise or to one of the authors of this paper. (Weatherwise, 28:156-160, 1975)

## REMARKABLE HAILSTONES IN INDIA, FROM MARCH 1851 TO MAY 1855

**Buist, George; *Reports of the British Association,* part 2, 34, 1855.**

There are four occasions on which remarkable masses of ice, of many hundred pounds in weight, are believed to have fallen in India. One near Seringapatam, in the end of last century, said to have been the size of an elephant. It took three days to melt. We have no further particulars, but there is no reason whatever for our doubting the fact.

In 1826, a mass of ice nearly a cubic yard in size, fell in Khandeish.

In April, 1838, a mass of hailstones, 20 feet in its larger diameter, fell at Dharwar.

On the 22nd of May, after a violent hailstorm 80 miles south of Bangalore, an immense block of ice, consisting of hailstones cemented together, was found in a dry well.

These masses of ice, like many of those considered hailstones of the largest size, have, in all probability, been formed by violent whirlwinds or eddies, and seem to have reached the monstrous dimensions in which we find them, either on their approach to or their impingement on the ground; and the same thing will apply to those of much more moderate bulk, and which are commonly considered hailstones, though when examined they turn out to be a number of these aggregated together. Many of the masses doubtless owe their origin to being swept, like that of 1852 near Belgaum, into hollows or cavities---in this particular case into a dry well---where they become almost immediately congealed into a mass.

Since 1850 two hailstorms of much greater magnitude, and more disastrous consequences, have occurred than any here made mention of, that in the Himalayas north of the Peshawur on the 12th of May, 1853, when eighty-four human beings and 3000 oxen were killed, and that which occurred at Nainee Tal, a Sanitarium on the lower Himalayas, on the 11th of May, 1855. Of the Peshawur storm we have few details beyond the fact that the ice masses were very hard, compact, and spherical, many of them measuring 3-1/4 inches in diameter, or nearly a foot in circumference; and this fact seems to have been given from measurement, not by guess.

The description of the Nainee Tal storm, from the pen of an eye-witness of intelligence and information, is the best we possess. The approach of the storm was heralded in by a noise as if thousands of bags of walnuts were being emptied in the air. At first the hail was of comparatively small size, about that of pigeons' eggs; it gradually increased in magnitude, till it reached the dimensions of cricket-balls. Pieces, picked up at all parts of the station, were carefully weighed and measured, and the results will be found further on. (Reports of the British Association, 34, 1855)

## GREAT MASS OF ATMOSPHERIC ICE

**Anonymous; *Edinburgh New Philosophical Journal*, 47:371, 1849.**

A curious phenomenon occurred at the farm of Balvullich, on the estate of Ord, occupied by Mr. Moffat, on the evening of Monday last. Immediately after one of the loudest peals of thunder heard there, a large and irregular-shaped mass of ice, reckoned to be nearly 20 feet in circumference, and of a proportionate thickness, fell near the farm-house. It had a beautiful crystalline appearance, being nearly all quite transparent, if we except a small portion of it which consisted of hailstones of uncommon size, fixed together. It was principally composed of small square, diamond-shaped, of from 1 to 3 inches in size, all firmly congealed together. The weight of this large piece of ice could not be ascertained; but it is a most fortunate circumstance, that it did not fall on Mr. Moffat's house, or it would have crushed it, and undoubtedly have caused the death of some of the inmates. No appearance whatever of either hail or snow was discernible in the surrounding district. (Edinburgh New Philosophical Journal, 47:371, 1849)

# DUST, THUNDERSTONES, AND SUNDRY INORGANIC MATTER

The winds of the high atmosphere can carry dust for many hundreds of miles. Dust from the western states frequently falls in normal precipitation in states farther east. The famous "dry fogs" at sea off Cape Verde and many European dust falls have their origins in the Sahara. Other dust falls are harder to explain for no handy nearby sources exist. Selected examples in all categories are recorded below.

The scientific and popular literature have noted many falls of yellowish substances. Investigations usually point to air-borne pollen. Other reports insist that real sulphur has fallen from the skies. Sulphur is a common enough element but it does not exist in elemental form where winds can pick it up and transport it elsewhere. "Sulphur rains," if true, have no conventional explanations.

The thunderstone problem cannot be avoided. Many credible witnesses have reported the luminous descents of objects accompanied by claps of thunder. Upon retrieval the purported fallen objects turn out to be not meteorites (these also may fall with considerable noise) but ordinary terrestrial stones. Some of these events are ultimately identified as hoaxes or human errors. But, as with UFOs, some bona fide events seem to persist. If thunderstones truly exist, what are they? Can they be reentering terrestrial material propelled long ago into orbit by meteorite impacts or some other catastrophes? An obvious parallel exists between thunderstones and the fall of large ice chunks. Are both phenomena real and, if so, are common terrestrial mechanisms adequate?

## DUST

### WIDESPREAD DUST HAZE
**Anonymous; *Nature*, 125:256, 1930.**

Feb. 16-19, 1898. Dust Haze. A dense haze occurred over a large part of the eastern Atlantic off West Africa, extending for at least 1500 miles north and south and a great but unknown distance east and west. The haze was caused by very fine red dust, so fine that it was impossible to sweep it up, and so dense that the sun and stars were completely obscured for two days. When visible the sun was generally red, but one observer described it as "a perfect blue ball" and another as greenish. At Teneriffe the occurrence was preceded by a strong and very hot southerly wind, but during the haze there was no wind. Many insects were observed, of species not generally found on the island. The dust evidently originated in Africa, for it was much coarser near the coast, and was thrown overboard from ships in large quantities. (Nature, 125:256, 1930)

## ATMOSPHERIC DUST
**Hepworth, M. W. Campbell; *Royal Meteorological Society, Quarterly Journal*, 28:68, 1902.**

In the Monthly Pilot Chart of the North Atlantic and Mediterranean, issued by the Meteorological Office, for the month of January 1902 there appeared the following remark among others under the heading Fog:---"On the African Coast, from Cape Verde southward, the land is often obscured by a low-lying treacherous haze, which calls for caution in navigation. There is also a great deal of misty haze, if not fog, about the Cape Verde islands; and haze in some seasons very dense, due to 'Red Dust,' extends far out on the eastern side of the Atlantic southward from the 35th parallel. In February 1898 the R. M. S. Briton entered a cloud of red dust 80 miles south of Madeira, and for a distance of over 2000 miles was dependent for her position on dead reckoning. She emerged from the haze when between 400 and 500 miles south of Cape Verde. In consequence of the density and persistency of this haze vessels were delayed as much as two or three days, one steamer occupying two and a half days in running the short distance between Teneriffe and Las Palmas, the run usually being accomplished in about 5 hours."

Since the publication of the foregoing a gale is reported to have visited the Canary Islands on January 17 or 18, and a dust storm to have been experienced in some districts.

On Friday, January 17 in the early morning the mail steamer Lagos outward bound, went ashore on one of the Dezertas islands, near Madeira, during what is described as a dense fog. She became a total wreck.

The mail steamer Raglan Castle, which arrived at Las Palmas from London early in the morning of January 18, reported that from 33° N. to 29° N. she had met with a strong South-south-east wind, and that the atmosphere was thick with fine sand causing a fog-like appearance. The Lisbon correspondent of a London newspaper, writing on January 25 mentions, in connection with the wreck of the steamer Lagos, "an unexplained phenomenon observed in different localities in Portugal. The leaves of the trees and the vegetables were thickly covered with a brownish yellow dust resembling cinnamon."

The deposit of this atmospheric dust so far north as Lisbon is unusual. Later reports show that it has recently occurred in higher latitudes.

The Secretary of the Meteorological Office has received a communication from the Secretary to Lloyds, Colonel H. Hozier, enclosing an extract from a letter (dated January 23) received by him from the Signalman in charge of Lloyds Signal Station at Barry Island, together with a specimen of the salmon-coloured dust to which it refers. This dust closely resembles that which fell at Fiume in the night of March 10-11, 1901, but appears to be of a somewhat coarser grain.

The extract from the report of the Signalman is as follows:---"I also beg to inform you that a very queer thing happened here last evening; there was a light wind from West with a wet fog. This morning I noticed the iron railings were coated with a salmon-coloured dust, although rather wet, I did not take much notice of it. Later in the day I found that the base of the railings was completely coated, and I have scraped a small portion off and am sending it to you. Probably you will send it to the Meteorological Office. We have not had any length of dry weather for some time, rain falling nearly every evening for the last three weeks, so it could not be dust. I called the attention of Mr. Waite (the Water Engineer at Barry) to it. He tells me it is something remarkable. I cannot account for it. Some call it pink rain. I found myself that my overcoat and cap, after being out last evening, had a red appearance this morning." (Royal Meteorological Society, Quarterly Journal, 28:68, 1902)

## DUST SHOWERS IN THE SOUTH-WEST OF ENGLAND

**Anonymous; *Symons's Monthly Meteorological Magazine*, 37:1-4, 1902.**

The "blood-rains" of March last in southern Europe produced so much excitement where they fell and were commented on so fully in the daily press, that it is somewhat strange to have to record a very similar occurrence in our own country which seems to have called forth no comment whatever in any London newspaper. On January 22nd or 23rd, many people in the extreme south-west of England had their attention attracted by a peculiar deposit of reddish or yellowish dust, which lay on exposed objects or the surface of the ground, and appeared to have fallen from the sky. Such a phenomenon is not unprecedented, but it is so rare as to be worth investigating, and we publish this preliminary notice in the hope of obtaining additional information as to the extent and the nature of the showers. It would be a favour if every reader who has observed the phenomenon would write a short account, stating exactly what was seen, with date and place and any other facts bearing on the subject, and send it to Dr. H. R. Mill, 62, Camden Square, London, N. W.

The accompanying map shows by a large black dot every place at which we have ascertained that the appearance was noticed. It will be seen that the dust fell over

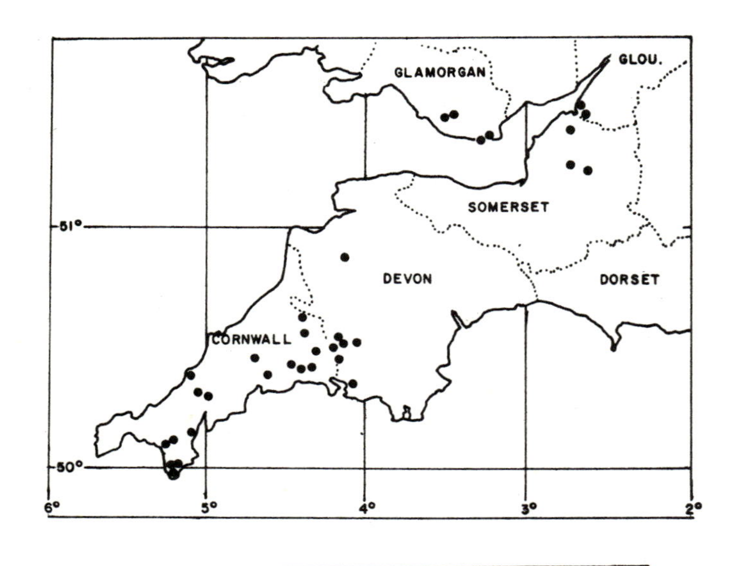


Dust falls in England, February 1902

practically the whole of Cornwall, but only attracted attention near the western border of Devon, reappearing again at a few points in Somerset, the extreme south of Gloucester and Glamorgan. There are thus two areas in which the dust fell abundantly---the Cornish area, measuring roughly 80 by 40 miles, and the Bristol Channel area, measuring perhaps 60 by 30 miles. The most northerly and easterly point in the former area, Black Torrington, is 50 miles from the nearest points in the latter area, Bridgend and Barry. Of course, the dust may have fallen in the intervening districts and have escaped notice, a probability which will appear the greater when we remember that the large unpeopled tracts of Dartmoor and Exmoor lie there. But, on the other hand, we have ascertained by enquiries specially addressed to rainfall observers, who are not likely to have overlooked so interesting an appearance, that nothing of the kind was noticed at Barnstaple, Taunton, Newton Abbott, Torquay, Portland, or North Cadbury (near Wincanton). These places are marked on the map by a small circle. Hence, there is some evidence for believing that two different falls took place, but we cannot at present say whether they were simultaneous or successive. With regard to this question, notes as to the hour when the dust was first observed would be of value.

An enquiry addressed to Mr. Dorrien Smith, of Tresco Abbey in the Scilly Islands, elicited the fact that splashes of muddy rain had been observed there on January 21st, but no general fall of dust.

An examination of the Daily Weather Charts shows that, from the 20th to the 23rd, the weather of the south of England was controlled by a high pressure area, the centre of which lay over the west of France and the Bay of Biscay until the 23rd, when it moved a little eastward. The wind over the whole of the west and south of England was westerly, with a southerly component, and light; and the rainfall was very slight, a little snow lying in some places.

The actual appearances observed may be judged from the following extracts from local newspapers, and from our correspondence:---

A correspondent of the Western Morning News, writing from Bere Alston, says: "A labourer here speaks of a slight hail shower on Thursday last [23rd], which was accompanied by a dust fog. His account somewhat resembles the descriptions we have of those fine dust showers which, at a distance of 400 or 500 miles from the coast of Africa, envelop vessels in a thick fog."

Rev. J. A. Wix, of Quethwick, speaks of a washing hung out to dry on the night of the 23rd being found next morning, "splashed to such an extent by some yellowish 'mess' that all had to be re-washed," while the cabbages were covered with a dust resembling Peruvian guano.

Mr. E. W. Waite, the Waterworks' Engineer of the Barry Urban District Council, writes us:---

"The fall of rain, ending on Thursday the 23rd at 9 a. m., was a light one, only .03 in., the wind very light from W. N. W. The deposit, like a very fine dust, coated the iron railings and fences with a pale salmon-pink colour. I was very much struck with some ordinary wire-netting which appeared to have been painted with a brush. The deposit was particularly noticeable on Barry Island, although I found it to a lesser extent at Barry itself."

Earl Waldegrave, writing from Chewton Priory, to the Western Daily Press, says:---

"Wednesday the 22nd was with us very warm, with wet mist, only measuring .02 in. of rain. Afterwards the glass and woodwork of the greenhouses and frames were covered with a rust-coloured dust, which has left stains on the paint."

Sir Edward Fry, writing to Nature of February 6th, from Failand, near Bristol, says:---"My men here noticed on Thursday last, the 23rd instant, that the leaves, glasses of the frames and iron-work of the gates were smeared with a reddish mud; one hedge in particular they describe as almost covered with the substance, and the pinafores of a cottager's children which were hanging out to dry were so stained with the deposit that they had to be re-washed. When the substance fell no one here knows, nor is it clear whether it fell as dust or mud; from the firm way in which it has attached itself to the iron-work, I should think that it fell as mud." (Symons's Monthly Meteorological Magazine, 37:1-3, 1902)

## ALANDROAL METEOR
**Anonymous; Cornell, James, and Surowiecki, John; *The Pulse of the Planet*, 19, 1972.**

It was nearly dark when a Portuguese man taking a walk heard a loud noise like a sonic boom. [On November 14, 1968] Suddenly, an eerie glow lit up the landscape as though it were daytime. Seconds later, a 55-pound meteorite slammed into the ground about 100 feet away.

The meteor had traveled from southwest to northeast across Portugal descending at a 50- to 60-degree angle. When it struck near the Portuguese village of Jurmenha, it caused a "rain of sand." It also formed a depression in the ground some three inches deep upon impact. (The Pulse of the Planet, 19, 1972)

## SUPPOSED METEORITE
**Anonymous; *American Journal of Science*, 2:24:449, 1857.**

We have received from Mr. Thomas Bassnett of Ottawa, Illinois, specimens of scoria, with a description of their supposed fall on the 17th of June last, about ten miles south-southwest from that place. The account of the fall, as written out by L. H. Bradley on whose farm the scoria was found, states, that the time it occurred was fifteen or twenty minutes before 2 p. m.; the wind blew west by south. He says, "The cinders fell in a northeasterly direction in the shape of the letter V. The weather had been showery, but I heard no thunder and saw no lightning. There appeared to be a small, dense black cloud hanging over the garden in a westerly direction, or a little to the south of west. The cinders fell upon a slight angle within about three rods of where I was at work; there was no wind at the moment, or none perceptible. My attention was called first to the freak the wind had in the grass, and the next moment to a hissing noise caused by the cinders passing through the air. The longer ones were considerably imbedded in the earth, so much as only to show a small part of it, while the smaller ones were about one-half buried. I noticed at the time that the ground where I afterwards picked up the cinders showed signs of warmth, as there was quite a steam or fog at that particular point. I thought it singular, as the ground had been very cold previously."

The scoria is in rounded inflated pieces, like what have been called volcanic bombs, the exterior being glassy, and the interior very cellular. They are little over an inch in the longest diameter. Color black.

The paper, called the "Sunny South," of Aberdeen, Miss., of Sept. 17, 1857, describes a mass of lava as large as a barrel, "which fell near the farm of Mr. John Fortson, ten miles west of Aberdeen, on the 8th of July, 1856, and which excited a good deal of wonder and speculation at the time for hundreds of miles around."

The Illinois scoria is unlike any meteorite, and suggests the idea of a terrestrial origin. We know nothing about the Mississippi "lava." (American Journal of Science, 2:24:449, 1857)

# SULPHUR

## RAIN OF BRIMSTONE
**Anonymous; *Nature,* 125:729, 1930.**

May 16, 1646. Rain of Brimstone.---Worm relates that "the whole city of Hasnia [? Asniara Is. off Italy] and all its streets were inundated with a mighty rain so that men were prevented from walking and the air was infested with a sulphureous odour; when the waters had subsided somewhat it was possible in some places to collect a sulphureous powder, some of which would pass for real sulphur as regards taste, colour, smell, etc.". (Nature, 125:729, 1930)

## SHOWER OF SULPHUR
**Anonymous; *Symons's Monthly Meteorological Magazine,* 2:130, 1867.**

The inhabitants of the village of Thames Ditton, Surrey, were, on Friday night, October 18th, 1867, a good deal startled at witnessing a very strange phenomenon, which had the appearance of a shower of fire. The shower lasted about ten minutes, and during its continuance afforded a brilliant light. Next morning it was found that the waterbutts and puddles in the upper part of the village were thickly covered with a deposit of sulphur. Some of the water has been preserved in bottles. (Symons's Monthly Meteorological Magazine, 2:130, 1867)

## MASS OF BURNING SULPHUR FALLS
**Glaisher, James, et al; *Reports of the British Association,* 272, 1874.**

[June 17, 1873; Hungary and Austria] It is remarkable that perfectly authentic statements were received of the deposition, soon after, or about the time of, the meteor's explosion over Zittau and its neighbourhood, of a mass of melted and burning sulphur the size of a man's fist, on the roadway of a village, Proschwitz, about 4 miles south of Reichenberg, where the meteor exploded nearly in the zenith. It was stamped out by a crowd of the villagers, who could give no other explanation of its appearance on the spot than that it had proceeded from the meteor; on examination at Breslau some remnants of the substance proved to be pure sulphur. With regard to the calculated course, the meteor must, however have passed quite 12 or 14 miles south-westwards from the place where this event is said to have occurred; and its questionable connexion with the fireball is occordingly rendered very doubtful from the great distance of the locality from immediately below the meteor's course. In

Chladni's work on Fiery Meteors and Stonefalls, only one similar instance is recorded, from ancient chronicles, where burning sulphur fell at Magdeburg, of the size of a man's fist, on the castle-roof at Loburg, 18 miles from Magdeburg, in June of the year 1642. The fact of this large fireball having deposited any stony or other aerolitic matter cannot therefore yet be regarded as decidedly established. (Reports of the British Association, 272, 1874)

### SULPHUR RAINS
**Anonymous; *Monthly Weather Review,* 26:115, 1898.**

The Cincinnati Enquirer of March 22 reports that a "sulphur rain" fell at Mount Vernon, Ky., early on the morning of March 21, as also at several other places in Rockcastle County; the stuff burned and gave out fumes of sulphur.

Those who are not seeking after mysteries may rest assured that such a rain of sulphur simply brings down to the ground some pollen from the pine woods, or some other light substance that has only a short time before being carried up by a strong gust of wind. It saddens one to think that any superstition should attach to such an ordinary phenomenon, one that occurs every day of the year at some place on the globe. Still more is it a pity that our daily press should repeat, and apparently indorse, any of the popular errors regarding these and other meteorological phenomenon. It is quite as easy for a popular journal to present the best thoughts of the best people as it is to merely diffuse and strengthen the errors of the ignorant. The past century has witnessed the banishment from our text-books of innumerable erroneous ideas that were accepted by our ancestors. Why can not the daily press assist in the work of educating the public and resolutely refuse to print such nonsense as "the people generally consider this a sure harbinger of war," or such headings as "a red sun; bloody omen," or again, "great drought: belief that the world is drying up and that its end is drawing near"? If any one thing is more clearly taught than another by all our teachers, both religious and secular, it is that the future is not and cannot be revealed by signs and omens. (Monthly Weather Review, 26:115, 1898)

## THUNDERSTORMS AND "THUNDERSTONES"

### THUNDERBOLTS
**Carus-Wilson, C.; *Knowledge,* 8:320, 1885.**

I have frequently been struck by the fact of thunderbolts falling during violent storms, and of their simultaneous appearance with flashes of lightning.

That masses of matter occasionally accompany the electric fluid when it reaches the earth there is no question; they have been seen and described as "balls of fire," and I have in my possession one that fell some time ago at Casterton, in Westmoreland.

It was seen to fall during a violent thunderstorm, and, killing a sheep en route, buried itself about six feet below the surface, and when dug out shortly after it was still hot. In appearance it much resembles a volcanic bomb; it is about the size of a large cocoa-nut, weighs over 12 lbs., and seems to be composed of a hard ferruginous quartzite. I have not yet submitted it to a technical analysis, but hope to do so shortly. There is an external shell of about an inch in thickness, and this contains a nucleus of the same shape and material as the shell, but is quite independent of it, so that the one is easily separated from the other; I attribute this separation of the parts to an unequal cooling of the mass.

I have often wondered from whence thunderbolts were derived, and why they always fell as "balls of fire" accompanied by flashes of lightning.

It occurs to me that if ten millions of meteorites enter our atmosphere daily (vide Knowledge, Feb. 8, 1884), and become dissipated into vapour on coming into contact with our air, that some of these bodies might chance to survive the ordeal of passing through the more rarified zones and reach the earth, if only an airless passage could be obtained during the latter part of their journey. (Knowledge, 8:320, 1885)

## GLOBE LIGHTNING

**Capron, J. Rand; *Symons's Monthly Meteorological Magazine,* 18:57, 1883.**

It is strange what vagaries lightning will sometimes play. I had recently brought to me a ball evidently largely composed of iron, spherical in shape, 14.38 oz. in weight, specific gravity 6.81. This had been found after a storm near the apparent place of fall of lightning, in a heap of refuse bark, thrown out from a barn door. In this heap was a round hole with charred sides, and at the bottom of this hole, 10 inches deep, was found the ball in question. As the finder was certain he had found a "meteorite," every means were taken by analysis, comparison, section, magnetic examination, &c., to ascertain its true character, and these proved it incontestably to be a charcoal smelted cast iron ball, probably of considerable antiquity, as charcoal furnaces have long ceased. There yet remained the charred hole to be explained, and upon seeking advice from Mr. Preece, he came to the conclusion that an offshoot from a flash of lightning, falling near even so small a mass, may have penetrated to it, and thus led to its discovery, and to the mistake about it. (Symons's Monthly Meteorological Magazine, 18:57, 1883)

## OBSERVED FALL OF AN AEROLITE NEAR ST. ALBANS

**Bullen, G. S.; *Nature,* 89:34 and 89:62, 1912.**

Observed Fall of an Aerolite near St. Albans. During a heavy thunderstorm which ensued on Monday, March 4, between 2.30 p.m. and 4.15 p.m. an aerolite was observed to fall at Colney Heath, near St. Albans. The observer, who has placed the specimen in my hands for examination, stated that the stone fell within a few feet from where he was standing, and that it entered the ground for a distance of about 3 ft. Its fall was accompanied by an unusually heavy clap of thunder. The example weighs 5 lbs. 14-1/2 oz., and measures 6-3/4 in. x 5-5/8 in. at its greatest length and breadth respectively. The mass is irregularly ovate on the one side, and broken in outline on the other. The actual surface throughout is fairly deeply pitted, and under magnification exhibits the usual chondritic structure of the crystalline matter with interspersed particles of what appears to be nickeliferous iron. (Nature, 89:34, 1912)

Observed Fall of an Aerolite near St. Albans. Under the above heading in the issue of Nature for last week I reported upon the circumstances and other details of a supposed fall of a meteorite during the storm of March 4, as described to me by an observer, Mr. H. L. G. Andrews, at Colney Heath, near St. Albans. I have now submitted the stone for examination to Dr. George T. Prior, of the British Museum (Natural History), who informs me that it is not of meteoric origin. (Nature, 89: 62, 1912)

## OBSERVING THE FALL OF A METEORITE
**Anonymous; *Symons's Meteorological Magazine*, 38:117-118, 1903.**

Referring to a note in our February number, stating that meteoric stones had been seen to fall only four times in the British Isles, the Rev. W. C. Plenderleath, of Mamhead Rectory, writes stating that he had heard of another case, and sending an account of it written by the lady who saw the phenomenon. In the absence of a mineralogical examination of the stone it is impossible to say positively that it was a meteorite and not a piece of ironstone struck by the flash of lightning which fused the sand. The description runs:---

"With pleasure I write all I remember of the incident referred to in Mr. Plenderleath's letter, and I will make it as clear as I can. In 1887 (I think early in August) there was a very heavy thunderstorm at Harrogate, and in the surrounding districts. We were then living at Starbeck, a hamlet between Harrogate and Knaresborough. I was watching the storm from the drawingroom window, when suddenly (simultaneously with a sharp, cracking clap of thunder, immediately overhead) I saw a bright white light fall, in a flash of lightning. As it touched the earth I heard a loud fizz---like an enormous match being struck. When the rain ceased I went to the spot where it fell, in a meadow, at the bottom of the garden, about 50 yards distant. I saw a disturbance in the ground, and on digging at a little below the surface, I found a hard, heavy substance, about the size of a guinea fowl's egg, surrounded by pieces of light grey and white substance, like slate. I collected it all and sent it to Professor Tyndal; by return of post I had a kind note from Mrs. Tyndal, saying that the Professor was from home, but that she had notified him. In a few days I heard from Professor Tyndal, thanking me for the specimen, and saying that it was 'ironstone,' and the fragments 'fusion of sand by the lightning.' Dr. Tyndal aid not return the specimens, and I do not know what became of them. My husband's health was very bad, and we left for the east, and Japan, immediately afterwards, and the meteorite passed from my mind, interesting though it was. I was seven years absent from England. It is a long time ago now, nearly sixteen years, and so much has happened since---but I always remember the interesting circumstance, and feel rather proud to have seen and found it. I wish I could tell what became of it." (Symons's Meteorological Magazine, 38: 117-118, 1903)

## METEORITE FALLS DURING THUNDERSTORM
**Anonymous; *Nature,* 81:134, 1909.**

Mr. H. Garrett, writing from Greensted Rectory by Ongar to the Times (July 28), says:---"During the severe thunderstorm on the 13th inst. a meteoric stone fell in the stable yard here with a terrific explosion when within a few feet of the ground, embedding itself in the gravel about 8 inches, the ground around for several feet being perforated with small holes caused by the fragments. The main part and fragments which we could collect weighed 1 lb. 13 oz. The fall was witnessed by my daughters, who were sheltering about eight yards away." (Nature, 81:134, 1909)

## THE SUPPOSED FALL OF METEORIC STONES IN THE "BLACK" COUNTRY
**Hall, Townshend M.; *Symons's Monthly Meteorological Magazine,* 4:184, 1869.**

With reference to the supposed shower of meteorites at Wolverhampton on Tuesday, May 25, it is perhaps worthy of notice that a precisely similar fall was recorded by Mr. Thomas Plant, F. M. S., as having taken place at Birmingham on Friday, May 29---only four days later. The following is an extract from Mr. Plant's letter, published in the Birmingham Daily Post for Saturday, May 30, 1869:---

"The thunderstorm this morning was remarkable for the immense quantity of rain which fell; also for the long duration of the tempest; likewise the strange shower of meteoric stones referred to at the close of this report. . . . There was an extraordinary phenomenon during the deluge of rain. From nine to ten meteoric stones fell in immense quantities in various parts of the town. The size of these stones varied from about one-eighth of an inch to three eights of an inch in length, and about half those dimensions in thickness. They resembled in shape broken pieces of Rowley ragstone. A similar phenomenon visited Birmingham ten years ago. On the 12th of June, 1858, during a severe thunderstorm, there fell a great quantity of meteoric stones, in every respect like those discharged this morning."

In a catalogue of all the known British meteoric descents, I find only twelve instances on record of the fall of meteorites in England, four in Scotland, and a like number in Ireland. But here we have not an isolated instance of this phenomenon, but two "showers" in "immense quantities."

The letter published in your last number gives, not one atom of evidence in this matter, save perhaps the fact that the author had placed traps, consisting of a sheet and a large tin pan, in his garden, but had failed in catching any of the meteorites. What we require is, however, positive evidence. If the meteoric stones descended in such numbers as Mr. Plant describes, surely there must be some forthcoming for a crucial examination. Let them be properly analyzed, and the results will prove either that the Birmingham and Wolverhampton meteorologists were very much mistaken, or that the "black" country had the unparalleled good fortune to be visited by two showers of meteorites within four days.

I may add, that it has fallen to my lot, on several occasions, to conduct examinations on reputed meteoric stones, and the result in general has been by no means favourable to their authenticity. One proved to be a piece of scoriae from an iron foundry; another was a fragment of iron pyrites; and a third was evidently derived from some neighbouring glass works. (Symons's Monthly Meteorological Magazine, 4:184, 1869)

## THE EFFECT OF LIGHTNING ON A HAYSTACK
**Anonymous; *Meteorological Magazine,* 69:195-196, 1934.**

When lightning strikes sandy soil, the intense heat generated by the passage melts the sand and forms a cylindrical mass of fused silica penetrated by a small hole, which is known as a "fulgurite". Colonel C. Buckle, of 9, Walbrook, has described a somewhat similar process which occurs when lightning strikes a haystack. He writes that "the enamel on the stems of the grasses is, I believe, largely silica from the soil. Ordinarily, of course, if a stack burns the quantities of silica in any one place are too minute to coagulate and they remain in the soft ash. When a stack has been struck by lightning however, you may find masses of fused matter which are hard and ring like a tile and which sometimes take the form of walls of a rough tube, the walls being about an inch thick and the tube up to about a yard in diameter.

The only explanation I can think of is that the heat of the discharge instantaneously consumes the hay or straw in its path and the gaseous products of combustion create an outward pressure against the remaining part of the stack which brings all the silica against the side of the tubular hole. Unfortunately the tube usually collapses with the ultimate destruction of the stack and vou only find pieces. I have a fragment here, however, which gives a good idea of the shape and size of the whole.

It is several years since I formed this opinion but I have only recently been able to confirm it by getting a specimen from a stack which was undoubtedly struck by lightning as a man was killed whilst sheltering by it. The School of Mines fulgurite has a small hole down it and it seemed to me vastly interesting that the same sort of thing can happen in a haystack but that, owing to the very small quantities of silica present, the hole is vastly bigger and the fused walls very much thinner. If we could get such a "thunderbolt" whole it would look very much like the bark of the trunk of a tree minus the rest of the tree.

Colonel Buckle kindly lent a specimen of part of one of these silica tubes to Dr. G. C. Simpson. The specimen has the appearance on one side of the effect of great heat, and the curvature of the fragment showed that the diameter of the whole tube was about one metre, which is approximately the diameter of the lightning channel in the air. (Meteorological Magazine, 69:195-196, 1934)

## HAY CONVERTED INTO A VITREOUS GLASS BY LIGHTNING
**Anonymous; *American Journal of Science,* 1:19:395, 1831.**

In the summer of 1827, a rick of hay, in the parish of Dun near Montrose, was set on fire by lightning, and partly consumed. When the fire was extinguished by the exertions of the farm-servants who were on the spot, there was observed in the middle of the stack a cylindrical passage, as if cut out by sharp instruments. This passage extended down the middle of the stack to the ground, and at the bottom of it there was found a quantity of vitrified matter, which, there is every reason to think, is the product of the silex contained in the hay which filled up the cylindrical passage. The existence of silex in the common grasses is well known, and the color of the porous and vesicular mass is very like that which is obtained from the combustion of siliceous plants. We have been indebted for a specimen of the substance to Captain Thomson of Montrose, who examined the spot almost immediately after the accident had taken place. (American Journal of Science, 1:19:395, 1831)

# MISCELLANEOUS SOLIDS

## PECULIAR HAILSTONES
**Kanhaiyalal; *Nature*, 48:248, 1893.**

A friend of mine writes me from Peshawar about a very curious phenomenon which I think is worth notice in your columns. The monsoon has set in this season earlier than for some years past. A few days ago in a village named Daduzai (a tehsil in the Peshawar district) rain fell, preceded by a wind storm, and with the rain came a shower of hailstones which lasted for a few minutes. The most curious part of this occurrence is that the hailstones when touched were not at all cold, and when put in the mouth (as is the custom in this hot country) tasted like sugar. I am further told that these hailstones were extremely fragil, and as soon as they reached the ground they broke in pieces. These pieces when examined looked like broken sticks of crystallised nitre. My informant tasted them, and was struck with their purity and sweetness. A few pieces were also sent to the Deputy Commissioner of the district. The phenomenon has been duly reported in the leading newspapers of the province, and the Akhbar-i-Am has noted it in its leading columns. (Nature, 48:248, 1893)

## METEORIC PHENOMENON
**Anonymous; *Annual Register*, 75:447-448, 1832.**

The "St. Petersburg Academical Gazette" contains the following account of an extraordinary phenomenon, from a letter, dated Moscow, May 2:---

"In March last, there fell, in the fields of the village of Kourianof, thirteen versts from Volokolamsk, a combustible substance of a yellowish colour, at least two inches thick, and covering a superficies of between 600 and 700 square feet. The inhabitants at first thought it was snow, but on examination it appeared to have the properties of cotton, having, on being torn, the same tenacity; but on being put into a vessel filled with water, it assumed the consistence of rosin. On being put to the fire in its primitive state, it burnt and sent forth a flame like spirits of wine; but in its resinous state it boiled on the fire without becoming inflamed, probably because it was mixed with some portion of the snow from which it had been taken. After a more minute examination, the rosin had the colour of amber, was elastic like Indian rubber, and smelt like prepared oil, mixed with wax." (Annual Register, 75:447-448, 1832)

## AN OLD-TIME SALT STORM
**C., H.; *Science,* 7:440, 1886.**

Can any of your readers tell me the exact date of the so-called 'salt-storm' which came upon the coast of Massachusetts about 1815? As described by old inhabitants, there was a high wind and heavy rain, and the houses and all objects within a mile of the water were coated with salt. Are such storms of frequent occurrence, and what is their explanation? (Science, 7:440, 1886)

## NINE YEARS OF CONTINUOUS COLLECTION OF BLACK, MAGNETIC SPHERULES FROM THE ATMOSPHERE
**Crozier, W. D.; *Journal of Geophysical Research,* 71:603, 1966.**

Abstract. Data on the deposition of black, magnetic spherules from the atmosphere, at two stations in central New Mexico, have accumulated for the 9-year period 1956-1964. These spherules, which appear to be similar to those which I have recovered from various sedimentary rocks and to those recovered by others from ice in Antarctica and Greenland and from Paleozoic salts, are believed to be of extraterrestrial origin. The 9-year average of the annual rate of mass accretion to the earth, for spherules in the diameter range of 5 to 60 u, was approximately $1.04 \times 10^{11}$ grams. The period included 2 years in which the annual infall was nearly $2.5 \times 10^{11}$ grams, however, and therefore the long-term average remains somewhat uncertain. The size distributions show fluctuations but, in general, are similar to those previously reported. The rate of deposition decreased from 1956 to 1962, and high peaks occasionally occurred. Interesting seasonal and biennial patterns appear in the rate of deposition, and there seems to be a degree of correlation between the spherule rates and radar meteor rates, the spherule rates leading the radar rates by about 6 weeks. In 1963, high rates of spherule deposition coincided with unusually high radar meteor rates. (Journal of Geophysical Research, 71:603, 1966)

# Chapter 8
# MAGNETIC DISTURBANCES

## INTRODUCTION

The magnetic anomalies described here are transitory---sudden deviations of the compass and jostles of the magnetometer. Permanent magnetic anomalies caused by local distortions of the earth's magnetic field (some of which are quite remarkable) are found in the companion handbook on geology.

The transitory magnetic anomalies are of two types: (1) those where no source can be identified; and (2) those that appear associated with specific phenomena but where it is difficult to understand how the magnetic effects are produced.

Careful observers of the compass needle noted long ago that it quivered during sun-induced magnetic storms. Possibly some of the disturbances noted below are of this type, but generally they are much larger and of shorter duration than those due to magnetic storms. Their origin(s) remain mysterious.

Auroras and thunderstorms are manifestly electrical in nature, and it is conceivable that electricity flows during these phenomena in such a way that it sets up a disturbing magnetic field. The actual mechanism may be obscure, but the possibility is there. Earthquakes are not obviously electrical, although magnetic effects have been reported and hotly controverted for over a century. The piezoelectric and piezomagnetic effects may be operating in these instances. Most obscure of all are the supposed magnetic effects of meteors. The rare reports of electric shocks on the ground under meteors fall in this class, too. It may be that intense meteor activity upsets the electrical equilibrium of the atmosphere and the induced flow of terrestrial electricity creates the observed magnetic effects.

# UNEXPLAINED MAGNETIC DISTURBANCES

## ABNORMAL MAGNETIC VARIATION
**Sell, R.; *Marine Observer,* 23:148, 1953.**

M. V. Britannic. Captain R. Sell, R. D., R. N. R. Liverpool to New York.

30th September, 1952, 1932-2052 G. M. T. Abnormal magnetic variation was experienced during this period. At 1932 the deviation was 3°E, at 2024 it was 1°W. The deviation then increased to 4-1/2°W by 2040 and decreased to the normal deviation of 3°E at 2052. Numerous bearings of celestial bodies were obtained having both E'ly and W'ly hour angles and the gyro compass remained correct throughout. No abnormal static was experienced.

Position of ship: at 1932, 50° 52'N, 18° 21'W; at 2052, 50° 50'N, 18° 55'W.

Note. The above observation was forwarded to the Admiralty Compass Observatory, Slough, who sent us the following comments:

"Various lines of investigation have been pursued in the hope of providing an explanation for the phenomena reported by the Britannic. The Department of Scientific and Industrial Research, however, have stated that their records give no indication of any magnetic storm which might have accounted for abnormal variation of the magnitude referred to on 30th September, 1952. Moreover, the Marine Superintendent of the Cunard White Star Line states that the Britannic is no longer fitted with degaussing coils, temporary and inadvertent energisation of which might have introduced a change in deviation.

"It can only be supposed therefore that this was an instance of a temporary change in deviation, brought about by some local agency such as movement of magnetic material in the neighbourhood of the compass, or a temporary fault in the finnacle lighting circuit if the latter were not non-inductively wound." (Marine Observer, 23:148, 1953)

## MAGNETIC DISTURBANCE
**Wilde, H. J.; *Marine Observer,* 15:6, 1936.**

The following is an extract from the Meteorological Record of S. S. Middlesex. Captain H. J. Wilde. Liverpool to Brisbane. Observer, Lieut. A. H. Martin, R. N. R., 3rd Officer.

30th March, 1937, 0217 G. M. T., in Latitude 9° 45' 30"S., Longitude 139° 45' W., between the islands of Hiva Oa and Ua Pow (Marquesas), a magnetic disturbance was experienced. The compass suddenly became very erratic, at one time pointing as much as 20° to the east of magnetic north. Within two minutes the compass-card settled down, and an azimuth showed the error to be normal again. (Marine Observer, 15:6, 1936)

## MAGNETIC DISTURBANCE
**Gordon, A. L.; *Marine Observer*, 6:261, 1929.**

The following is an extract from the Meteorological Log of S. S. Elpenor, Captain A. L. Gordon, Liverpool to Cape Town. Observer, Mr. A. R. Pearson, 2nd Officer.

"3. 55 p. m., December 23rd, 1928, in Latitude 13° 51' S., Longitude 1° 21' E. Steering by Standard Compass S. 17° E. (Error 21° W. S. 38° E. True), Steering Compass S. 22° E., both Compasses were suddenly deflected 90°, North Point swinging to East, and then swung freely back to normal. Period of disturbances about three minutes." (Marine Observer, 6:261, 1929)

## REGULAR PULSATIONS IN ORBIT
**Anonymous; *Nature*, 222:415-416, 1969.**

Remarkably regular pulsations have been found in the magnetograms from the geostationary satellite ATS 1 (Cummings et al., J. Geophys. Res., 74, 778; 1969). The pulsations are nearly monochromatic and usually nearly linearly polarized, the polarization ellipse being almost perpendicular to the steady magnetic field. The sensitivity of the magnetometer is about 1$\gamma$ ($10^{-5}$ gauss) and amplitudes up to 10 have been observed. The pulsations are observed only during magnetically quiet periods (Kp$<$ 3) and chiefly between 0400 and 1600 local time. With these conditions, they are detectable for a few per cent of the time.

The pulsation events lasted for up to 460 minutes and the pulsation period usually varied little during an event. A histogram for the periods shows two groups, one centred on 190 seconds and the other on 102 seconds. The longer period occurred in very quiet conditions and never during local night-time. Cummings et al. suggest that these pulsations are related to the "giant micropulsations", observed on the ground near the auroral zones, discussed in the 1930s, and this seems plausible. Following earlier suggestions, their notion is that quasi-transverse hydromagnetic standing waves can occur for which the disturbance is concentrated near one line of force, in line with the observed localization of giant micropulsations. When the disturbance in the magnitude of the field is neglected, so that the disturbance is purely in direction, the hydromagnetic equations for weak disturbances simplify in such a way that only spatial variation along a field line is involved so that normal modes can be calculated for a field line. Cummings et al. have carried out normal mode computations for the first six harmonics and several density models. They point out that the odd harmonics have no magnetic disturbance at the equator and that it is most plausible that the pulsations they observed on ATS 1 are the second harmonic. For giant micropulsations, the phase relations between conjugate points also require even modes.

The periods agree satisfactorily with the assumption that the plasmapause is outside the synchronous orbit when the longer period is observed; the plasmapause is known to move out in the very quiet conditions corresponding to the longer period.

The computations also show that the amplitude should be an order of magnitude larger at the reflecting surface than at the equator, while the pulsations observed at the ground seldom exceed 10$\gamma$. It is likely, however, that the amplitude at the ground is much reduced by the ionosphere, and large disturbances have been found above the auroral ionosphere, though their nature cannot be elucidated from a single rapidly moving satellite.

It is most unlikely that such pure, even modes would be excited by any hydromagnetic mechanism. The alternative explanation is a "velocity space instability, involving energetic particles", and an instability driven by particles in bounce resonance has been proposed (Southwood et al., Nature, 219, 56; 1968). This mechanism requires the phase of the pulsations to vary rapidly with longtude and it is hoped that this phase variation can be tested by combining the data from ATS 1 with those from DODGE, the orbit of which is not quite synchronous. The mechanism would account for pulsations in the radial (or vertical) component of the magnetic field, but the observed direction is typically 30$^{o}$ away from this direction. (Nature, 222:415-416, 1969)

# MAGNETIC EFFECTS OF AURORAS

## REMARKABLE EFFECT OF AN AURORA BOREALIS UPON AN ELECTROMAGNETIC CLOCK
**Anonymous; *Popular Astronomy,* 12:288, 1904.**

In A. N. 3932 Dr. Ernst Hartwig writes of the influence of an aurora borealis, on the night of Oct. 31, 1903, upon an electro-magnetic clock in his study at Bamberg. The pendulum of the clock receives an electro-magnetic impulse from two accumulator cells every minute and, when the cells are in order, has a constant rate through the year. On this night, however, the pendulum was accelerated in an extraordinary manner, as if the accumulators were over charged. On the morning of Nov. 1 Dr. Hartwig found the clock violently disturbed, the pendulum striking on both sides of the case, and the hand pointing several hours and perhaps over a whole revolution ahead, while because of the too violent swing several seconds-intervals were skipped at a time. The accumulators had not been charged for several days past, so that it must have been an earth-current resulting from the aurora which produced the disturbance. (Popular Astronomy, 12:288, 1904)

# MAGNETIC EFFECTS OF EARTHQUAKES

## MAGNETIC DISTURBANCES PRECEDING THE 1964 ALASKA EARTHQUAKE

**Moore George W.; *Nature*, 203:508-509, 1964.**

Through a fortunate circumstance, a recording magnetometer was operating in the city of Kodiak, 30 km north-west of the surface trace of a fault zone along which movement occurred at the time when the earthquake occurred in Alaska on March 27, 1964. Fortunately, too, the instrument was on such high ground that it was not reached by the subsequent seismic sea wave which virtually destroyed the city. The magnetometer recorded the fact that the largest of several magnetic disturbances briefly increased the intensity of the Earth's magnetic field by 100$\gamma$ at Kodiak, 1h 6min before the earthquake.

..........

One possibility is that the magnetic events which preceded the Alaska earthquake resulted from piezomagnetic effects of rocks undergoing a change in stress. Why such abrupt disturbances occurred in advance of the earthquake is not known; but a causal relation is indicated by the fact that Breiner has recently reported similar positive magnetic disturbances prior to minor earthquakes in Nevada and California. These observations, taken together, suggest that magnetic monitoring may provide a means of predicting a major earthquake in time to save lives and property. (Nature, 203:508-509, 1964)

## A NOTE ON TWO APPARENT LARGE TEMPORARY LOCAL MAGNETIC DISTURBANCES POSSIBLY CONNECTED WITH EARTHQUAKES

**Chapman, S.; *Terrestrial Magnetism and Atmospheric Electricity*, 35:81-83, 1930.**

The two reports from ships, reproduced and discussed in this Note, were communicated to me by Professor H. H. Turner (President of the Section of Seismology of the International Union for Geodesy and Geophysics), with the suggestion that they were worth bringing to the notice of workers on terrestrial magnetism.

I---(From Seismological Despatches, Georgetown, U. S. A., 1926, August 3, S. S. West Holbrook.)

"While the above-named vessel was entering the Gulf of Tokyo, 6:40 p. m. A. T. S. (apparent time of ship), on the above-named date, a slight tremor was felt, as if vessel was grounding. Vessel shivering and a noise was heard, as if touching a rocky ledge, but no reduction in speed could be noticed.

"Vessel heading at the time 328° P. S. C., 319° true. Deviation 3.°6 west on course.

"Bearings were immediately taken of Suno Saki Light bearing north 26° east true, and Merano Hana Light bearing north 93° east true; which placed vessel in latitude 34° 55' north, longitude 139° 43' 30" east, and by chart in from 46 to 50 fathoms of water.

"From the above position to Yokohama, the ship's compasses were acting queerly, having on northerly courses a deviation of 0°. 5 west: where before there had been 3° west.

"Upon arrival in Yokohama learned that an earthquake had taken place.

"The following morning, August 4, azimuths were taken for recording deviation. The compasses were still found to be out. On August 5 azimuths were again taken and compasses were now found to be back to normal."

The following report is from a ship which is provided with two entirely independent master gyro compasses situated low down in the ship. These communicate with gyro repeaters in the navigating position for the Officer of the Watch and the steering position for the Quarter-Master. In each position there is also a magnetic compass, the one in the navigating position being known as the standard compass and the other as the steering compass. In order to check one master gyro against the other, the repeaters in the two positions are run by different masters. The report is as follows:

II---"On April 22, 1928 at 18:20 G. M. T. in latitude 33° 38' north, longitude 24° 04' east, course 286° true---north 71° west (Standard)---a difference of 5° was noticed in the comparison between gyro and magnetic compass-courses. As both standard and steering compasses were similarly affected, it was at first supposed that gyro compasses were at fault, but investigation showed both masters to be running normally, and they remained in agreement throughout---nor had any electrical changes taken place in the ship.

"It had therefore to be assumed that for some reason the north point of the magnetic compasses had been deflected to the westward. After 10 minutes the difference between gyro and magnetic courses began to decrease, until at 19:10 G. M. T. the standard compass-course became normal again---to be followed shortly after by the steering compass.

"As both magnetic compasses were deflected similarly it is doubtful whether a vessel not fitted with gyro compasses would have been aware of the disturbance. Unfortunately no azimuths could be taken during the phenomenon.

"It is noteworthy that the time of the occurrence was practically coincident with the earthquake which destroyed the town of Corinth, and that the above position is on the same magnetic meridian as that place."

The occurrence described in I appears to be a quite definite instance of a large local magnetic disturbance of a temporary kind, apparently associated with an earthquake. The fact that after two days the compass-direction returned to the normal value seems to render it unlikely that the disturbance of direction was due merely to some mechanical action associated with the shock experienced by the vessel.

It would be interesting to know how far away, and how local, was the epicentre of the earthquake referred to, and what was its intensity. It is desirable also to know whether the records of the neighboring Tokyo magnetic observatory showed any disturbance at the time; if not, then the disturbance must have been either only apparent, or, if real, extremely local, since the observatory is only a degree or two away from the position of the ship.

As regards II, it should be noted that the Corinth earthquake occurred at 20h 13m50s on April 22, 1928, that is, about two hours after the magnetic disturbance was noted. Hence, the latter can have nothing to do with the actual earthquake, though it might conceivably be due to the causes leading up to the quake. The ship itself experienced no shock, and its officers only learned of the earthquake next morning.

The position of Corinth is 37°.9 north, 22°.9 east, and the epicentre of the earthquake of April 22, 1928, is within a degree of this; the ship must therefore have been within 4° or 5° from the epicentre.

In another letter the ship's Commander states that there was "no lightning or any signs of electrical activity in the atmosphere which might have explained a temporary deflection of the compass." He also says:

> "At the time of the occurrence there were, therefore, the following checks on the various compasses: (i) Comparison between gyro repeater in navigating position and standard compass; (ii) comparison between gyro repeater in steering position and steering compass; (iii) comparison between gyro repeater in navigating position and gyro repeater in steering position; and (iv) comparison between standard compass and steering compass.
>
> "The discovery was originally made by the Officer of the Watch, who noticed that when the ship was on her course by gyro repeater in navigating position the course by standard compass was 5° different from what it had been for some hours. Investigation showed the same difference in comparison (ii), while comparisons (iii) and (iv) remained unchanged. In other words, both magnetic compasses were in agreement and both gyro compasses were in agreement, but with 5° difference between the course by magnetic compass and the course by gyro compass.
>
> "Any disturbance sufficient to cause a wander of 5° would be quite clearly indicated on one or the other various electrical instruments or spirit levels attached to the master gyros.
>
> "Inspection showed both these to be running perfectly normally and that the repeaters were both still in step with their respective master.
>
> "In these circumstances it was only possible to conclude that the magnetic compasses were disturbed, and a careful investigation was made to find out if this could be due to any cause within the ship, such as starting up any electrical machinery near the compass. No cause whatever could be found and in fact, being Sunday evening, with all workshop-machinery etc. stopped, the electrical state of the ship was probably as stable as it ever could be. In any case, a disturbance due to ship causes would be unlikely to affect both compasses by exactly the same amount. One could only infer therefore that the disturbance was an external one."

The remarkable features of the magnetic disturbance mentioned in report II are its large amount and its short duration; its amount is paralleled, in such latitudes, only during the most intense world-wide magnetic storms, which last for many hours. No such storm occurred on that date. Unfortunately there is no continuously-recording magnetic observatory within several hundred miles of Corinth, the nearest being at Helwan; at my request Mr. H. E. Hurst kindly furnished me with a tracing of the horizontal-force magnetograph for Helwan for the day in question; it showed no trace of any unusual disturbance at the time, and Mr. Hurst states that the other magnetic records were equally quiet.

Hence, if the magnetic disturbance near Corinth was real, as seems scarcely open to doubt, it must have been much more local than are most magnetic disturbances, even when much less intense than this one. One possibility is that, in the region where the ship then was, there is a large permanent local irregularity in the Earth's field, but it seems unlikely that so large an irregularity in such frequented waters could have escaped notice.

The two occurrences raise the question whether, if such local temporary disturbances occur from time to time, they would be likely to attract attention. This naturally depends on their frequency, duration, and spatial extent; the latter is probably the most important factor in determining whether they would be noticed. In Great Britain there are four continuously-recording magnetic observatories, at

Abinger, Stonyhurst, Eskdalemuir, and Lerwick, separated by distances of from one hundred to several hundred miles; but in most parts of the world observatories are much more sparsely spread, and a disturbance concentrated within an area of linear extent 100 miles, or even much more, would have a good chance of escaping notice at any continuously recording observatory. Further, save when magnetic or other special surveys involving the use of the compass are in progress, a large temporary local magnetic disturbance might easily pass unrecognized except possibly through its effect on local cables; probably general wireless transmission would be but little affected by a local magnetic disturbance of the stated extent. Further, according to the ship's Commander, ships not fitted with gyro compasses might not become aware of such a disturbance. It may therefore be that, if they occur, they would be noticed only rarely, and even then might well pass unreported; if, however, they occur in thickly populated and industrial regions, the chance of their detection would be greater, though to set against this there is the greater likelihood, in such places, of artificial disturbances by electric currents.

Our present ignorance of the cause of the Earth's main field and its secular change makes it difficult to exclude a possible connection between earthquakes and local temporary magnetic disturbances, and if the latter occur at all, the fact that they have not been detected at the magnetic observatories of Britain and similar countries may be due to the rarity of serious earthquakes there. This, however, does not apply to such an observatory as that at Tokyo, and if its records give no indication of any really local disturbances associated with earthquakes, the suggested connection would be rendered doubtful, and the occurrences here reported on would remain completely mysterious. (Terrestrial Magnetism and Atmospheric Electricity, 35:81-83, 1930)

# VOLCANO MAGNETISM

## MAGNETIC DISTURBANCE AT TIME OF ERUPTION OF MOUNT PELEE

**Bauer, L. A.; *Science*, 15:873, 1902.**

Coincident, as far as can be at present ascertained, with the time of eruption of Mont Pelee on May 8, a magnetic disturbance set in which was registered on the self-recording instruments of the two U. S. Coast and Geodetic Survey magnetic observatories, the one at Cheltenham, Md., seventeen miles southeast of Washington, and the other at Baldwin, Kansas, seventeen miles south of Lawrence. The preliminary reports received from Mr. L. G. Schultz, in charge of Cheltenham observatory and Mr. W. C. Bauer, in charge of Baldwin Observatory are sufficient to indicate that the disturbance began at practically the same instant of time at both observatories, viz., at 7h. 54m. St. Pierre local mean time. According to the newspaper reports the catastrophe befell St. Pierre about 8 A. M. of May 8 and it has been stated that the town clock was found stopped at 7h. 50m.

Purely mechanical vibrations caused by earthquakes are often recorded by the delicately suspended magnetic needles, as for instance the Guatemalan one which was felt at the Cheltenham Observatory on April 18 from about 9h. 20m. to 9h. 50m. P. M.

75th meridian mean time.

The disturbance on May 8, however, was distinctively a magnetic and not a seismic one and hence was not recorded on seismographs. The Cheltenham magnetograms exhibit magnetic disturbances amounting at times to .00050 to 0.00060 c.g.s. units (about 1/350 of the value of the horizontal intensity) and from 10' to 15' in declination, beginning at the time stated and continuing until midnight of the 9th. Even on the 10th tremors were still discernible. (At the time of writing the subsequent curves had not yet been received.)

Until further information has been received from other observatories, it cannot be determined definitely whether this magnetic disturbance was due to some cosmic cause or came from within the earth's crust and was associated with the Martinique eruption. The coincidence in time is however a remarkable fact. (Science, 15:873, 1902)

## MAGNETIC NOISE PRECEDING THE AUGUST 1971 SUMMIT ERUPTION OF KILAUEA VOLCANO

**Keller, George V., et al; *Science,* 175:1457, 1972.**

Abstract. During the course of an electromagnetic survey about Kilauea Volcano in Hawaii, an unusual amount of low-frequency noise was observed at one recording location. Several weeks later an eruption occurred very close to this site. The high noise level appeared to be associated in some way with the impending eruption. (Science, 175:1457, 1972)

# THUNDERSTORM MAGNETISM

## COMPASS DEVIATION IN THUNDERSTORM

**Elliott, J.; *Marine Observer,* 34:11, 1964.**

m.v. Dartwood. Captain J. Elliott. Botwood to London. Observers, Mr. B. L. Bass, and Officer and Mr. J. Roberts, Apprentice.

13th February 1963. At 0440 GMT, the vessel encountered a violent squall, accompanied by large hailstones approx. 3/8 in. in diameter. The wind increased to force 11-12 from W and there was a great deal of forked lightning and heavy thunderclaps: visibility was almost nil. As the storm was passing overhead, the magnetic compass was seen to be deviated 4° to the W: it returned to its normal position after the storm moved away. Air temp. 38°F at 0440: 40° at 0500.

Position of ship: 49° 53'N, 30° 34'W.

Note. The following comment has been received from the Superintendent, Eskadalemuir Observatory, Dumfriesshire, Scotland:

"The magnetic field was moderately disturbed on 13th February with a range of 20' in declination at Eskadalemuir.

The deviation experienced by m. v. Dartwood seems much too large to be part of this disturbance and it appears possible that it may have been due to a point discharge from the mast head. The current up the mast would be increased by the abnormal potential gradient due to the thunderstorm overhead, and by the high wind, but if this is the explanation it is rather surprising that St. Elmo's fire was not observed: the bad weather may explain this." (Marine Observer, 34:11, 1964)

# POSSIBLE MAGNETIC EFFECTS OF METEORS

## THUNDERBOLT AFFECTS COMPASSES
**Campbell, Alex; *Marine Observer*, 5:157, 1928.**

The following is reported by Captain Alex Campbell, S. S. Sai On:---

"At 0. 50 p. m., on August 2nd, 1927, whilst steaming across the mouth of Deep Bay and having Black Point 2 points on our port bow, a thunderbolt fell on our port beam, and on striking the water burst with a terrific report.

"It was about 100 yards from the ship when it struck the water, and after the ship had ceased to vibrate we found the standard compass with an error of 2 points left and the steering compass 1 point left. Before this occurrence they were correct."

The Harbour Master, Hong Kong, obtained the following additional particulars from Captain Campbell:---

"There are absolutely no signs of the ship having been struck.

"The course at the time was S. E. 1/2 S.

"The deviation introduced was westerly in the N. semicircle and easterly in the south.

"The error appears to be permanent and it will be necessary to readjust his compasses." (Marine Observer, 5:157, 1928)

## MAGNETIC MICROPULSATIONS ACCOMPANYING METEOR ACTIVITY
**Campbell, Wallace H.; *Journal of Geophysical Research*, 65:2241, 1960.**

Abstract. Increased activity of magnetic micropulsations with periods of 5 to 30 seconds and magnetic flux densities of 20 to 320 m$\gamma$ was found to accompany the $\eta$ Aquarid, $\delta$ Aquarid, and Perseid meteor showers in 1958. Conflicting reports are discussed. (Journal of Geophysical Research, 65:2241, 1960)

## OBSERVED MAGNETIC EFFECTS FROM METEORS
**Jenkins, Alvin W., et al; *Journal of Geophysical Research*, 65:1617, 1960.**

A correlation between geomagnetic fluctuations and meteoric activity was reported by Kalashnikov, who used sensitive fluxmeters and a photographic recording technique. In his work, he noted an increase in the number of pulses in the vertical component over the dates of meteor showers. Hawkins, using more sensitive equipment also sensitive to the vertical component, attempted to correlate pulses with visual meteors. His results were negative, indicating only such correlation as might be expected statistically. A real discrepancy thus exists between the results of these two workers. Hawkins has pointed out, however, that Kalashnikov's results may not be significant, since the correlation he noted is not much greater than that expected to occur accidentally. A preliminary analysis of data recently available from the IGY program concerned with subaudio fluctuations in the geomagnetic field seems to indicate that meteoric activity and the average level of the fluctuations are related. (Journal of Geophysical Research, 65:1617, 1960)

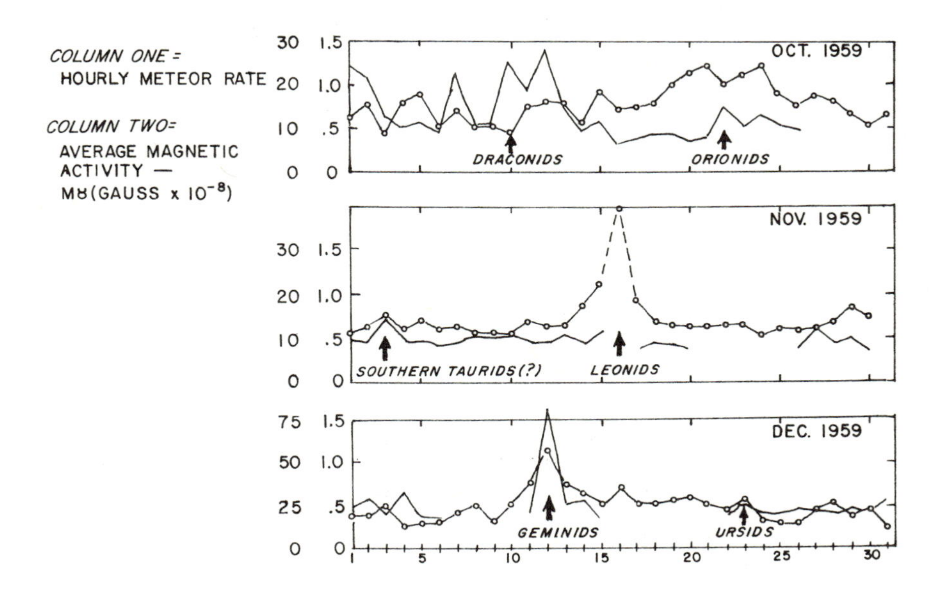


Correlation of meteors and magnetic activity

# INDEXES

## SUBJECT INDEX

# SOURCE INDEX

3 2311 00062 599 9